3401388188

Fundamentals of Digital Logic Design: with VLSI circuit applications

D0528816

PRENTICE HALL
SILICON SYSTEMS ENGINEERING SERIES

Editor: Kamran Eshraghian

Haskard & May	*Analog VLSI Design: nMOS and CMOS*
Pucknell & Eshraghian	*Basic VLSI Design: Systems and Circuits*

Fundamentals of Digital Logic Design: with VLSI circuit applications

Douglas A Pucknell

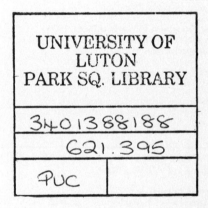

UNIVERSITY OF
LUTON
PARK SQ. LIBRARY
3401388188
621.395
PUC

PRENTICE HALL

New York London Toronto Sydney Tokyo

© 1990 by Prentice Hall of Australia Pty Ltd

All rights reserved. No part of this publication may be reproduced, stored in a retrieval system, or transmitted in any form or by any means, electronic, mechanical, photocopying, recording, or otherwise, without the written permission of the publisher.

Prentice Hall, Inc., *Englewood Cliffs, New Jersey*
Prentice Hall of Australia Pty Ltd, *Sydney*
Prentice Hall Canada, Inc., *Toronto*
Prentice Hall Hispanoamericana, SA, *Mexico*
Prentice Hall of India Private Ltd, *New Delhi*
Prentice Hall International, Inc., *London*
Prentice Hall of Japan, Inc., *Tokyo*
Prentice Hall of Southeast Asia Pty Ltd, *Singapore*
Editora Prentice Hall do Brasil Ltda, *Rio de Janeiro*

Typeset by: Keyboard Wizards, Harbord, NSW.
Printed and bound in Australia by:
Impact Printing Pty Ltd, Brunswick, Victoria.

Cover design by: Philip Eldridge

2 3 4 5 93 92 91
ISBN 0 7248 0432 3 (paperback)
ISBN 0-13-332693-4 (hardback)

National Library of Australia
Cataloguing-in-Publication Data

Pucknell, Douglas A. (Douglas Albert), 1927 -
Fundamentals of digital logic design: with VLSI circuit applications.

Includes index.
ISBN 0 7248 0432 3.

1. Integrated circuits - Very large scale integration - Design and construction.
2. Logic design. I. Title.

621.395

Library of Congress
Cataloguing-in-Publication Data

Pucknell, Douglas A., 1927 -
Fundamentals of digital logic design: with VLSI circuit applications / Douglas A. Pucknell.
p. cm.
Includes index.
ISBN 0-13-332693-4

1. Integrated circuits - Very large scale integration - Design and construction - Mathematical models. 2. Logic design.
I. Title.

TK7874.P85 1989
621.395--dc19

89-3455
CIP

PRENTICE HALL

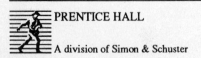

A division of Simon & Schuster

Contents

Preface xiii

Chapter 1 **An introduction to digital systems engineering, semiconductors and basic MOS technology** 1

1.1 General considerations 2
1.2 An introduction to semiconductor technology 5
 1.2.1 Basic semiconductor properties 5
 1.2.2 The effect of doping 9
 1.2.3 p-n (or n-p) junctions 13
1.3 Basic nMOS and CMOS technologies 16
 1.3.1 Basic MOS transistors 16
 1.3.2 Summary of nMOS fabrication processes 19
 1.3.3 CMOS fabrication 20
 1.3.4 Latch-up in CMOS circuits 22
1.4 MOS layers 22
1.5 Stick diagrams 23
1.6 Design rules and layout 23
1.7 Electrical properties 23
 1.7.1 Parameters for MOS transistors 25
 1.7.2 Sheet resistance, R_S 33
 1.7.3 Area and peripheral capacitances 34
 1.7.4 Properties of pass transistors and transmission gates 35
 1.7.5 Inverters 38
1.8 Observations 47
1.9 Worked examples 47
 1.9.1 Semiconductor technology 47
 1.9.2 MOS fabrication 50
 1.9.3 MOS layout and design 50
 1.9.4 MOS electrical parameters 54
1.10 Tutorial 1 56

2 **Number systems and arithmetic** 60

2.1 Introduction 61
2.2 Characteristics of positionally weighted number systems 61
 2.2.1 Number representation 61
 2.2.2 Some implications of the choice of radix 62
2.3 The binary number system 63
 2.3.1 Some factors 63
 2.3.2 Some disadvantages of the binary system 63
 2.3.3 Other systems related to binary 64
2.4 A general process for radix conversion 66
2.5 Operations on signed numbers 67

2.5.1 The representation of signed numbers 67
2.6 An overview of binary multiplication and division processes 74
2.6.1 Multiplication of unsigned or sign/magnitude numbers 75
2.6.2 Signed (twos complement) multiplication—Booth's algorithm 76
2.6.3 Some division algorithms 78
2.6.4 Some observations 80
2.7 Some other number systems of interest 80
2.7.1 Ternary arithmetic 80
2.7.2 Quadernary arithmetic 81
2.7.3 Residue arithmetic 81
2.8 Floating point arithmetic 83
2.8.1 Format and representation 83
2.8.2 Floating point arithmetic operations 85
2.9 Summary 86
2.10 Worked examples 86
2.10.1 Radix conversion 86
2.10.2 Twos and ones complement arithmetic 89
2.10.3 Multiplication and division 91
2.10.4 Residue arithmetic 92
2.10.5 FP representation and arithmetic 93
2.11 Tutorial 2 94

3 Some basic techniques for handling problems in designing logic circuitry 96

3.1 Introduction to switching algebra 97
3.2 Boolean algebra and logic functions 98
3.2.1 Negation: The *Not* function (bar ‾ or prime ') 98
3.2.2 Logical sum: The *Or* function (+) 99
3.2.3 The *Nor* (*Not Or*) function 100
3.2.4 Logical product: The *And* function (".") 100
3.2.5 The *Nand* (*Not And*) function 101
3.2.6 The *Xor* (*Exclusive Or*) function (⊕) 102
3.2.7 The *Xnor* (*Exclusive Nor* or *Equality*) function 103
3.2.8 Logic functions of two variables 103
3.3 Further aspects of switching theory and Boolean algebra 103
3.3.1 Variables 103
3.3.2 Literals 103
3.3.3 Theorems & aids to simplification of Boolean algebra 104
3.3.4 Duality 107
3.3.5 Product terms 107
3.3.6 Sum terms 107
3.3.7 Normal terms 107
3.3.8 Sum of products (SOP) form of expression 107
3.3.9 Product of sums (POS) form of expression 107
3.3.10 Canonic (standard) sum and product terms 107

3.3.11 Maxterms and minterms 108
3.3.12 Maxterm notation 108
3.3.13 Minterm notation 109
3.3.14 Expansion of simplified expressions 110
3.3 15 Conversion from one form to another 110
3.3.16 The Karnaugh map 112
3.4 Simplification 115
3.4.1 Algebraic simplification 115
3.4.2 Graphical methods 115
3.4.3 Karnaugh map based simplification 116
3.4.4 A tabular approach to simplification: The Quine-McCluskey method 130
3.5 Summing up the introductory chapters 137
3.6 Worked examples 138
3.7 Tutorial 3 146

4 The design of combinational logic 148

4.1 Introduction 149
4.2 "Random" logic 150
4.3 Using read only memories (ROM) [or non-volatile random access memories(RAM)] to realize combinational logic 152
4.3.1 General memory configurations 152
4.3.2 Mapping combinational logic expressions into a ROM or PROM chip 153
4.4 Using multiplexers (MUXs) to realize combinational logic 155
4.4.1 Problems with four variables or less 155
4.4.2 Using MUXs for problems with large numbers of variables 157
4.5 Programmable logic arrays (PLAs) for combinational logic 158
4.6 Symbols for and a brief discussion of "active Lo" logic 162
4.7 Summary 162
4.8 Worked examples 163
4.9 Tutorial 4 167

5 Combinational logic in silicon 170

5.1 Implementing combinational logic functions in silicon 171
5.2 nMOS and CMOS custom design of logic circuits 171
5.2.1 Switch-based logic 171
5.2.2 Complementary switch-based logic 173
5.2.3 A general procedure for the design of CMOS complementary logic 176
5.2.4 The complementary (CMOS) inverter and inverter-based logic 181
5.2.5 nMOS and pseudo-nMOS inverter-based logic (ratio logic) 182

5.2.6 The use of bridging switches 187
5.2.7 PLAs in silicon 189
5.2.8 Precharged and clocked logic 190
5.2.9 Selector switch (multiplexer and demultiplexer) based logic 198
5.3 Gate arrays (uncommitted logic arrays-(ULAs)) and related semi-custom implementations of logic circuitry in silicon 204
5.3.1 User programmable logic arrays 207
5.4 The adder 207
5.4.1 The adder equations 211
5.4.2 A CMOS implementation of a "ripple through" adder 214
5.4.3 A carry look-ahead approach to the parallel adder 214
5.4.4 The serial adder 216
5.5 Summary 218
5.6 Worked examples 220
5.7 Tutorial 5 226

Color Plates

1 Color encoding schemes
2(a) Design rules for width and separation of wires (nMOS & CMOS)
2(b) Transistor design rules (nMOS, pMOS & CMOS)
3(a) Particular rules for p-well CMOS process
3(b) Simple contacts (nMOS & CMOS)
4 Example layouts
5 One possible mask layout for Example 2
6 Possible stick diagrams and mask layouts for typical CMOS gates
7 Possible stick and mask layout for 3 I/P *Nor*
8 Two to four line decoder

6 Asynchronous (fundamental mode) sequential logic 227

6.1 A basic sequential logic circuit 228
6.2 A general model for sequential logic circuits 229
6.3 Simple analysis of asynchronous sequential logic circuitry 230
6.3.1 A further, more complex, example of analysis 231
6.3.2 The RS flip-flop (reset-set flip-flop) 233
6.4 Synthesis of simple asynchronous sequential circuits 235
6.4.1 A simple asynchronous design problem 236
6.4.2 A more difficult example, demonstrating further aspects of the asynchronous sequential circuit design processes 238
6.4.3 On the importance of words 244
6.4.4 Outline of a procedure for designing asynchronous sequential logic 245
6.4.5 Further aspects of asynchronous sequential circuits 246

6.5 Realization of asynchronous sequential circuits in silicon 254
6.5.1 PLA based asynchronous sequential circuits 254
6.6 Concluding remarks on asynchronous sequential circuits 256
6.7 Worked examples 256
6.8 Tutorial 6 260

7 **Clocked sequential circuits and memory—basic techniques 262**

7.1 Some common types of flip-flop 263
7.1.1 The asynchronous RS flip-flop 263
7.1.2 The synchronous (clocked) JK flip-flop 263
7.1.3 The D (data) flip-flop or latch 265
7 1.4 The T (toggle) flip-flop 267
7.1.5 A summary of some common characteristic equations 268
7.1.6 Designing an edge sensitive flip-flop 269
7.2 Clocked sequential circuit design 271
7.2.1 Example 1: Design of a clocked serial parity detector 271
7.2.2 Summarized design procedure 276
7.2.3 Example 2: A clocked sequence detector circuit using (i) JK and (ii) D flip-flops 278
7.2.4 VLSI based realizations 281
7.2.5 Mealy and Moore (finite state) machines 287
7.3 Memory elements 289
7.3.1 General considerations 290
7.3.2 Some basic arrangements 291
7.3.3 Some static storage circuits 298
7.4 Remarks and observations 303
7.5 Worked examples 303
7.6 Tutorial 7 313

8 **Some commonly applied clocked sub-systems 316**

8.1 Introductory remarks 317
8.2 Counters 317
8.2.1 A ripple-through 4-bit binary counter 317
8.2.2 An up/down synchronous 4-bit binary counter 323
8.2.3 An up/down counter (incrementer/decrementer) for VLSI 329
8.3 Registers 331
8.3.1 A static 4-bit parallel register, D flip-flop and VLSI based 332
8.3.2 Shift registers 334
8.3.3 A successive approximation register (SAR) for an A/D converter 340
8.4 Observations 346
8.5 Tutorial 8 346

9 Characteristic and application equations—a difference equation approach to clocked sequential circuit design and analysis 348

9.1 Introduction 349
9.2 A summary of some common characteristic equations 349
9.3 Application equations 350
9.4 Application equation based design procedure for clocked elements 352
9.5 Examples of the design of clocked flip-flop based circuits using the application equation approach 353
9.5.1 Example 1: A binary coded decimal (8421) up/down synchronous counter 353
9.5.2 Example 2: The design of a base 5 counter for up counts only 357
9.5.3 Example 3: A 6-bit serial code detector 362
9.6 Application equation based design for clocked VLSI circuits 365
9.6.1 A PLA based version of a binary coded decimal (BCD) (8421) synchronous up/down counter 366
9.6.2 A further discussion of PLA based design—dimension reduction 368
9.7 Analysis of clocked sequential circuits using an application equation based approach 371
9 7.1 Analysis of JK flip-flop based designs 371
9.7.2 Analysis of VLSI (PLA based) designs 373
9.8 Conclusions 377
9.9 Tutorial 9 378

10 Basic microprocessor architecture and organisation with interfacing techniques 381

10.1 Introduction 382
10.2 General microcomputer architecture and organization 383
10.2.1 Main memory 384
10.2.2 The I/O facilties 385
10.2.3 The arithmetic and logical unit (ALU) 385
10.2.4 The control unit 385
10.2.5 The buses 386
10.2.6 Key registers 386
10.3 Interfacing with a microprocessor 388
10.3.1 The software interface 389
10.3.2 The Z80 programmer's model—registers, instruction set and addressing modes 389
10.3.3 The 68000 programmer's model—registers, instruction set and addressing modes 397
10.3.4 Reduced instruction set computer (RISC) concepts 405
10.3.5 The hardware interface 407
10.3.6 Typical parallel and serial I/O packages 409

10.3.7 Memory and I/O space maps 412
10.3.8 The Z80 hardware interface 414
10.3.9 The 68000 hardware interface 420
10.4 Interrupts 426
10.4.1 The Z80 interrupts 429
10 4.2 The 68000 interrupts 432
10.5 Concluding remarks 436
10.6 Tutorial 10 436

Appendix 1 The design of phase detectors using the logic signal flow graph (by C.J. Kikkert) 439

A1.1 Introduction 440
A1.2 Phase detectors 440
A1.2.1 *Exclusive Or* gate 440
A1.2.2 Basic sequential logic phase detector 441
A1.2.3 Combination phase detector 442
A1.2.4 Phase detector for negative pulses 443
A1.2.5 Phase and frequency detector 443
A1.2.6 Phase only detector 444
A1.2.7 Double-edged phase detector 444
A1.3 Phase detector design 445
A1.4 Conclusions 446

Appendix 2 Transition equations (TEs) and TE based design 447

A2.1 Introduction 448
A2.2 Transition characteristic equations 448
A2.3 Example: The TE design of an 8421 BCD counter (up count only) 451
A2.3.1 Using the J and K inputs only 452
A2.3.2 Using the clock inputs "T" alone 452
A2.3.3 A ripple through approach using T and also Clr inputs 454
A2.4 Application equations revisited 455
A2.5 A particular case study—the design of a successive approximation register (SAR) for an A/D converter 457
A2.5.1 A TE design approach 457
A2.6 Conclusions 460
A2.7 Further reading 461

References for general reading 463

Index 465

Acknowledgements

Much of the material in this book is based on coursework presented to third and fourth year undergraduate students in electrical and electronic engineering at the University of Adelaide. The fact that this book has been written is largely due to the enthusiastic way in which they have received and digested the topics covered here. I have benefitted greatly from their constructive criticism and ideas, and also from discussion and general interaction with members of staff in this department.

Thus many have been involved in the creation of this book but there are some direct contributions which must be acknowledged. A former colleague, Associate Professor Keith Kikkert of the James Cook University of Queensland, kindly provided some of the material for Chapter 6 and also Appendix 1 in its entirety. A current colleague, Mike Leibelt, was most helpful in double-checking the chapter on microprocessors. I must also acknowledge the assistance provided by Capilano Computing of Canada, in particular, Chris Dewhurst and Neil McKenzie. They kindly provided up-to-date versions of LogicWorks™, the Capilano software package used to simulate various logical arrangements in the text. I am also indebted to series editor Kamran Eshraghian for the useful discussions we had on the content of the book.

In having this work published I have been greatly helped and encouraged first by Prentice Hall Australia's Senior Editor Ted Gannan and lately by his successor Andrew Binnie. I am also very much indebted to the production team of Prentice Hall in Sydney of whom Ian McArthur and Fiona Marcar rate a special mention. They have been a pleasure to work with and have set and maintained the highest standards in getting this book into print. I also benefitted greatly from the comments and criticisms of the unnamed but nevertheless revered reviewers of the manuscript.

Finally I must express my gratitude to my wife Ella. She has been a tower of strength, patience and understanding throughout the long period it has taken to bring this work to fruition.

Douglas A. Pucknell
University of Adelaide, SA
August 1989

Preface

Since the early 1960s, the widely available standard ranges of packaged integrated circuits (ICs), have provided the vast majority of all system elements and components for digital designers. This technology has proved to be so convenient that, following the initial introduction of packaged gates and flip-flops, various logic functions and digital subsystems of increasing complexity have been integrated into single silicon chips. The technology has ranged through small scale (SSI), medium scale (MSI) and large scale (LSI) integration densities.

The IC technology of the 1980s has now advanced to very large scale integration (VLSI) density. This increase in "on chip" complexity has provided an abundance of sophisticated "off the shelf" packaged digital subsytems, but has also created an awakening of interest in the possibilities presented by custom design in silicon. In this respect, VLSI has to some extent, "turned back the clock" as far as design "know how" is concerned. The designer, using custom digital chips, is now faced with the task of circuit as well as subsystem and system design. No longer will all designs be implemented from standard packages for which a knowledge of the terminal characteristics, and interconnection rules alone is required.

To effectively implement any design using ICs, and for custom logic design in particular, it is essential to understand the characteristics of semiconductor devices, and of the particular technology to be used. MOS technology is by far the most widely used for ICs, and almost without exception, all the devices we utilize or design in this or the next decade, will be realised in nMOS or CMOS technology. Thus, a good working knowledge of MOS circuitry is essential for custom design and highly desirable for the user of packaged logic.

Naturally, the creation of good and effective digital systems also depends on the designer having a good grasp of the appropriate fundamental aspects of digital logic design, and of design methods for logic subsystems.

This text combines a treatment of digital logic, with a complementary treatment of semiconductors and MOS VLSI technology. Where logic circuits and subsystems are developed, then the text mostly illustrates their implementation in silicon as well as in logic diagram form. The characteristics of circuits in silicon are also explained and illustrated so that the reader will appreciate the performance limits for nMOS and CMOS technologies, and also why such limitations are present.

It is hoped that the links between theory and practice are thus adequately established. It is the author's considered opinion that good design stems from a sound knowledge of available technologies, together with a proper understanding of the nature and behavior of the "components" of the chosen technology.

Appropriate design procedures should then be utilized to effect system design.

Apart from the topics already discussed, and since many digital systems are concerned with "number crunching", the text also deals with number systems, and with microprocessors. Microprocessors are also treated as digital system components with emphasis on the architectural and interfacing aspects. The Zilog Z80™ (8bit) and Motorola 68000™ (16bit) microprocessors are used to illustrate concepts.

The text is organized in such a way that the first three chapters comprise introductory material on semiconductors, MOS VLSI design, number systems and the basic aspects and techniques of logic circuitry. Appropriate examples are included in the text, as well as worked examples at the end of each of these chapters. The instructor and/or the reader may utilize as much or as little of this material as necessary. A knowledge of these matters will be assumed in the remaining chapters. The next six chapters comprise the main body of material on digital logic design. They contain appropriate design and analysis techniques and procedures for combinational logic and asynchronous and clocked sequential digital logic. This part of the text is also reinforced throughout with worked examples, and is well illustrated with figures and diagrams.

Wherever possible, alternative realizations are dealt with and, in many cases, realizations in silicon are also discussed and demonstrated. The design of many of the commonly required digital subsystems is dealt with, and it is hoped that these designs will provide a useful source of applications material for the designer. The last of these six chapters illustrates a *difference equation*–based approach to design and analysis, which the author has found to be useful, convenient and well appreciated by students. Other lesser known, but useful aspects of design are discussed in the two appendices.

The final chapter is devoted to a treatment of microprocessors with emphasis on the interfacing aspects.

Tutorial work follows each chapter, and this is designed to exercise the student in the appropriate area of study before proceeding to the next chapter. Learning is more effective if this is done and, ideally, the tutorial sessions should be supervized by an instructor.

The main body of material presented here is taught,in this University, in the penultimate year of a four–year degree course, but the inclusion of the introductory material of Chapters 1 to 3 extends the usefulness of this text to earlier years. The matters covered in the final chapters of the book may well find application in the final year of a degree level course, as is the case in Adelaide.

The material is presented in a concise manner, together with carefully chosen figures and diagrams so that it is readily adapted to a series of lectures.

It is hoped that this book will find ready application in undergraduate Electrical/ Electronic Engineering and Computer Science courses. It is further hoped that practicing professionals will also find it useful and readable for "brushing up", or in updating in the Digital Logic area.

Finally, in the author's experience, the essence of good engineering lies in simplicity. In achieving this, one must first, carefully define the task to be performed, second, choose and utilize the appropriate technology, and third, select the appropriate design procedures. If these things are done, then design becomes a relatively simple and straightforward process. This text is intended to help in achieving these aims.

"*Our life is frittered away by detail ... simplify, simplify*".
Henry David Thoreau.

Douglas Pucknell
University of Adelaide
April 1989

1. An introduction to digital systems engineering, semiconductors and basic MOS technology

1.1 General considerations.

As far as the electronics engineer is concerned, digital systems design, as a significant branch of electrical engineering, dates from the mid to late 1950s. At that time digital computers were just emerging as viable commercial entities. Previously, any ambitious form of electronic calculator or computer had been confined to the "back-room" areas of a few research and development groups and had not been easy to develop; nor had they proved reliable. However, advances in technology overcame these problems and, since then, digital systems engineering and computer design have advanced interdependently.

Forms of computer are not new. Some 3000 or more years have passed since the abacus was developed and it is interesting to note that it has stood the test of time so well that it is still in widespread use today particularly in Asia (including the home of the electronic calculator—Japan). It was also in widespread use in Europe prior to the twelfth century. At that time Arabic numbers came into general use and so complete was the transition that one of Napoleon's officers claimed a startling new discovery when he came across an abacus in use in a remote part of Russia.

Nearer our own time, but nevertheless over 300 years ago, Blaise Pascal invented a computing engine. He was spurred on by a good measure of self-interest arising from a need to do the many repetitive calculations imposed on him by his father. Times haven't changed—history records that his father was a tax collector! The invention was made in 1642 when Blaise was 19 years old. His invention was subsequently improved upon in 1670 by another notable man, Baron Gottfried Wilhelm von Leibnitz. His device was known as a stepped reckoner and could add, subtract, multiply, divide and find square roots.

More than 150 years ago, the principles upon which the modern digital computer is based were anticipated in an invention by Charles Babbage. Babbage, a British mathematician, set out the design of an "analytical engine" which encompassed program storage in memory, iterative instruction sequences, branching, an arithmetic unit and, to top it all off, the equivalent of punched cards—plates with pins inserted adapted from the Jacquard weaving loom of the times. Babbage's engine, proposed in 1833, would have been capable of processing up to 100 decimal numbers each up to 25 digits long. However, the technology of the time was inadequate and his creativity was thwarted. Although a limited form of his machine was finally completed by his son, Babbage was largely discredited and his creativity went unrecognized until more than 100 years later.

It is interesting to reflect on the consequences had the computer been successfully invented at that time. The Industrial revolution and a computer revolution would have taken place simultaneously in both Europe and the USA and there is no doubt that the course of history would have been significantly altered.

A major use of computers is for processing data and in that context a most important name from the past is that of Hollerith. More than 100 years ago he developed the punched card form of data representation, a development which once again was born of necessity, necessity to improve the handling and processing of data gathered at the USA census taking which occurred every five years. Prior to the use of punched cards it was taking seven years to complete the results and that was clearly a "no-win" situation. Hollerith's invention cut the time by two-thirds and was a major milestone in data handling.

Coincidentally, one of the main mathematical tools of digital design originated about the same time as the work of Babbage. This was due to George Boole whose work in the area of logic and reasoning laid the foundations of Boolean algebra and switching theory which are so widely used today. Engineers are not always "quick on the uptake"and the relevance of Boole's work, completed in 1847, was not appreciated for nearly 100 years. Then it was first applied by C.E.Shannon in 1938 for solving problems in telephone switching networks.

The first really useful digital circuit appeared in 1919 rejoicing in the title "Eccles-Jordan trigger relay". This type of circuit is now known by the more familiar name flip-flop. Once again the potential of this type of circuit was not realized for some time; it was not until the 1930s that it was first used and then it appeared in counters.

Through much of the first half of this century, power, rotating machinery, radio communications, and telephone networks represented the main areas of activity of the electrical engineer. In consequence, apart from some switching operations, it was an entirely analog world and electronics in consequence was concerned with linear circuits and the handling of analog signals.

World War 2 brought a departure into pulse techniques for radar and sonar applications but the constraints of military secrecy inhibited the rapid dissemination of knowledge and technology.

The first digital computers began to appear in the 1940s. For example, a computer development designed to solve up to 30 simultaneous equations and using vacuum tube technology (and the Eccles-Jordan flip-flop) was developed at Iowa State College by Atanasoff and Berry. This project was pursued over several years but was never finished. Another development "Eniac" (electronic numerical integrator and computer) was developed by Eckert and Mauchly at the University of Pennsylvania for the US Army and was completed in 1946.

Again, the circuitry was based on the Eccles-Jordan flip-flop using vacuum tubes—some 18,000 in fact! It was a formidable example of engineering technology in many respects. It weighed 30 tons and dissipated about 100 kilowatts. It was also very difficult to program, programs being set up via 6000 multi-position switches (any one of which could be wrongly set!). It did work however and was capable of some 5000 addition operations per second and constituted an important link in the chain of development.

It is interesting to digress at this point to speculate on the "portability" of computers had this been the technology of today. A typical personal computer of today if realised in vacuum tube technology would weigh more than 1000 tons and would require a special connection to the electricity supply due to its large dissipation (several megawatts). The extremely high purchase price and running costs would have prevented any possibility of personal computing and all its consequences.

Closely on the heels of Eniac, but in another country—Britain—"Edsac" was developed at Cambridge University and completed in 1949. This machine used binary arithmetic and a mercury delay line based memory in which both program and data were stored (up to 512 numbers, each of 17 digits). This machine is also important since it is the first recorded instance of the use of the stored program concept.

The first really commercial computer was developed between 1950 and 1952. This machine was "Univac" and it was built by the Eckert-Mauchly corporation (later part of Sperry Rand). This machine was designed for data processing as well as computation, thus the "general purpose digital computer " was born. To prove its worth, it was used to process results of the 1950 US census and to predict the outcome of the 1952 USA presidential elections. It tipped Eisenhower to win—which he did! By1955, 15 Univacs had been sold and the computer age (and the age of computer-based predictions) had begun.

It is fortunate indeed that semiconductors emerged as a viable technology following the invention of the transistor in 1947. Early computers using vacuum tubes experienced serious reliability problems but a fatal defect became apparent in their use in the binary world of 1's and 0's. Vacuum tubes had never before been used with anode current suppressed for long periods (as is the case when storing a 0). It was found that when the grid was returned again to a positive potential (a logic 1) then anode current was not re-established. The problem was tracked down to an insulating interface which formed on the cathode during a long period of current suppression. The solution was to develop special vacuum tubes which, of course, attracted high costs. Further development of the computer might well have been seriously hindered had it not been for the emergence of semiconductor based technology at the critical time. Subsequent developments have progressed rapidly, particularly since the invention of planar processing and the consequent development of the integrated circuit(IC) in the early 1960s.

To date, one may distinguish six generations of computer hardware and associated digital circuitry:

- *1st generation* based on vacuum tubes, relays, etc.
- *2nd generation* based on individual transistors and semiconductor diodes with discrete passive components.
- *3rd generation* based on small scale integrated **(SSI)** logic circuits (gates and flip-flops) with some discrete passive components.
- *4th generation*—the introduction of medium scale ICs **(MSI)** providing simple subsystems and more complex logic functions in single packages.
- *5th generation*—the introduction of large scale ICs **(LSI)** providing complete subsystems and complete simple systems in single or a few packages.
- *6th generation*—the introduction of very large scale ICs **(VLSI)** providing large complex subsystems and complete systems in single packages.

The development of VLSI technology and the associated design techniques have brought about some interesting and significant changes in design philosophy for the system designer.

Following the developments in semiconductor technology of the 1950s and 1960s two major trends could be observed.

The first was that the area of electronic design split into two major subdisciplines—analog design and digital design. So marked was the divergence that analog and digital designers wrestled with quite different problems, used different tools and techniques, and regarded each other's discipline (and each other!) with a great deal of suspicion.

The second, particularly in the area of digital design, was an almost complete departure from circuit design. The digital designer became almost exclusively a designer of systems, small or large, by interconnecting blocks (or packages) of logic circuits—SSI, MSI and some LSI.

Systems were mostly configured around "standard" ranges of "off the shelf" logic (e.g. 74 series TTL). A similar trend was also, to a lesser extent, evident in analog design where standard packaged operational amplifiers and other common analog functions were readily available.

A third, less obvious but very significant trend was that most detailed circuit design became the task of relatively few IC designers within the semiconductor industry. The design of circuits in silicon became a highly specialized function performed by a self-styled elite using closely guarded rules and techniques, the whole process being only vaguely understood by the majority of "general practitioners". IC design relied on a detailed "in-house" knowledge of particular processes and parameters.

The advent of VLSI has changed the emphasis. To take full advantage of these advances it is essential to mate application and technology. This means, at the very least, that the system designer must fully understand the essential features, characteristics and constraints of the technology. For these reasons, knowledge of the basic behavior of semiconductor devices and design procedures and of the dominant technologies (currently nMOS and CMOS) is essential. An introduction to these topics now follows which will allow us to consider circuit realizations in silicon as we deal with various aspects of digital technology in the following chapters.

1.2 An introduction to semiconductor technology

If we are to design digital systems and have the greatest freedom in so doing, then it is necessary to acquire a basic understanding of integrated circuits and the skills necessary for the design of circuits in silicon, as well, of course, as being able to design systems by interconnecting logic and subsystems using custom and/or standard packaged logic. Although various technologies are available it is nevertheless true to say that the bulk of current digital designs are well served by MOS (metal oxide semiconductor) technology which embraces nMOS, pMOS and CMOS circuits and their derivatives.

However, it is useful to first establish a background in fundamental aspects of silicon-based semiconductor devices.

1.2.1 Basic semiconductor properties

As the name implies, an intrinsic semiconductor is neither an insulator nor a good conductor, Most often the semiconductor devices we wish to consider are based on the properties of silicon crystals. A silicon atom is represented diagramatically in Figure 1.1 and will be seen to consist of a nucleus and planetary electrons which are arranged in shells. A single crystal of silicon (monocrystalline silicon) has a regular lattice structure in which individual atoms are arranged in a regular array and in which, due to overlapping of the outer shell electron orbits, there is a consequent trading of outer (so called valence) electrons between atoms. Due to this trading process there is a strong bond between neighboring atoms and the crystal is in the solid state. This type of bonding is referred to as covalent bonding; the valence electrons involved in the bonding process occupy a band of energy levels referred to as the valence band. The fact that there is a band and not a single energy level is due to the large numbers of electrons involved (in the silicon crystal there are 10^{22} atoms/cm^3, each atom having 14 planetary electrons).The distance between atoms is determined by the radius of the outer shell electron orbit, and for silicon this radius is 1.17 Å which results in a nucleus to nucleus separation (center to center) of twice this value. Electrons in the valence band are not available to contribute to current conduction but if some can be given sufficient energy to cross an energy gap, denoted E_g , then they will become available for conduction from a higher energy band known as the conduction band (see Figure 1.2). For silicon, the energy

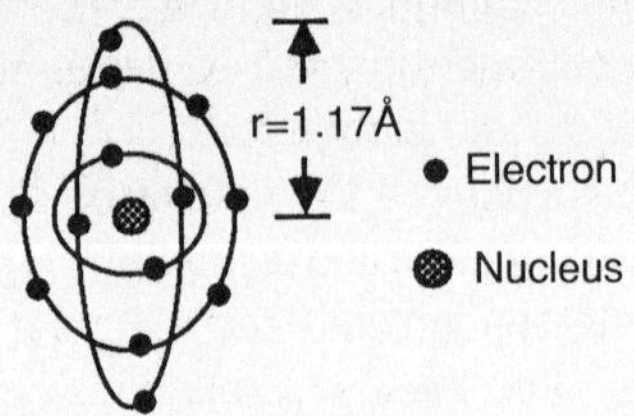

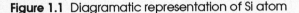

Figure 1.1 Diagramatic representation of Si atom

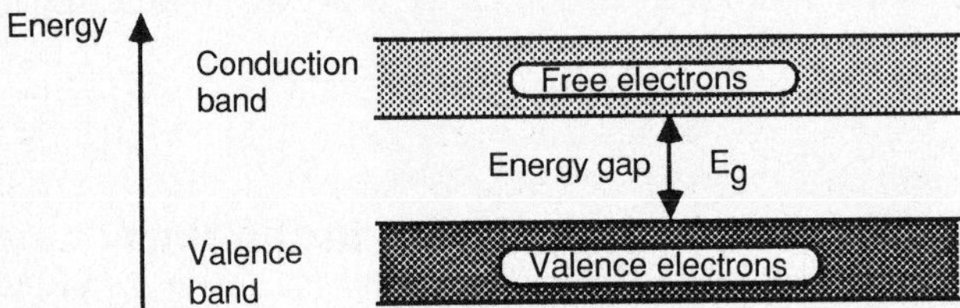

Figure 1.2 Band structure of a semiconductor

gap E_g between the valence and conduction band is approximately 1.1 e.V (electron.volts) at room temperature (300°K).

At any given temperature, atoms in the lattice vibrate around their lattice positions, the higher the temperature the more pronounced the effect. Due to random interactions there are some atoms which will vibrate more violently than others and "packets" of oscillations will exist at various locations in the lattice. Both the number of packets and the amplitude of vibration increase with temperature and at any given temperature above 0°K some electrons in the valence band will gain sufficient energy E_g to cross the energy band gap and become available in the conduction band. Such electrons are then referred to as free electrons. At room temperature the number of such free electrons is relatively small—appproximately $10^{10}/cm^3$ compared with the total of 14 X$10^{22}/cm^3$ (14 electrons per silicon atom arranged in three shells as indicated in Figure 1.1). In consequence, the intrinsic resistivity of silicon is large (ρ_i = 230 kΩ.cm at 300°K). At elevated temperatures, many electrons will cross the gap and silicon becomes a good conductor.

An atom consists of a positively charged nucleus which is neutralized electrically by the sum of the negative charges of the planetary electrons. Thus an un-ionized atom is space charge neutral so that, in the case of silicon, the nucleus has a positive charge of $+14q$ to balance the 14 planetary electrons (total charge $-14q$ where $q = 1.6 \times 10^{-19}$ coulomb). The shells defining the electron orbits are denoted 1st, 2nd, 3rd, etc., in terms of their closeness to the nucleus, the 1st being the closest.

The maximum number of electrons which can occupy a shell is given by:

Limiting number per shell = $2\,n^2$

where n is the shell number. Thus for silicon:

- the 1st shell is full and contains 2 electrons,
- the 2nd shell is full and contains 8 electrons, and
- the 3rd(outermost)shell is not full—containing the balance of 4 electrons.

Figure 1.1 attempts to illustrate this arrangement.

The force of attraction binding an electron to the atom is given by:

$$F = 9 \times 10^9 q^2 / r^2$$

where F is the force in newtons and r (meters) is the separation—nucleus to electron.

Clearly, electrons in the inner shells are strongly bound to the nucleus but outer shell electrons are more readily persuaded to break away from the atom. If we integrate to find the work done in removing an outer electron at radius $r1$ from one atom in isolation, we find that:

$$E_\infty = \int_{r1}^{\infty} F.dr = 9 \times 10^9 q^2 / r1$$

where $r1$ is the separation from the nucleus to the outer shell ($r1$ =1.17 Å for silicon). Substituting for q we find that the energy required is in the region of 10 e.V where e.V (electron.volt) is a unit of energy such that 1 e.V is the energy required to lift a charge q (one electron) through a potential of 1 volt. We have also seen that the band-gap energy E_g is 1.1 eV and that thermal agitation at 300°K is such that there are relatively few free electrons and the material will be a poor conductor (ρ_i =230 k .cm). However, our discussion of intrinsic conductivity is as yet incomplete; one important concept must be introduced at this point, and that is the concept of positively charged "**holes**". An electron crossing the band gap leaves the valence band and leaves a positively ionized silicon atom (one electron short). This positively charged atom may in turn capture an electron from a neighboring atom, and so on; in effect then the location of the positively charged ion moves through the crystal lattice. Thus a positively charged hole ($+q$) moves through the valence band of the lattice. It may be shown that a hole has an effective mass approximately equal to that of an electron. If an electric field is now applied across the crystal then free electrons will move in one direction in the conduction band and the corresponding holes in the other direction in the valence band. Both electrons and holes will contribute to the total current flow and clearly there will be an equal number of each. Thus at any given temperature in an intrinsic semiconductor there will be equal numbers of free electrons and holes, i.e. free electron-hole pairs.

For equilibrium in a semiconductor, both the temperature and the Fermi level must be constant throughout. The Fermi energy level, E_F, is a fundamental parameter describing the distribution of electron (and hole) energies and is such that at 0°K no electrons can be in states with energies higher than E_F. In an intrinsic semiconductor, it is readily shown that E_F lies halfway between the top of the valence and the bottom of the conduction bands as shown in Figure 1.3.

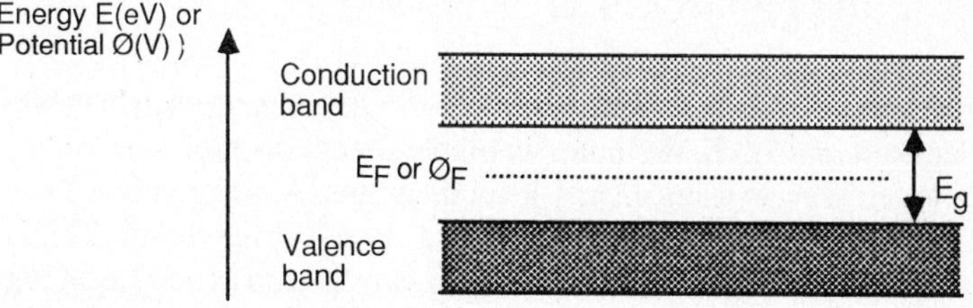

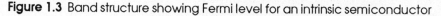

Figure 1.3 Band structure showing Fermi level for an intrinsic semiconductor

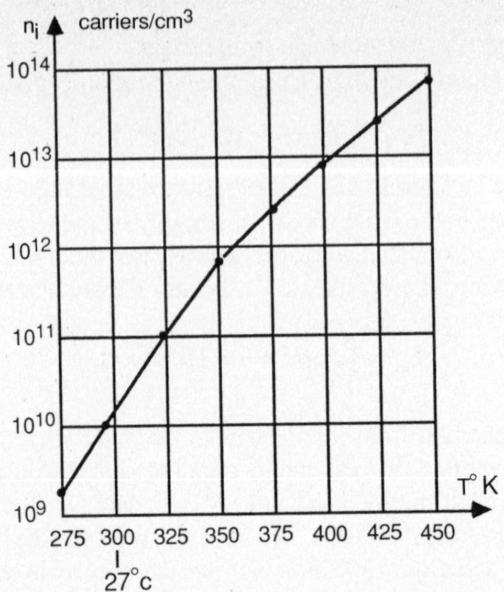

Figure 1.4 Intrinsic carrier concentration for silicon as a function of temperature

Again, for the intrinsic case, at all but very low temperatures, free electron-hole pairs are generated and both are current carriers having charges of $-q$ and $+q$ respectively. Electron or hole concentration, n_i may be calculated as follows:

$$n_i = \frac{4\sqrt{2}}{h^3}(\Pi m_e kT)^{\frac{3}{2}}.e^{\frac{-E_g}{2kT}}$$

h is Planck's constant = 4.14 X 10^{-15}e.V.sec (or 6.62 X 10^{-34}joule.sec)
m_e is electron mass (= m_h (hole mass)) = 9.11 X 10^{-31} kg
k is Boltzmann's constant = 8.61 X 10^{-5}e.V/°K (or 1.38 X 10^{-23}joule/°K
T is absolute temperature °K
E_g the band-gap energy (for silicon E_g =1.1 e.V approx.)
e is the base of natural logarithms = 2.7183

The relationship between n_i and T for silicon is plotted in Figure1.4 and may be restated as follows:

$$n_i(T) = 3.88 \times 10^{16}.T^{\frac{3}{2}}.e^{\frac{-7000}{T}}\ cm^{-3}$$

At room temperature (300°K) n_i evaluates to 1.5 X 10^{10} for silicon which means that there are 1.5 X 10^{10} electrons *and* 1.5 X 10^{10} holes available for conduction .

We have illustrated aspects using energy level diagrams. Alternatively we could have used potential rather than energy by dividing the energy levels by the charge. This is readily done without rescaling for energy levels in e.V which may be read as volts as in Figure 1.3. Thus energy-related parameters may be readily re-expressed, e.g. E_F, the Fermi level in electron volts may be given as $Ø_F$ the Fermi potential in volts.

1.2.2 The effect of doping

The type of conduction we have discussed up to now has been the intrinsic conductivity of a pure semiconductor crystal, such as silicon, in which equal numbers of electrons and holes contribute to the current flow but it is obvious that the range of devices which could be fabricated from intrinsic silicon would be severely limited due to its high resistivity.

However, the introduction of controlled amounts of the right types of impurities into the crystal lattice of a semiconductor will drastically modify its electrical properties. The materials thus formed will also exhibit "extrinsic conductivity" which can be drastically different from the poor intrinsic conductivity even for relatively low impurity doping levels. When an impurity atom enters the crystal lattice it may replace one of the silicon atoms in its position in the lattice or it may occur interstitially, i.e. between silicon atoms in the lattice. It is the former situation which is of interest in the fabrication of MOS circuits in silicon.

Silicon (and germanium) are group IV elements in the atomic table (see Table 1.1). Group IV means that their chemical valence is 4, or in other words, they each have four electrons in the outer (valence) shell as already discussed for silicon. Consider now what will happen if we replace a silicon atom in the lattice by, say, a group III element atom. Group III elements have only three valence electrons and when such an atom is located in place of the silicon atom it will pick up an additonal electron in the trading process. This will have two effects: first, it will negatively ionize the group III impurity atom since it will now have one more electron than it needs for neutrality. Secondly, a "hole" must be left somewhere else in the lattice since another (silicon) atom must be an electron short. This hole is then free to move in the lattice; however, the crystal as a whole remains electrically neutral.

The first effect is not important from the conduction point of view since the position of the ionized impurity atom is fixed in the lattice. However, the hole produced is free to move through the trading process. If an electric field is applied across the crystal, then free holes thus created will contribute to the resultant current flow (the ionized impurity atom will not move). Thus the doping process results, in this case, in extra holes (in addition to free electron-hole pairs due to intrinsic conductivity).In consequence of the interaction discussed, group III atoms are referred to as *acceptors* since they each *accept* an additional electron. Semiconductor material doped with acceptor or "p type" impurities is referred to as p type material in which the *majority* current carriers will be holes—majority because electron-hole pairs will also be generated due to the intrinsic conductivity of silicon. Clearly, doping levels can range from zero upwards to, say, 1 percent. In this range the number of impurity atoms is small compared with the number of silicon atoms. We may see that at the 1 percent level, 1 in every 100 silicon atoms will be displaced by an impurity atom of the chosen type. Remembering that there are 10^{22} silicon atoms per cm^3 then clearly there will be 10^{20} impurity atoms per cm^3.

Consider the effect of relatively small doping levels, say an impurity concentration $N_A=10^{16}/cm^3$ (i.e. 1 impurity in every 10^6 silicon atoms). In this case there will be 10^{16} free holes per cm^3 due to the impurities. Compare this with the intrinsic free electron-hole pair concentration of $10^{10}/cm^3$. It may be seen that even this relatively low doping level will increase conductivity by a factor of 10^6 compared with the intrinsic case (and of course the resistivity decreases by the same factor).

Consider now the effect of doping with a group V element (see Table 1.1). A group V atom has five valence electrons in the outer shell, one more than is needed in the trading (bonding) process in the silicon lattice. Thus this type of impurity atom sheds an electron which becomes available for conduction and the impurity atom will consequently become positively ionized in its fixed position in the lattice. By similar reasoning, the majority carriers will now be electrons. Group V elements are known as *donors* (because they give up electrons) and material thus doped is referred to as "*n*" *type*. The situations applying in doped silicon are illustrated in Figure 1.5 and for either "p" or "n" type doping levels N_A or N_D there is a direct relationship between doping level and resistivity as in Figure 1.6.

As a memory jog it is useful to remember that electro*n*s, do*n*ors, "*n*" type and *n*egative charge carriers are all interrelated and all contain the letter *n*. To reinforce this we may note that acce*p*tors, "*p*" type, and *p*ositive charge carriers all contain the letter *p*—unfortunately this does not extend to include holes.

The properties of silicon and common doping elements are summarized in Table 1.1 and it may be shown that the Fermi energy level E_F (and hence the Fermi potential $Ø_F$) will shift with doping, the direction of shift depending on the doping type (n or p) and the amount of shift being related to the doping level. This is indicated diagrammatically in Figure1.7 and for strongly n-type material it may be shown that:

$$E_F = \text{intrinsic level} + kT \ln (N_D/n_i)$$

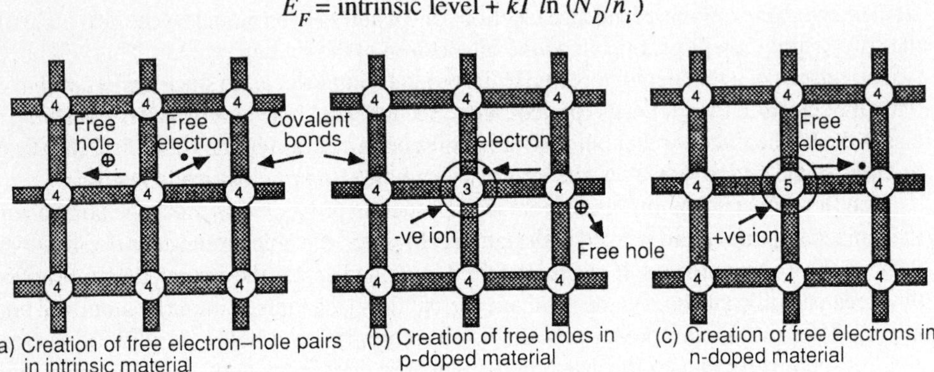

Figure 1.5 Diagramatic representation of effect of doping in crystal lattice

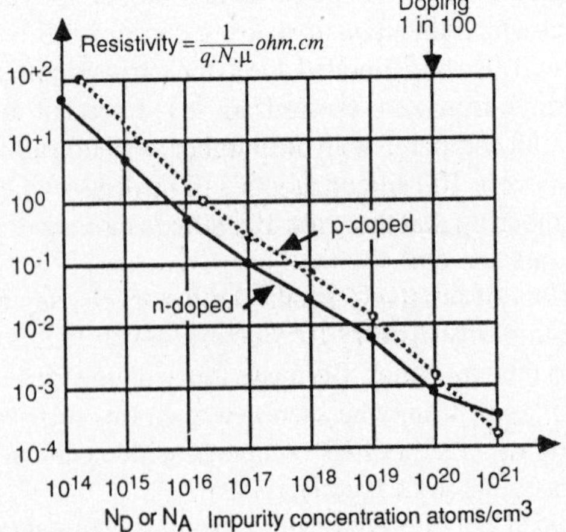

Figure 1.6 Resistivity of silicon at room temperature

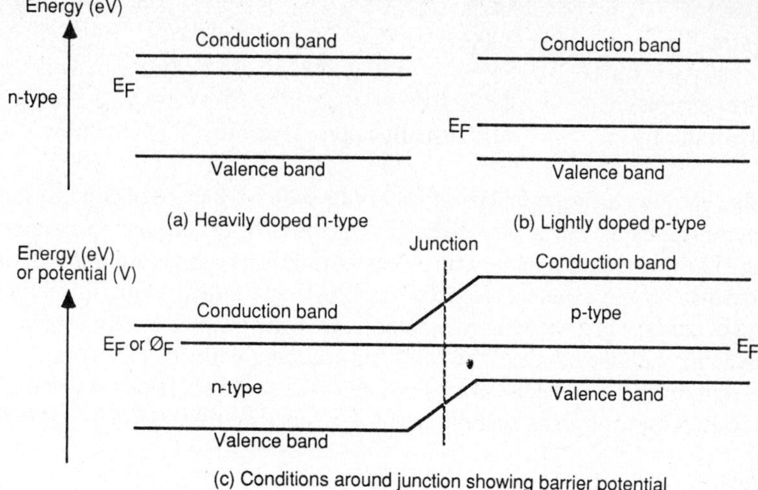

Figure 1.7 Position and alignment of Fermi levels ($N_D >> N_A$)

and for strongly p-type material:

$$E_F = \text{intrinsic level} - kT\ ln\ (N_A / n_i)$$

N_D and N_A are donor and acceptor doping concentrations respectively.

Table 1.1 Atomic table

	p-type Group III				*intrinsic* Group IV		*n-type* Group V		
	Boron (B)	Aluminium (Al)	Gallium (Ga)	Indium (In)	Silicon (Si)	Germanium (Ge)	Phosphorus (P)	Arsenic (As)	Antimony (Sb)
Number of valence electrons	3	3	3	3	4	4	5	5	5
Atomic number	5	13	31	49	14	32	15	33	51
Ionization energy to free carrier(e.V)	.045	.057	.065	.16	1.205- $2.8X10^{-4}T$	0.782- $3.9X10^{-4}T$	.012	.0127	.0096
Radius in lattice Å	.81	1.26	1.26	1.44	1.17	1.22	1.13	1:18	1.36

Some further relevant properties of silicon:

Atomic number	14
Atomic weight	28.66
Density kg/ m^3	$2.33 X 10^3$

Melting point °C	1420
Boiling point °C	2600
Thermal condvty watt/°C	84
Coeff. of expansion /°C	4.2 X 10^{-6}
Relative permittivity	11.7 (usually approximated as 12)

In digital circuitry an important parameter is the switching speed of circuitry, and this in turn is determined by the rate at which charges can move from one point to another through the lattice. The movement of charge with respect to time is, of course, current and it is readily appreciated that the movement of electrons (or holes) in the lattice is quite different from the movement of charges in a vacuum; for example, we can visualize electrons being accelerated, interacting with the atoms of the solid and colliding with other carriers.

Let us start by evolving a general expression for the current between two points l meters apart in a solid of uniform cross-section, area A. Considering electrons in n-type material we have:

$$\text{Total charge per unit volume} = -q.n$$

where n is the current carrier concentration. Thus:

$$\text{charge in region of interest} = -q.n.A.l.$$

Now the time t required for an electron to move distance l is:

$$t = l/v_n$$

where v_n is the velocity of the electron.

Thus, conventional current I_n (opposite in direction to the movement of electrons) is given by:

$$I_n = +q.n.A.l.\ v_n /l = +q.n.A.v_n$$

[Similarly, if we were considering holes in a p-type region:

$$I_p = -q.n.A.l.\ v_p /l = -q.n.A.v_p$$

where v_p is the hole velocity .]

Now we know that current must be proportional to voltage or electric field, but in the expression for current we may see that q, n and A are all fixed. However, if we determine v_n (or v_p) we find that:

$$v_n = \mu_n.E \text{ or } v_p = \mu_p.E$$

where E is the electric field and μ is known as *mobility* and is a factor determined by the ease with which the charge carriers can move through the lattice (the subscripts indicating that there are differences between electrons and holes). Mobility is not a constant but depends both on temperature (approx. as $T^{-3/2}$) and on transverse electric fields. Mobility also varies with doping levels as indicated in Figure 1.8. In consequence it is only possible to quote typical figures for mobility, for example for *bulk mobility* we may use the following:

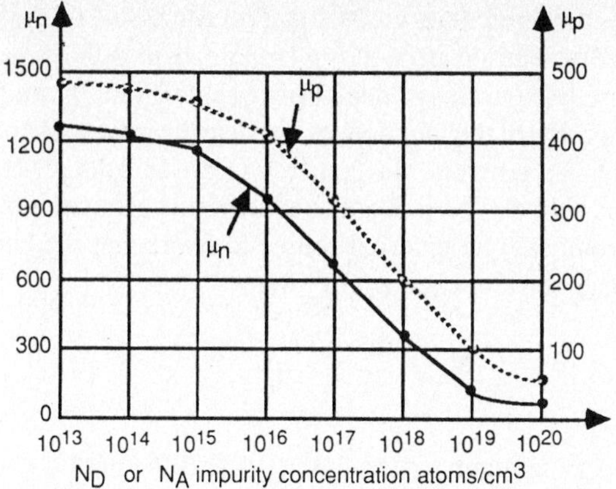

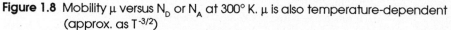

Figure 1.8 Mobility μ versus N_D or N_A at 300° K. μ is also temperature-dependent (approx. as $T^{-3/2}$)

$$\mu_n = 1250 \text{ cm}^2 / \text{V.sec}$$
$$\mu_p = 480 \text{ cm}^2 / \text{V.sec}$$

Note the approximate 2.5 : 1 ratio between μ_n and μ_p.

However, in a consideration of MOS circuits it is the *surface mobility* which is relevant and it has been found that *surface figures are roughly half the bulk values* and in consequence we will use:

$$\mu_n = 650 \text{ cm}^2 / \text{V.sec}$$
$$\mu_p = 240 \text{ cm}^2 / \text{V.sec.}$$

All the figures given are for relatively lightly doped regions at 300°K.

Resistivity, ρ, is clearly also dependent on mobility and in fact it is readily established that:

$$\rho_n = \frac{1}{q.N_D.\mu_n} \quad \Omega.\text{m.}$$

$$\rho_p = \frac{1}{q.N_A.\mu_p} \quad \Omega.\text{m.}$$

1.2.3 p-n (or n-p) junctions

If p-type and n-type regions are in direct contact, a junction is established which has interesting electrical properties, acting as a diode and as a voltage-dependent capacitor. Almost all semiconductor devices include p-n (or n-p) junctions and usually the doping levels on either side of the junction are unequal; for example, we will consider a junction for which $N_D >> N_A$ (Figure 1.9). When a free electron-rich region is brought into contact with a free hole-rich region, there will be a momentary flow of electrons in one direction and holes in the other, as indicated in Figure 1.9, and a subsequent recombination of electrons with holes in the regions in the immediate vicinity and on either side of the junction. Since both regions were electrically neutral before they were joined, then the depletion of charges due to the recombination will leave charge depletion regions (as in Figure 1.10) with immobile ionized atoms on either side of the junction, positive in the n-type and negative in the *p*-type depletion regions. Thus a potential difference (or barrier voltage) now exists across the junction, and it acts as a diode and will not conduct until sufficient voltage of the opposite polarity is applied from, say, an external source to overcome the barrier potential. The barrier voltage, V_B, may be calculated as follows:

$$V_B = \frac{kT}{q} . \ln \frac{N_A . N_D}{n_i^2}$$

where:
- k is Boltzmann's constant = 1.4 X 10^{-23} joule/°K
- T is temperature °K
- q is electron charge = 1.6 X 10^{-19} coulomb
- N_A = Acceptor concentration p-region
- N_D = Donor concentration n-region
- n_i = intrinsic carrier concentration

[Note that kT/q evaluates to 25.9 mV at 300°K.]

The barrier potential can be represented diagrammatically as shown in Figure 1.7, noting the alignment of the Fermi voltage through the junction. From our considerations it is apparent that the barrier voltage is +ve on the n side, so that in order to forward bias the diode we must apply an external voltage > V_B in such polarity that the p side is positive with respect to the n side. The diode will present a high impedance to voltages of the same polarity as the barrier voltage.

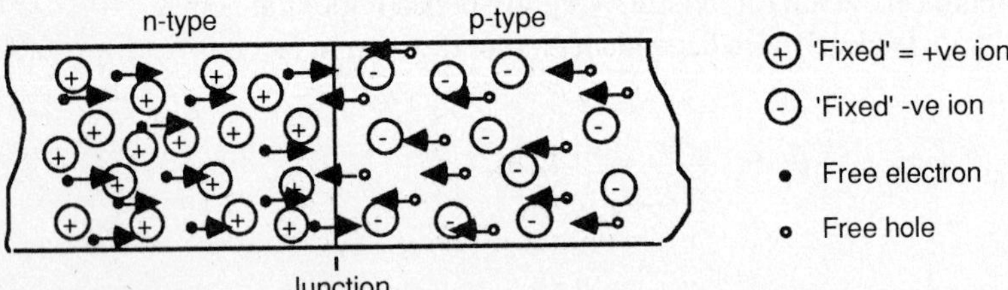

Figure 1.9 Diffusion of carriers across abrupt junction

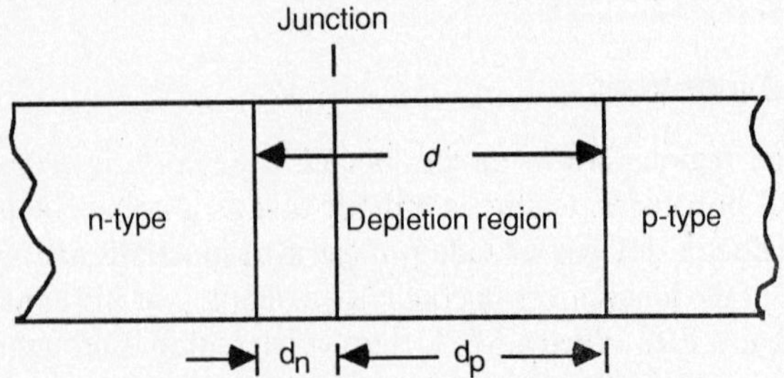

Figure 1.10 Charge depletion region around abrupt junction

Apart from its diode characteristics, the junction also exhibits capacitance due to the charge depletion regions (Figure1.10) which act as a dielectric between the adjoining semiconductor regions which, in turn, act as the plates of a parallel plate capacitor structure. The capacitance C_{pp} of a parallel plate structure is determined from:

$$C_{pp} = \frac{\varepsilon_0 \varepsilon_r A \times 10^{12} pF}{d}$$

where: ε_0 is the permittivity of free space = 8.85 X 10^{-14} $F/_{cm}$
ε_r is relative permittivity =11.7 for Si (often taken as 12)
A is the area of the "plates" (m^2)
d is the plate separation (m)

Alternatively, we may write:

$$C_{pp} = \frac{\varepsilon_0 \varepsilon_r}{d} \frac{pF}{\mu m^2}$$

Now, if we assume that $N_D > N_A$ (the situation reflected in Figure 1.10)) then $d_n < d_p$. It is readily shown that:

$$N_D . d_n = N_A . d_p .$$

and if $N_D >> N_A$ then $d_n << d_p$.

The total depletion region width $d = d_n + d_p$; but since $d_p >> d_n$ we may take $d = d_p$ as a good approximation (or $d = d_n$ if $N_A >> N_D$).

Thus the effective dielectric width, d, may be taken as:

$$d = \sqrt{\frac{2\varepsilon_0 . \varepsilon_r . V_T}{q . N_A}} \qquad \text{when } N_D >> N_A$$

or,

$$d = \sqrt{\frac{2\varepsilon_0 . \varepsilon_r . V_T}{q . N_D}} \qquad \text{when } N_A >> N_D$$

All parameters have already been defined except for V_T which is the total voltage across the junction. A little thought will establish that:

$$V_T = V_B + V_{App.}$$

where: $V_{App.}$ is an externally applied voltage acting across junction
V_B is thejunction barrier voltage as already defined.

If N_A and N_D are not widely different, the full expression for d is:

$$d = \sqrt{\frac{2\varepsilon_0 . \varepsilon_r . V_T}{q . (N_A + N_D)} . \left(\frac{N_A}{N_D} + \frac{N_D}{N_A} \right)}$$

Combining this with the expression for parallel plate capacitance, the junction capacitance C_j is given by:

$$C_j = A\sqrt{\frac{\varepsilon_0.\varepsilon_r.q.(N_A+N_D)}{2(V_B+V_{APP.})}}.\left(\frac{N_A}{N_D}+\frac{N_D}{N_A}\right)$$

Or for the case previously considered when N_A and N_D are very different,

$$C_j = A\sqrt{\frac{\varepsilon_0.\varepsilon_r.q.N}{2(V_B+V_{APP.})}}\times 10^{12}\,pF$$

For silicon, and taking $\varepsilon_r = 12$, we may write:

$$\text{junction capacitance per unit area } = 2.9\times 10^{-12}\sqrt{\frac{N}{V_T}}\frac{pF}{\mu m^2}$$

Where N is the impurity concentration on the lightly doped side.

Junction and depletion region capacitances are important features in determining the characteristics of MOS circuits.

In discussing the production of useful electrical features in silicon we have covered the underlying principles of making diodes, resistances of known value (doping levels determine resistivity) and capacitors using the junction effect. Clearly we must have effective means for making transistors and for connecting these circuit elements to form circuits, subsystems and systems as the need arises.

1.3 Basic nMOS and CMOS technologies

1.3.1 Basic MOS transistors

Let us first turn our attention to basic nMOS transistor structures with reference to Figure 1.11. The nMOS circuits are formed in a p-type substrate of moderate doping level. Transistor source and drain regions are formed by diffusing (or Implanting) n-type impurities through masks and clearly the source and drain of an enhancement mode transistor (Figure1.11(a)) are isolated one from another by the depletion regions which are formed round the n-type areas in a p-type substrate(diodes are formed at each junction). In order to make a useful transistor we must have means for establishing and controlling current flow between source and drain. This is accomplished by a polysilicon gate on thin gate oxide insulation over the region between the two. If a suitably positive voltage V_{gs} is applied between the gate and source(which is shown connected to the substrate) then, if $V_{gs} > V_t$ a charge inversion region, n type in this case, is formed under the gate oxide, and source and drain are joined through this channel. V_t is known as the threshold voltage and for typical nMOS processes has the value of +20% V_{DD}, where V_{DD} is the positive supply rail voltage, source and substrate being connected to GND (0V). V_t is typically +1 V for V_{DD} = +5 V. A current I_{ds} may now flow between source and drain if a positive voltage V_{ds} is applied between them.

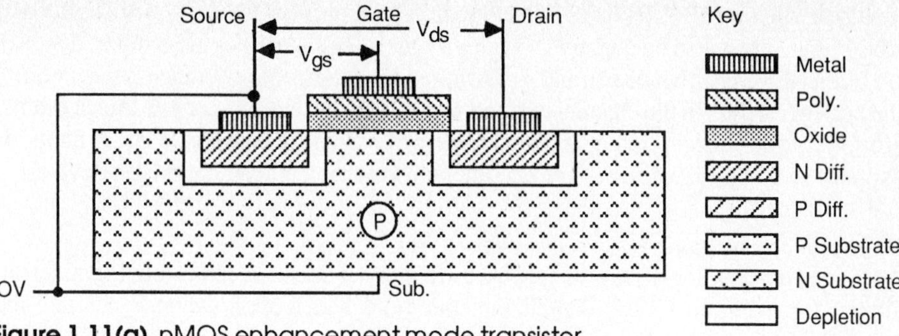

Figure 1.11(a) nMOS enhancement mode transistor

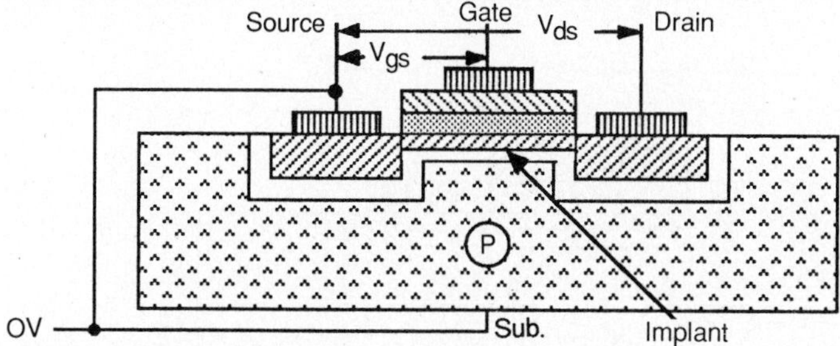

Figure 1.11(b) nMOS depletion mode transistor

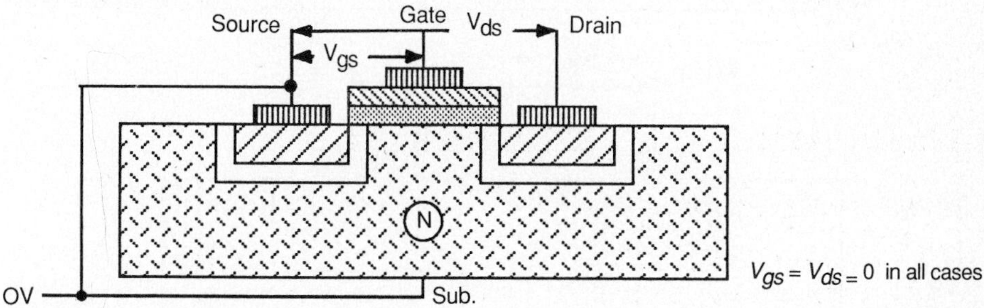

Figure 1.11(c) pMOS enhancement mode transistor

It is also possible to fabricate depletion mode transistors as in Figure 1.11(b). The structure is similar to the enhancement mode device except that an additional processing step is used to implant n-type impurities in the channel region so that the source and drain are connected through an n-type channel even in the absence of any applied voltage V_{gs}. If, however, a suitable negative value of V_{gs} is applied then holes attracted to the underside of the gate insulation will neutralize the n-type implant and the channel will cease to exist. The threshold voltage is thus negative and typically has a value of -80% V_{DD} (i.e.-4 V for V_{DD} = + 5 V). The characteristics of the two types of transistor are similar except for the difference in the value of V_t.

We will also be making use of enhancement mode p-type transistors in forming CMOS circuits. The basic pMOS structure as in Figure 1.11(c) is similar to the nMOS enhancement mode device except that p and n regions are interchanged and consequently all voltage and current polarities are also changed. For example, V_t is typically -1 V for a -5 V supply and V_{gs} and V_{ds} etc. must also be negative. Otherwise their characteristics are similar. There is also a further difference between n and p-type transistors due to the difference in the mobility of the current carriers in the channel regions. For n-type, the current carriers are electrons, but for p-type the carriers are holes which have lower mobility than electrons. A factor of 2.5 applies so that p-type devices are 2.5 times slower than comparable n-type devices.

nMOS enhancement mode transistor action

To gain an understanding of, say, the nMOS transistor, let us consider the conditions set out in Figure 1.12. Figure 1.12(a) depicts the conditions which apply when $V_{gs}>V_t$ and the channel is established but no voltage is applied between source and drain so that there will be no current flow. Now, with reference to Figure1.12(b), let a modest voltage $V_{ds}<V_{gs}-V_t$ be

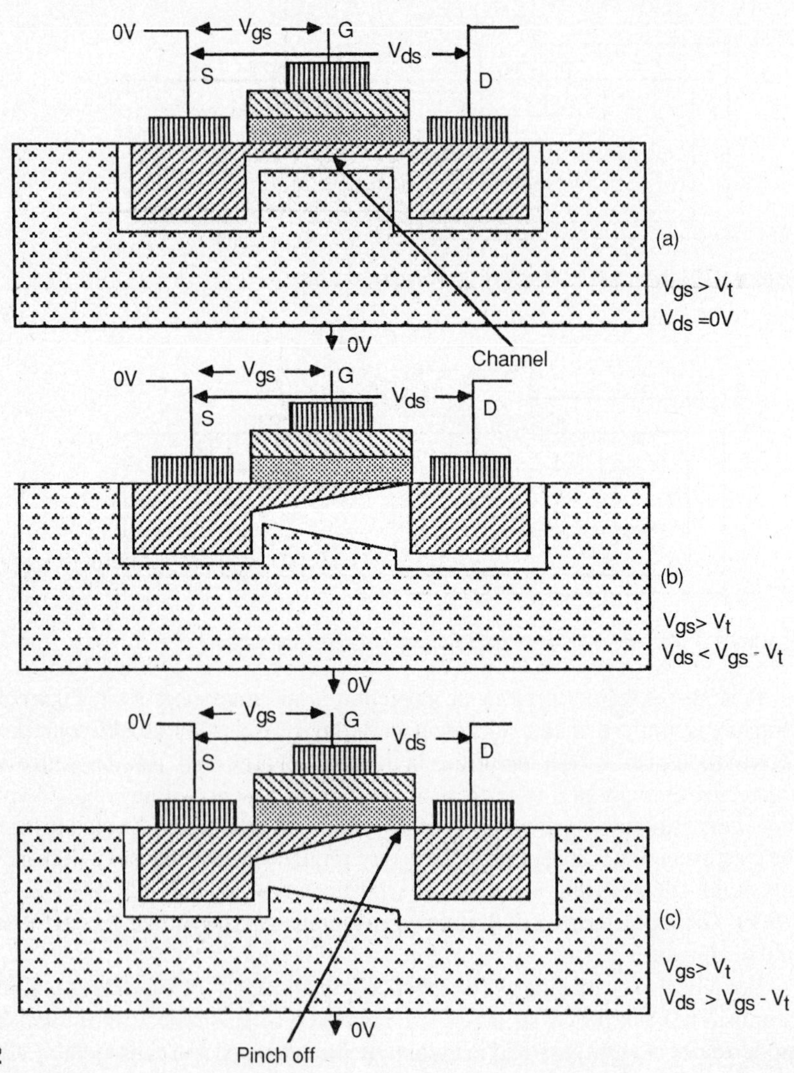

Figure 1.12

applied which results in a current flow I_{ds} with a consequent IR drop in the channel as shown. In this region the channel behaves resistively, the resistance being controlled by V_{gs}. Now if a larger value of V_{ds} is applied, such that $V_{ds}>V_{gs}-V_t$ (Figure1.12(c)), then an IR drop of $V_{gs}-V_t$ takes place over less than the entire channel length so that, at the right-hand end, there is insufficient gate to channel voltage to establish the channel in that region; the channel becomes "pinched off" as shown. In this region of operation, known as saturation, current flow is completed by diffusion currents and the device then exhibits a high resistance which is almost independent of V_{ds}, that is, the transistor now behaves as a constant current source. Thus there are three distinct regions of operation:

1. cut off when $V_{gs}<V_t$,
2. the linear or resistive region where $V_{gs}>V_t$ and $V_{ds}< V_{gs}-V_t$,
3. saturation where $V_{gs}>V_t$ and $V_{ds}> V_{gs}-V_t$. Here the device behaves as a constant current source.

nMOS depletion mode transistor action.
The action is similar to that described in the preceding section except that the threshold voltage is now negative.

pMOS enhancement mode transistors
The deliberations for the nMOS enhancement mode transistor apply but with a reversal of voltage and current polarities.

Commonly used symbols for nMOS and pMOS transistors are set out in Figure 1.13.

1.3.2. Summary of nMOS fabrication processes

An overview of fabrication may be obtained from a very brief summary :

- Processing takes place on a thin (e.g. 0.1 mm) wafer cut from a single crystal (up to 150 mm diameter) of p doped silicon .
- Mask 1: Thin oxide mask is used to produce a patterned layer of thick oxide (e.g.,1 μm thick for 5 μm technology) with the thick oxide *not* present where we have diffusion regions and transistor channels.

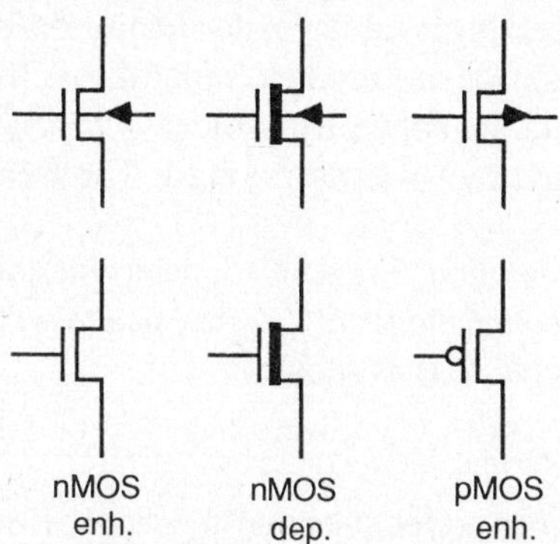

Figure 1.13 Transistor circuit symbols

- Thin oxide (about 1/10th of thick oxide) is then grown over the regions where the thick oxide is not present.
- Mask 2: Implant mask, is "Anded" with mask 1 to allow n-type ion implantation in depletion mode transistor channels regions .
- Mask 3: Polysilicon mask is used to pattern polycrystalline silicon (polysilicon or poly.) which is n doped to reduce resistance and deposited overall to about the same thickness as the thick oxide. It is then removed except for the areas defined by this mask. This forms the gate structures of transistors and some interconnections.
- Remove thin oxide not covered by polysilicon.
- Diffuse or implant n-type impurities into all regions not covered by thick oxide or poly. This process is *self-aligning* since there is no way that a transistor source and drain regions can misalign to the gate.
- Grow thick oxide overall.
- Mask 4: Contact cuts are made through the thick oxide to expose the underlying poly. or diffusion in the areas where contacts to the metal layer (or from poly to diffusion) are desired.
- Mask 5: Metal mask is used to pattern metal (aluminum) which is deposited overall(about same thickness as thick oxide), metal being etched away outside those areas defined by this mask.
- Mask 6: Overglassing is applied overall except for areas defined by this mask (e.g.where pads to the outside world are present).
- A substrate connection is made in fabricating and packaging nMOS chips.

1.3.3 CMOS fabrication

CMOS circuits utilize both n-type and p-type transistors so that, in effect, two substrates are needed, a p substrate for the n devices and an n substrate for the p devices.

This problem is usually solved by fabricating a substrate within a substrate and there are several ways of doing this.

The CMOS p-well process

In this popular process the wafer is n-type in which all p-type devices are formed. The second (p-type) substrate is provided by a deep p-type diffusion forming a p-well in which all n-type devices are formed. The processing is similar to that for the nMOS devices of the preceding section but two extra masking stages are needed: one mask to define the p-well boundary and a second new mask to differentiate between those areas where n-type and p-type diffusions are required. This is most often known as the p + mask. The process may be appreciated by considering Figure 1.14.

An inverter circuit cross-section, Figure 1.15, demonstrates the fact that *we must now make deliberate substrate connections to V_{DD} (the positive rail) and deliberate p-well connections to V_{SS} (the negative rail)* as shown.

The CMOS n-well process

Clearly, we may also start with a p-type substrate in which we form the n-type components. The p-type components are then accommodated in an n-well. The same extra masking stages are needed but an n+ mask replaces the p + mask.

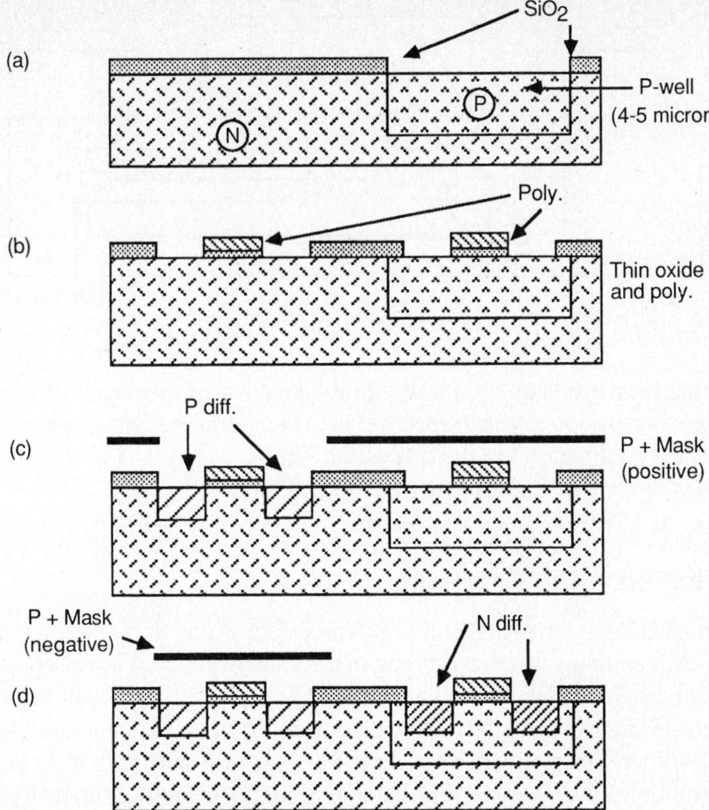

Figure 1.14 CMOS p-well process steps

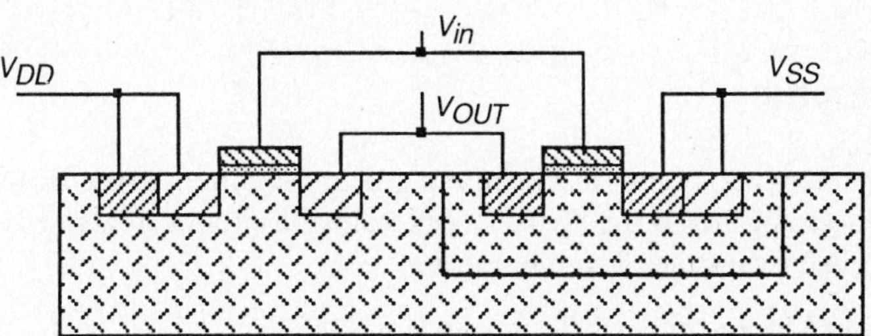

Figure 1.15 P-well CMOS inverter

N-well processes are also popular and are often established as retrofits to existing nMOS lines.

Both n-well and p-well CMOS processes must be carefully carried out to avoid latch-up problems.

The CMOS twin-tub process

The latch-up problems and associated compromises in fabrication led to the development of this process. The substrate upper layer is now formed of intrinsic silicon which, as we have

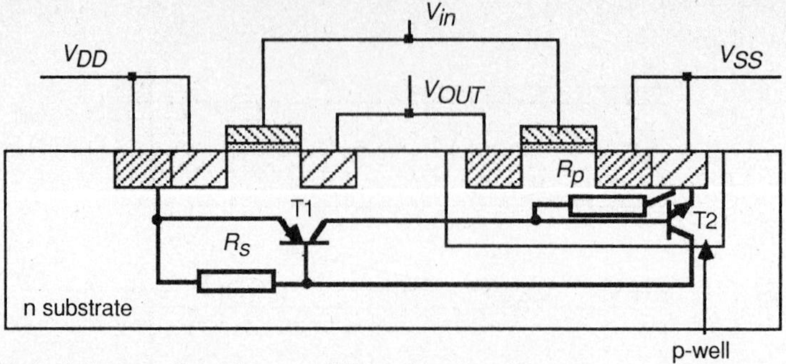

Figure 1.16 Latch-up effect in p-well inverter structure

seen, is virtually an insulator. Two wells or tubs are diffused into this layer, one p and one n-type. The n and p components respectively are then formed in these tubs.

The penalty paid here is in extra processing steps.

In this text we will concentrate on the p-well process.

1.3.4 Latch-up in CMOS circuits.

A problem inherent in the p-well and n-well processes is due to the parasitic components formed by the numerous junctions present in a typical structure. The parasitic component configuration leading to latch-up is indicated in Figure 1.16. If sufficient voltage is developed due to substrate current flow in R_s (due to the substrate region resistivity) then both transistors will turn on, giving rise to a disastrous low resistance path from rail to rail.

This condition is avoided by careful parameter control in fabrication and by the designer placing sufficient substrate (V_{DD}) and p-well (V_{SS}) connections in the designs.

1.4 MOS layers

The design of integrated circuits in silicon is a process of turning a specification into mask layouts for processing the silicon. We have seen that MOS circuits are formed of conducting layers on and in the substrate. These layers are:

1. (Diff.)—n diffusion with p diffusion also for CMOS,
2. (Poly.)—polysilicon, and
3. Metal—usually aluminum.

During fabrication, these layers are insulated from each other by thick or thin silicon dioxide (referred to as oxide). We may also deduce that wherever the polysilicon and thin oxide masks intersect (that is wherever poly crosses diff.) a transistor is formed. Layers may also be joined by contacts.

In design, therefore, we need a simple way of representing layer and topology so that we can do initial design work with relatively simple diagrams if we so desire. A simple expedient in this respect is the use of *stick diagrams.*

1.5 Stick diagrams

Stick diagrams convey layer information through the use of a colour code (Color plate1) - red for poly., blue for metal, green for n diffusion, yellow for p diffusion (CMOS) or implant (nMOS), and black for contacts. Monochrome encoding is also used where color is not available (e.g. using a monochrome graphics terminal or when material has to be reproduced by a copying machine) one such scheme being set out in Figure1.17. The figures also show that color and monochrome encoding schemes carry through to mask layout encoding. (These encoding schemes are commonly used and are set out in full, for example, in Pucknell and Eshraghian, *Basic VLSI Design*, 2nd edition, Prentice Hall '88).

By way of example, two stick diagrams are included in Figure 1.17 while Figure 1.18 indicates the ready translation from stick diagram to mask layout for simple structures.

1.6 Design rules and layout

Design rules are necessary to provide a working interface between the designer and the fabricators and reflect the capability of the fabrication process to be used. Thus, design rules in absolute terms will vary from one fabricator to another, but thanks initially to the work of Mead and Conway*, a widely acceptable set of rules for both nMOS and CMOS has been developed. The set is known as lambda (λ) based rules and is based on a single dimensional parameter, λ, which is subsequently allocated an absolute value in microns (μm) for fabrication.

In general, λ will have a value equal to half the *feature size* of the fabrication technology to be used. For example, a 5 μm feature size implies a minimum line (or path) width of 5 μm with a consequent value of $\lambda = 2.5$ μm. A 2 μm feature size implies a minimum line width of 2 μm with a consequent value of $\lambda = 1.0$ μm, etc..

The design rules then are all specified in terms of λ and are fixed with the possible effects of mask misalignments and fabrication imperfections in mind.

Design rules therefore specify minimum allowable line widths, minimum separations between separate paths in the same layer and paths in different layers, minimum acceptable overlaps for contact areas and for transistor geometry, and acceptable overlaps and clearances for other features such as p-wells and p+ masks, etc.

The design rules are summarized here as figures 1.19 (a) - (e) (in monochrome form) and colored versions in color plates 2 and 3.

To illustrate the use of the design rules, some stick diagrams and associated mask layouts are set out in Figure 1.20 (monochrome) and in color plate 4, but a full exposition may be sought in texts covering basic VLSI design.

1.7 Electrical properties

In order to assess the likely performance of circuits being designed for implementation in silicon, either by hand calculations or by computer-aided modelling or simulation, we need

* Mead C.A., Conway L.A. *Introduction to VLSI systems*, Addison & Wesley, U.S.A., 1980.

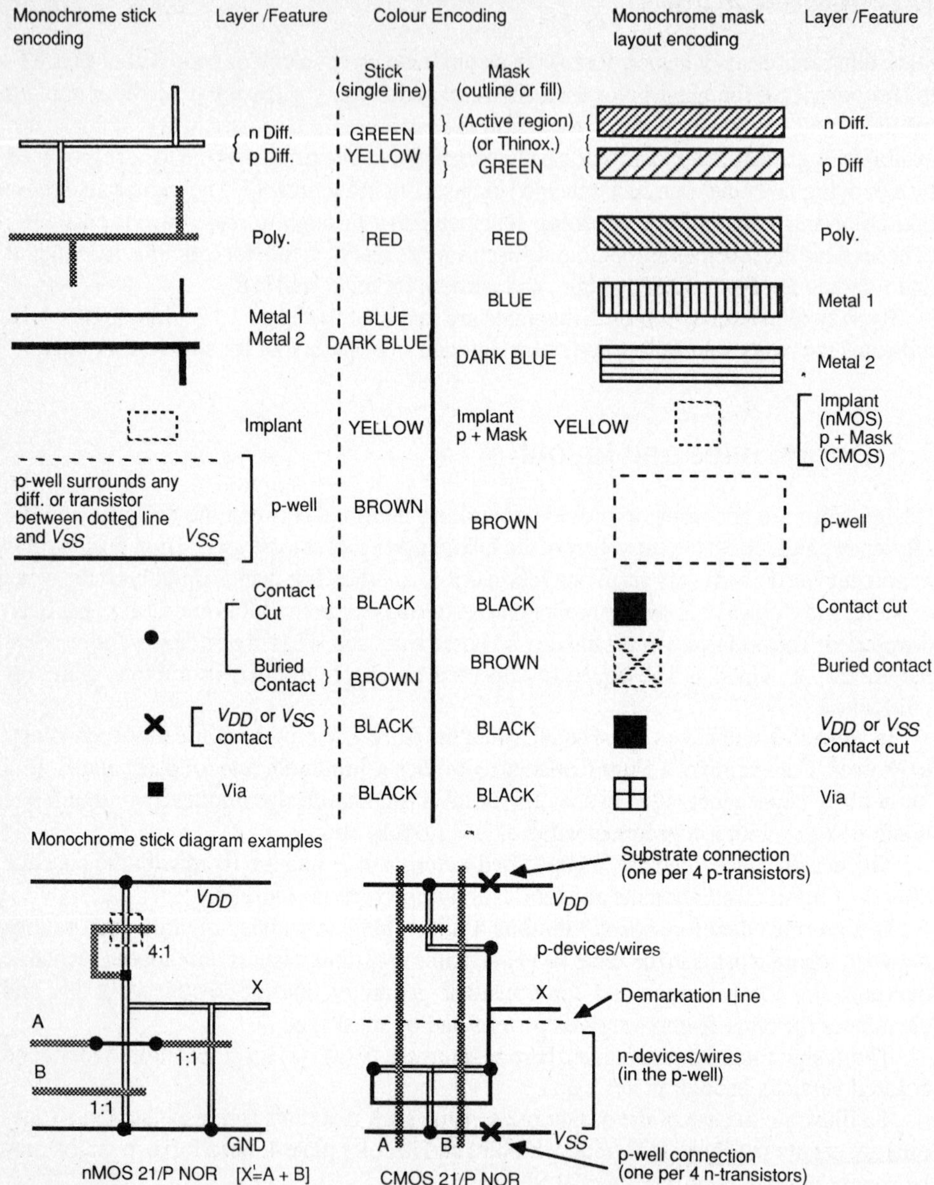

Note: Layers are insulated from each other unless deliberately joined by a contact. However, wherever Poly crosses diffusion, a transistor is formed.

Figure 1.17 Layer and feature encoding schemes (see also Color Plate 1)

a knowledge of the parameters determining the behavior of layers and transistors. In particular we need to know the resistance and capacitance associated with the layers and transistor channels as well as other parameters associated with the operation of transistors. All must apply to the particular fabrication process to be used.

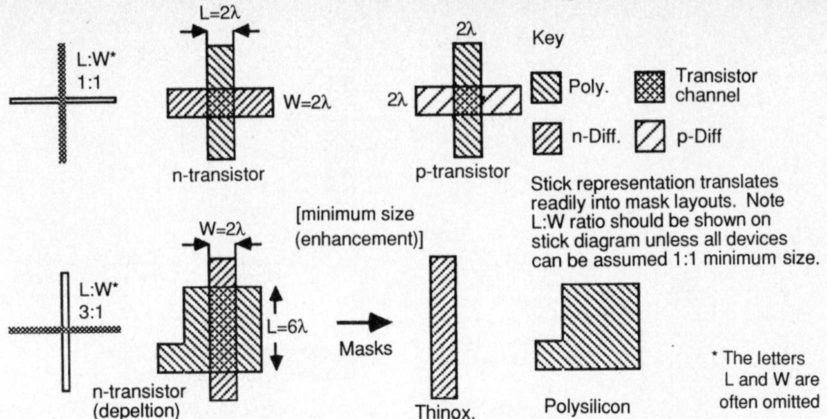

Note: The width of the channel is determined by the Thinox width under poly.

Figure 1.18 Translation of stick diagrams (transistors) into mask layouts

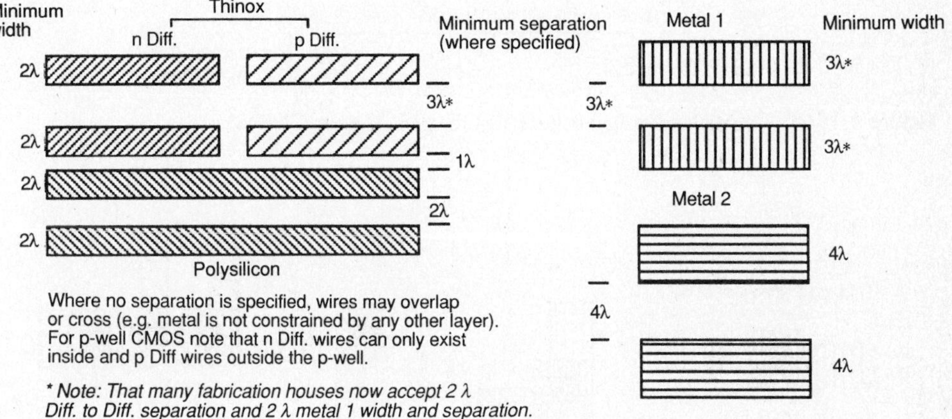

Figure 1.19(a) Design rules for wires (nMOS and CMOS)

1.7.1 Parameters for MOS transistors

We have seen that *nMOS* transistors exhibit the following *positive* voltages and current in operation:

$$V_{ds}\,;\,V_{gs}\,;\,V_t\ \text{(effective gate voltage} = V_{gs} - V_t\text{)}$$

(substrate and source assumed connected to $0V$ as in Fig. 1.21(b)).

I_{ds} - current is *electron flow source to drain* (i.e. conventional current is $+I_{ds}$ from drain to source).

For *pMOS* transistors all the above voltages are *negative* and I_{ds} is also negative since current is *hole flow from source to drain*.

For depletion mode transistors, the threshold voltage V_{td} takes on the opposite sign to V_t and is usually of larger magnitude.

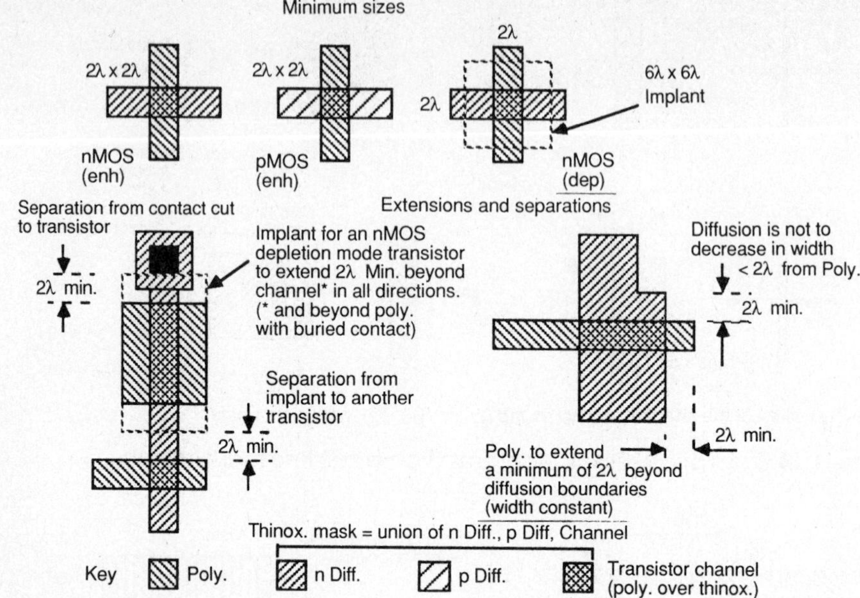

Figure 1.19(b) Transistor design rules (nMOS, pMOS and CMOS)

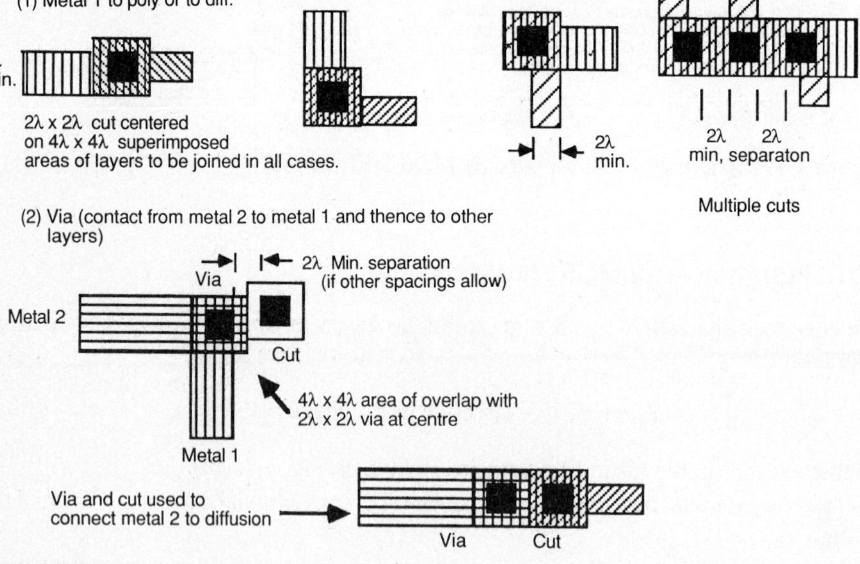

Figure 1.19(c) Contacts (nMOS and CMOS)

(i) Buried contact usually nMOS only: Basically, layers are joined over a 2λ x 2λ area with the buried contact cut extending by 1λ in all directions around the contact area, except that the contact cut extension is increased to 2λ in diff. paths leaving the contact area. This is to avoid forming unwanted transistors. See following examples.

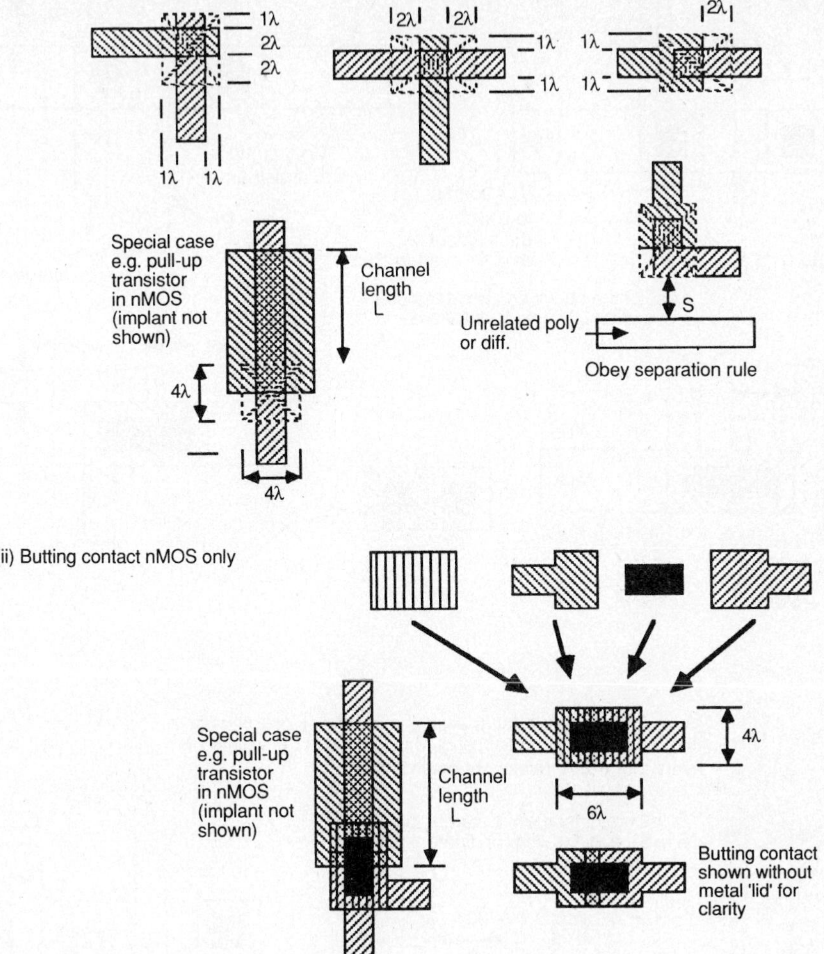

Figure 1.19(d) Contacts poly. to diff.

Now the I_{ds} v V_{ds} relationship is deduced from considering Figure 1.21(a). *First for n-type in the resistive (non-saturated) region of operation* (for which the voltage conditions are as set out in Figure 1.21(b)).

$$\textit{Current } I_{ds} =^{+} \left(\frac{\varepsilon_0 \cdot \varepsilon_{SiO2} \cdot \mu_n}{D}\right) \cdot \frac{W}{L}\left[\left(V_{gs}-V_t\right)V_{ds} - \frac{V_{ds}^{2}}{2}\right] \qquad (1)$$

and for p-type,

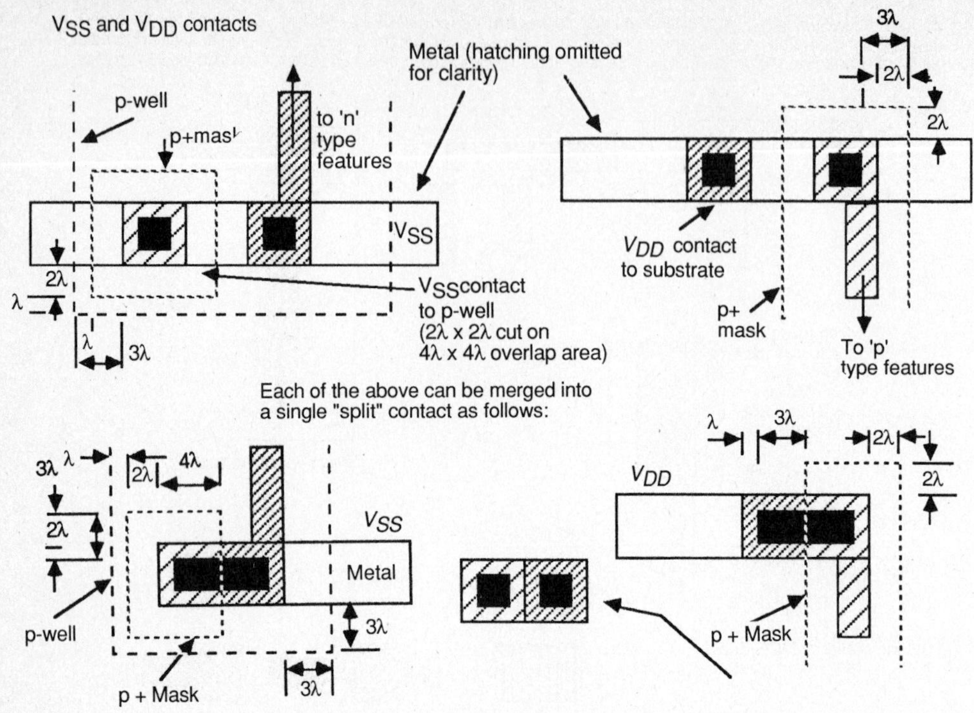

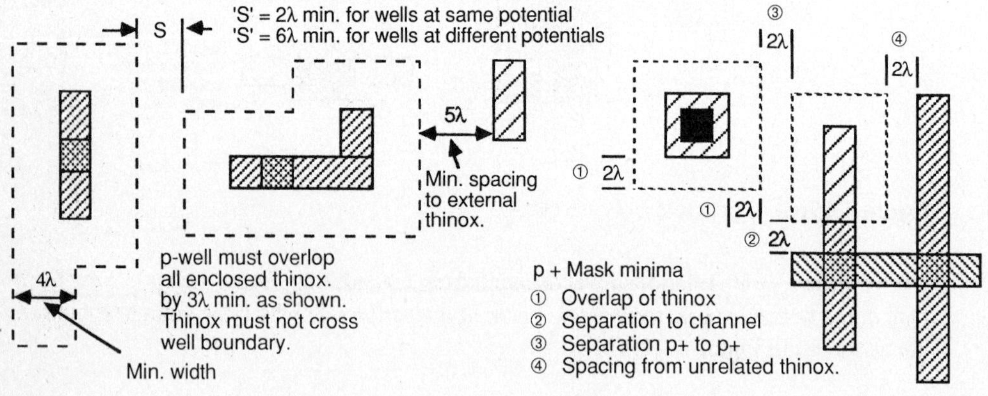

Figure 1.19(e) Particular rules for p-well CMOS process

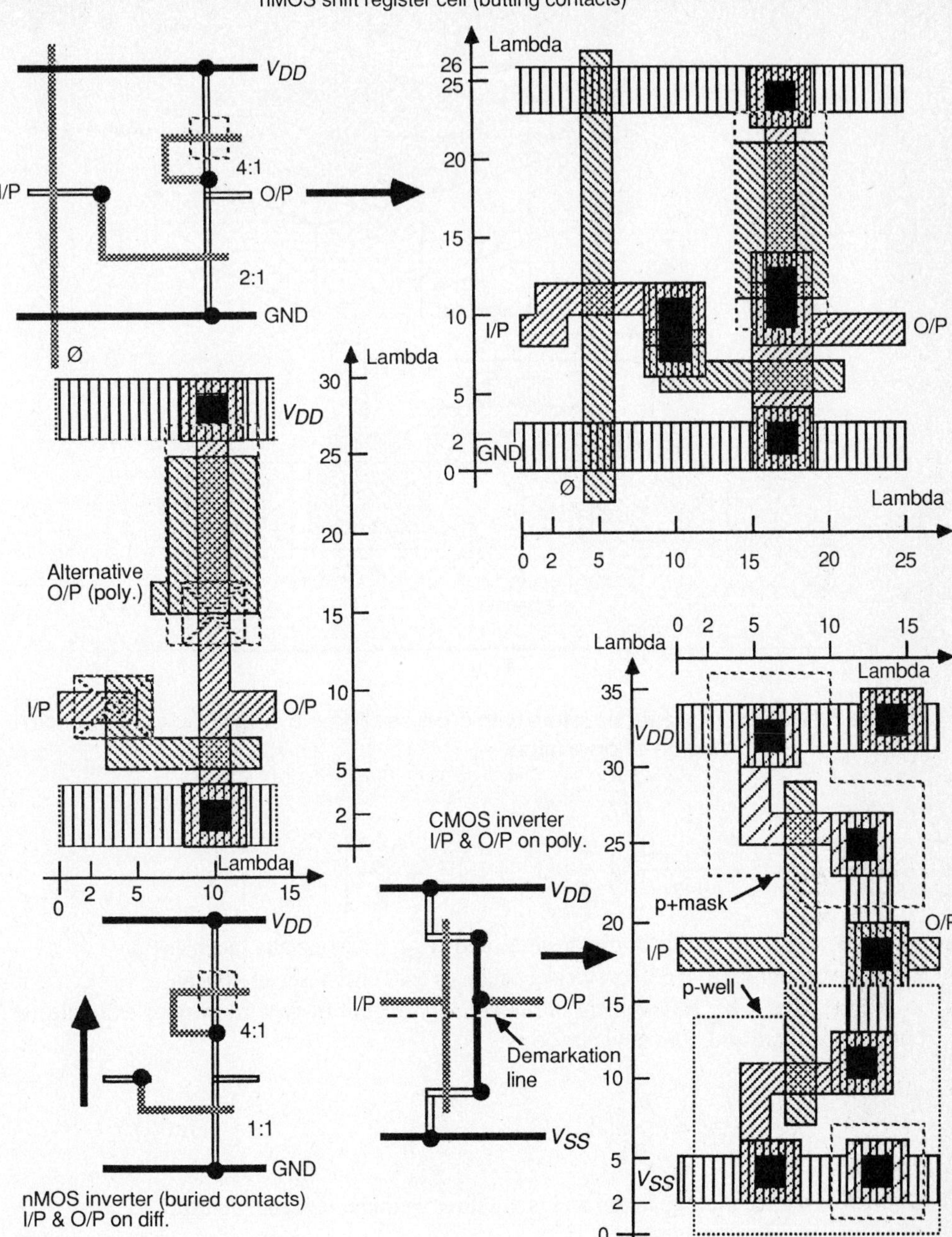

Figure 1.20 Example layouts

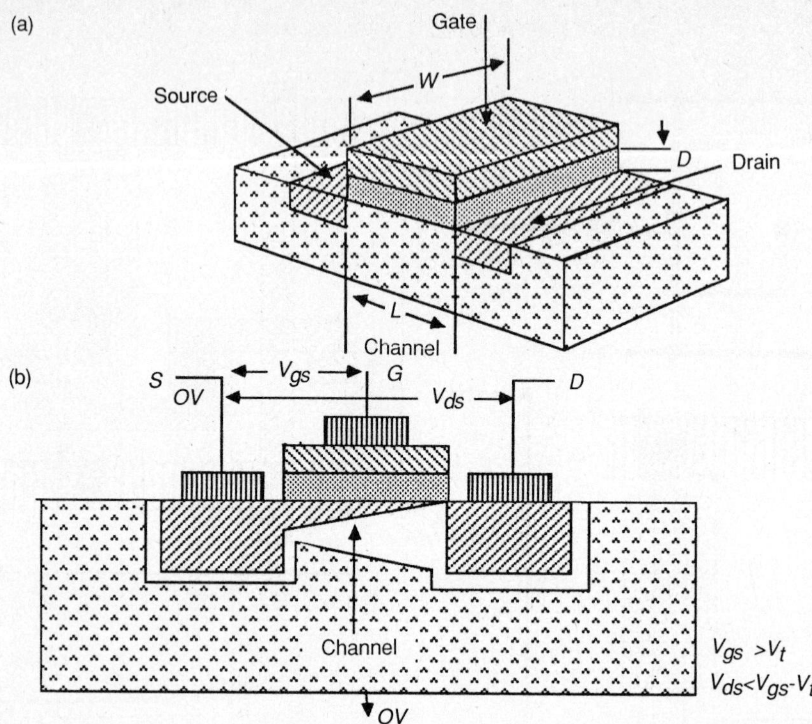

Figure 1.21 (a) (b) Transistor structure with cross-section showing conditions in resistive region of operation

$$\text{Current } I_{ds} =^{-} \left(\frac{\varepsilon_0 \cdot \varepsilon_{SiO2} \cdot \mu_p}{D}\right) \cdot \frac{W}{L}\left[(V_{gs} - V_t)V_{ds} - \frac{V_{ds}^{2}}{2}\right]$$

where W and L are transistor dimensions, and, ε_{SiO2} =3.9 (usually taken as 4). Typically, mobility μ_n = 650 cm²/V.sec and μ_p = 240 cm²/V.sec at 300°K.

The I_{ds} versus V_{ds} *relationship in the resistive region* (n-type by way of example) is commonly expressed for convenience as:

$$I_{ds} =^{+} K.\frac{W}{L}.\left[(V_{gs} - V_t)V_{ds} - \frac{V_{ds}^{2}}{2}\right] \quad (1a)$$

where *K* is the technology factor and is not directly under designer control;

or

$$I_{ds} =^{+} \beta\left[(V_{gs} - V_t)V_{ds} - \frac{V_{ds}^{2}}{2}\right] \quad (1b)$$

where ß is the Beta factor, and $ß = K.\frac{W}{L}$

or, $$I_{ds} = {}^{+}\mu . C_0 . \frac{W}{L} . \left[\left(V_{gs} - V_t\right) V_{ds} - \frac{V_{ds}^{\ 2}}{2} \right] \quad (1c)$$

where C_0 is the gate capacitance per unit area $= \dfrac{\varepsilon_0 . \varepsilon_{SiO2} .}{D}$

Alternatively, total gate capacitance C_g may be used, where $C_g = C_0 . W.L.$ Suitable sign reversals yield the equations applicable to pMOS transistors.

Channel resistance $$R_{ch} \text{ or } R_{ds} = \frac{V_{ds}}{I_{ds}}$$

and from eqn. *(1b)* for example,

$$R_{ch} \text{ or } R_{ds} = \frac{V_{ds}}{\beta \left[\left(V_{gs} - V_t\right) V_{ds} - \frac{V_{ds}^{\ 2}}{2} \right]} \quad (2)$$

and if V_{ds} is small,

$$R_{ch} \text{ or } R_{ds} = \frac{1}{\beta \left(V_{gs} - V_t\right)} \quad (2a)$$

thus, when $V_{ds} < V_{gs} - V_t$ the channel acts as a resistance controlled by V_{gs}.

Secondly, for current I_{ds} in the saturation region:

$V_{ds} = V_{gs} - V_t$ defines the onset of saturation which continues for $V_{ds} > V_{gs} - V_t$.

Substituting $V_{ds} = V_{gs} - V_t$ in equations *(1)* and allowing for both n-channel and p-channel devices we have:

$$I_{ds} = {}^{+/-} \left(\frac{\varepsilon_0 . \varepsilon_{SiO2} . \mu}{D} \right) . \frac{W}{L} \left[\frac{\left(V_{gs} - V_t\right)^2}{2} \right] \quad (3)$$

or, $$I_{ds} = {}^{+/-} \frac{K}{2} . \frac{W}{L} . \left(V_{gs} - V_t\right)^2 \quad (3a)$$

or, $$I_{ds} = {}^{+/-} \frac{\beta}{2} . \left(V_{gs} - V_t\right)^2 \quad (3b)$$

The transistor behaves as a voltage controlled current source in saturation.

The behavior of the nMOS transistor under consideration is clearly apparent in the characteristics set out in Figure 1.22. The characteristics of a p-channel transistor are similar but with polarity changes.

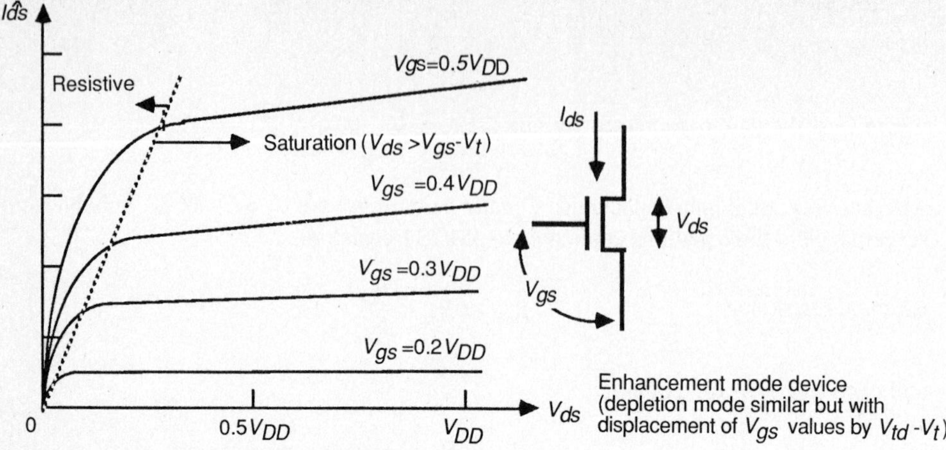

Figure 1.22 MOS transistor characteristics

Channel resistance at the onset of saturation:

$$R_{ch} \text{ or } R_{ds} = \frac{2}{\beta.\left(V_{gs} - V_t\right)}$$

and thereafter in saturation:

$$R_{chsat} \text{ or } R_{dssat} = \frac{2V_{ds}}{\beta.\left(V_{gs} - V_t\right)^2}$$

Up to now, we have assumed that substrate and p-well connections, asappropriate, are made to V_{DD}or to V_{SS}respectively, and this is usually the case. However, it is possible to return the substrate and well connnections to voltages other than V_{DD} or V_{SS} and thus have an effect on threshold voltages (V_t or V_{td}) since any bias voltages will change the effective potential of the channels. The effect of substrate bias V_{SB}, may be calculated from the following approximate expressions:

$$V_t = \gamma\,(\,V_{SB}\,)^{1/2}$$

where V_t is the change from the Zero bias condition $V_{t(0)}$, and γ is a process dependent parameter, sometimes denoted as B_E (Body effect).

Typical values for γ are in the range 0.15 to 0.8 $V^{1/2}$ and this parameter will be specified by the fabricator for a particular process. Note that different values of γ for n and p transistors will most likely apply. In assessing this effect, typical values for a 5 µm nMOS process are:

Typical values (zero bias)

$$V_{t(0)} = +0.2\,V_{DD}$$

$$V_{td(0)} = -0.7\,V_{DD}$$

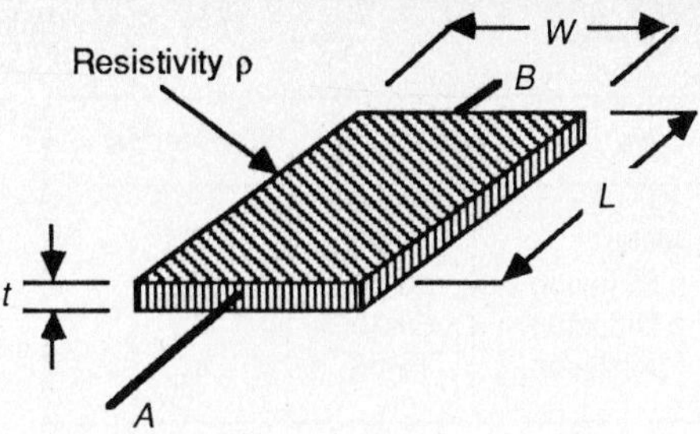

Figure 1.23 Sheet resistance R_s

Typical values ($V_{SB} = -V_{DD}$)

$$V_t = +0.3\ V_{DD}$$
$$V_{td} = -0.6\ V_{DD}$$

In all cases, a proper assessment of effects requires a knowledge of the parameters associated with the process to be used for fabrication.

1.7.2 Sheet resistance R_s

It is useful to recognize the concept of sheet resistance for the various layers and with regard to path A to B in a layer as in Figure 1.23:

$$R_{AB} = \frac{\rho . L}{W . t .}$$

and for $W=L$,

$$R_{AB} = R_s = \frac{\rho}{t}$$

that is, R_s ohm per square is a constant value for a given thickness t and resistivity ρ.

Since t and ρ are fixed by the process, the designer may conveniently work in terms of R_s and paths or regions in any layer can be regarded as being composed of squares in series and/or in parallel. For example, a metal wire 60 λ long and 3 λ wide in, say 5 μm technology, may be regarded as 20 squares in series giving an overall resistance from end to end = $20R_s$ = 20X .03 Ω = 0.6 Ω, since R_s for metal is .03 Ω in 5 μm technology. In order to put the relative resistance values in perspective, the range of typical values set out in Table 1.2 cover those to be expected in 2 μm to 5 μm CMOS(n-well or p-well) and nMOS technologies.

It will be seen that the most significant resistance values are those of transistor channels. Since transistors are often designed with $W=L$ (and with minimum dimensions) then the values given in the table are those typically associated with an "on" transistor (remembering that R_{ch} and thus R_{CH} depends on applied voltages, see Expressions (2) and(2a)). Other sheet resistance values are less significant but care must be taken when running wires of any significant length in polysilicon or diffusion since they may present sufficient resistance to generate large delays when driving normal circuit C values. Contact resistances (Table 1.2(b))are generally negligible, but the designer should be aware that contacts are not perfect.

Table 1.2(a) Typical R_s Values

Layer	R_s Ω/□	Denoted by
Metal	.03	R_{SM}
n Diffusion	20-40	R_{SDn}
p Diffusion	50-100	R_{SDp}
Polysilicon	15-50	R_{SP}

Channel R_s

n Transistor	10k-20k	R_{CHn}
p Transistor	25k-50k	R_{CHp}

Table1.2(b) Typical contact resistance values

Metal/Poly.	1.5 - 15Ω
Metal/Diff.	5 - 35Ω
Poly./Diff.(Butting)	25 - 100Ω

1.7.3 Area and Peripheral capacitances

It is easy to see that metal and poly regions, which are separated from each other and from the substrate by layers of insulation, will form parallel plate capacitor structures. Such capacitances are dependent on the areas in question and on the relevant capacitances per unit area.

For diffusion regions the structure is less obvious until one recognizes that the depletion region associated with a diffusion region acts as a dielectric(see Figure 1.24). However, it will be seen that area capacitance is only part of the total C presented by a diffusion region.

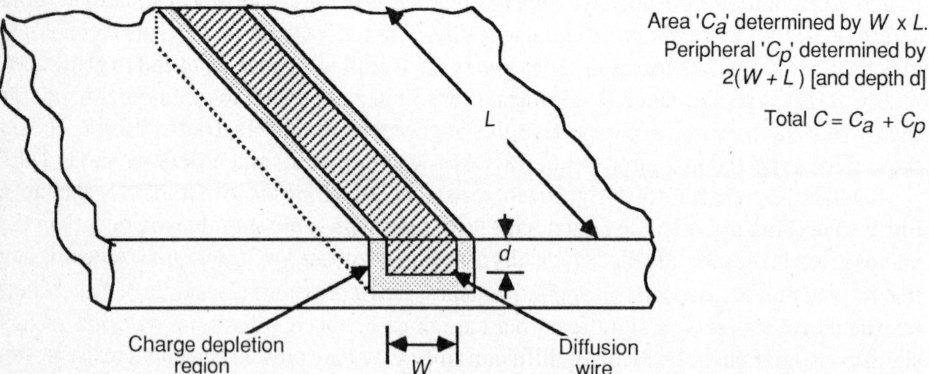

Figure 1.24 Determination of total capacitance to substrate for "deep" diffusion wire

Considerable "sidewall" or peripheral capacitance is also present for diffusions of any depth (see Figure 1.24) and this must be allowed for in calculating total C. In the case of implanted regions, the depth of the region is usually very small and peripheral effects are negligible. Significant capacitive effects are also associated with gate/channel regions of transistors since oxide is thin in these areas.The capacitance presented does not vary widely over the three possible states of the channel, that is:

1. on, in resistive region,
2. on, in saturation, and
3. off.

A convenient measure is the gate C per unit area, denoted C_0or, alternatively, the C_g value for a minimum size transistor ($W = L = 2\,\lambda$) denoted □C_g. Typical C values are set out in Table1.3.

Table 1.3(a) Capacitance values per unit area

Feature	*Denoted*	*Typical value* (10^{-4} pF/μm^2) 5 μm features	2 μm features
Gate/channel C	C_0	4	8-10
Poly./substrate	C_P	0.4	0.6
Metal/substrate	C_M	0.3	0.3
Diff./substrate	C_{DAn} or C_{DAp}	1.0	1.75
Metal1/Poly.	C_{MP}	0.3	0.6
Metal1/Diff.	C_{MD}	0.3	0.6

□ Cg = .01pF (5μmfeatures) or .004pF (2μm features).

Table 1.3(b) Peripheral capacitance value n or p diffusion

Feature	*Denoted*	*Typical value* (10^{-4} pF/μm) 5 μm features	2 μm features
Peripheral C	C_{DPn} or C_{DPp}	8	2 to 4 or negligible*

* If a shallow Implant is used for the diffusion areas.

1.7.4 Properties of pass transistors and transmission gates

Given switches which are operated by and which can be used to switch logic levels, and which can be connected in series and/or in parallel, we have the capability of producing *And* and *Or* type logic circuits. Add inverters to the switches and we then have all the necessary elements to put together all possible combinational and sequential logic circuits. We will now evaluate these elements in a CMOS environment, noting that nMOS and pMOS are also covered as subsets of CMOS.

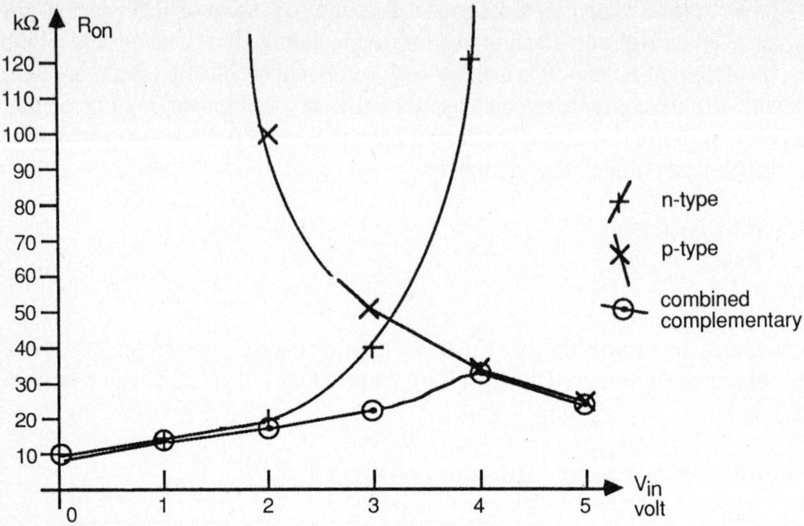

Figure 1.25 "On" resistance of pass transistor switches

Pass transistor switches

The basic nMOS and pMOS enhancement mode transistors find ready application in switch-based logic arrangements. Used in this way they are referred to as *pass transistors*. In order to assess their usefulness we may set out their basic properties in an environment in which logic level 0 = 0 volt (V_{SS}); logic level 1 = $+V_{DD}$ (e.g. +5 volt).

For pass transistors, we must also note that threshold voltage V_t effects will degrade one or other output logic level and that each pass transistor, when "on", presents a series resistance R_{on} (or R_{ch}) as well as gate/channel capacitance C_g at all times. The properties are summarized in Table 1.4 and in Figure 1.25.

Table 1.4 Properties of pass transistor switches

Type	*Symbols*	*Switching signals to gate*	*Threshold effects*	*Typical Series 'Ron'*	*Gate/Channel 'Cg'*
n-type	G, S D; G, S D	ON: Logic 1; OFF: Logic 0 (single input G)	1, 0; '1'; poor 1, good 0	10kΩ min. big increase with V in.	1□Cg e.g. .01pF for min. size device.
p-type	G', S D; G', S D	ON: Logic 0; OFF: Logic 1 (single input G')	1, 0; '0'; good 1, poor 0	25kΩ min. big increase as V_{in} drops.	1□Cg e.g. .01pF for min. size device.

Limitations on the use of pass transistors

Due to the threshold-voltage-induced loss of logic levels, we must place the following restriction on the use of pass transistors:

The input to the gate of a pass transistor must not come from the output of a pass transistor of the same type. The reasons are apparent when one considers the arrangement as in Figure 1.26.

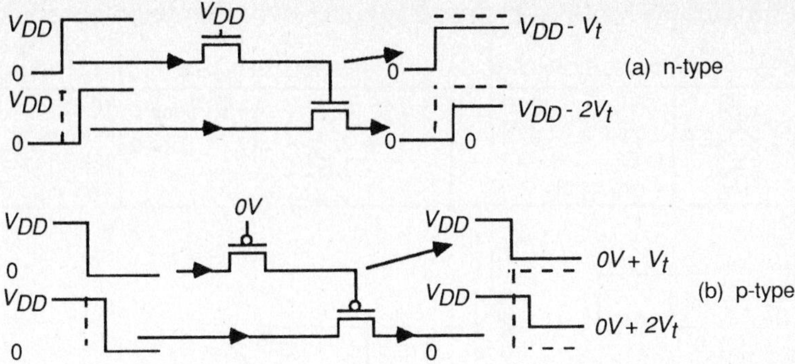

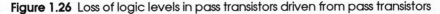

Figure 1.26 Loss of logic levels in pass transistors driven from pass transistors

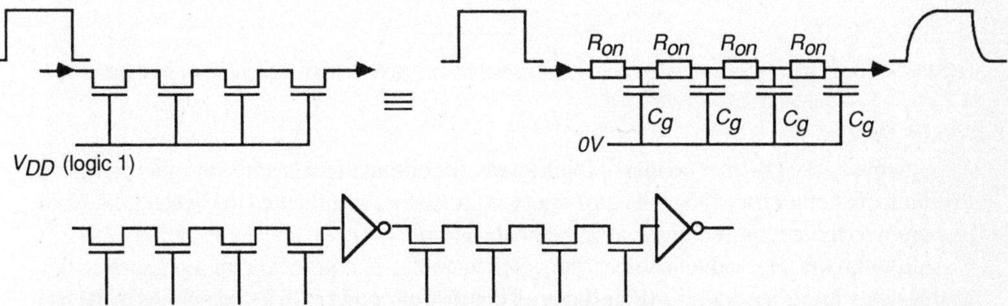

Figure 1.27 Delay (proportional to n²) through n-pass transistors in series. This effect also applies to transmission gates in series

A second limitation is due to the finite "on" resistance in association with the gate/channel capacitance C_g which gives rise to delays when pass transistors are connected in series as in Figure 1.27. It will be noted that the value of C_g may be taken as $C_g = C_0.W.L$ under all conditions.

Such considerations lead to the following restriction:

The number of pass transistors to be directly connected in series should not exceed four if excessive delays are to be avoided. If more than four are necessary then they should be sectionalized in fours with inverter/buffers as shown.

Transmission gate (complementary) switches

By combining the n and p pass transistors in parallel we may overcome the loss of logic levels and produce a switch which faithfully transmits good logic 1 and logic 0 levels. In addition, the use of two switches in parallel also produces a lower and more uniform R_{on} than is the case for simple pass transistor switches (see Figure 1.25). The properties of the transmission gate are summarized in Table 1.5.

Limitations on the use of transmission gates

Due to the use of two transistors in parallel, the transmission gate occupies more than double the area of a pass transistor switch and also presents twice the gate/channel capacitance. It also needs two (complementary) switching inputs rather than one.

Table 1.5 Properties of complementary switches (transmission gates)

TYPE	SYMBOLS	SWITCHING SIGNALS TO GATE		THRESHOLD EFFECTS	TYPICAL VALUES SERIES 'R_{on}'	GATE/CHANNEL 'C_g'
		ON	OFF			
n-type and p-type	G, in, out, G' (two symbols)	Logic 1 to G	Logic 0 to G	good 1 good 0	10kΩ min. increasing thru' 33kΩ to 25kΩ with V_{in}	2□Cg e.g. .02pF for min size devices.
		(two inputs G, G')				

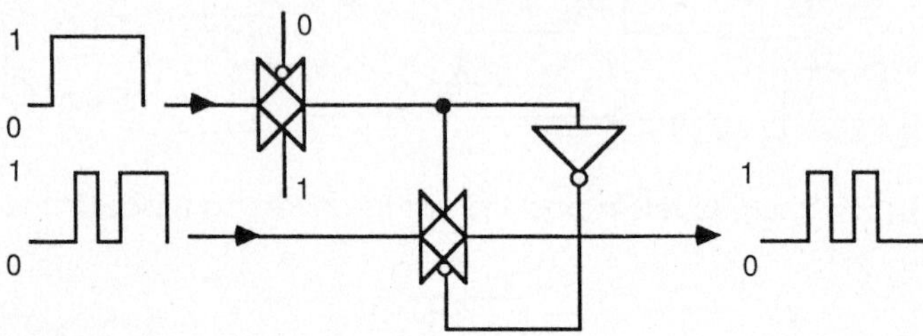

Figure 1.28 Transmission gates may be used to drive the inputs of other transmission gates (note inverter)

However, since there is no loss of logic levels, the output of a transmission gate can drive the input of another transmission gate (or a pass transistor) as indicated in Figure 1.28. Note that an inverter is required in the arrangement shown.

Although the R_{on} value is lower the gate/channel C is double that of a pass transistor so that delay line effects are still significant. Thus the following restriction applies in the use of transmission gates:

The number of transmission gates to be directly connected in series should not exceed four if excessive delays are to be avoided. If more than four are necessary then they should be sectionalized and buffered using inverters in a similar manner to the pass transistors considered earlier.

1.7.5 Inverters

The simplest form of inverter would comprise an enhancement mode transistor in series with a resistor across the supply rails (Figure 1.29). In order to limit the current drawn, and hence dissipation, when the transistor is "on", the resistor should have a reasonably high value (e.g. several 10s of k). However, the technology does not lend itself to the production of high values of resistors in a reasonable area.

The nMOS (or pMOS) inverter

Use is made of a depletion mode transistor (Tr2), connected as shown in Figure 1.30(a) so that it is always "on", to form a load resistance for the inverter. In view of the way in which each of the two transistors is connected to the output point, Tr1 is known as the "pull-down" (pd) transistor and Tr2 is known as the "pull-up" (pu) transistor.

When a logic 0 is presented at the input, Tr1 is turned off and the output is connected (i.e. pulled-up) to logic 1 = V_{DD}. However, when a logic 1 is present at the input then both Tr1 and Tr2 will be on and the output will take up a nominal logic 0, the actual output voltage being determined by the potential divider formed by Tr1 and Tr2 in series.

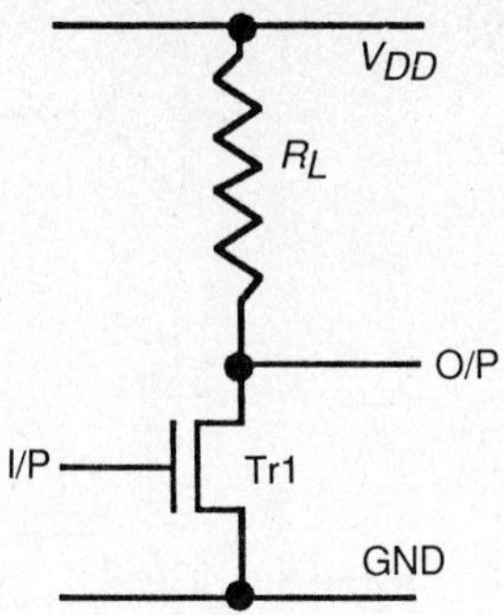

Figure 1.29 A simple inverter

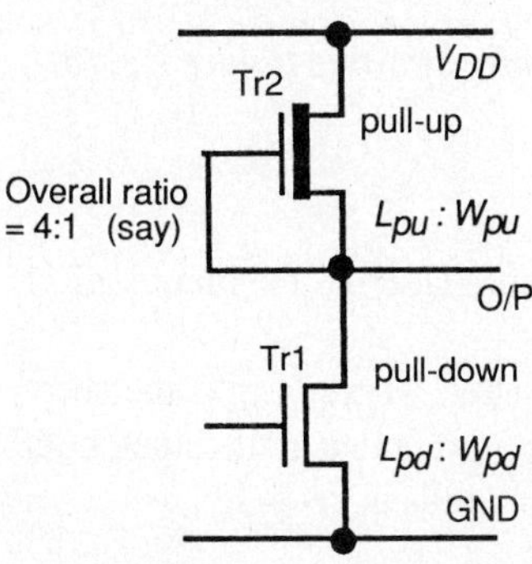
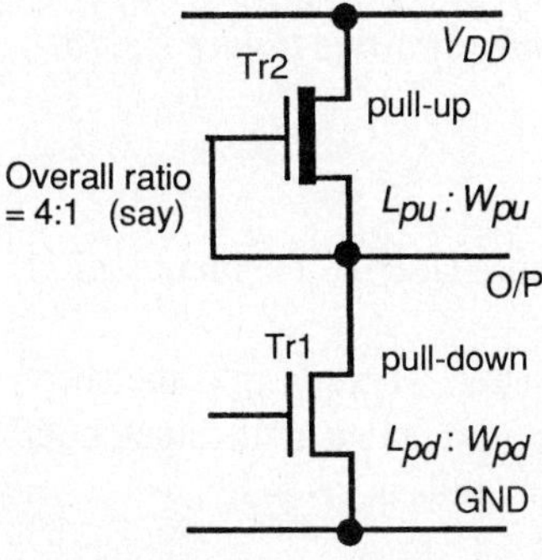

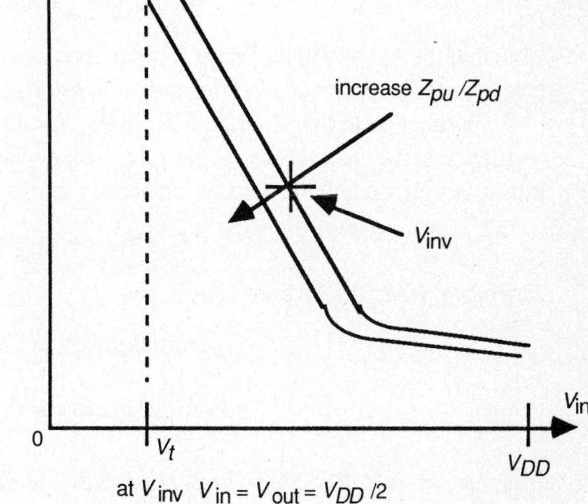

Figure 1.30

If several inverters are to be connected in series, then output logic levels must not be progressively degraded; thus the ratio between the value of the resistances of Tr1 and Tr2 must be properly chosen. Thus *ratio rules apply in designing nMOS, or pMOS, (or pseudo-nMOS) inverters.*

nMOS ratio rules

1. *For an inverter driven directly from the output of another nMOS inverter (as in Figure 1.31(a)), the required overall ratio is to be 4 : 1,* that is:

$$Z_{pu}/Z_{pd} = 4/1$$

where, $Z_{pu} = L_{pu}/W_{pu}$ and $Z_{pd} = L_{pd}/W_{pd}$

(a) Inverter driven directly from the output of another inverter

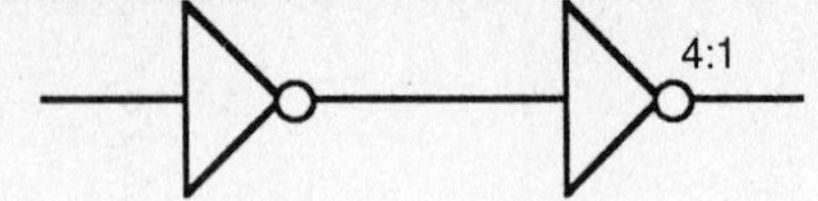

(b) Inverter driven through one or more pass transistors (n-type)

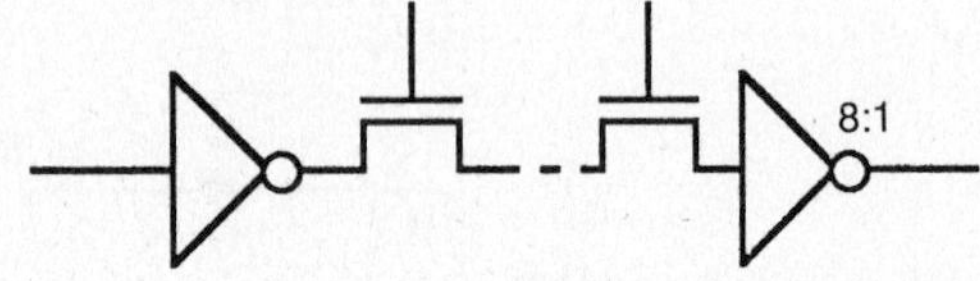

Figure 1.31 nMOS ratio rules

2. *For an inverter driven through one or more n type pass transistors as in Figure 1.31(b)), the required overall ratio is to be 8 : 1*, that is:

$$Z_{pu}/Z_{pd} = 8/1$$

Note that the use of a higher ratio is to compensate for the degradation of the input logic 1 level due to the presence of series pass transistor(s).

In passing, note that $Z . R_{CHn} = R_{ch} = R_{on}$ for either transistor, where R_{CHn} is the sheet resistance of the channel (see Table 1.2) so that, when a logic 1 is applied at the input, both transistors are on and the total resistance R_{on} across the supply rails is given by:

$$R_{on} = (Z_{pu} + Z_{pd}) . R_{CHn}$$

Therefore, when the inverter is on:

$$\text{current drawn} = V_{DD} / R_{on}$$

$$\text{power dissipation} = V_{DD}^2 / R_{on}$$

Both current and dissipation present problems in designing nMOS systems

The transfer characteristic (V_{out} versus V_{in}) for a typical nMOS inverter is illustrated in Figure 1.30(b) and it will be noted that an increase of the overall ratio (Z_{pu}/Z_{pd}) shifts the transfer characteristic to the left and lowers the output logic 0 level at maximum V_{in}.

These ratio rules also apply to logic circuitry based on the nMOS inverter.

Pseudo-nMOS ratio rules

A form of CMOS logic emulates the nMOS inverter and associated logic circuitry but uses a permanently turned on p transistor for the pull-up as indicated in Figure 1.32.

The ratio rule for this type of circuit is that the required overall ratio is to be 3: 1, that is:

$$Z_{pu}/Z_{pd} = 3/1$$

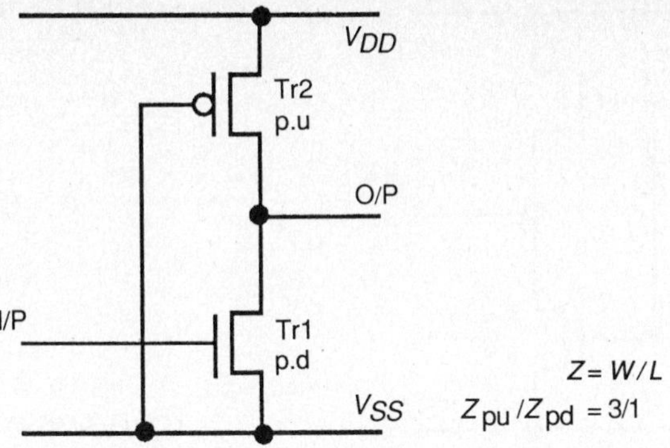

Figure 1.32 Pseudo-nMOS inverter ratio rules

This rule applies irrespective of the input signal source, provided that any series switching is by transmission gate(s) and not pass transistors.

These ratio rules also apply to logic circuitry based on the pseudo-nMOS inverter. This type of circuit also has current and dissipation problems.

Some problems associated with nMOS (and pMOS) and pseudo-nMOS inverters and associated logic circuitry

The inverter is the parent element from which the *Nand* and *Nor* gates and other related logic is derived. The main problems presented may be summarized as follows:

1. The need to obey ratio rules, which, for nMOS also depends on the source of the input signal at any input. This leads to difficulties in design due to the differing ratios and the consequent loss of regularity in the mask layout. This results in some inefficiency in the use of area.
2. There is a static flow of current in the "on" state requiring the supply rails to provide typically 0.05 to 0.1 mA per "on" inverter/gate. The supply of several thousand or tens of thousand inverter/gates (half assumed to be "on") places a heavy demand for current and poses problems in the design of V_{DD} and *GND* (V_{SS}) rails due to the current density limitations in metal wires.
3. There is a significant power dissipation; for example, each "on" inverter /gate may dissipate (typically) 0.1 to 0.5 mW which presents problems in the thermal design and packaging of chips of any complexity.
4. Logic 0 levels (for nMOS) are not good and the output "0" level from an inverter/gate can be almost $+0.2V_{DD}$ in some cases.

The complementary (CMOS) inverter

It would be nice to think that we could develop an inverter arrangement which overcomes all the problems discussed in the previous section. In fact, the arrangement commonly used in CMOS designs and set out here in Figure 1.33, does just that! The reason why the arrangement is so good is that only one or other of the two transistors is "on" at any time (except during switching) and this yields the following desirable characteristics:

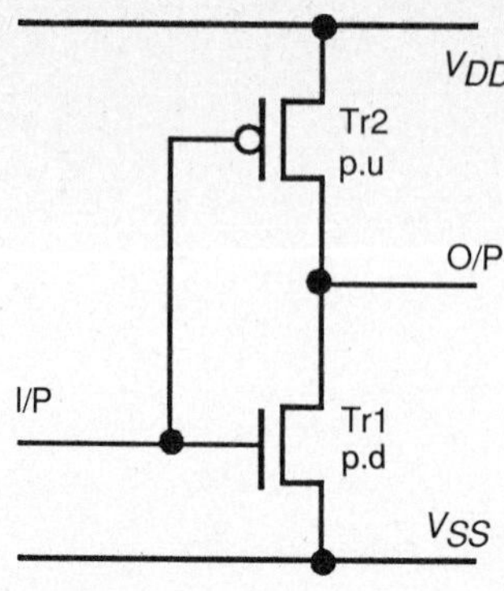

Z_{pu}/Z_{pd} = 1/1 usually, or
= 0.4 /1 for equal rise and fall times

Figure 1.33 Complementary CMOS inverter

1. With a logic 0 input, Tr1 is off but Tr2 is on due to the negative V_{gs} which is applied in this condition. The output point is thus connected by Tr2 to V_{DD} giving a full logic 1. No current* is drawn from the supply.
2. With a logic1 input, Tr2 is off but Tr1 is on due to the positive V_{gs} which is applied in this condition. The inverter output point is thus connected by Tr1 to V_{SS} giving a good logic 0 output. No current* is drawn from the supply rails.
3. In either of the two static states there is no current flow and thus no power dissipation. However, there will be power dissipation associated with switching between states which will depend on the load capacitance and the frequency of switching. This effect is also present with other types of inverter but is usually masked by the large static dissipation. It will also be applicable to logic circuits.
4. The n and p transistors can be made the same size (minimum W = L usually) and thus the complementary arrangements are regular in geometry. However, equal size transistors will yield unequal output rise and fall times since $\mu_n = 2.5\mu_p$.

Inverter delays

We will now make some observations on the delays associated with the basic inverters since this will to a large extent determine the speed of logic circuitry. Consider the general arrangement of Figure 1.34(a), from which we may see that the load on the output of inverter 1 comprises the input capacitance of similar inverter 2 (=C_g) plus the capacitance of the 'wiring' between them (C_w often assumed to be of the same order as C_g). We may thus write:

$$C_L = C_g + C_W.$$

We may now envisage the two possible switching operations (∇V_{in} or ΔV_{in}) performed by the inverter as in Figure 1.34(b) or (c) respectively. The various circuit and performance parameters are conveniently set out in Table 1.6.

It will be noted that, using all minimum size transistors and obeying ratio requirements, the output rise time is inherently slower than the fall time. This is due to the fact that the ratio

* There is a small temperature dependent leakage current in "off" transistors which gives rise to a small static power dissipation.

Table 1.6 Inverter delay summary
[assuming all minimum size (1:1) transistors and $C_W = 1\square C_g$ in all cases*]

Parameter	nMOS 4:1	nMOS 8:1	Pseudo-nMOS (CMOS)	Complementary (CMOS)	
C_g	$1\square C_g$	$1\square C_g$	$1\square C_g$	$2\square C_g$	
C_L*	$2\square C_g$	$2\square C_g$	$2\square C_g$	$3\square C_g$	
Typical value C_L(pF)	.02	.02	.02	.03	
R_{pu}	$4R_{CHn}$	$8R_{CHn}$	$3R_{CHp}$	$1R_{CHp}$	$\{R_{CHp}=2.5R_{CHn}\}$
Typical value R_{pu} kΩ	40	80	75	25	
Turn off T.C.	8T	16T	15T	7.5T	$\{T= R_{CHn}.\square C_g\}$
Typical value (n.sec)	0.8	1.6	1.5	0.75	
R_{pd}	$1R_{CHn}$	$1R_{CHn}$	$1R_{CHn}$	$1R_{CHn}$	
Typical value R_{pd} kΩ	10	10	10	10	
Turn on T.C.	2T	2T	2T	3T	$\{T=R_{CHn}.\square C_g\}$
Typical value (n.sec)	0.2	0.2	0.2	0.3	

*C_L is taken as $C_g + C_W$

requirements and/or the use of p transistors always makes the pull-up resistance higher than the pull-down resistance. The best performer in this respect is the complementary inverter but, at the same time, it should be noted that it presents twice the input C_g and is thus slower on fall times.

The complementary inverter can be designed to have symmetrical rise and fall times by reducing the resistance of the pull-up transistor by a factor of 2.5 (since $\mu_n/\mu_p = 2.5$). This may be accomplished by increasing the width W_p by a factor of 2.5 (e.g. if $W_n = L_n = L_p = 2\,\lambda$, then we make $W_p = 5\,\lambda$). In doing so we increase the C_g of the p transistor to $2.5\square C_g$

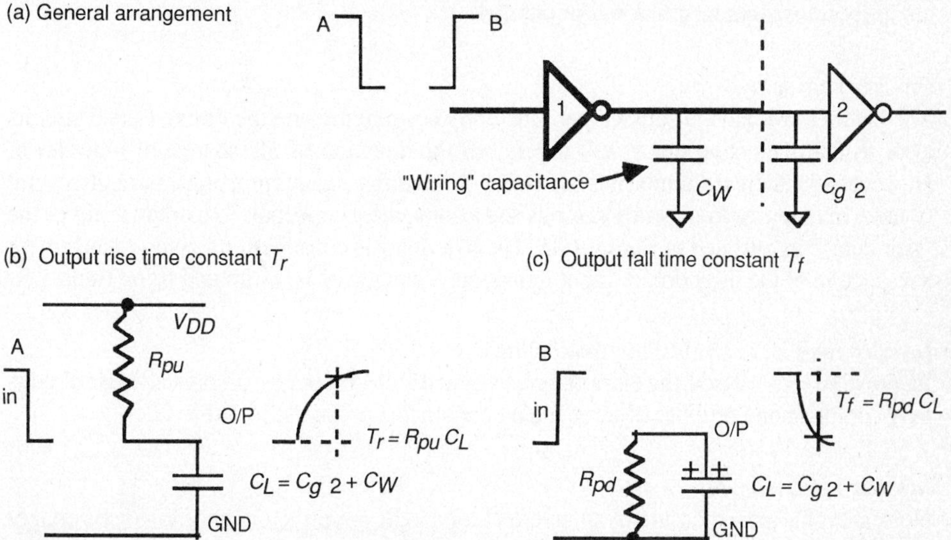

Figure 1.34 Switching time estimation for inverters

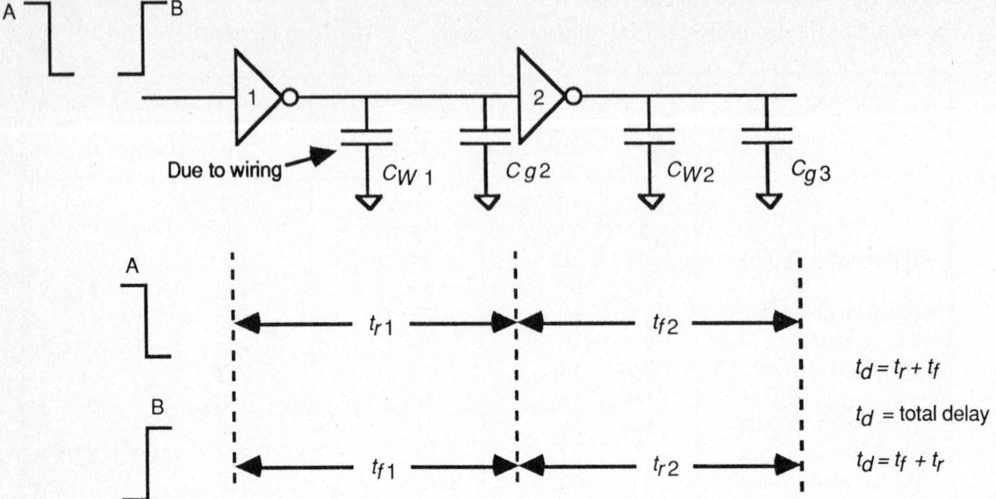

Figure 1.35

and the total inverter input capacitance to 3.5 □C_g and thus both rise and fall times will be affected in consequence.

The same approach of increasing the width of pull-up (and pull-down) transistors can be taken for nMOS or pseudo nMOS circuits to reduce pull-up and pull-down resistances (e.g. both W_{pu} and W_{pd} can be increased by the same factor to keep overall ratios the same). This will, however, increase the value of C_g in each case by the same factor *and* increase the static power dissipation, but there will generally be a reduction in area. The same considerations apply to logic circuits based on the various basic arrangements, although the pull-up and or pull-down structures are not likely to be single transistors in all cases and will often comprise transistors in series and/or parallel.

Inverter pair delays

The inverters we have discussed are inherently asymmetric and the derived logic circuits also, so that the actual delay will depend on the direction of the change of input level. However, in practical circuits, it is most often the case that signals propagate through several or more inverters or logic gates in series and a convenient concept is to work in terms of the "pair delay" as outlined in Figure 1.35. The pair delay is constant for a given arrangement irrespective of the direction of input transition. Values may be estimated from Table 1.6.

Inverter (and logic circuit) dissipation factors

As previously discussed, the dissipation associated with inverters or associated logic circuits may comprise one or more of three main components as set out in Table 1.7.

Inverter noise margins

Noise margins are a measure of an inverter's or a logic circuit's tolerance to noise voltages

Table 1.7 Dissipation factors

(In the table "on" refers to the state of inverter (or logic circuit) in which the output is at logic 0; "off" is the complementary state)

Static dissipation	*nMOS*		*Pseudo-nMOS*		*Complementary*	
	on	off	on	off	on	off
$V_{DD}^2/(R_{pu}+R_{pd})$ Due to rail to rail current	yes	no	yes	no	no	no
$V_{DD} \times I_L$ Due to leakage current	no	yes	no	yes	yes	yes
Switching dissipation	nMOS		Pseudo-nMOS		Complementary	
$V_{DD}^2.C_L.f$	yes*		yes*		yes*	

* Dissipation is calculated for complete on/off cycles where "*f*" is the frequency of such cycles.

when in either of the two logic states. It defines by how much the input voltage can change without disturbing the current logic output state. In order to examine this it is convenient to consider a pair of inverters (nMOS or CMOS) and derive the noise margins for signals applied to the input of the second inverter, inverter 2, which is driven from the output of a similar inverter (inverter 1) as in Figure1.36(a). We may then infer the noise performance of associated logic.

Referring now to Figure 1.36.(b), we see the transfer characteristics (V_{out} v V_{in}) for a pair of CMOS Inverters set out in such a way that the output voltage of inverter 1 is applied as the input voltage to inverter 2. By first considering the point at which output1 starts to enter the transition region (the unity gain point *A*) and calling this voltage "V_{OHmin}", and then considering the input voltage level "V_{IHmin}" (point *B*) at which the transition of the output of Inverter 2 commences, we are able to define the "high level" noise margin of Inverter 2 as NM_H where:

$$NM_H = V_{OHmin} - V_{IHmin} \quad \text{(a positive voltage).}$$

Similarly, a consideration of the low logic level conditions gives:

$$NM_L = V_{OLmax} - V_{ILmax} \quad \text{(a negative voltage).}$$

A similar approach will yield noise margins for the nMOS inverter as shown in Figure 1.36.(c). It may be seen that, generally, the CMOS inverter will have better noise margins than the nMOS inverter, particularly for the "low" condition. This also applies to the derived logic circuitry in both cases.

In both cases, symmetry about V_{inv} is assumed (where V_{inv} is at $V_{out} = V_{in} = V_{DD}/2$). This assumes that $\beta_p = \beta_n$ for CMOS and that the correct ratio of Z_{pu} to Z_{pd} has been observed for nMOS.

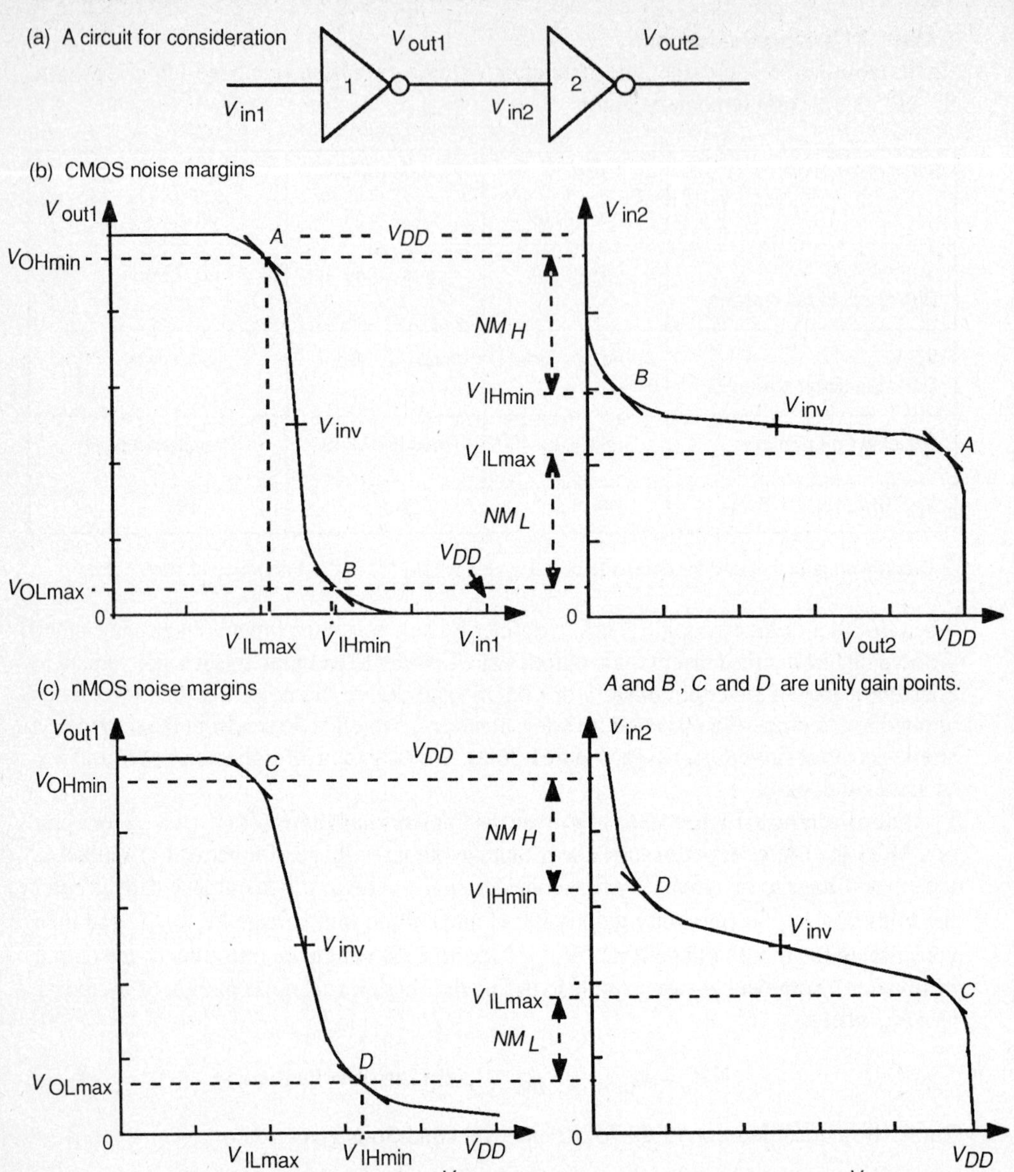

Figure 1.36 Inverter noise margins

Changes in the β_n/β_p ratio for CMOS or to the Z_{pu}/Z_{pd} ratio for nMOS will result in a shift in the V_{out} v V_{in} characteristics (see Figure 1.30(b) for nMOS, for example) and consequent degradation of one or the other noise margin in each case.

Thus the effect of ratios on noise margins performance must be taken into account in design.

1.8 Observations

This chapter has been devoted to the task of introducing semiconductor technology and to an overview of the characteristics and design processes for MOS circuits in silicon. The reasons for this are apparent when one considers the fact that logic circuitry is employed in an extremely wide range of engineering applications and, almost without exception, all logic circuitry is realized in semiconductor technology. MOS technology is already widely employed, particularly in LSI and VLSI technologies and the full potential of systems design in silicon using a VLSI approach has yet to be realized over the next decade. It is therefore most appropriate that engineers should become fully conversant with the matters so far set out and with the principles and practice of the design of logic circuitry and digital systems and their realization in silicon.

It is also true that digital logic is employed widely in data processing and in "number crunching" generally. To fully appreciate the possibilities in this direction, it is desirable to have a working knowledge of number systems and arithmetic processes and of binary arithmetic in particular.

Such is the purpose of Chapter 2 which will put in place the background for work which follows later on registers, counters and adders, and so on.

1:9 Worked examples

1.9.1 Semiconductor technology

1. (a) Calculate the bulk resistivity ρ for intrinsic silicon at 350°K .
   (b) What is the value of the *surface* mobility at this temperature ?

Solution:

(a) Resistivity $$\rho = \frac{1}{q.\mu.n}$$

where: $q = 1.6 \text{ X } 10^{-19}$coulomb, n is carrrier concentration.

To calculate μ we note that room temperature (300°K) figures are taken as μ_n = 1250cm²/Vsec; μ_p = 480cm²/Vsec and that μ varies as $T^{-3/2}$ approx.

In this case at 350°K we have:

$$\mu_{350}/\mu_{300} = (300/350)^{3/2} = 0.86^{3/2} = 0.8.$$

whence bulk mobilities at 350°K:

$$\mu_n = 1000\text{cm}^2/\text{Vsec}; \ \mu_p = 384\text{cm}^2/\text{Vsec}$$

Value of n (the carrier concentration) at 350°K is obtained from Figure 1.4, say 5 X 10^{11} per cm^3. Note that there will be this number of electrons *and* same number of holes.

Therefore, resistivity $= \dfrac{1}{qn(\mu_n + \mu_p)}$

$$= \frac{1}{1.6 \text{ X } 10^{-19} \text{ X } 5 \text{ X } 10^{11} \text{ X } 1384}$$

$$= \frac{10^8}{11072} = 90.3k\Omega.cm$$

(b) The surface mobility is roughly half the bulk values, so that surface mobilities at 350°K:

$$\mu_n = 500\ cm^2/Vsec;\ \mu_p = 192\ cm^2/Vsec$$

(2) Determine the resistance of an n-doped 1 cm long silicon rod of rectangular cross section 2 mm wide by 1 mm thick. The n-doping concentration is $10^{17}/cm^3$ and $T = 27°C$.

Solution:

From Figure 1.6, resistivity $\rho = 10^{-1}\Omega.cm$.

Resistance is given by: $R = \dfrac{\rho l}{A}$

where A = cross section area cm^2
therefore,

$$R = \frac{10^{-1} \text{ X } 1}{0.2 \text{ X } 0.1} = 5\Omega$$

3. Assuming the value for surface mobility given in the text, calculate the transit time:

(a) for an electron in an n type channel, and
(b) for a hole in a p type channel,

where the channel length is 5 µm in both cases and 5 volts of appropriate polarity is applied between drain and source.

Solution:

Transit time = channel length l /velocity v.

Now $v = \mu E$, where E is the electric field strength drain to source.

(a) For the electron in a 5 μm n channel, and with $V_{ds} = +5V$:

$$l = 5\mu m\ ;\ \mu_n = 650\ cm^2/V\ sec\ ;\ E = V_{ds}/l = 5V/5\mu m$$

Velocity $\quad v = \mu_n.E$

$$= (650\ cm^2/V\ sec) \times (5 \times 10^4 V/5\ cm)$$

$$= 650 \times 10^4\ cm/sec$$

Thus:

$$\text{transit time } (l/v) = \frac{5}{10^4 \times 650 \times 10^4}\ sec = 0.077 nsec$$

(b) For the hole in a 5 μm p-channel, and with $V_{ds} = -5V$:

$$l = 5\mu m\ ;\ \mu_p = 240\ cm^2/V\ sec\ ;\ E = V_{ds}/l = 5V/5\mu m$$

Velocity $\quad v = \mu_p.E$

$$= (240\ cm^2/V\ sec) \times (5 \times 10^4 V/5\ cm)$$

$$= 240 \times 10^4\ cm/sec$$

$$\text{Thus transit time } (l/v) = \frac{5}{10^4 \times 240 \times 10^4}\ sec = 0.21 nsec$$

4. Calculate the barrier voltage V_B for a p-n junction for which $N_A = 10^{16}$ and $N_D = 10^{14}$ atoms/cm³ when the temperature is 300°K. Which side is positive?

Solution:

$$V_B = (kT/q)\ ln.\ (N_A N_D / n_i^2)$$

Substituting values for $k.T$ and q we find that:

$$kT/q = 25.9mV \text{ at } 300°K.$$

Therefore :

$$V_B = 25.9mV \times \ln.(N_A N_D / n_i^2)$$

$$= 25.9 \times \ln.(10^{30}/10^{20})$$

$$V_B = 25.9mV.\ \ln(10^{10}) = 596\ mV = 0.596V$$

The n side is positive.

5. Calculate the junction capacitance per mm² for the junction described in question 4.

Solution:

$$C = \varepsilon_0\ \varepsilon_{si}/d\ pF/\mu m^2$$

where d is the depletion region width on the lightly doped side of the junction, and, in this case,

$$d = \sqrt{2\varepsilon_0\varepsilon_{Si}V_T / qN_D}$$

Therefore:

$$C = \sqrt{\varepsilon_0\varepsilon_{Si}.qN_D / 2V_T}$$

$$= 2.9 \times 10^{12}.\sqrt{N_D / V_T}\ pF / \mu m^2.$$

that is:

$$C \text{ per } \mu m^2 = 2.9 \text{ X } 10^{-12}.\ \sqrt{10^{14}/.596} = 2.9 \text{ X } 10^{-12} \text{X } 6.87 \text{ X } 10^7$$

Thus (for 1 mm^2) we have:

$$C = 10^6 \text{X } 2.9 \text{ X } 10^{-12} \text{X } 6.87 \text{ X } 10^7 = 199\ pF/mm^2.$$

1.9.2 MOS fabrication

1. Draw a cross-section of an n-well CMOS inverter structure and indicate the parasitic components leading to possible latch-up.

   Solution: See Figure 1.37.

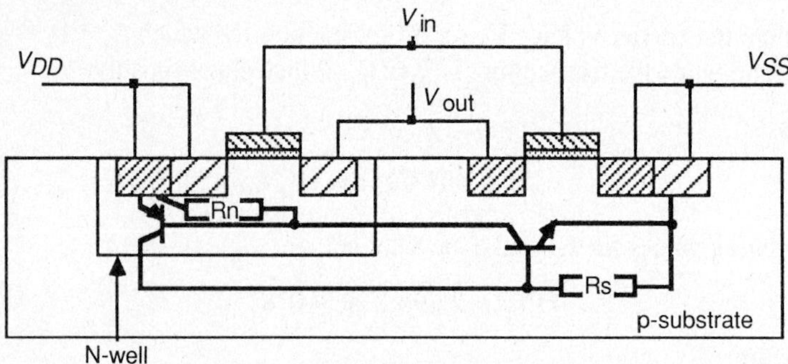

Figure 1.37 Latch-up configuration for n-well CMOS inverter

2. With reference to Figure 1.14, draw a similar set of diagrams for the fabrication of an n-well CMOS inverter.

   Solution: See Figure 1.38.

1.9.3 MOS layout and design

1. Draw stick diagrams and then mask layouts for the circuits set out in Figure 1.39(a) and (b).

   Solution: See Figure 1.40(a) and (b) and 1.41(a) and (b).

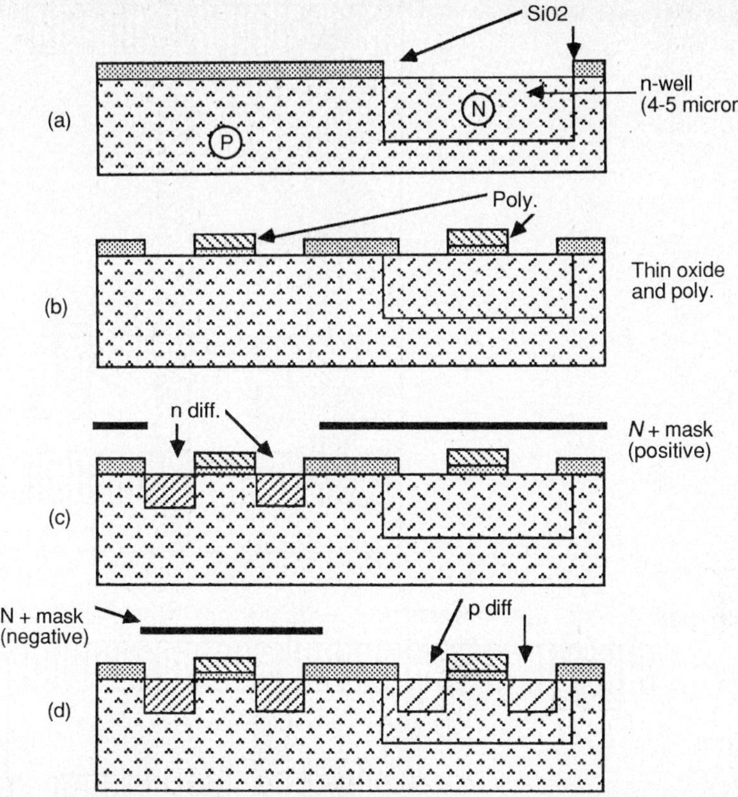

Figure 1.38 CMOS n-well process steps (dimensions, where shown, are for 5μm feature size)

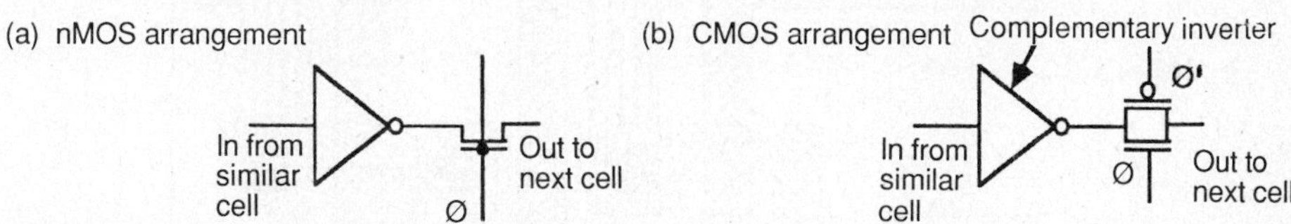

Figure 1.39(a) nMOS arrangement (b) CMOS arrangement

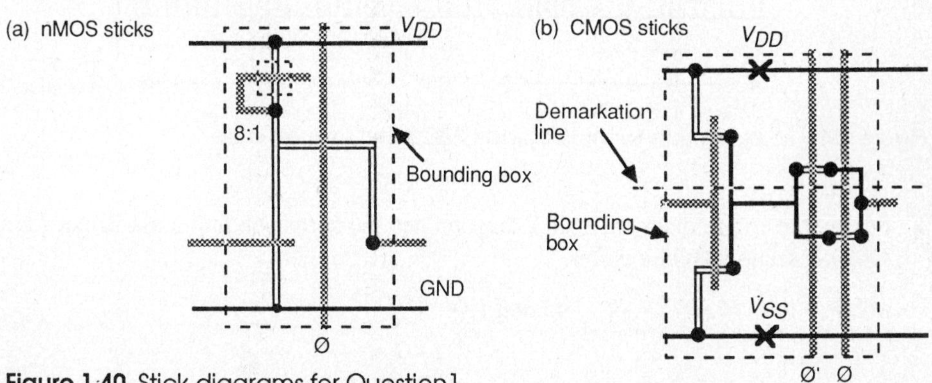

Figure 1.40 Stick diagrams for Question1

(a) nMOS masks (buried contacts)

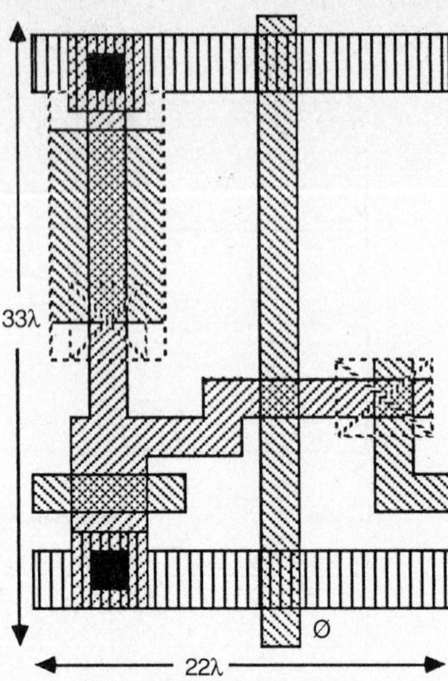

(b) CMOS masks

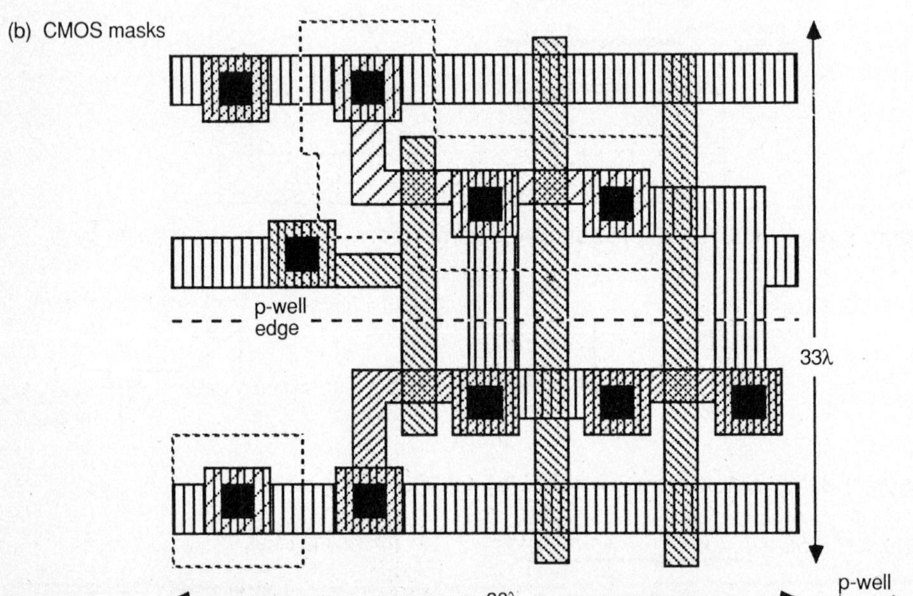

Figure 1.41 Mask layouts for nMOS and CMOS shift register cells

2. Set out the circuit diagram, a stick diagram and the corresponding mask layout for a CMOS pseudo-nMOS inverter.

   Solution : See Figures 1.42 , 1.43 and 1.44.

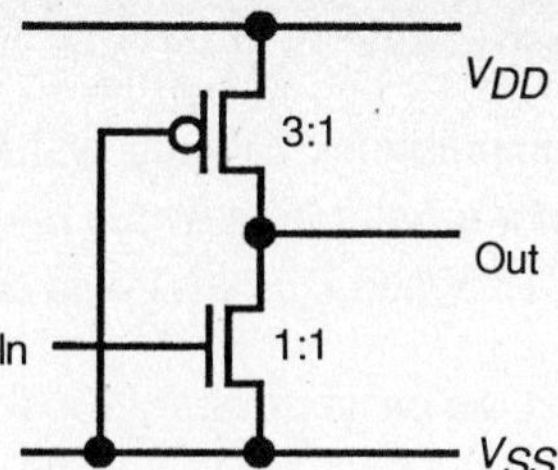

Figure 1.42 Psuedo-nMOS inverter circuit

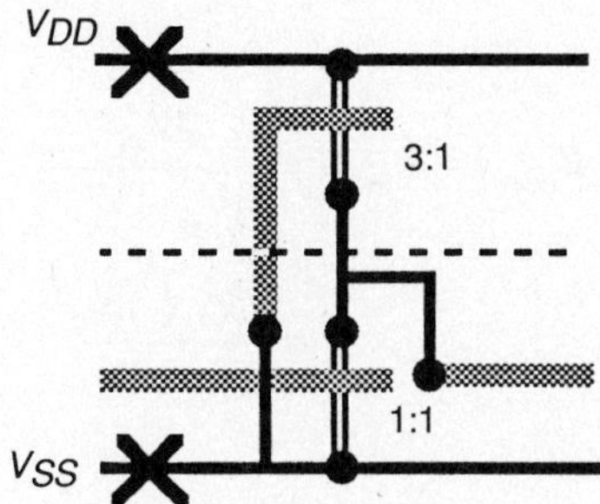

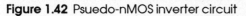

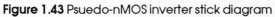

Figure 1.43 Psuedo-nMOS inverter stick diagram

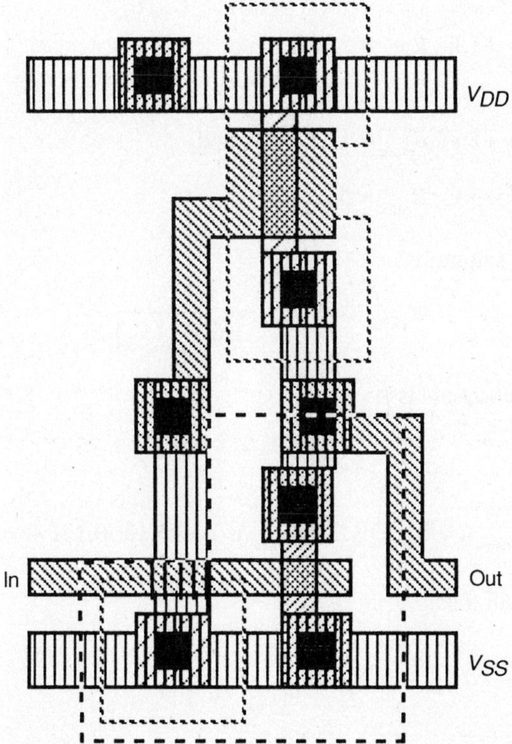

Figure 1.44 Psuedo-nMOS inverter mask layout

1.9.4 MOS electrical parameters

1. (a) Using the approximate expressions, calculate values for R_{chsat} $(=R_{dssat})$ for an n type enhancement mode transistor in saturation over a range of applied V_{gs} from +2 V to +5 V. You may assume V_{ds} = +5 V and V_t = +1 V and that ß = 0.1 mA/V².

(b) What is the resistance at the onset of saturation for V_{ds} = +4 V?

(c) Assuming V_{ds} is small, calculate R_{ds} (non sat) for same V_{gs} values used in (a).

Solution:

(a) In saturation:

$$R_{chsat}\,(=R_{dssat}) = \frac{2V_{ds}}{ß(V_{gs} - V_t)^2}$$

$$= \frac{100 \text{ X } 10^3}{(V_{gs} - 1)^2}$$

Therefore,

for V_{gs} = +2 V , R_{chsat} = *100 kΩ*

and,

for V_{gs} = +3 V , R_{chsat} = *25 kΩ*

and,

for V_{gs} = +4 V , R_{chsat} = *11.11 kΩ* ; and,

for V_{gs} = +5 V , R_{chsat} = *6.25 kΩ*

(b) At onset of saturation:

$$R_{chsat}\,(=R_{dssat}) = \frac{2}{ß(V_{gs} - V_t)}$$

and onset of saturation is at:

$$V_{ds} = V_{gs} - V_t = +4 \text{ V for } V_{gs} = +5 \text{ V}$$

thus:

$$R_{chsat}\,(=R_{dssat}) = 2\,/ß(V_{gs} - V_t) = 2\text{X}10^3/0.1 \text{ X } 4 = 5\,k\Omega$$

(c) If V_{ds} is small then:

$$R_{ds} = \frac{1}{ß(V_{gs} - V_t)}$$

$$= \frac{10 \text{ X } 10^3}{(V_{gs} - V_t)}$$

Therefore,

$$for\ V_{gs} = +2\ V,\ R_{ds} = 10\ k\Omega$$

and,

$$for\ V_{gs} = +3\ V,\ R_{ds} = 5\ k\Omega$$

and,

$$for\ V_{gs} = +4\ V,\ R_{ds} = 3.33\ k\Omega$$

and,

$$for\ V_{gs} = +5\ V,\ R_{ds} = 2.5\ k\Omega$$

2. *(a)* Calculate the total series resistance R_{AB} and total capacitance to substrate (C_{total}) presented by a 5 μm technology p-type diffusion path having the geometry of Figure1.45.
   (b) Find the new value of C_{total} if the entire area is overlaid by metal 1 (at 0 Volts).
   (c) For *(a)*, if *A* is driven from, and *B* drives a minimum size CMOS complementary inverter, find the delays associated with Δ and ∇ signal edges at *B*.

Solution:

(a) From Table 1.2 (a) a mid range R_s value for p-diffusion is 75Ω/□ . Ignoring the small effect of current crowding at corners, we have a mean path length of (12.5 + 35 + 12.5) λ [= 60 X 2.5 μm = 150 μm].

Now path width is 5 λ, therefore there are 60/5 =12 squares in series.

Thus: $R_{AB} = 12 \times 75\Omega = 900\Omega.$

Total $C = C_{area} + C_{periph}$

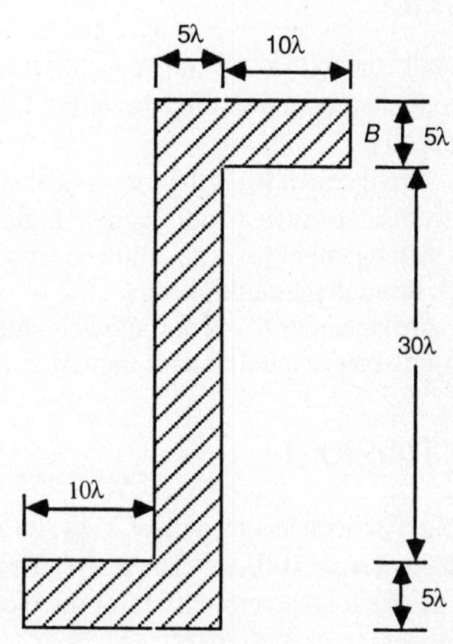

Figure 1.45 Diffusion "wire" geometry

From Table 1.3(a),

$$C_{area} = 1.0 \times 10^{-4}pF/\mu m^2 \times (150 \times 12.5)\mu m^2 = 0.1875 \text{ pF}.$$

and, (from Table 1.3(b)),

$$C_{periph} = 8 \times 10^{-4}pF/\mu m \times (2 \times 65 \times 2.5)\mu m = 0.26pF.$$

$$C_{total} = 0.4475pF$$

(b) If metal 1 is overlaid, then additonal area C, is present.

From Table 1.3(a)

$$C_{diff/metal} = 0.3 \times 10^{-4} pF/\mu m^2 \times (150 \times 12.5)\mu m^2 = 0.05625pF.$$

This will add to the original value to give *new* $C_{total} = 0.50375pF$

(c) Driving source is a CMOS inverter where the pull up resistance R_{pu} is 25 kΩ and the pull down resistance R_{pd} is 10 kΩ.

An additional C must be added to the value calculated in *(a)* due to the input $C = 2\square C_g$ of the inverter driven from B. Note that $1 \square C_g$ is the gate to channel C of a minimum size transistor [= $25\mu m^2 \times 4 \times 10^{-4}pF/\mu m^2$] = 0 .01pF.
Thus total C is now 0.4475 + 0.02 = 0.4675pF.
The total series R for Δ edges $= R_{AB} + R_{pu} = (0.9 + 25) = 25.9$ kΩ
The Δ edge time constant is (0.4675 X 25.9) nsec = 12.1nsec.
The total series R for ∇ edges $= R_{AB} + R_{pd} = (0.9 + 10) = 10.9$ kΩ
The Δ *edge time constant* is (0.4675 X 10.9) nsec = *5.1nsec.*

(3) Which of the four arrangements of Figure 1.46 are allowable and which are not? Give reasons.

Solution:

Arrangement (a) is not allowed since pass transistors Tr1 and Tr2 both have gates driven through a pass transistor. Therefore, logic "1" levels at X, (i.e. $V_{DD} - 2V_t$), will be unacceptable.

Arrangement (b) is the transmission gate equivalent of (a) and is thus allowed since transmission gates do not degrade logic levels.

Arrangement (c) is not allowed since there are five pass-transistors in series in each path through the multiplexer.

Arrangement (d) is not allowed since, again, there is a pass transistor gate driven through one or another pass transistor.

1.10 Tutorial 1

1. A bar of silicon has a length of 2 cm and a cross-sectional area of 1 cm². It is doped n type such that $N_D = 10^{15}/cm^3$. The resistance from end to end is measured and is found to be 10 Ω. What is the mobility μ for the electrons? Hence deduce the temperature.

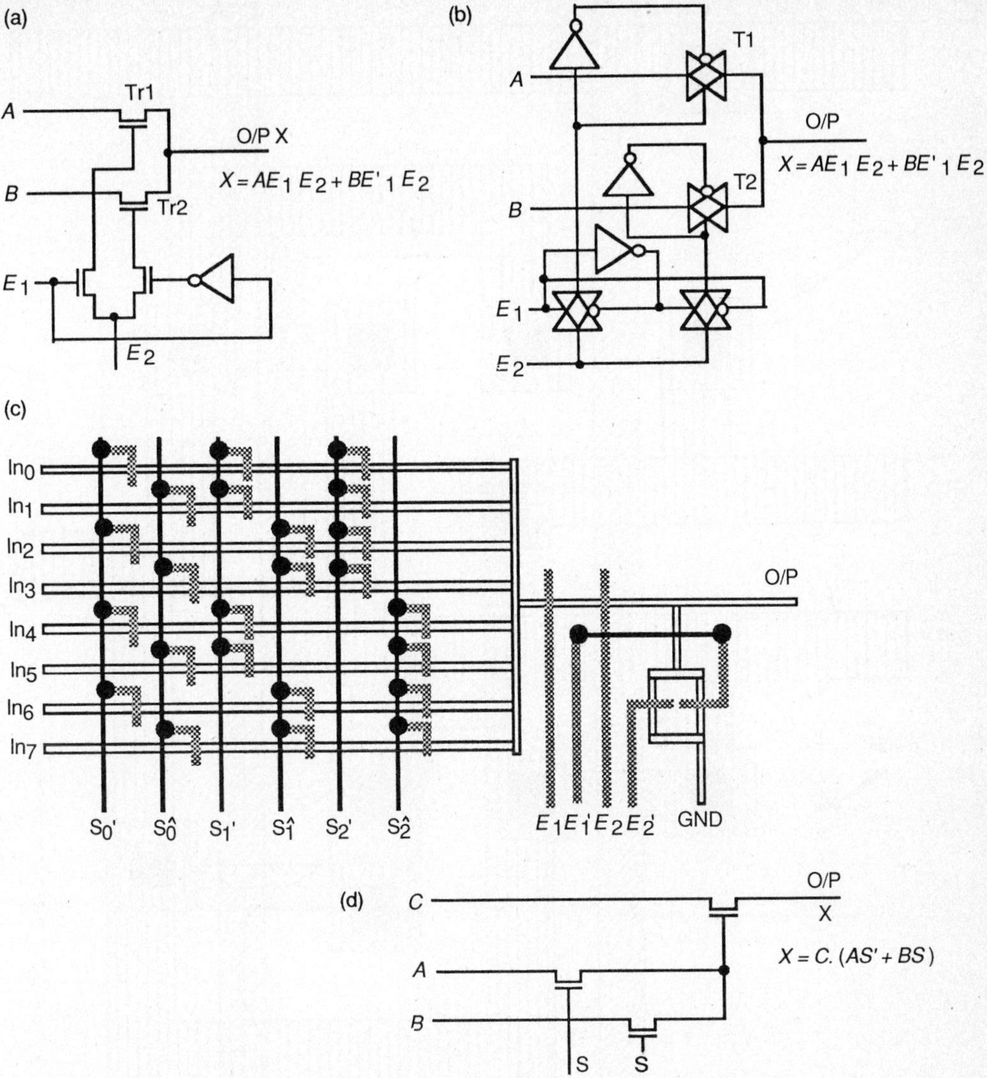

Figure 1.46 Some pass transistor and transmission gate arrangements

2. Derive an expression for the bulk resistivity of an intrinsic semiconductor.

3. Assuming room temperature and the surface mobility figures given in the text, calculate the transit time, (i) of an electron in an n transistor channel, and (ii) of a hole in a p transistor channel, where the geometry in both cases is such that the channel length = 2 μm.

   You are to examine two situations:

   (a) where $V_{ds} = 3$ volt; and
   (b) where $V_{ds} = 5$ volt

   in the correct polarity in both cases for conduction to take place.

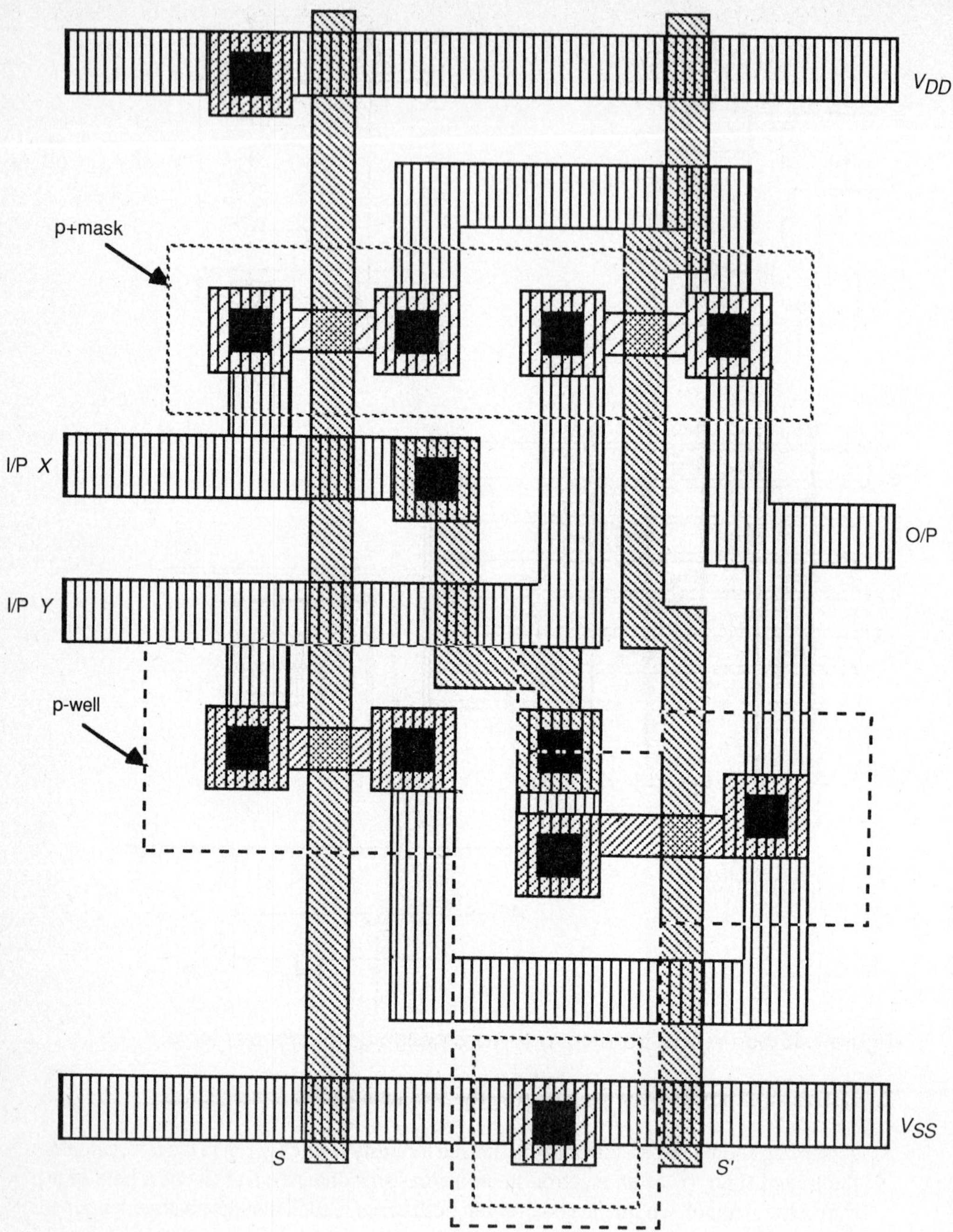

Figure 1.47 A mask layout

4. (a) Calculate the power dissipated by (i) a 5 μm 4:1 ratio nMOS inverter having a minimum size pull down and a 5V supply rail, and (ii) a 5 μm CMOS pseudo-nMOS inverter having a minimum size pull down and a 5V supply rail.
(b) Which circuit would give the fastest output rise and fall times when driving a load of 2 □C_g?

5. Draw circuit and stick diagrams for the mask layout given as Figure 1.47. Comment on the nature and purpose of the circuit. Why are the V_{DD} and V_{SS} rails shown in the layout?

2. Number systems and arithmetic

2.1 Introduction

Digital systems are used in a wide range of processors including, of course, the ubiquitous digital computer. The number system employed in a computer or processor has a most significant impact on speed and the complexity of the hardware, and thus a discussion of this topic is relevant to the everyday needs of the digital designer.

The choice of a number system depends not only on the efficiency of the algorithms associated with its arithmetic processes but also on the characteristics of available technology. For example, a pure decimal system is hard to implement in hardware because of the lack of any simple electronic circuit elements with 10 clearly defined states. On the other hand binary arithmetic is easy to realize in hardware because there are many simple devices having two clearly defined and reliable states, e.g. a switch "on" or "off", a transistor "conducting" or "cut off", etc.

The number systems we will discuss are associated with positionally weighted representation, that is to say, the significance of a digit depends on the position (or column) in which it appears. Our everyday decimal system, which originally evolved from the fact that we have 10 digits (fingers), is interpreted as having the most significant digit on the left working through all others in sequence to the least significant on the right. Convention has dictated that it should be so. We will also be dealing with number systems having a symbol to represent zero. These factors are not present in, say, the Roman number system. We would have been restricted, to say the least, if the Roman system was still in use today, and engineering calculations of any complexity would be difficult if not impossible. On the other hand, the Romans were good engineers and some of their works have endured well through twenty centuries or more.

However, we will look at the commonly used systems and in particular at the binary system in some detail. We will also take an overview of other systems such as ternary and residue arithmetic which may have a part to play in processor design.

2.2 Characteristics of positionally weighted number systems

2.2.1 Number representation

Number systems may be developed to any reasonable Integer base (or radix) and, in general, any integer number, N, is the sum of a series of digits, thus:

$$N_q = a_n q^n + a_{n-1} q^{n-1} + a_{n-2} q^{n-2} + \dots + a_2 q^2 + a_1 q^1 + a_0 q^0$$

where:

q is the radix (base) and is an integer, and
a is a positive integer such that $0 \le a \le q-1$.

When writing this number we omit the weighting factors and merely write:

$$a_n \, a_{n-1} \, a_{n-2} \, \dots \, a_2 \, a_1 \, a_0$$

where the position of a digit infers its weight.

We may also deal with fractional quantities, and any fraction, M, may also be represented as a sum of a series, thus:

$$M_q = a_{-1}q^{-1} + a_{-2}q^{-2} + a_{-3}q^{-3} + \ldots + a_{-m+1}q^{-m+1} + a_{-m}q^{-m}$$

When writing this fraction we omit the weighting factors but insert a "point" and write:

$$.a_{-1}\, a_{-2}\, a_{-3}\ \ldots\ a_{-m+1}\, a_{m}.$$

In everyday arithmetic this "point" is, of course, the decimal point which separates the integer and fractional parts . For example: $(975.84)_{10}$* represents the sum of:

$$9 \text{ X } 10^2 + 7 \text{ X } 10^1 + 5 \text{ X } 10^0 + 8 \text{ X } 10^{-1} + 4 \text{ X } 10^{-2}$$

Note now that a general characteristic of such systems is that the number of symbols - a - required to represent all possible digits = q.

For the decimal system, q = 10 and the ten symbols are:

0 1 2 3 4 5 6 7 8 9

that is: $[0 \leq a \leq 9]$

For the binary system, q = 2 and the two symbols required are: 0 and 1 only.

that is: $[0 \leq a \leq 1]$

2.2.2 Some implications of the choice of radix

Arithmetic operations

Digital computers and processors must be "taught" arithmetic just as we ourselves were. The hardware and software must be configured in such a way as to implement the common operations of add, subtract, multiply and divide. This involves the "learning" of tables: the addition and multiplication tables, etc.

In the case of decimal arithmetic, for example, the addition tables require a knowledge of 55 separate relationships: 0 + 1 = 1, 1 + 2 = 3, 5 + 6 = 11 (or 1 carry 1), 9 + 9 = 18 (or 8 carry 1), etc. This assumes that the machine recognizes commutativity, e.g., that 2 + 3 gives the same result as 3 + 2.

The multiplication tables are similarly proportioned and again we must remember a minimum of 55 relationships.

A general expression sums up the requirements as follows: For radix q we need to remember S relationships, where:

$$S = \frac{q}{2}(1+q)$$

For the decimal case this evaluates to $S_{10} = \frac{10.}{2}(10+1) = 55$

* Note that we should get into the habit of indicating the radix when more than one is in use.

and for the binary case this evaluates to $S_2 = \frac{2}{2}.(2+1) = 3$ relationships.

To reinforce this, the binary addition and multiplication tables may be stated thus: For add,

$$0 + 0 = 0\ ;\ 0 + 1 = 1\ ;\ 1 + 1 = 0 \text{ carry } 1.$$

For multiply,

$$0 \text{X } 0 = 0\ ;\ 0 \text{X } 1 = 0\ ;\ 1 \text{X } 1 = 1$$

a very simple set of things to remember.

We will show later that subtraction and division may also be covered by this simple set. The same simplicity would not be the case for higher radix systems.

2.3 The Binary number system

2.3.1 Some factors

We have seen that simple arithmetic processes and small radices go hand in hand. Clearly then there is a strong inducement to adopt the binary system when designing arithmetic processors. Further inducements when considering hardware are:

1. The number of symbols, therefore states, which must be represented is two only—0 and 1—which is readily done in a wide range of electronic devices.
2. Binary arithmetic operations and logical operations are closely akin and readily implemented as we shall see in later chapters.

2.3.2. Some disadvantages of the binary system

Clearly, life wasn't meant to be this easy and a little thought will reveal some penalties for adopting the binary system. If when you next apply for a position, you are offered an annual salary of $1 000 000, you might well be pleased—unless the potential employer is binary minded and the offer is for $\$(1000000)_2$. Evaluating this you will see that the offer is for $\$(1 \text{ X } 2^6) = \$(64)_{10}$ which is not quite so encouraging!

This brings out two key factors:

1. Even small numbers require lots of digits for small radix number systems.
2. In interpreting a number you need to know the radix(base) and should get into the habit of indicating it whenever more than one radix is in use.

A third factor can be added:

3. We have been conditioned to think in decimal terms, and the visual and mental interpretation of other radix numbers can be difficult.

2.3.3 Other systems related to binary

In order to reduce the number of digits to be written, or input through a keyboard, etc., we may adopt "shorthand" methods of expressing binary numbers. To do this we can look to number systems which have radices which are a power of the binary radix 2. We therefore choose systems based on radix q where, $q = 2^n$ and where n is a positive integer. This will reduce the number of digits to be written by a factor n. Two systems are commonly used, octal and hexadecimal.

The octal number system

For the octal system, the radix is 8 (2^3) so that three **binary digits** (known as **bits**) can be represented by one octal digit.

The general characteristics are:

Radix: $q = 8$
Digits (a): 0 1 2 3 4 5 6 7 $[0 \leq a \leq 7]$

A binary number can be written in octal form by dividing the binary number into groups of three bits from the binary point outwards in each direction, converting each group (add 0s as needed) to a single octal digit (Table 2.1).

Table 2.1 Binary to single digit octal conversion

Binary	*Octal*
000	0
001	1
010	2
011	3
100	4
101	5
110	6
111	7

For example:

$$(10111011.1101)_2$$

converts to: $(2 \quad 7 \quad 3 \quad . \quad 6 \quad 4)_8$

The 3-bit groups are: *0*10 111 011.110 1*00*
The added bits are shown in italics.

The octal number $(273.64)_8$ may, in turn, be interpreted in decimal form by summing up the weighted digits as shown:

$$2 \times 8^2 + 7 \times 8^1 + 3 \times 8^0 + 6 \times 8^{-1} + 4 \times 8^{-2} = (187.8125)_{10}$$

Arithmetic operations can be carried out in octal arithmetic but this is awkward for those used to the decimal system. For example:

$$
\begin{array}{r}
(352)_8 \\
\times \quad (52)_8 \\
\hline
724 \\
22220 \\
\hline
(23144)_8
\end{array}
$$

However, our interest in octal is usually for representing binary numbers, not for arithmetic.

The Hexadecimal number system

This is the most popular "shorthand" form for binary quantities and in this case the radix $q = (16)_{10} = 2^4$ so that $n = 4$. Hexadecimal (hex.) numbers are also widely used for representing machine language (binary) for microprocessors and computers generally.

The characteristics then are:

Radix: $q = (16)_{10}$
Digits(a): 0 1 2 3 4 5 6 7 8 9 A B C D E F $[0 \leq a \leq F]$

Note that we have had to "invent" six numeric symbols in this case.

Conversions from binary are now carried out in groups of four bits from the binary point outwards in each direction. Conversions are made according to Table 2.2.

Table 2.2 Binary to single hexadecimal digit conversion

Binary	*Hex*	*Binary*	*Hex*
0000	0	1000	8
0001	1	1001	9
0010	2	1010	A
0011	3	1011	B
0100	4	1100	C
0101	5	1101	D
0110	6	1110	E
0111	7	1111	F

Note that each conversion is made by weighting the binary digits 2^3, 2^2, 2^1, and 2^0 from left to right respectively. As an example of conversion:

$$(11011001.11)_2$$

converts to: $(D \quad 9 \quad . \; C)_{16}$

or, D9.CH which is an accepted form.

Conversions are equally readily made in the other direction, for example:

3A6.7CH

converts to, $(1110100110.011111)_2$.

Clearly calculations can be carried out in hexadecimal arithmetic but the process is daunting to say the least.

Conversions to decimal form can be made without too much trauma, for example

$$(3\,A\,6\,.\,7\,C\,H)_{10}$$

is given by:

$$3 \text{ X } 16^2 + 10 \text{ X } 16^1 + 6 \text{ X } 16^0 + 7 \text{ X } 16^{-1} + 12 \text{ X } 16^{-2}.$$

2.4 A general process for radix conversion

If we wish to convert from radix (base) p to radix (base) q, then we have already seen that simple processes are involved if the radices are directly related.

1. If $p = q^n$ or if $p^n = q$ then subdivide the number with the smallest radix into groups of n digits and convert each group separately to the new radix. We have already seen this demonstrated for binary/octal and binary/hex. conversions.
2. If $p = r^m$ and $q = r^n$ then carry out the conversion through the common radix, r. For example to convert from octal to hexadecimal we may use base 2. Say we wish to convert $(6341.26)_8$ to hexadecimal form:

First, convert to binary: $(1\,1\,0\;0\,1\,1\;1\,0\,0\;0\,0\,1.0\,1\,0\;1\,1)_2$

$$[(1\,1\,0\,0\;1\,1\,1\,0\;0\,0\,0\,1\,.\,0\,1\,0\,1\;1)_2]$$

now to hex (via groups of 4 bits): $(C \quad E \quad 1\;.\;5 \quad 8)_{16}$.

If the radices are not related in a simple way then a general process is as follows:

(a) *General case for N_p to N_q where N is an integer*

$$N_p = N_q = a_n q^n + a_{n-1}q^{n-1} + a_{n-2}q^{n-2} + \dots a_2 q^2 + a_1 q^1 + a_0 q^0$$

The conversion algorithm is: *Using the arithmetic of the old radix, p,* divide the number to be converted by the new radix, q, i.e. N_p/q. Note the remainder in base q; this is the least significant (LS) digit of the converted number as follows:

$$N_p/q = N'_q = a_n q^{n-1} + a_{n-1}q^{n-2} + a_{n-2}q^{n-3} + \dots a_2 q^1 + a_1 q^0 \text{ rem } a_0$$

where a_0 in this case is the remainder (rem) and the first (LS) digit.

Repeat now with the integer quotient N'_q to obtain the next digit:

$$N'_p/q = N''_q = a_n q^{n-2} + a_{n-1}q^{n-3} + a_{n-2}q^{n-4} + \dots a_3 q^1 + a_2 q^0 \text{ rem } a_1$$

a_1 in this case is the remainder and the second(LS) digit.

Repeat with the integer quotient N''_q to obtain the next digit, and so on until all digits are obtained.

(b) *General case for M_p to M_q where M is a fraction, i.e. M<1.*

$$M_p = M_q = a_{-1}q^{-1} + a_{-2}q^{-2} + a_{-3}q^{-3} + ... + a_{-(m-1)}q^{-(m-1)} + a_{-m}q^{-m}$$

The conversion algorithm is: *Using the arithmetic of the old radix, p,* multiply the number to be converted by the new radix, i.e. $M_p X q$. Note the overflow in base q; this is the most significant(MS) digit of the converted number as follows:

$$M_p X q = M'_q = a_{-1}q^0 + a_{-2}q^{-1} + a_{-3}q^{-2} + ... + a_{-(m-1)}q^{-m} + a_{-m}q^{-m+1}$$

where a_{-1} in this case is the overflow and the first (MS) digit. Note that overflow here is a digit which moves to the other side of the point.

Repeat now with the fraction M'_q to obtain the next digit:

$$M'_p X q = M''_q = a_{-2}q^0 + a_{-3}q^{-1} + a_{-4}q^{-2} + ... + a_{-(m-1)}q^{-m+1} + a_{-m}q^{-m+2}$$

a_{-2} in this case is the overflow and the second (MS) digit. Repeat with the fraction M''_q to obtain the next digit, and so on until all digits (in base q form) are obtained or the required degree of accuracy achieved.

The process of radix conversion is best demonstrated by example and worked examples are included in section 2.10. Unworked questions appear in Tutorial 2 at the end of this chapter.

2.5 Operations on signed numbers

Everyday decimal arithmetic makes use of two additional symbols, + and - , to represent the sign of a number. By convention, the sign is written to the left of the most significant digit (MSD) and may thus appear in any integer column. Also, the sign of a positive number may be omitted altogether. Examples show these features:

```
     + 1
  - 7 6 5 4
1 4 9 6 3 2
   + 9 9 9
      2 2 . 3 4
       - 0 . 4 5 6
  2 3 4 5
       - . 8 7 6 5 etc.
```

Thus, in any of the integer columns, any one of $q + 2$ symbols may appear. For decimal there could be any one of twelve: 0,1,2,3,4,5,6,7,8,9,+,-; and for binary, any one of four: 0,1,+,-. In the binary case, this approach would destroy the inherent two-symbol (two-state) simplicity of the representation.

2.5.1 The representation of signed numbers

Sign/magnitude representation

As a first attempt at overcoming this problem we can consider the dedication of one column to sign alone. Let this be the most significant column, but for this to be sign alone and never a digit, we must fix a *word length*, eight bits for example, to which all signed numbers conform. The leftmost bit will then only convey sign information and this is normally done by having 0 represent + and 1 represent -.

For example, for an 8-bit word:

0 0 1 1 0 0 1 1 represents + 0110011 (i.e. $+51_{10}$)
1 0 1 1 0 0 1 1 represents - 0110011 (i.e. -51_{10})
1 1 1 1 0 0 1 0 represents - 1110010 (i.e. -114_{10}) etc.

This approach could also be carried through to decimal arithmetic, where 0 could represent + and 9 represent -. Then for, say, a four-digit word length, we could write:

0 0 5 1 for + 5 1
9 0 5 1 for - 5 1
9 1 1 4 for - 1 1 4 etc.

A characteristic of the representation is that the maximum magnitude is given by $q^{N-1}-1$ (where N is the word length in digits) and the range of numbers which can be represented is $2(q^{N-1})$ disposed equally in the negative and positive ranges (allowing for zero in each range).

For the 8-bit binary case, the range is from 0 to +/- $2^{8-1}-1$, i.e. +127 ... 0 ... -127.

However, if we are seeking simplicity of arithmetic and hardware, the sign/magnitude form has a further drawback in that the processes of addition and subtraction are different and two sets of rules and two hardware configurations would be needed, for example, in a digital arithmetic processor. Therefore, it is necessary to seek a signed number representation which allows for subtraction through the addition process. Such a representation may be based on the *radix complement* or *diminished radix complement* form of signed numbers.

The radix complement representation

To dispel any mystery right away, this approach, in the binary system, leads to the widely used *twos complement* representation of signed numbers.

In general terms, the radix complement N'_q of integer N_q is given by:

$$N'_q = q^n - N_q$$

where, n is the number of integral digits of N_q
q is the radix (base).

This general form applies to any radix in a positionally weighted system.

For example, in the decimal system we may form the *tens complement*, say of 125_{10}, assuming the word length to be three digits:

$$N'_q = q^n - N_q$$

$$125'_{10} = 10^3 - 125$$

$$= 875 \text{ (the tens complement of +125)}$$

Similarly, the tens complement of 475 = 525.
We can now use 125 to represent +125
and 875 to represent -125
and 475 to represent +475
with 525 to represent -475, etc.

Such a system allows positive and negative numbers to be *added directly* with the sign indication being presented as the most significant digit. In this case a leading digit between 0 and 4 inclusive indicates a positive number, whilst a leading digit between 5 and 9 inclusive indicates a negative number. Tens complementation works in either direction (from positive to negative form or vice versa) and, for example, the tens complement of 747 (-253) is given by:

$$10^3 - 747 = 253 \text{ (i.e. +253)}$$

Just to demonstrate that the system works, take some simple examples using the few numbers we have dealt with above:

$$+125 - 125 = 125 + 875 = (1)000$$

i.e. zero, since the carry (in brackets) beyond the MS digit is discarded.

$$+253 - 475 = 253 + 525 = 778$$

i.e., a negative number, the positive form of which is obtained by tens complementation (to check its value):

$$10^3 - 778 = 222$$

The result (778) is the representation of -222 as one would expect.

The range of numbers in this case for a three-digit representation is from the maximum positive number, [499], through zero [000] and into a negative range from [999] (-1) to [500] (-500). Note that there is only one representation of zero and that it is in the positive range.

Twos complement representation

Although the consideration of tens complement representation is not directly of use, it has served to illustrate some key features of radix complement forms. Our main interest is in the binary system and the radix complement form is of course *twos complement* representation.

Using the general expression given above and *taking an 8-bit environment*, we may, for example, form the twos complement of 00011010, [= $+26_{10}$] (taking the MSB as indicating sign (0 = +, 1 = -)). Twos complement = 2^8 - 00011010 = 100000000 - 00011010 = *11100110 which is the representation of* -26_{10} in twos complement form.

Thus we may represent a range of positive and negative binary numbers for any given word length. For example, for an 8-bit word the representable range of numbers is from $+127_{10}$ through 0 to -128_{10} as shown in Table 2.3.

Note the unique zero and that there is one extra value in the negative range (-128, there is no +128), the extra value being due to the fact that zero is in the positive range only. Note also that the twos complement of a positive number gives the negative form of that number, and vice versa. Further, the process of subtraction now takes place by addition, which is the object of the exercise.

Table 2.3 8-bit twos complement numbers

Twos complement form	Decimal equiv.
0 1 1 1 1 1 1 1	+127
0 1 1 1 1 1 1 0	+126
0 1 1 1 1 1 0 1	+125
...	...
0 0 1 0 1 1 1 0	+ 46
...	...
0 0 0 0 0 0 1 0	+ 2
0 0 0 0 0 0 0 1	+ 1
0 0 0 0 0 0 0 0	zero
1 1 1 1 1 1 1 1	- 1
1 1 1 1 1 1 1 0	- 2
...	...
1 1 0 1 0 0 1 0	- 46
...	...
1 0 0 0 0 0 1 0	-126
1 0 0 0 0 0 0 1	-127
1 0 0 0 0 0 0 0	-128

For example, to subtract, say, the binary representation of 46_{10} from127_{10}, we add the representation of +127 to the twos complement of $+46_{10}$ (i.e. -46_{10}) as follows:

In twos complement arithmetic, form $(127\text{-}46)_{10}$ as follows:

	0 1 1 1 1 1 1 1	+ 127
Add	1 1 0 1 0 0 1 0	- 46
result (1)	0 1 0 1 0 0 0 1	+ 81

discarding the (carry) beyond the 8-bit word.

As a further example, add $+127_{10}$ to -128_{10}:

	0 1 1 1 1 1 1 1	+ 127
	1 0 0 0 0 0 0 0	- 128
result	1 1 1 1 1 1 1 1	- 1

or subtract $+3_{10}$ from $+1_{10}$:

0 0 0 0 0 0 0 1	+ 1
1 1 1 1 1 1 0 1	- 3
1 1 1 1 1 1 1 0	- 2

Overflow will occur when two numbers being "added" exceed the available range. This is readily demonstrated as follows:

Using an 8-bit twos complement representation add:

0 1 1 1 0 0 0 0 + 0 1 0 0 0 0 1 = 1 0 1 1 0 0 0 1

which tells us that $(+112+65=-79)_{10}$, which is clearly incorrect in both sign and magnitude. The correct result, +177, is outside the 8-bit range.

Rules for forming the twos complement

Other than using the general expression, there are two simple rules which allow the generation of the positive from the negative, or negative from the positive form, of an n-bit number in twos complement form.

1. Complement all n-bits of the number to be twos complemented and then add 1.
   or,
2. Scanning the number to be twos complemented from right to left, reproduce all bits up to and including the first 1 and then complement all the remaining bits to the left of the first 1.

For example, form the (8-bit) negative form of $+46_{10}$

Rule1	0 0 1 0 1 1 1 0	+ 46
Complement all bits:	1 1 0 1 0 0 0 1	
Add 1:	1	
Giving:	1 1 0 1 0 0 1 0	- 46

Or for *Rule 2*	0 0 1 0 1 1 1 0	+ 46

Scan and reproduce all bits up to first 1 (in **bold**), then complement others, giving:

1 1 0 1 0 0 **1 0**   - 46

Or again for *Rule 2*	0 1 1 0 0 0 0 0	+ 96

Scan and reproduce all bits up to first 1 (in **bold**), then complement others, giving:

1 0 **1 0 0 0 0 0**   - 96   etc.

Rules of twos complement arithmetic

1. Add positive and negative numbers directly. The correct sign will be generated if range is not exceeded.
2. To form A-B, form the twos complement of B and add to A.
3. Discard any carry beyond the most significant bit.

The diminished radix complement representation

This approach, in the binary system, leads to the widely used *ones complement* representation of signed numbers.

In the most general form, the diminished radix complement $N''q$ of a number N_q is given by:

$$N''_q = q^n - q^{-m} - N_q$$

where, n is the number of integral digits of N_q
m is the number of fractional digits of N_q and,
q is the radix (base).

This general form applies for any radix in a positionally weighted system.

For example, in the decimal system we may form the *nines complement,* say of 125_{10} assuming the word length to be three digits:

$$N''_q = q^n - q^{-m} - N_q$$

$$125''_{10} = 10^3 - 1 - 125 = 874$$

Similarly, the nines complement of 475 = 524.

We can now use	125 to represent +125
and	874 to represent - 125
and	475 to represent +475
with	524 to represent - 475 etc.

We may also illustrate the nines complement of a decimal number with a fractional part, for example:

Form the nines complement of 136.21:

$$N''_q = q^n - q^{-m} - N_q$$

where $q^n = 10^3$; $q^{-m} = 10^{-2}$; $N_q = 136.21$; whence:

$$N''_q = 863.78$$

Such a system also allows positive and negative numbers to be *added together directly*, including the sign indication carried in the most significant digit. Again, a leading digit between 0 and 4 inclusive indicates a positive number, while a leading digit between 5 and 9 inclusive indicates a negative number. Nines complementation works in either direction (from positive to negative form or vice versa). For example, the nines complement of 747 (-252) is given by:

$$10^3 - 1 - 747 = 252 \text{ (i.e.+252).}$$

Zero is not unique in this system, there being a positive range zero (e.g. 000 for three digits) and a corresponding negative range zero (999).

Just to demonstrate that this system also works, we may use the numbers we have dealt with above:

$$+125 - 125 = 125 + 874 = 999 \quad \text{i.e. negative zero .}$$
$$+253 - 475 = 253 + 524 = 777$$

That is, a negative number, the positive form of which is obtained by nines complementation:

$$10^3 - 1 - 777 = 222$$

Thus the result (777) is the representation of -222 as one would expect.

The range of numbers in this case for a three-digit representation is from the maximum positive number [499] through zero [000] and into a negative range from [999] (-0) to [500] (-499). Note that the positive and negative ranges cover the same range.

Ones complement representation

Our main interest in the diminished radix form is in the binary system and the form is of course the ones complement (ones comp.) representation.

Using the general expression given above and *taking an 8-bit environment* we may, for example, form the ones complement of 00011010 which, taking the MSB as indicating sign (0 = +, 1 = -), $= +26_{10}$.

$$\text{Ones complement} = 2^8 - 00000001 - 00011010$$
$$= 100000000 - 00011011 = \underline{11100101}$$

which may be taken as a representation of -26_{10} *in ones comp. binary form.* Thus we may represent a range of positive and negative binary numbers for any given word length. For example, for an 8-bit word, the representable range of numbers is from $+127_{10}$ through 0 to -127_{10} as shown in Table 2.4.

Table 2.4 8-bit ones complement numbers

Ones complement form	*Decimal equiv.*
0 1 1 1 1 1 1 1	+127
0 1 1 1 1 1 1 0	+126
0 1 1 1 1 1 0 1	+125
...	...
0 0 1 0 1 1 1 0	+ 46
...	...
0 0 0 0 0 0 1 0	+ 2
0 0 0 0 0 0 0 1	+ 1
0 0 0 0 0 0 0 0	+ 0
1 1 1 1 1 1 1 1	- 0
1 1 1 1 1 1 1 0	- 1
1 1 1 1 1 1 0 1	- 2
...	...
1 1 0 1 0 0 0 1	- 46
...	...
1 0 0 0 0 0 1 0	- 125
1 0 0 0 0 0 0 1	- 126
1 0 0 0 0 0 0 0	- 127

Note the double zero and the equal ranges. Note also that the ones comp. of a positive number gives the negative form of that number and vice versa. Subtraction can again be effected through addition.

For example, to subtract, say, the binary representation of 46_{10} from 127_{10}, we add the representation of +127 to the ones comp. of $+46_{10}$ (i.e. -46_{10}) as follows:

```
        0 1 1 1 1 1 1 1      +127
        1 1 0 1 0 0 0 1      - 46
     (1)0 1 0 1 0 0 0 0      (carry in brackets)

Add any carry into LSB             1
        0 1 0 1 0 0 0 1      + 81
```

As a further example, add $+127_{10}$ to -127_{10}

01111111	+ 127
10000000	- 127
11111111	- 0

or subtract $+3_{10}$ from $+1_{10}$:

00000001	+ 1
11111100	- 3
11111101	- 2

Overflow will again occur when two numbers being "added" exceed the available range. This is readily demonstrated as follows:

Using an 8-bit ones complement representation add:

$$01110000 + 0100001 = 10110001$$

which tells us that $(+112 + 65 = -78)_{10}$, which is clearly incorrect in both sign and magnitude. The correct result, +177, is outside the 8-bit range.

Rules for forming the ones complement

A very simple rule applies to generate the positive from the negative or negative from the positive form of an n-bit number in ones complement form. The rule is:

Complement all n-bits of the number to be ones complemented. For example, to form the ones comp. of the 8-bit form of $+46_{10}$:

	00101110	+ 46
Complement all bits	11010001	- 46

Rules of ones complement arithmetic

1. Add positive and negative numbers directly. The correct sign will be generated if range is not exceeded.
2. To form *A-B*, form the ones comp. of *B* and add to *A*.
3. Add any carry beyond the most significant bit into the LSB.

Further worked examples in twos and ones complement arithmetic are included in section 2.10.

2.6 An overview of binary multiplication and division processes

We have seen that binary *addition* and *subtraction* operations are easy to conceive in terms of algorithms and hardware, needing only an adder and a complementer (twos or ones). However, a complete kit of arithmetic operations must include the *multiplication* and *division* operations. If we are to maintain the inherent simplicity of approach then ideally we would like to be able to multiply and divide using these same facilities and very little else. To that end we will examine some possible approaches.

2.6.1 Multiplication of unsigned or sign/magnitude numbers

Consider the basic process of long multiplication, and for binary numbers, we may recognize that each bit of the multiplier can only have a value of 1 or 0. So each multiplication is affected either by reproducing and adding the (shifted) multiplicand or adding zero respectively. The process is thus one of shift and add, as illustrated in the following example which assumes two 5-bit unsigned binary numbers.

Multiplicand	1 0 1 1 1
Multiplier	1 0 1 0 1
{*multiplier bits*	5 4 3 2 1}

	Major product	Minor product	
Multiply by bit 1 to form 1st PP	1 0 1 1 1		i.e. reproduce multiplicand same
Shift 1st PP right one place	0 1 0 1 1	1	effect as left shift to next product
Multiply by bit 2 and add to form 2nd PP	0 0 0 0 0		
2nd PP	0 1 0 1 1	1	
Shift 2nd PP right one place	0 0 1 0 1	1 1	
Multiply by bit 3 and add to form 3rd PP	1 0 1 1 1		
3rd PP	1 1 1 0 0	1 1	
Shift 3rd PP right one place	0 1 1 1 0	0 1 1	
Multiply by bit 4 and add to form 4th PP	0 0 0 0 0		
4th PP	0 1 1 1 0	0 1 1	
Shift 4th PP right one place	0 0 1 1 1	0 0 1 1	
Multiply by bit 5 and add to form 5th PP	1 0 1 1 1		
5th PP	1 1 1 1 0	0 0 1 1	
Shift to adjust *final Product*	0 1 1 1 1	0 0 0 1 1	

Note that the major and minor products are accumulated as two 5-bit words. This is conveniently accommodated in hardware by using two 5-bit registers. For sign/magnitude numbers, the sign bits are compared first and then removed prior to multiplication.

A suitable hardware arrangement for multiplication is set out in Figure 2.1 and the algorithm for multiplication would then be as follows.

An algorithm for unsigned binary multiplication
This algorithm is for use with a hardware configuration as in Figure 2.1 (8-bit words and registers are assumed for example.)

1. Load multiplier into MR register and the multiplicand into the N register. The multiplier least significant bit (LSB) is now in position to control *And* gate A1; if it is a 1, the gate is open, otherwise it is closed.
2. The multiplicand is now allowed to enter the adder if A1 is open; otherwise all 0's enter the adder when A1 is closed. The output of the adder is stored in the accumulator AC to form the first partial product (1st PP).

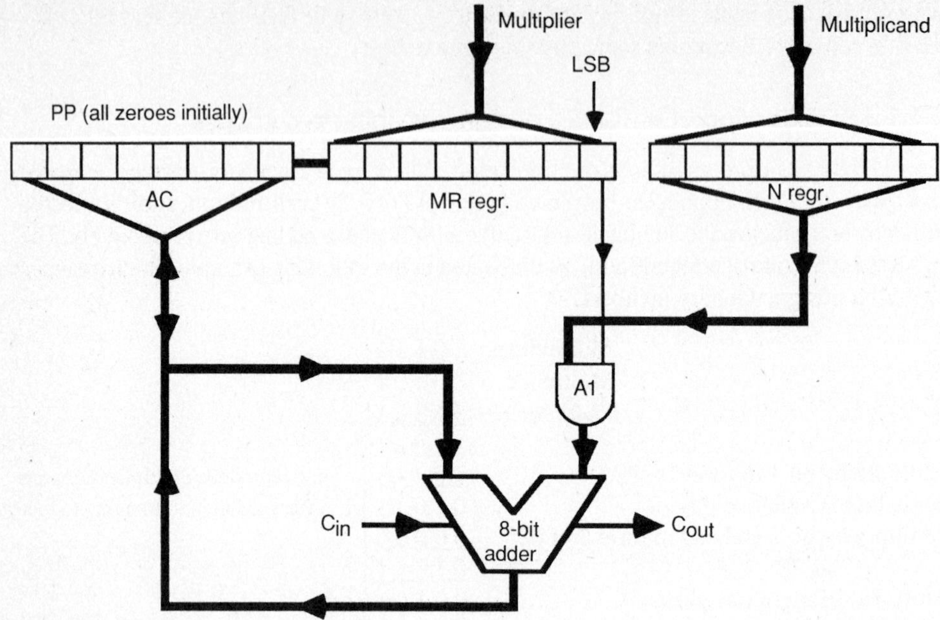

Figure 2.1 Multiplier architecture

3. We have now finished with the multiplier LSB, so that we may now shift register AC and MR contents right one place, so that the next LSB of the multiplier enters the control position and the LSB of register AC moves into the vacated most significant bit (MSB) of register MR.
4. The contents of N is now allowed (or not) to enter the adder together with the contents of AC, and the 2nd PP is thus formed and stored in the accumulator AC.
5. We have now finished with the second LSB of the multiplier, so that we may now shift register AC and MR contents right one place. The next bit of the multiplier enters the control position and the current LSB of register AC moves into the again vacated MSB of register MR.
6. The contents of N is now allowed (or not) to enter the adder together with the contents of AC and the 3rd PP is thus formed; etc.
7. At the end of the process when all bits of the multiplier have been used and shifted out of register MR, the16-bit product is presented in registers AC (major product) and MR (minor product).

   The muliplier is lost but the multiplicand is still present in the register N.

2.6.2 Signed (twos complement) multiplication-Booth's algorithm*

We will not delve into the derivation or the proving of Booth's algorithm but examine its application in multiplying twos complement numbers.

* Booth A.D. "A signed binary multiplication algorithm", *Journal of Mechanics and Applied Mathematics*, Vol 4, pt. 2, 1951.

Let Y be an $(n + 1)$ bit multiplicand written as:

$$y_n\, y_{n-1} \cdots y_i \cdots y_2\, y_1\, y_0$$

and X be an $(n + 1)$ -bit multiplier written as:

$$x_n\, x_{n-1} \cdots x_i \cdots x_2\, x_1\, x_0$$

where y_n and x_n are the sign bits.

Booth's algorithm is fast and deals with the multiplier bits in pairs, starting with the LSB x_0 which is paired with an imaginary 0 (in bit position x_{-1}). The execution of the algorithm is best set out as a short table, Table 2.5.

Table 2.5 Booth's algorithm

x_i	x_{i-1}	*Action*
0	0	Shift PP (arithmetic shift) right
1	1	one place to form next PP.
1	0	PP - *Y*, then shift result as above.
0	1	PP + *Y*, then shift result as above.

PP is initially set to all zeroes

In an arithmetic shift, the sign bit is reproduced as right shifts take place.

An example, showing the multiplication of two 5-bit twos complement numbers will serve to illustrate the process as follows:

Example: Multiply $+15_{10}$ X -13_{10} in 5-bit twos complement form.
Note that $+15_{10}$ = 0 1 1 1 1 and,
-13_{10} = 1 0 0 1 1 in twos comp. form.

Y	0 1 1 1 1		$+15_{10}$ [-Y = 1 0 0 0 1]
X	1 0 0 1 1	*0*	-13_{10}
	0 0 0 0 0	0 0 0 0 0	Initial PP
(x_0 =1: x_{-1}=0) PP-*Y*	1 0 0 0 1	0 0 0 0 0	
Arithmetic shift rt.	1 1 0 0 0	1 0 0 0 0	1st PP
(x_1 =1: x_0=1) shift rt.	1 1 1 0 0	0 1 0 0 0	2nd PP
(x_2 =0: x_1=1) PP+*Y*	0 1 1 1 1	0 0 0 0 0	+*Y*
	0 1 0 1 1	0 1 0 0 0	2nd PP + Y
Arithmetic shift rt.	0 0 1 0 1	1 0 1 0 0	3rd PP
(x_3 =0: x_2=0) shift rt.	0 0 0 1 0	1 1 0 1 0	4th PP
(x_4 =1: x_3=0) PP-*Y*	1 0 0 0 1	0 0 0 0 0	-*Y*
	1 0 0 1 1	1 1 0 1 0	4th PP - *Y*
Arithmetic shift rt.	1 1 0 0 1	1 1 1 0 1	Final product.

The result is a negative twos complement number = 1 1 0 0 1 1 1 1 0 1 [Check value by twos complementing to positive form = 0 0 1 1 0 0 0 0 1 1.

Evaluating, we have (128 + 64 + 2 + 1) = 195, i.e correct.]

2.6.3 Some division algorithms

There are numerous algorithms for division but we have no brief in this text to carry out a complete review. However, we will look at algorithms which can be implemented using a register structure similar to that in Figure 2.1. We will deal with unsigned numbers in this case.

A possible algorithm for fractions (i.e. binary point on left of MSB)
Assume numbers A and B which are both fractional and such that $A < B$. Further assume that we wish to form their quotient $Q = A/B$, which will also be fractional since $A < B$.

Thus Q (quotient), A (dividend) and B (divisor) are all less than 1.
Further, we assume Q to be of the form:

$$Q = a_{-1}.2^{-1} + a_{-2}.2^{-2} + a_{-3}.2^{-3} + a_{-4}.2^{-4} + \ldots \text{ etc } \ldots + a_{-m}.2^{-m}$$

where m is the number of bits needed for an exact result or for the desired or allowed degree of accuracy.

The division process is as follows:

1. Form: $A' = 2A - B$ {Shift A and add - B}
$= 2BQ - B$ (since $A = BQ$)
$= B(2Q - 1)$

i.e., $A' = B(-1 + a_{-1}(2^0) + a_{-2}.2^{-1} + a_{-3}.2^{-2} + a_{-4}.2^{-3} + \text{etc.}$

|- - I_n - -|- - - - - -Q' always <1- - - - - -|

$I_n = 0$ or 1 since 2^0 implies a non-fractional value.

2. Test sign of result to determine a_{-1}, (+ implies 1, - implies 0).

3. (a) If sign was positive:

then, $A' = BQ' = B(a_{-2}.2^{-1} + a_{-3}.2^{-2} + a_{-4}.2^{-3} + \text{etc})$

Now form: $A'' = 2A' - B = 2BQ' - B$.

$A'' = B(-1 + a_{-2}.(2^0) + a_{-3}.2^{-1} + a_{-4}.2^{-2} + \text{etc})$

|- - - - - Q'' - - - - -|

(b) If sign was negative;

then, $A' = BQ' - B = B(Q' - 1)$

Now form: $A'' = 2A' + B \; = B(-2 + 1 + Q')$

$$A'' = B(-1 + a_{-2}.(2^0) + a_{-3}.2^{-1} + a_{-4}.2^{-2} + \text{etc})$$

$$\qquad\qquad |\text{------}\; Q'' \;\text{------}|$$

4. Test Sign of result to determine a_{-2}, (+ implies 1, - implies 0).

5. Repeat the process with A'' etc., until the desired accuracy or the allowed number of bits are evaluated.

An algorithm for an unsigned number

This algorithm is based on the pencil and paper long division process and may be set out as follows:

Assume Y = dividend; X = divisor; Q = quotient.

1. Store dividend Y in a double length register (Y in LS half).
2. Shift Y left one place; compare MS half (Y(MS)) with divisor X (can be done by subtracting divisor X from Y(MS)).
3. (a) $Y(\text{MS}) < X$: subtract 0 from Y(MS) and record 0 for 1st bit of Q.
   (b) $Y(\text{MS}) \geq X$: subtract X from Y(MS), record 1 for 1st bit of Q.
4. Shift result of 3 left one place and compare.
5. Repeat 3 and 4 to give desired number of bits.

Example: Divide Y = 10011 by X = 00101 (both unsigned binary numbers).

Dividend Y	1 0 0 1 1
Divisor X	0 0 1 0 1

	Y(MS)	Y(LS)	
Store Y double length	0 0 0 0 0	1 0 0 1 1	
Shift Y one place left	0 0 0 0 1	0 0 1 1 0	
Compare: $Y(\text{MS})<X$	0 0 0 0 1	0 0 1 1 **0**	**(0 recorded as shown)**
Subtract 0 from Y(MS)	0 0 0 0 1	0 0 1 1 **0**	(result)
Shift Y one place left	0 0 0 1 0	0 1 1 **0** 0	
Compare: $Y(\text{MS})<X$	0 0 0 1 0	0 1 1 **0 0**	**(0 recorded as shown)**
Subtract 0 from Y(MS)	0 0 0 1 0	0 1 1 **0 0**	(result)
Shift Y one place left	0 0 1 0 0	1 1 **0 0** 0	
Compare: $Y(\text{MS})<X$	0 0 1 0 0	1 1 **0 0 0**	**(0 recorded as shown)**
Subtract 0 from Y(MS)	0 0 1 0 0	1 1 **0 0 0**	(result)
Shift Y one place left	0 1 0 0 1	1 **0 0 0** 0	
Compare: $Y(\text{MS})>X$	0 1 0 0 1	1 **0 0 0 1**	**(1 recorded as shown)**
Subtract X from Y(MS)	<u>0 0 1 0 1</u>		
	0 0 1 0 0	1 **0 0 0 1**	(result)
Shift Y one place left	0 1 0 0 1	**0 0 0 1** 0	
Compare: $Y(\text{MS})>X$	0 1 0 0 1	**0 0 0 1 1**	**(1 recorded as shown)**
Subtract X from Y(MS)	<u>0 0 1 0 1</u>		
	0 0 1 0 0	**0 0 0 1 1**	

Quotient Q is now in Y(LS) register; remainder in Y(MS) register.

Some worked examples are included in section 2.10.

2.6.4 Some observations

Even in an overview such as this, it is apparent that multiplication and division are long processes involving many sequential steps. For this reason, they are slow and often constitute bottlenecks in arithmetic processes in a digital system or computer. Also, we have seen that binary numbers require many digits to express large magnitudes or to represent a number to a high degree of accuracy. Other number systems are sometimes considered to limit these constraints.

2.7 Some other number systems of interest

To limit the number of digits, we require a radix higher than 2. On the other hand, we require a number system for which simple electronic devices may be used to represent all possible digits of the system and for which the arithmetic is simple. Two immediate possibilities which present themselves are base 3 and base 4, which are candidates because of their simple arithmetic and for which future developments may well yield simple devices with three or even four reliable and clearly defined states.

2.7.1 Ternary arithmetic

In this case the radix q is 3 and suitable symbols for the digits a are 0,1,2. Any number, N, may then be represented as:

$$N_3 = a_n.3^n + a_{n-1}.3^{n-1} + \dots + a_1.3^1 + a_0.3^0 + a_{-1}.3^{-1} + a_{-2}.3^{-2} + \dots + a_{-m}.3^{-m}$$

For example, $(2\,1\,2\,.\,1\,2)_3$ represents:

$$2.3^2 + 1.3^1 + 2.3^0 + 1.3^{-1} + 2.3^{-2} = 18 + 3 + 2 + \frac{1}{3} + \frac{2}{9}$$

Taking the integer part $(212)_3$, the binary equivalent is:

$$(1\,0\,1\,1\,1)_2$$

i.e. five digits rather than three so that the adoption of the ternary system reduces the number of digits quite considerably, by 40 percent in this example. Other numbers show more or less reduction but, on average, a 45 percent reduction is the order of the gain.

The arithmetic is simple. For example, addition requires the learning of $\frac{q\,(q+1)}{2} = \frac{3\,(4)}{2}$ = 6 relationships as follows:

0 + 0= 0 ; 0 + 1= 1 ; 0 + 2= 2 ; 1 + 1= 2 ; 1 + 2= 0 carry 1; 2 + 2= 1 carry 1.

Multiplication is equally straightforward.

2.7.2 Quadernary arithmetic

In this case the radix, q, is 4 and suitable symbols for the digits a are 0,1,2,3. Any number, N, may then be represented as:

$$N_4 = a_n.4^n + a_{n-1}.4^{n-1} + ... + a_1.4^1 + a_0.4^0 + a_{-1}.4^{-1} + a_{-2}.4^{-2} + ... + a_{-m}.4^{-m}$$

For example, $(3\,2\,1\,.\,3\,2)_4$ represents:

$$3.4^2 + 2.4^1 + 1.4^0 + 3.4^{-1} + 2.4^{-2} = 48 + 8 + 1 + \frac{3}{4} + \frac{2}{16}$$

Taking the integer part, $(321)_3$, the binary equivalent is:

$$(1\,1\,1\,0\,0\,1)_2$$

i.e. six digits rather than three so that the adoption of the quadernary system reduces the number of digits by 50 percent in this example.

The arithmetic is still reasonably simple. For example, addition requires the learning of $\frac{q}{2}(q + 1) = \frac{4}{2}(5) = 10$ relationships as follows:

0 + 0= 0 ; 0 + 1= 1 ; 0 + 2= 2 ; 0 + 3= 3 ; 1 + 1= 2 ; 1 + 2= 3 ;1 + 3 = 0 carry 1 ; 2 + 2 = 0 carry 1 ; 2 + 3 = 1 carry 1; 3 + 3 = 2 carry 1.

Multiplication is equally straightforward.

The adoption of either radix 3 or 4 reduces the number of digits but does not eliminate lengthy carry propagation effects nor the effects of long sequential multiplication and division operations. What is required is a number system in which all digits can be added, subtracted, multiplied or divided simultaneously. Such, for add subtract and multiply, are the properties of residue arithmetic.

2.7.3 Residue arithmetic

It is only possible to give a brief introduction to this topic in this text. For those readers wishing to read in more depth an excellent treatment is provided by N.S. Szabo, and R.I. Tanaka, in "*Residue arithmetic and its application to computer technology*", McGraw-Hill,1967.

Let us consider a simple residue system based on two moduli which must be mutually prime—say 7 and 9. Any number N within a range determined by the choice of moduli may then be represented as a two-digit number, the first digit being the residue (remainder) when N is divided by 7 and the second digit being the residue when N is divided by 9.

Take, for example, two values of N, $N_1 = 8$ and $N_2 = 6$. The representations of $N_1 = 8$ and $N_2 = 6$ in the 7, 9 residue system are:

$$N_1 = 1, 8$$

and,
$$N_2 = 6, 6$$

arrived at by dividing by 7 and then by 9 and writing the remainders.

Addition: To add $N_1 + N_2$ we *add each column separately* and write the sum to base 7 in the first column and the sum to base 9 in the other column and *disregard any carry.* For example:

$$\begin{array}{rl} N_1 = & 1, 8 \\ + \; N_2 = & 6, 6 \\ \hline (N_1{+}N_2) = & 0, 5 \quad = 14_{10}. \end{array}$$

Subtraction: To form $N_1 - N_2$ we *subtract each column separately* and write the difference to base 7 in the first column and the difference to base 9 in the other column and disregard any borrow. For example:

$$\begin{array}{rl} N_1 = & 1, 8 \\ - \; N_2 = & 6, 6 \\ \hline (N_1 - N_2) = & 2, 2 \quad = 2_{10}. \end{array}$$

Thus any adders using this arithmetic would have no carry (or borrow) propagation delay effects.

Multiplication: To multiply N_1 X N_2 we *multiply each column separately* and write the product to base 7 in the first column and the product to base 9 in the other column and disregard any carry. For example:

$$\begin{array}{rl} N_1 = & 1, 8 \\ \text{X} \; N_2 = & 6, 6 \\ \hline (N_1 x N_2) = & 6, 3 \quad = 48_{10} \end{array}$$

Division: This is a more complex and slower process than those described above and can be implemented using a look-up table. For those readers who wish to study division, the reference given earlier and the additional references given at the end of this chapter should provide a starting point.

The *range of numbers* which can be represented is determined by the number and value of the moduli and for two moduli is determined by:

$$\text{largest number} = (m_1 \text{ X } m_2) - 1$$

where m_1 and m_2 are the moduli of a two-modulus system. For the example used so far, m_1 and $m_2 = 7$ and 9 respectively, so that the largest number = (7 X 9)-1 = 62_{10} and the range is from 0 to 62_{10}.

A better choice for a two-modulus system from the range point of view would be to choose moduli 15 and 16, which also fit in well with 8-bit representation of the residue numbers in this range. Clearly, the range of numbers is now from 0 to (15 X 16) -1 = 0 to 239_{10}.

It is also possible to split the range to generate positive and negative numbers, and *larger ranges* can be accommodated by using more mutually prime moduli. For example, if three are used, then the range is from zero to (m_1 X m_2 X m_3) -1. For example, chosing16,15,14 would give 0 to 3359, etc.

2.8 Floating point arithmetic

Clearly, for any fixed word length, there will be a limit to the range of integers or fractions which can be represented. This is determined by the number of available digits (or bits) and whether a signed or unsigned number is being represented. In any event for any reasonable number of bits in a word the limits are unacceptable for many of the calculations we may wish to do. For this reason, an alternative form of representation, floating point representation, is in widespread use to extend the range available.

2.8.1 Format and representation

A floating point number "*N*" comprises three main components:

1. a *sign* for the overall number,
2. a magnitude component *M* called the *mantissa* which is represented to some chosen radix, *q*, and
3. a *signed exponent, e*.

The value of *N* is then given by:

$$N = +/- \; (M.\, q^{+/-e}) \quad \{\text{where +/- indicates + or -}\}$$

M may be either a fraction or an integer, and the number of bits allocated to *M* and the choice of radix *q* determine the accuracy of the representation. The radix *q*, together with the number of bits allocated to the exponent *e*, determine the overall range of numbers which can be covered. The format of a typical representation in one or more computer words is as follows:

Sign	Mantissa *M*	Signed exponent *e*

To give a satisfactory range and acceptable accuracy, the number of bits allocated to represent *M* and *e* are critical factors and, in general, a satisfactory floating point(FP) representation will require at least 32 bits per number. In an 8-bit microcomputer, this is achieved by using four words, while in a 16-bit machine there will be a need for two words per number.

Accuracy

For a given number of bits allocated to representing the mantissa, the best accuracy is achieved when all bits of the mantissa are significant, i.e. when there are no leading zeroes.

Taking the resolution represented by the LSB of M when all other bits are 1, we can evaluate the best accuracy for a binary weighted mantissa as follows:

Number of bits allocated to M	Best accuracy
16 bits	1 in 65 536, i.e. approx. 1 in $10^{4.8}$
20 bits	1 in 1 048 576, i.e. approx. 1 in 10^{6}
24 bits	1 in 16 777 216, i.e. approx. 1 in $10^{7.2}$

We may see that acceptable accuracy accompanies the use of a 20- or 24-bit mantissa.

Normalization

In order to maintain the accuracy of representation, we use a process of *normalization* which *adjusts the exponent so that a significant digit appears in the MS digit position of the mantissa*. For example, normalisation would be effected to adjust a 32-bit binary result such as:

$$\cdot 00101110001011100010111000101110 \times 2^{+12}$$

to give a normalized result of:

$$.10111000101110001011100010111011 \times 2^{+10}$$

which may well allow for the inclusion of extra digits in the LS bits as suggested here.

When the chosen radix q is not 2, normalization is carried out to put a *digit* in the *MS digit position*. Since other radices are invariably encoded in binary form, then normalisation may not put a 1 in the MS *bit* position of the mantissa and the accuracy of representation will be affected. For example, if a hexadecimal weighting is used, then each hex digit of the mantissa will occupy four bits and normalization may well result in a 1 in the MS digit position which will be represented as 0 0 0 1 so that there will be three leading 0s in the binary representation of the normalized mantissa.

Range of numbers

This is determined by the choice of radix q and by the number of bits allocated to expressing the signed exponent e.

Given a fixed number of bits for e, then the range is determined by the choice of radix q. Clearly the choice of a binary radix will give the smallest range and the choice of another radix, say 4 or 8 will give a progressively larger range since $2^e < 4^e < 8^e$, etc.

For small word length computers it is often necessary to choose a radix of 8 (octal) or 16 (hexadecimal) to achieve the required range. The larger radices will result in a loss of accuracy as illustrated in the following example of a typical 32-bit word format:

Sign (1 bit)	Mantissa M (24 bits)	Signed exponent e (7 bits)

For the allocation of bits shown, and interpreting the mantissa as a fraction, and the exponent e as a signed twos complement binary number, and allowing for the effects of normalization we have:

Chosen radix	*Range covered*	*Accuracy after normalization*
2	$1 \times 2^{+63}$ to 1×2^{-64}	(1 bit in 24), 1 in 16×10^6
4	$1 \times 4^{+63}$ to 1×4^{-64}	(1 bit in 23), 1 in 8×10^6
8	$1 \times 8^{+63}$ to 1×8^{-64}	(1 bit in 22), 1 in 4×10^6
16	$1 \times 16^{+63}$ to 1×16^{-64}	(1 bit in 21), 1 in 2×10^6

Choice of radix

There is obviously a trade-off between range, implying a large radix, and accuracy, which implies a small radix. Studies have shown that the radix giving the best compromise is 4, but many computer FP arithmetic formats use binary or hexadecimal weighting.

2.8.2. Floating point arithmetic operations

A summary of the outline algorithms will serve to illustrate the way in which FP operations are realised.

Addition and Subtraction

1. Subtract exponents to determine the largest.
2. Adjust the number with the smaller exponent so that both have the same exponent.
3. Add (or subtract) the mantissas.
4. Normalize the result and adjust the exponent accordingly.
5. Return the result with exponent and sign.

Example: Add $A = +0.111010 \times 2^7$

$+ \ B = +0.101010 \times 2^5$

1. Compare exponents. *A* has the larger.
2. Adjust *B* to exponent 7.

   $B = .001010 \times 2^7$
3. Add aligned mantissas.

   Sum = +1.000100
4. Normalize mantissa and adjust exponent of sum.

   FP Sum = 0.100010×2^8
5. Return result = +, M = 0. 100010, e = +8.

This is more complex than ordinary arithmetic since both the exponents and the mantissas must be processed.

Multiplication and Division

These processes are easier than addition or subtraction.

Multiplication

1. Multiply the mantissas.
2. Add the exponents.
3. Normalize.
4. Return result.

Division

1. Divide the mantissas.
2. Subtract the exponents.
3. Normalize.
4. Return result.

In some computers, a small additional adder of *e* bits (where *e* is the number of exponent bits) is provided to operate on the exponents in parallel with the processing of the mantissas.

In any event, floating point (FP) operations are so important in everyday computing that computer power is often assessed in terms of the number of FP arithmetic operations that can be performed in a second. This leads to such delightful terminology as "2 megaflops" implying that a particular computer can process 2 million floating point operations per second.

2.9 Summary

This chapter has given an overview of the various number systems, representations and arithmetic of interest to the digital system designer. It is by no means a complete coverage but it is hoped that most of the essential facts, factors and processes are at least introduced.

Like the preceding chapter and the following chapter, it is intended to supply necessary background material for the reader and should be read selectively to fill in the gaps where necessary.

The next chapter will establish the underlying principles in utilising logic and dealing with logical expressions. Boolean algebra and other basic techniques are also covered.

2.10 Worked examples

2.10.1 Radix conversion

Convert the following numbers as stated, fractional parts of converted numbers to be limited to four significant figures where necessary:

1. Convert $(28.625)_{10}$ to binary.

   Solution:

   Taking the integer and fractional parts separately and noting that the arithmetic is to be that of base 10 and that the new base q is 2:

   Integer part—successive division by 2   Fractional part—successive multiplication by 2

```
   2|28                              .625
   2|14  r 0(LSB)          overflow   X 2
   2| 7  r 0               (MSB)1   .250      (continued)
```

```
2|__3_ r 1                    X2
2|__1_ r 1                 0 .500
    0  r 1(MSB)               X2
                     (LSB) 1 .000
```

Thus, binary representation is: $(11100.101)_2$

2. Convert $(329.85)_{10}$ to octal.

Solution:

The arithmetic is to be that of base 10 and the new base q is 8:

Integer part-successive division by 8 Fractional part-successive multiplication by 8

```
8|329                              .85
8|__41_ r 1(LSD)          overflow  X8
8|___5_ r 1              (MSB)6 .80
     0  r 5                          X8
                                 6 .40
                                     X8
                                 3 .20
                                     X8
                           (LSB)1 .60
```

Thus, octal representation is ... $(511.6631)_8$.

Note: The next digit will be 4 so that the last digit could be rounded to 2.

3. Convert $(645)_8$ to decimal form.

Solution:

(a) Applying the general algorithm we would approach the conversion thus. The arithmetic is to be that of base 8 so that new base q of 10_{10} is 12_8.

Integer only, therefore successive division by 12_8. All numbers in the calculation are to base 8.

```
12|645
12|__52_ r 1(LSD)
12|___4_ r 2
      0  r 4
```

Thus the conversion yields $(421)_{10}$.

(b) An alternative way to approach this conversion is to take the base 8 number and evaluate each digit (in base 10) by using the weighted series, so that:

$$
\begin{aligned}
(6\,4\,5)_8 &= 6 \text{ X } 8^2 + 4 \text{ X } 8^1 + 5 \text{ X } 8^0 \\
&= 6 \text{ X } 64 + 4 \text{ X } 8 \;\; + 5 \text{ X } 1 \\
&= 384 \; + 32 + 5 = \underline{(421)}_{10}
\end{aligned}
$$

4. Convert A B C 8 . 2 D H to octal form.

Solution:

Rather than go through the process of successive divisions for the integer part and then successive multiplications for the fraction *all in hexadecimal arithmetic*, a better approach is to convert through base 2.

(a) Convert A B C 8 . 2 D H to binary, giving:

$$(1\,0\,1\,0\,1\,0\,1\,1\,1\,1\,0\,0\,1\,0\,0\,0\,.\,0\,0\,1\,0\,1\,1\,0\,1)_2$$

(b) Now convert $(1\,0\,1\,0\,1\,0\,1\,1\,1\,1\,0\,0\,1\,0\,0\,0\,.\,0\,0\,1\,0\,1\,1\,0\,1)_2$ to octal forming groups of 3 bits:

$$\mathit{00}1\;\;010\;\;101\;\;111\;\;001\;\;000.001\;\;011\;\;01\mathit{0}$$

giving: $(1\,2\,5\,7\,1\,0\,.\,1\,3\,2)_8$

5. Convert $(6\,4\,5)_8$ to binary, thence convert to decimal.

Solution:

$$(6\,4\,5)_8 = (1\,1\,0\;\;1\,0\,0\;\;1\,0\,1)_2$$

Now to convert to decimal we could sucessively dividě the binary number by $(1\,0\,1\,0)_2$ [$= 10_{10}$]. However, this would be cumbersome and it is easier to evaluate each bit and sum up as follows:

$$
\begin{aligned}
(1\,1\,0\;\;1\,0\,0\;\;1\,0\,1)_2 = (&1 \text{ X } 2^8 + 1 \text{ X } 2^8 + 0 \text{ X } 2^8 + 1 \text{ X } 2^8 + 0 \text{ X } 2^8 + 0 \text{ X } 2 \\
&+ 1 \text{ X } 2^8 + 0 \text{ X } 2^8 + 1 \text{ X } 2^8)_{10} \\
= (&256 + 128 + 32 + 4 + 1)_{10} \qquad = \underline{(421)}_{10}
\end{aligned}
$$

The result may be checked with question 3.

6. Convert $(241)_{10}$ to radix 3 and also to radix 2 and compare.

Solution:

Both conversions will be carried out in decimal arithmetic:

3	241		2	241	
3	80	r 0(LSD)	2	120	r 1(LSB)
3	26	r 2	2	60	r 0

```
3|___8_  r2                    2|__30_  r0
3|___2_  r2                    2|__15_  r0
     0   r2(MSD)               2|___7_  r1
                               2|___3_  r1
                               2|___1_  r1
                                    0   r1(MSB)
```

Thus, $(241)_{10} = (22220)_3 = (11110001)_2$
Note: The choice of radix 3 rather than radix 2 considerably reduces the number of digits (i.e. by about 40%).

7. Convert $(11100011010.1)_2$ to radix 5.

Solution:

First convert to decimal form:

$$1X2^{10}+1X2^{9}+1X2^{8}+0X2^{7}+0X2^{6}+0X2^{5}+1X2^{4}+1X2^{3}+0X2^{2}+1X2^{1}+0X2^{0}+1X2^{-1}$$
$$= 1024 + 512 + 256 + 16 + 8 + 2 + 0.5 = (1818.5)_{10}$$

The arithmetic is to be that of base 10 and the new base q is 5:

Integer part-successive division by 5 Fractional part-successive multiplication by 5

```
5|1818                              .5
5|__363_  r3(LSD)      overflow    X5
5|___72_  r3           (MSB)2       .5
5|___14_  r2                       X5
5|____2_  r4                2       .5
      0   r2(MSD)                  etc 2 recurring.
```

Thus, radix 5 representation is, $(24233.2222)_5$
Many other conversions are possible but the essential features of radix conversion are well enough illustrated by the preceding examples.

2.10.2 Twos and ones complement arithmetic

1. (a) For an 8-bit word length, carry out the following arithmetic operations in (i) twos complement, and (ii) ones complement arithmetic.

$$-37_{10} - 91_{10} =$$

Solution:

Start with the positive forms of the numbers since this is the same for both twos and ones comp. forms:

```
+37 = 0 0 1 0 0 1 0 1;   +91 = 0 1 0 1 1 0 1 1
      sign 32    4 1           sign 64 16 8  2 1
```

	(i) Twos comp.	(ii) Ones comp.
-37	11011011	11011010
-91	10100101	10100100
	(1) 10000000	(1) 01111110
	i.e. -128(correct)	1 add carry
		overflow ... *01111111 (incorrect)*

Note: The result is out of range for ones complement and therefore incorrect.

(b) For an 8-bit word length form $(A - B + C) + D$ in (i) twos complement, and (ii) ones complement arithmetic where:

$$A = +58_{10}\,;\ B = +33_{10}\,;\ C = -25_{10}\,;\ D + 65_{10}\,.$$

Start with the positive forms of the numbers:

A=00111010 ; B=00100001 ; C=00011001 ; D=01000001

	(i) Twos comp.	(ii) Ones comp.
A	00111010	00111010
-B	11011111	11011110
A-B	(1) 00011001	(1) 00011000
		1 add carry
		00011001
A-B	00011001	00011001
+(-C)	11100111	11100110
(A - B + C)	(1) 00000000	11111111
+D	01000001	01000001
	01000001 = + 65	(1) 01000000
		1
		01000001 = + 65

2. For a 16-bit word length:

(a) Determine the range of numbers for (i) twos and (ii) ones complement representations.

Solution:

Allowing for the sign bit, we have a maximum positive range value of: $0111111111111111 = 2^{15} - 1 = 32{,}767_{10}$.

Thus, (i) Twos comp. range is from +32 767 through zero to -32 768.
and, (ii) Ones comp. range is from +32 767 through +/- zero to -32 767.

(b) Evaluate, 0100000000000011 and 1111111100110011 in decimal form for (i) twos and (ii) ones comp. representation.

Solution:

The positive form is the same in both cases, therefore for (i) and (ii): *0100000000000011* = + ($2^{14} + 2^1 + 2^0$) = 16,384 + 2 + 1 = *+16,387*$_{10}$.

For the negative numbers:

(i) 1111111100110011 twos complemented = 0000000011001101. Therefore in twos complement representation it represents:

$$-(2^7 + 2^6 + 2^3 + 2^2 + 2^0) = -(128 + 64 + 8 + 4 + 1)_{10} = -205_{10}.$$

(ii) 1111111100110011 ones complemented = 0000000011001100. Therefore in ones complement representation it represents:

$$-(2^7 + 2^6 + 2^3 + 2^2) = -(128 + 64 + 8 + 4)_{10} = -204_{10}.$$

(c) Using (ii) twos and (i) ones comp. arithmetic, subtract $+204_{10}$ from $-16,183_{10}$. Check the result by decimal arithmetic.

Solution:

$+204_{10}$ = 0000000011001100_2. Therefore -204_{10} is given by:

(i) twos comp.	(ii) ones comp.
1111111100110100	1111111100110011

$+16183_{10}$ = 0011111100110111_2. Therefore -16183_{10} is given by:

(i) twos comp.	(ii) ones comp.
1100000011001001	1100000011001000

Now by addition:

	(ii) twos comp.	(i) ones comp.
-204	1111111100110100	1111111100110011
-16183	1100000011001001	1100000011001000
	(1) 1011111111111101	(1) 1011111111111011
		1
		1011111111111100

Check by evaluating the positive form of the results:

= 0100000000000011 = 16387 in both cases i.e correct, which checks with the decimal calculation -16183 -204 = -16387.

2.10.3 Multiplication and division

1. Multiply $+15_{10}$ X $+13_{10}$ in 5-bit twos complement form.

X	0 1 1 1 1 |	$+15_{10}$	[*-Y* = 1 0 0 0 1]
Y	0 1 1 0 1 | 0	$+13_{10}$	

	00000|00000	Initial PP
($x_0 = 1{:}x_{-1}=0$) PP-Y	10001|00000	
Arithmetic shift rt.	11000|10000	1st PP
($x_1 = 0{:}x_0=1$) PP+Y	01111|00000	+Y
	00111|10000	1st PP + Y
Arithmetic shift rt.	00011|11000	2nd PP
($x_2 = 1{:}x_1=0$) PP-Y	10001|00000	-Y
	10100|11000	2nd PP - Y
Arithmetic shift rt.	11010|01100	3rd PP
($x_3 =1{:}x_2=1$) shift rt.	11101|00110	4th PP
($x_4 = 0{:}x_3=1$) PP+Y	01111|00000	+Y
	01100|00110	4th PP + Y (discard carry)
Arithmetic shift rt.	00110|00011	Final product.

The result is $+195_{10}$ *as (hopefully) expected.*

2. Divide Y = 11011 by X = 00011 (both unsigned binary numbers)

Dividend Y 11011 27_{10}
Divisor X 00011 3_{10}

	Y(MS) Y(LS)	
Store Y double length	00000|11011	
Shift Y one place left	00001|10110	
Compare: Y(MS)<X	00001|10110	**(0 recorded as shown)**
Subtract 0 from Y(MS)	00001|10110	(result)
Shift Y one place left	00011|01100	
Compare: Y(MS)=X	00011|01101	**(1 recorded as shown)**
Subtract X from Y(MS)	00011	
	00000|01101	(result)
Shift Y one place left	00000|11010	
Compare: Y(MS)<X	00000|11010	**(0 recorded as shown)**
Subtract 0 from Y(MS)	00000|11000	(result)
Shift Y one place left	00001|10100	
Compare: Y(MS)<X	00001|10100	**(0 recorded as shown)**
Subtract 0 from Y(MS)	00001|10100	(result)
Shift Y one place left	00011|01000	
Compare: Y(MS)=X	00011|01001	**(1 recorded as shown)**
Subtract X from Y(MS)	00011	
	00000|01001	**=9_{10}**

Quotient Q is now in Y(LS) register. Remainder in Y(MS) register. This is an exact division in this case.

2.10.4 Residue arithmetic

1. (a) Given a three-modulus residue number system, what are the ranges of numbers covered if the moduli are (i) 7,8,9 and (ii) 14,15,16?

(b) Assuming an unsigned environment, write the representations for systems (i) and (ii) of the following decimal numbers:

0 ; 7 ; 10; 100 ; 345 ; 500 ; 510.

Solution:

(a) Range is determined by maximum value = (m_1 X m_2 X m_3) -1
Therefore, for (i) range is 0 to (7X8X9)-1, i.e. *0 to 503*$_{10}$.
and for (ii) range is 0 to (14X15X16)-1 i.e. *0 to 3359*$_{10}$.

(b)	$Number_{10}$	System(i)	System(ii)
	0	0,0,0	0,0,0
	7	0,1,2	7,7,7
	10	3,2,1	A,A,A (hex symbols assumed)
	100	2,4,1	2,A,4
	345	2,1,3	9,0,9
	500	3,4,5	A,5,4
	510	out of range	6,0,E

2. Using two-modulus (m_1 = 15; m_2 = 16) residue arithmetic, carry out the following arithmetic operations (expressed in decimal form).

(a) Add 22 + 58.

Solution:

22 = 7,6 ; 58 = *D*, *A*.Adding residues separately...
7,6 + *D*,*A* = 5,0 (Check: residue form of 80 = 5,0 √)

(b) Subtract 58 - 22.

Solution:

22 = 7,6 ; 58 = *D*, *A*.Subtract residues separately...
D,*A* - 7,6 = 6,4 (Check: residue form of 36 = 6,4 √)

(c) Multiply 22 X 8.

Solution:

22 = 7,6 ; 8 = 8, 8. Multiply residues separately...
7,6 X 8,8 = *B*,0 (Check: residue form of 176 = *B*,0 √)

2.10.5 FP representation and arithmetic

1. Given a 24-bit word length, choose a suitable division of the word for a FP representation and, choosing a suitable radix, determine the range and accuracy.

Solution:
A possible choice is:

Sign(1-bit)	Mantissa M (16-bits)	Signed exponent e (7-bits)

and a reasonable range will be generated if a hex radix is adopted which gives four hex digits for the 16-bit mantissa. The 7-bit exponent will be regarded as a twos complement number.

Range (regarding the mantissa as a fraction) = +/- (1 X 16^{+63} to 1 X 16^{-64})

Accuracy (worst case after normalisation) = 1 bit in 13

2. For a representation comprising a 4-hex digit mantissa with a leading sign bit and a 7-bit twos complement exponent, add the two following FP numbers.

A = 1 1 0 0 0 1 1 1 1 0 0 0 1 1 0 1 0 0 0 1 0 1 0 0

B = 1 0 0 0 0 0 1 1 0 1 0 1 0 0 0 1 0 0 0 1 0 0 1 1

Solution:

The assumed format is:

Sign(1-bit)	Mantissa M (16-bits)	Signed exponent e (7-bits)

Therefore A = -.1000111100011010 exponent 0010100
and B = -.0000011010100010 exponent 0010011

that is: $A = (-.8F1A \text{ X } 16^{+20})_{10}$
and $B = (-.06A2 \text{ X } 16^{+19})_{10}$

Therefore A = -.1000111100011010 exponent 0010100
+ adjusted B = -.0110101000100000 exponent 0010100
= sum $A+B$ = -.1111100100111010 exponent 0010100

No normalisation is needed since MS hex digit is occupied.

Sum = $(-.F93A \text{ X } 16^{+20})_{10}$

2.11 Tutorial 2

1. (a) If we wished to configure a processor to carry out basic arithmetic operations in hexadecimal arithmetic, how many "table entries" would we need to teach the machine so that it could add, subtract and multiply? You may assume that commutativity is recognized.
(b) How many symbols would be needed to represent the digits for base 32_{10} arithmetic? Suggest a suitable set of symbols.
(c) Set out the advantages of using binary arithmetic in digital computers.

2. (a) Set out the general algorithms for the radix conversion of integers and also for fractions of numbers from base p to base q.

(b) Convert $(346.375)_{10}$ and $(492)_{10}$ to hexadecimal form.
Add these two hexadecimal numbers *using hexadecimal arithmetic.*
Convert their sum to decimal form and then check your results using decimal arithmetic.

3. Using 8-bit (a) twos complement and (b) ones complement arithmetic, perform the following arithmetic operations and check your results through binary to decimal conversion.

   (i) $+(48)_{10}$ - $(+(69)_{10})$
   (ii) $-(125)_{10}$ + $(45)_{10})$
   (iii) $+(125)_{10}$ - (128_{10})
   (iv) $+(68)_{10}$ + (60_{10})
   (v) $-(-(67)_{10})$ + $(-(31)_{10})$

4. Using Booths algorithm, and a 5-bit word length, form the product of -7_{10} and -12_{10}.

5. Assuming unsigned 6-bit binary integers, divide 57_{10} by 3. Check your result by binary to decimal conversion of the result.

6. (a) Calculate the range of numbers which could be represented using a two modulus residue system where the moduli are 14 and 15.
   (b) Using residue arithmetic, form the sum, difference, and product of A and B for the following values:

   (i) A = 21; B = 9
   (ii) A = 15; B = 13
   (iii) A = 10; B = 21
   (iv) A = 12; B = 12
   } All base 10

7. Floating point numbers are to be represented in a 30-bit word. Indicate the factors determining the allocation of the bits of the word and the interrelationship with the choice of mantissa radix. Make the choice of bit allocations and radix and then calculate the range and accuracy of the chosen representation. Compare with the choice of radix 2 (unless that is the radix you have chosen).
8. Devise a more detailed algorithm for floating point addition based on the outline algorithm in section 2.8.2 of the text.

3. Some basic techniques for handling problems in designing logic circuitry

The design of digital systems will generally comprise two classes of logic circuitry—combinational logic and sequential logic. In this chapter we will examine switching algebra and methods of representing and simplifying the expressions which describe the requirements to be met, such methods being general to both classes of circuits.

Switching algebra (Boolean algebra) is a convenient way of representing, in a logical sense, the characteristics of logic circuits and is based on relationships between two-state (logical 0 and logical 1) variables and three simple operators—"." (*And*), "+" (*Or*) together with " – " alternatively " ' ", (the *complement* or *Not* operator).

3.1 Introduction to switching algebra

We are to consider an algebra based on variables which can only take on the value of one or other of two logic levels—logic 1 or logic 0. The state of these variables is often conveniently represented by switches (closed or open) or by voltage levels (high or low), so that the following (Table 3.1) are often used, interchangeably, in the design of combinational logic:

Table 3.1 Logic level representation

Logic level 1	Logic level 0
On	Off
High (Hi)	Low (Lo)
1	0
True	False
+V	-V
+V	OV
-V	OV

The variables used are commonly letters (upper or lower case) but other characters or symbols are sometimes employed.

Consider the switch arrangements shown in Figure 3.1 in which switches *A* and *B* are used to operate the lamp *L*.

If either switch *A* or switch *B*, or both are operated then the input voltage source will be connected across the lamp, current will flow and the lamp will light. Note that one or more closed switch will operate the lamp.

To avoid any ambiguity, the circuit action may be described by entries in a Truth Table as in Table 3.2(a).

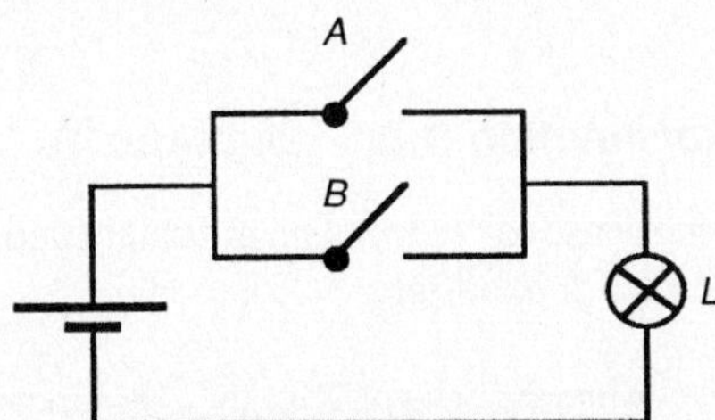

Figure 3.1 Switching circuit

Table 3.2(a) Truth table defining the operation of the circuit of Figure 3.1

A	*B*	*L*
Off	Off	Off
Off	On	On
On	Off	On
On	On	On

On = switch closed: Off = switch open

The entries in this table may be replaced by 1 (on), 0 (off) which results in the truth table of Table 3.2(b).

Table 3.2(b) Truth table for logic level representation for circuit of Figure 3.1

A	*B*	*L*
0	0	0
0	1	1
1	0	1
1	1	1

Truth tables express all possible relationships between the inputs (A and B in this case) and the output(s). A more concise representation is to write an equation in Boolean algebra:

$$L = A + B$$

a two-input *Or* function.

3.2 Boolean algebra and logic functions

The structure of this algebra is based on the work of George Boole* using variables in two states—0 and 1.

3.2.1 Negation: The *Not* function (bar $\overline{\ }$ or prime $'$)

The *Not* function changes the sense (or state) of an argument and is represented by a bar or prime operator as convenient. For example:

* George Boole, "An investigation into the Laws of Thought", 1854, reprinted by Dover Publications, USA, 1954.

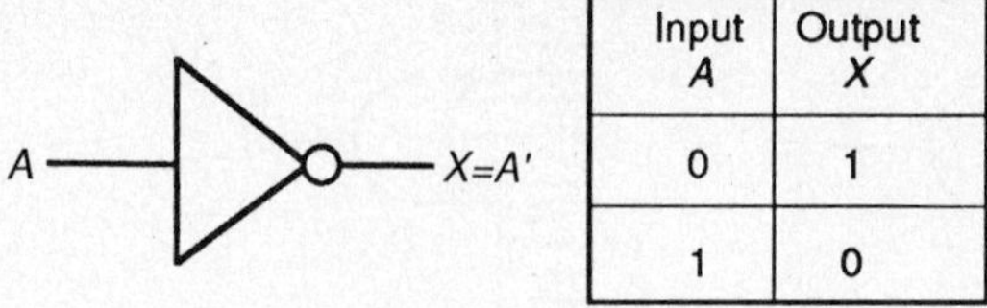

Input A	Output X
0	1
1	0

Truth table

Figure 3.2 The inverter

$$\text{Proposition} = A$$
$$\text{Negation or inverse } (\textit{Not } A) = \overline{A} \text{ or } A'$$

(Note that $\overline{\overline{A}} = (A')' = (\overline{A})' = A$ and so on.)

In this chapter, both bar and prime are used since both are in wide use in practice. However, in following chapters, the prime is mainly used as a matter of convenience.

A logic symbol and the truth table for the *Not* function (or inverter) are given in Figure 3.2.

3.2.2 Logical sum: The *Or* function (+)

The *Or* function of two or more variables is true (1) if one or more of the variables is true (1). The *Or* operators used in engineering applications is a plus sign.

Some basic properties of the *Or* function are:

$$A+A = A$$
$$A+A' = 1$$
$$A+1 = 1$$
$$A+0 = A$$
$$A+A+A+ \ldots A = A$$

Consider next, a two-input *Or* gate. The logic symbol and the truth table are illustrated in Figure 3.3.

A	B	F
0	0	0
0	1	1
1	1	1
1	0	1

Truth table

A
B
F

$F = A + B$

Figure 3.3 Two-input *Or* gate: Logic symbol and truth table

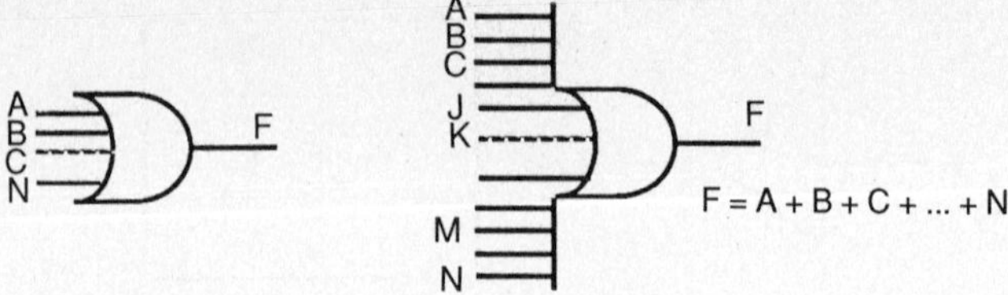

Figure 3.4 Logic symbols for multiple input *Or* gate

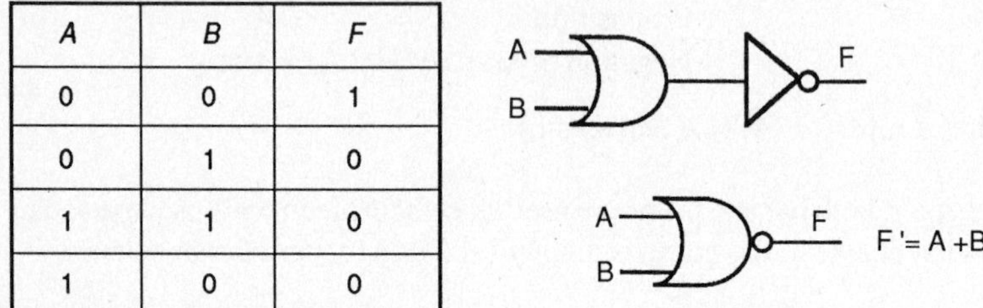

A	*B*	*F*
0	0	1
0	1	0
1	1	0
1	0	0

Figure 3.5 Two-input *Nor* gate: Logic symbols and truth table

Or gates with any number of inputs may be configured (Figure 3.4) subject, of course, to practical limitations.

3.2.3 The *Nor* (*Not Or*) function

With reference to Figure 3.3, negation of *F* results in the arrangement of Figure 3.5.

Multiple input *Nor* gates may be configured as in Figure 3.6 again subject to practical limitations.

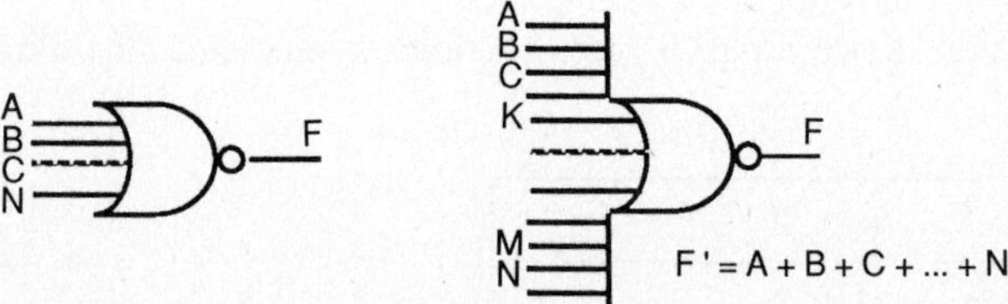

Figure 3.6 Logic symbols for multiple input *Nor* gate

3.2.4 Logical product: The *And* function (".")

The *And* function of two or more logical variables is true *if* and *only if* all the input variables are true. The operator is " . ".

Some properties of the *And* operation are:

$$A.A = A$$
$$A.A' = 0$$
$$A.1 = A$$
$$A.0 = 0$$
$$A.A \ldots A = A$$

Note: The *And* operator "." is often omitted for convenience, so that $X = ABC$ means the same as the more formal $X = A.B.C$, etc.

Figure 3.7 illustrates the symbol and truth table for a two-input *And* gate.

A	*B*	*F*
0	0	0
0	1	0
1	1	1
1	0	0

Truth table 2I/P *And*

A
B
F
F = A.B

Figure 3.7 Two-input *And* gate: Logic symbol and truth table

And gates with any practical number of inputs may also be represented as in Figure 3.8.

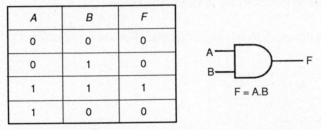

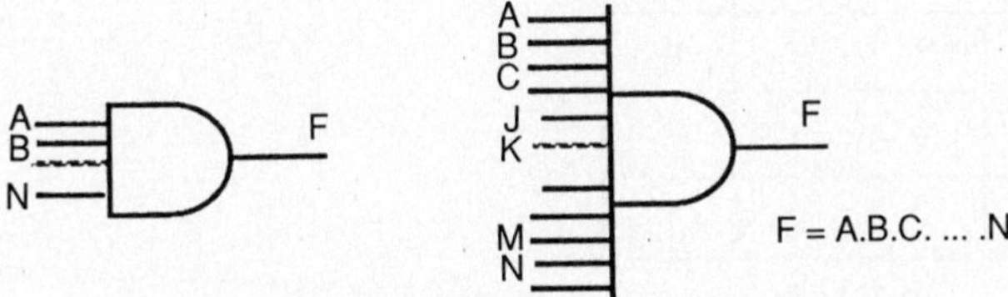

Figure 3.8 Logic symbols for multiple input *And* gate

3.2.5 The *Nand (not And)* function

Negation of F in Figure 3.7 results in the *Nand* gate (Figure 3.9).

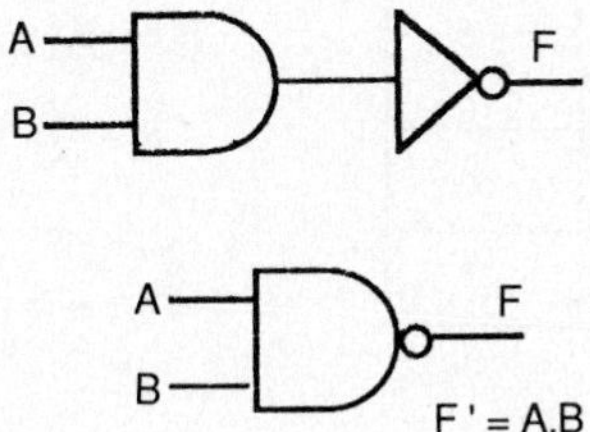

Figure 3.9 Two-input *Nand* gate

Multiple input *Nand* gates may also be represented as in Figure 3.10.

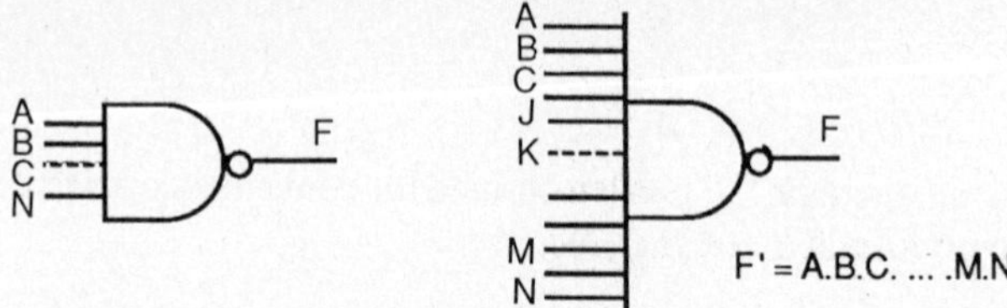

Figure 3.10 Logic symbols for multiple input *Nand* gate

Truth tables may be drawn up for any logic function. That for a 3 input *Nand* gate is given in Figure 3.11

A	*B*	*C*	*X*
0	0	0	1
0	0	1	1
0	1	1	1
0	1	0	1
1	1	0	1
1	1	1	0
1	0	1	1
1	0	0	1

Truth table

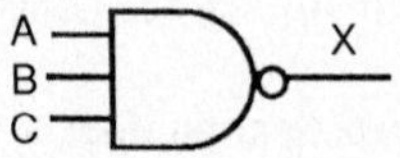

A
B
C
X

X' = A.B.C

Figure 3.11 Three-input *Nand*

3.2.6 The *Xor (exclusive Or)* function (⊕)

A	*B*	*X*
0	0	0
0	1	1
1	1	0
1	0	1

Truth table

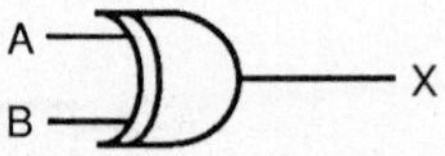

A
B
X

X = A'B + AB'

$X = A\overline{B} + \overline{A}B$ or, $X = AB' + A'B$
Sometimes written $X = A \oplus B$

Figure 3.12 Exclusive *Or* (*Xor*) gate

This gate is sometimes referred to as an inequality gate.

3.2.7 The *Xnor (exclusive Nor* or *equality)* function (Figure 3.13)

A	*B*	*X*
0	0	1
0	1	0
1	1	1
1	0	0

Truth table

$$X = A'B' + AB$$

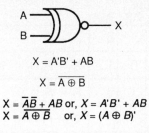

$$X = \overline{A \oplus B}$$

$$X = \overline{A}\,\overline{B} + AB \text{ or, } X = A'B' + AB$$
$$X = \overline{A \oplus B} \quad \text{or, } X = (A \oplus B)'$$

Figure 3.13 Exclusive *Nor* gate

3.2.8 Logic functions of two variables

Various functions may be generated by combinations of the logic functions *And*, *Or*, *Nand*, *Nor* and *Invert*. In order to assess the scope it is informative to consider the number of logical functions which relate two binary variables, *A* and *B*, to a single binary output *X* as in Figure 3.14 and Table 3.3

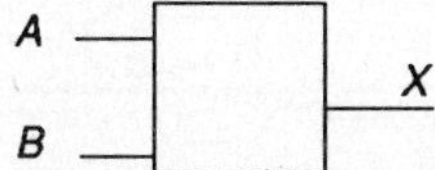

Figure 3.14 General function of two variables

Note the alternative logic symbols introduced in this table (see Figure 4.14).

Similar but larger tables may be drawn up for functions of three or more variables.

3.3 Further aspects of switching theory and Boolean algebra

In order to put matters in perspective it is first necessary to set out some definitions.

3.3.1 Variables

Variables are usually denoted by lower or upper case letters. For example, the following are each expressions in five variables:

$$ABE + CDE + ABCDE = 1 \qquad (1)$$

$$(p + q + r' + s) . (p + s' + t) . (p + q' + r' + s + t') = 0 \qquad (2)$$

3.3.2 Literals

Each appearance of a variable or its complement in expression is termed a literal.

Thus expression (1) above has 11 literals, and expression (2) has 12 literals.

Table 3.3 Functions of two variables

Inputs A'B'	A'B	AB	AB'	FUNCTION GENERATED	
0	0	0	0	X = 0	
0	0	0	1	X = A.B'	
0	0	1	0	X = A.B *(And)*	
0	0	1	1	X = A	A — X
0	1	0	0	X = A'.B	
0	1	0	1	X = A.B'+A'B *(Xor)*	
0	1	1	0	X = B	B — X
0	1	1	1	X = A + B *(Or)*	
1	0	0	0	X' = A + B *(Nor)*	
1	0	0	1	X = B'	
1	0	1	0	X = A.B + A'B' *(XNor)*	
1	0	1	1	X' = A'.B	
1	1	0	0	X = A'	
1	1	0	1	X' = A.B *(Nand)*	
1	1	1	0	X' = A.B'	
1	1	1	1	X = 1	

3.3.3 Theorems and aids to simplification of Boolean algebra

$0+0 = 0$	$0.0 = 0$	
$1+1 = 1$	$1.1 = 1$	
$0+1 = 1$	$0.1 = 0$	
$\overline{0} = 1 \quad \overline{\overline{A}} = (\overline{A})' = A$	$\overline{1} = 0$	
$A+0 = A$	$A.0 = 0$	
$A+1 = 1$	$A.1 = A$	
$A+A = A$	$A.A = A$	
$A+\overline{A} = 1$	$A.\overline{A} = 0$	
$A+AB = A$	$A(A+B) = A$	
$A+B = B+A$	$A.B = B.A$	(commutative law)
$A(B+C) = AB+AC$		(distributive law)

$(A+B)+C=A+(B+C)=(A+C)+B$ $\quad (AB)C=A(BC)=(AC)B$ (associative law)

$A+\bar{A}B= A+B$ $\quad A+AB+AC+ \ldots +AN = A$

Demorgan's theorem

The complement of any expression is formed by complementing all the literals and exchanging *And* (.) with *Or* (+) operations and vice versa.

For example, let function *T* be defined as:

$$T = a \,.\, b' \,.\, c + a' \,.\, b \,.\, c$$

Then the complement is:

$$\bar{T} = (a' + b + c') \,.\, (a + b' + c')$$

Similarly, if:

$$X = (p + q' + r' + s) \,.\, (p' + q' + r + s)$$

then:

$$\bar{X} = p' \,.\, q \,.\, r \,.\, s' + p \,.\, q \,.\, r' \,.\, s'$$

This is a most important theorem and is widely used to manipulate expressions involving gate logic, for example:

The *And* operation: $X = A \,.\, B$ can be written $X' = A' + B'$

The *Or* operation: $X = A + B$ can be written $X' = A' \,.\, B'$

The *Nand* operation: $X' = A \,.\, B$ can be written $X = A' + B'$

The *Nor* operation: $X' = A + B$ can be written $X = A' \,.\, B'$

*Huntington's Postulates**

All the relationships which we will use are based on a set of independant Postulates, each one of which is essential since none can be proved from the others.

1. There exists a set of K objects or elements, subject to an equivalence relationship, denoted "=", which satisfy the principle of substitution.
   For example, if $a=b$ then a may be substituted for b in any expression involving b without affecting the validity of the expression.

The next eight postulates are given in pairs.

2. (a) A rule of combination "+" is defined such that $a + b$ is in K whenever both a and b are in K.
   (b) A rule of combination "." is defined such that $a.b$ (and abbreviation ab) is in K whenever both a and b are in K.
3. (a) There exists an element 0 in K such that, for every a in K, $a + 0 = a$.
   (b) There exists an element 1 in K such that, for every a in K, $a.1 = a$.

* Huntington E.V., "Sets of independant postulates for the algebra of Logic", *Trans American Maths. Soc.*, 5, 266-305, 1904.

4. (a) $a + b = b + a$ (commutative law).
   (b) $a.b = b.a$ (commutative law).
5. (a) $a + (b.c) = (a + b).(a + c)$ (distributive law).
   (b) $a.(b + c) = (a.b) + (a.c)$ (distributive law).
6. For every element a in K there exists an element a' such that
$$a.a' = 0\,; \qquad a + a' = 1$$
7. There are at least two elements x and y in K such that $x \neq y$.

Note that the paired postulates are *duals*.

Shannon's Expansion Theorem

This is applied in the decomposition of switching expressions into sub-expressions which depend on subsets of the variables. It is often applied in dealing with problems in large numbers of variables.

It may be stated as follows:

Any switching expression $f(x_{n-1}, x_{n-2}, ..., x_1, x_0)$ can be decomposed as:

$$f(x_{n-1}, x_{n-2}, ..., x_1, x_0) = x_i f(x_{n-1}, ...\ x_{i+1}, 1, x_{i-1}, ..., x_0) + x_i' f(x_{n-1}, ...\ x_{i+1}, 0, x_{i-1}, ..., x_0)$$

or its dual:

$$f(x_{n-1}, x_{n-2}, ..., x_1, x_0) = x_i + f(x_{n-1}, ...\ x_{i+1}, 0, x_{i-1}, ..., x_0) + x_i' + f(x_{n-1}, ...\ x_{i+1}, 1, x_{i-1}, ..., x_0).$$

For example, applying the theorem first to x_0 then to x_1 ... we have:

$$\begin{aligned} f(x_{n-1}, x_{n-2}, ..., x_1, x_0) = {} & x_1 x_0 f(x_{n-1}, ...\ x_2, 1, 1) \\ & + x_1 x_0' f(x_{n-1}, ..._2, 1, 0) \\ & + x_1' x_0 f(x_{n-1}, ...\ x_2, 0, 1) \\ & + x_1' x_0' f(x_{n-1}, ...\ x_2, 0, 0) \end{aligned}$$

As an example consider the decomposition of a 4-variable relationship $T = f(x_3\, x_2\, x_1\, x_0)$ as defined by entries in the following table:

From table $f(x_3\, x_2\, 1,1) = (x_3'\, x_2')$

and $f(x_3\, x_2\, 1,0) = 1$

and $f(x_3\, x_2\, 0,1) = (x_3\, x_2)$

and $f(x_3\, x_2\, 0,0) = (x_3'\, x_2 + x_3\, x_2')$

Thus $T = x_1\, x_0\, x_2'\, x_3' + x_1\, x_0' + x_1'\, x_0\, x_3\, x_2 + x_1'\, x_0'\, (x_3' x_2 + x_3\, x_2')$

$x_3\ x_2\ x_1\ x_0$	T
0 0 0 0	0
0 0 0 1	0
0 0 1 0	1
0 0 1 1	1
0 1 0 0	1
0 1 0 1	0
0 1 1 0	1
0 1 1 1	0
1 0 0 0	1
1 0 0 1	0
1 0 1 0	1

(continued)

$x_3\ x_2\ x_1\ x_0$	T
1 0 1 1	0
1 1 0 0	0
1 1 0 1	1
1 1 1 0	1
1 1 1 1	0

3.3.4 Duality

For every theorem there is a dual which is also true, one being formed from the other by exchanging "." with "+" and "0" with "1" and vice versa.

For example:

$a + 1 = 1$ then $a \,.\, 0 = 0$ (duals)

3.3.5 Product terms

A product term is a series of literals related by the *And* operation. For example: $A \,.\, B \,.\, \bar{E}$; $P\bar{Q}R\bar{S}T$; $x \,.\, y' \,.\, z$ are all product terms

3.3.6 Sum terms

A sum term is a series of literals related by the *Or* operation. For example: $(p + s' + t)$; $(A + \bar{B} + C + \bar{E})$; $(X + Y')$ are all sum terms.

3.3.7 Normal terms

A normal term is a product or sum term in which no variable appears more than once. For example: $AB'CD'$is normal, $AB'BC'D$ is not.

3.3.8 Sum of products (SOP) form of expression

An expression comprising a series of product terms related by the sum operation. For example: $T = A \,.\, B' \,.\, E + C' \,.\, D' \,.\, E + B' \,.\, C \,.\, D \,.\, E' + \ldots$ etc.

3.3.9 Product of sums (POS) form of expression

An expression comprising a series of sum terms related by the product operation. For example: $F = (A + B' + E) \,.\, (B' + F' + G') \,.\, (A' + B' + D' + E + G') \,.\, \ldots$ etc.

Note: For every expression in SOP form there is an equivalent POS form, and vice versa.

3.3.10 Canonic (standard) sum and product terms

A canonic $\left\{\begin{matrix}\text{product}\\ \text{sum}\end{matrix}\right\}$ term is a normal $\left\{\begin{matrix}\text{product}\\ \text{sum}\end{matrix}\right\}$ term comprising as many literals as there are variables. For example: For 5-variable environments, the following are canonic sums:

$$(a+b'+c+d+e') \quad ; \quad (P+\bar{Q}+\bar{R}+\bar{S}+\bar{T}) \quad ; \quad (\bar{F}+\bar{G}+H+\bar{I}+J) \ldots \text{etc.}$$

Canonic sums are otherwise known as *maxterms*.

The following product terms are canonic in a 5-variable context.

$$a.b.c'.d'.e \quad ; \quad \bar{p}.\bar{q}.r.s.\bar{t} \quad ; \quad F.G.\bar{H}.\bar{I}.\bar{J} \ldots \text{etc.}$$

Canonic products are otherwise known as *minterms*.

3.3.11 Maxterms and Minterms

In order to illustrate the concepts, let us take a three-variable (ABC) environment in which we may show all possible maxterms and minterms as in table 3.4.

Table 3.4 Maxterms and minterms of three variables

		Specific values				
Maxterm no.	*Maxterm*	*A*	*B*	*C*	*Minterm*	*Minterm no.*
M_0	(A+B+C)	0	0	0	A'.B'.C'	m_0
M_1	(A+B+C')	0	0	1	A'.B'.C	m_1
M_2	(A+B'+C)	0	1	0	A'.B.C'	m_2
M_3	(A+B'+C')	0	1	1	A'.B.C	m_3
M_4	(A'+B+C)	1	0	0	A .B'.C'	m_4
M_5	(A'+B+C')	1	0	1	A .B'.C	m_5
M_6	(A'+B'+C)	1	1	0	A.B.C'	m_6
M_7	(A'+B'+C')	1	1	1	A.B.C	m_7

Canonic sums, each having a value of 0 for the specified values for *ABC*.

Canonic product terms each having a value of 1 for the specified values for *ABC*.

Clearly for the 3-variable situation there are eight minterms and eight maxterms. In general, for an n-variable case there are 2^n minterms and 2^n maxterms.

3.3.12 Maxterm notation

An expression may be written thus:

$$S = \Pi M\,(0,1,2,3,6)$$

which is interpreted as:

$$S = (M_0) \,.\, (M_1) \,.\, (M_2) \,.\, (M_3) \,.\, (M_6)$$

and if $S = f\,(xyz)$:

$$S = (x+y+z) \cdot (x+y+z') \cdot (x+y'+z) \cdot (x+y'+z') \cdot (x'+y'+z)$$

Interpreting maxterms in this way gives rise to an expression in *expanded POS form* (all canonic sum terms). The expression, in this case, could be simplified to:

$$S = x(y'+z) \text{ ... in POS form also.}$$

3.3.13 Minterm notation

This is the more widely used form since it is a "shorthand" form of the SOP form of expression and most engineering and other problems are expressed more readily in SOP like terms. Take, for example:

$$X = \Sigma m\,(0, 1, 14, 15)$$

interpreted as:

$$X = (m_0) + (m_1) + (m_{14}) + (m_{15})$$

Clearly this is a 4-variable expression since the highest minterm number is 15 ($>2^3 - 1$ but $\leq 2^4 - 1$) therefore needing a 4-variable vector.

Let $X = f\,(abcd)$

then $X = a'b'c'd' + a'b'c'd + abcd' + abcd$

This is the expression for X in expanded SOP form. In this case the expression can be simplified to:

$$X = a'b'c' + abc$$

Simplification is an important concept having very significant implication in an engineering situation. For example, Figure 3.15 compares the realizations of the previous expression for X in its unsimplified (a) and simplified (b) forms. The savings are clearly seen.

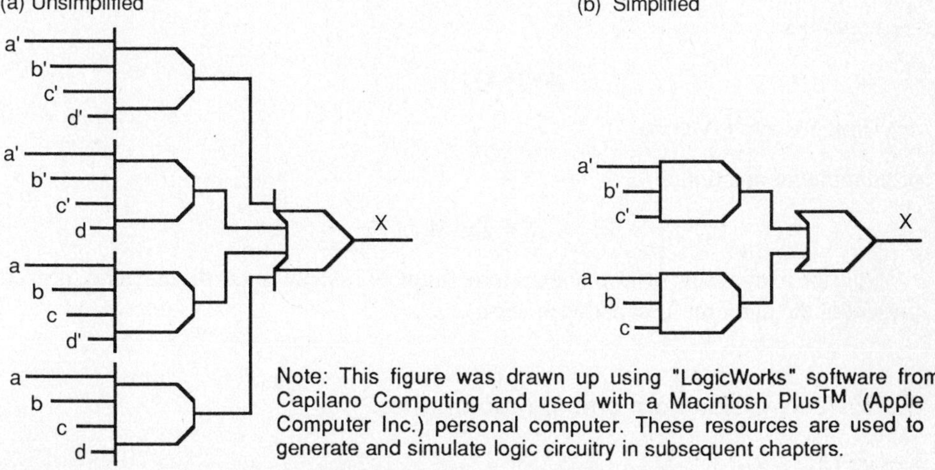
(a) Unsimplified
(b) Simplified
a'
b'
c'
d'
a'
b'
c'
d
a
b
c
d'
a
b
c
d
X
a'
b'
c'
a
b
c
X
Note: This figure was drawn up using "LogicWorks" software from Capilano Computing and used with a Macintosh Plus™ (Apple Computer Inc.) personal computer. These resources are used to generate and simulate logic circuitry in subsequent chapters.

Figure 3.15 Comparison of simplified and unsimplified logic

3.3.14 Expansion of simplified expressions

Having just discussed the merits of simplification it seems inappropriate to now consider the de-simplification or expansion of expressions, but this is sometimes a necessary process as we shall see in later sections (e.g. when implementing logic functions with read only memories [ROMs]).

For example, to expand:

$S = x\,.\,(y'+z)$ which is in POS form so that, x implies all maxterms containing x, namely:

$$(x+y+z)\,.\,(x+y+z')\,.\,(x+y'+z)\,.\,(x+y'+z')$$

while $(y'+z)$ implies both maxterms:

$$(x+y'+z)\,.\,(x'+y'+z)$$

Thus, the expanded form of S is:

$$S = (x+y+z)\,.\,(x+y+z')\,.\,(x+y'+z)\,.\,(x+y'+z')\,.\,(x'+y'+z)$$

or, in maxterm notation:

$$S = \Pi M\,(0, 1, 2, 3, 6)$$

The original expression for S, namely

$$S = x\,.\,(y'+z)$$

is readily converted (in this case) to SOP form by multiplying out:

$$S = xy'+xz$$

This can be expanded in SOP form:

xy' implies all minterms including xy' namely:

$$xy'z'+xy'z$$

xz implies:

$$xy'z+ xyz$$

Thus, $S = xy'z'+xy'z+xyz$

or, in minterm notation:

$$S = \Sigma m\,(4, 5, 7)$$

Note the relationship to the maxterm form (numbers *not* present in the maxterm form are present in the minterm form and vice versa).

3.3.15 Conversion from one form to another

From POS to SOP

Simple rules apply as follows:

1 Simplify if appropriate.
2. "Multiply out" all brackets.
3. Simplify if possible.

For example:

$$T = X(\bar{Y}+Z) \, . \, (X+Y+\bar{Z}) \qquad \text{(POS form)}$$

No obvious simplification, therefore multiply out:

$$T = X\bar{Y}+X\bar{Y}\bar{Z}+XZ+XYZ$$

Simplifying, we have:

$$T = X\bar{Y}+XZ \qquad \text{(SOP form)}$$

From SOP to POS
The rules are as follows:
1. Simplify if appropriate.
2. Complement the expression.
3. Multiply out all brackets.
4. Simplify if possible.
5. Complement back again.

For example:

$T = \Sigma m\,(0, 1, 2, 4, 6, 7)$ (minterm form)

If $T = f(abc)$, we have:

$$T = \bar{a}\bar{b}\bar{c}+\bar{a}\bar{b}c+\bar{a}b\bar{c}+a\bar{b}\bar{c}+ab\bar{c}+abc$$

Simplifying, we have:

$$T = \bar{c}+\bar{a}\bar{b}+ab$$

Complementing, we have:

$\bar{T} = c \, . \, (a+b) \, . \, (\bar{a}+\bar{b})$ (Using DeMorgan's theorem)

Multiply out:

$$\bar{T} = a\bar{b}c+\bar{a}bc$$

Simplification—none possible.

Complement back:

$T = (\bar{a}+b+\bar{c}) \, . \, (a+\bar{b}+\bar{c})$ (POS form)

or, in maxterm form:

$$T = \Pi M\,(3, 5)$$

Again, note the relationship between the minterm and maxterm forms of the same expression.

It will be seen that the numbers present in, say, the minterm form are those numbers which do not appear in the maxterm form and vice versa. Thus it is very easy to convert between the two expanded forms of an expression. However, a more general approach is to make use of a Karnaugh map for conversions since this allows for easy simplification.

3.3.16 The Karnaugh map

The Karnaugh map provides a concise way of expressing the information contained in a truth table or in an expression, in a manner which allows ready interpretation, manipulation and simplification.

Maps are readily drawn up for two, three and four variables and may also be set out for five and six variable problems as we shall see later in the section covering simplification processes.

The 2-variable map

Consider the function: $F = A'B + AB'$.

This is clearly a 2-variable function and it may be mapped as shown in Figure 3.16.

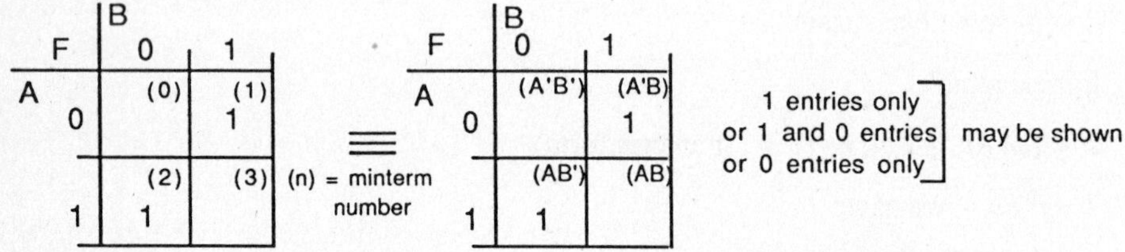

Figure 3.16 2-variable map

Terms present in the SOP form of expression are mapped as 1's. Other map cells must, by inference, be 0's and represent the equivalent maxterm form of the same expression. Each cell of the map represents a potential minterm, the minterm numbers being shown in each cell of Figure 3.16(a).

The 3-variable Karnaugh map

Consider the function: $F = A + A'B'C' + BC$.

Setting out the maps in a similar manner to Figure 3.16 we have alternative forms as in Figure 3.17(a) and (b), or we may, for example, omit the minterm number entries as in Figure 3.17(c); 0 entries are also often omitted.

Note that the minterm form of the expression is taken directly from the 1 entries:

$$F = \Sigma M(0, 3, 4, 5, 6, 7)$$

while the 0 entries give the maxterm form:

$$F = \Pi M(1, 2)$$

Note that either "tall thin" or "short fat" forms of the map are equally acceptable.

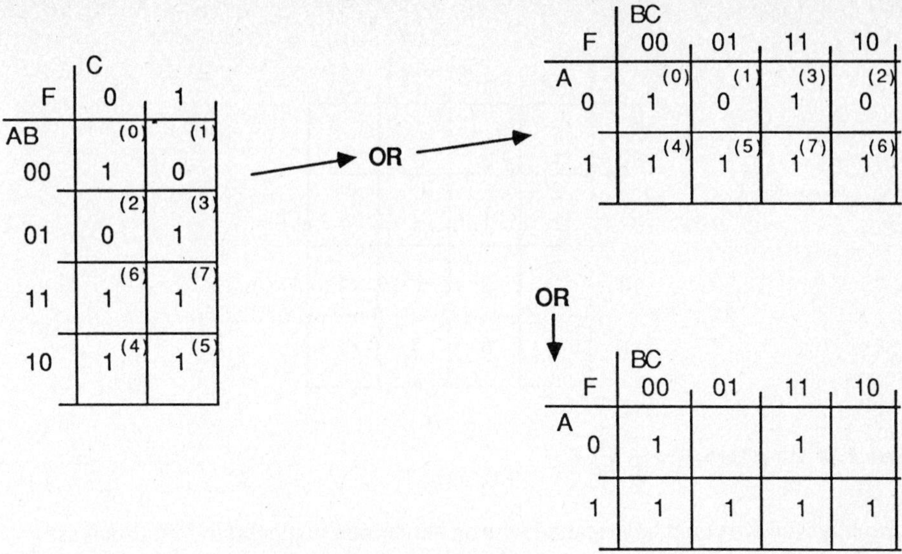

Figure 3.17 Forms of 3-variable Karnaugh maps

The 4-variable Karnaugh map

The form of the 4-variable map is readily deduced and is illustrated for the function:

$$T = \Sigma m(0, 3, 5, 6, 9, 10, 12, 15) \qquad \text{(assuming } T = f(PQRS)\text{)}$$

The function is mapped in Figure 3.18.

T (PQ \ RS)	00	01	11	10
00	(0) 1	(1)	(3) 1	(2)
01	(4)	(5) 1	(7)	(6) 1
11	(12) 1	(13)	(15) 1	(14)
10	(8)	(9) 1	(11)	(10) 1

Figure 3.18 4-variable map for T

Note that minterm numbers (n) for each cell are readily determined by "weighting" the variables, $PQRS$, as follows: $P = 8, Q = 4, R = 2, S = 1$ and hence determining the value of the vector $PQRS$ for each cell.

X (AB \ CD)	00	01	11	10
00	1	0	1	1
01	1	1	1	1
11	1	0	0	1
10	1	0	1	1

Figure 3.19 4-variable map for X

Equally well we could have started with an expression in algebraic form, such as:

$$X = A'B + B'C + BCD' + D'$$

This will map directly as in Figure 3.19.

The 0's on the map occupy those cells not containing 1's. Clearly the 0 entries indicate the conditions under which $X = 0$. In other words, the 0's represent the complementary function X'.

In this case the complementary function is:

$$X' = ABD + B'C'D$$

From the expression for X', the POS form is readily obtained by complementation:

$$X = (A' + B' + D') \,.\, (B + C + D') \text{ (POS form.)}$$

Expanding, we have:

$$X = (A'+B'+C'+D') \,.\, (A'+B'+C+D') \,.\, (A'+B+C+D') \,.\, (A+B+C+D')$$

In the maxterm form:

$$X = \prod M(1, 9, 13, 15)$$

This can be checked from the map and can be seen to comprise the 0 entries on the map. Thus, the minterm form of expression is entered as 1's while the maxterm form is entered as 0's on the map. Clearly the map is a convenient way of converting between minterm and maxterm representation.

The maps, of course, may be differently presented by re-ordering the variables so that, for example, the map of Figure 3.19 may be presented as in Figure 3.20. As with all maps, the function may be readily converted to minterm form.

X DC \ BA	00	01	11	10
00	(0) 1	(8) 1	(12) 1	(4) 1
01	(2) 1	(10) 1	(14) 1	(6) 1
11	(3) 1	(11) 1	(15) 0	(7) 1
10	(1) 0	(9) 0	(13) 0	(5) 1

Figure 3.20 Alternative mapping of X

It will be seen then that the Karnaugh map provides a convenient way of manipulating functions for, say, conversions from POS to SOP or vice versa; for deriving the complementary function; for conversion between minterm or maxterm and algebraic forms; and, as we shall see later, for simplifying functions.

3.4 Simplification

The simplification of logic expressions is obviously a desirable process in most situations. In the design of engineering systems, for example, simplification can lead to:

1. lower complexity;
2. lower assembly and test costs;
3. lower power consumption ; and
4. higher reliability.

Several simplification techniques are commonly available.

3.4.1 Algebraic simplification

Uses the rules and theorems of Boolean algebra, success relying on an informed and sometimes intuitive approach. It is not recommended as the best approach except for simple expressions in no more than two or three variables and in some cases for simple expressions in large numbers of variables.

3.4.2 Graphical methods

Three-dimensional representation

A 3-variable problem, say, a function of *xyz*, can be visualized in three dimensions with

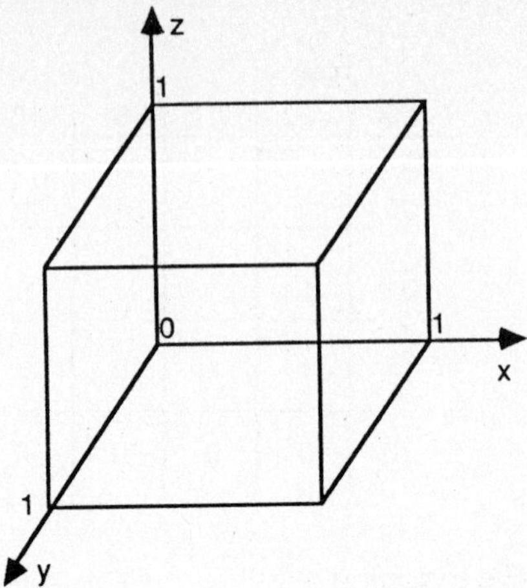

Figure 3.21 Cube representing f(xyz)

respect to three mutually perpendicular axes. Each axis represents a variable which can only have the value 0 or 1 on that axis as shown in Figure 3.21. A little thought will reveal that the problem space will map out a cube as shown. Further thought will reveal that each of the eight vertices represents a 3-literal term (minterm) ($x'y'z'$ ... xyz), each of the twelve edges represents a 2-literal term ($x'y'$ ... $x'z$ ... yz) and each face represents a 1-literal term (x, x', y, y' z, z').

Consider the expression:

$$T = xz + xy'z' + xyz'$$

Enter terms as shown in Figure 3.22. Note that all vertices of a complete face (x) are covered by these entries so that the simplified representation of the expression is: $T = x$.

Simplification is thus a process of reducing the number of terms and the number of literals (by forming larger groups).

Venn diagrams
A 3-variable environment, for example, is now represented by three intersecting circles as shown in Figure 3.23. The area inside each circle represents the "true" or "1" state of a particular variable. The area outside any circle is the complement of that variable. The whole problem falls within a square or rectangular outline as shown. Let us take the problem from the previous section and map it on the Venn diagram as in Figure 3.24.

3.4.3 Karnaugh map based simplification

The use of the map is a most useful concept and it is a widely used approach to simplification.

The form of Karnaugh maps
Consider again 2-variable, 3-variable and 4-variable Karnaugh maps (Figure 3.25). The

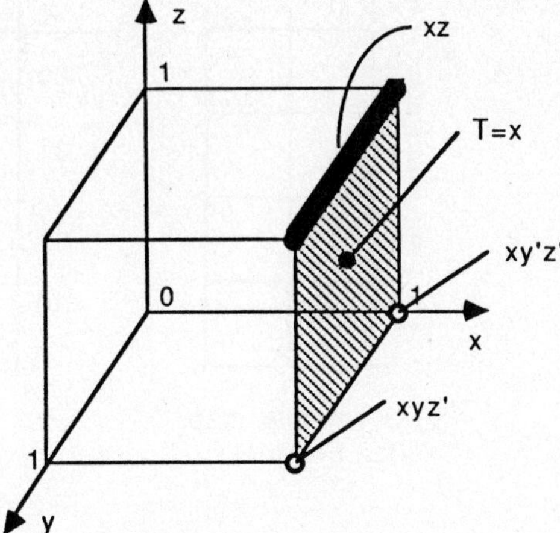

Figure 3.22 T = xz+xy'z'+xyz' mapped onto cube

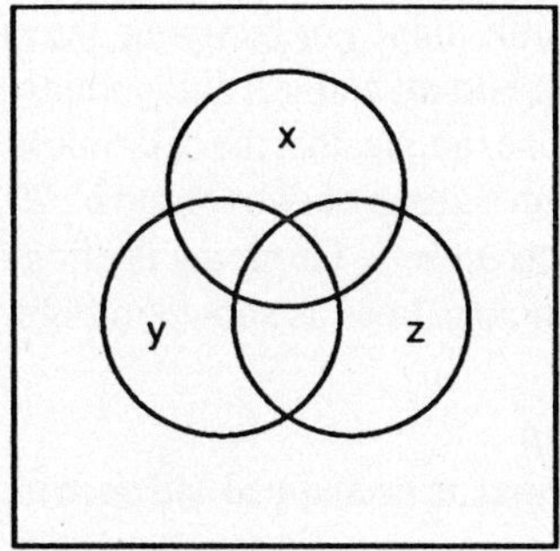

Figure 3.23 Venn diagram (3-variables)

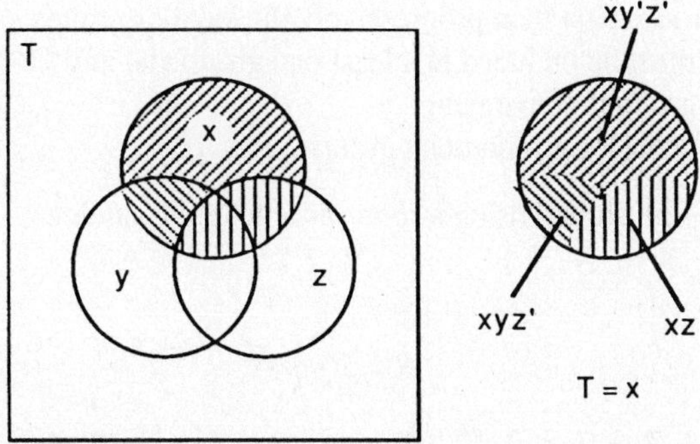

Figure 3.24 Venn diagram for problem

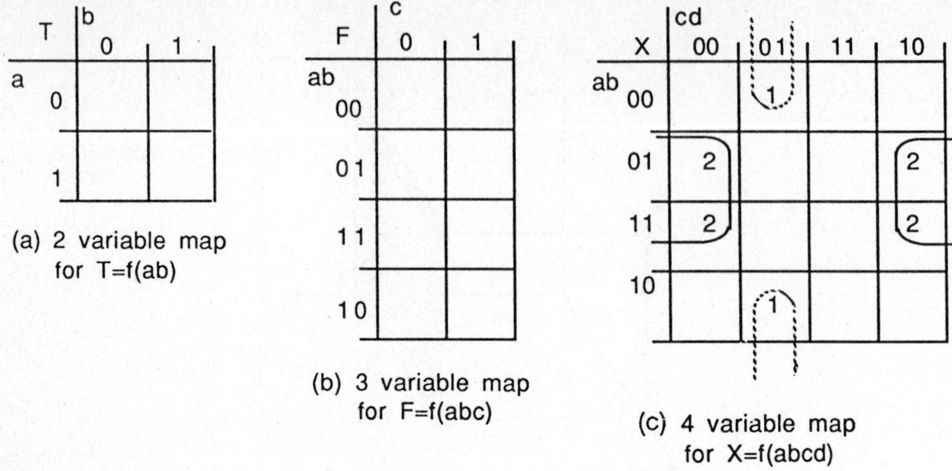

Figure 3.25 Forms of Karnaugh map

maps are set out in such a way that the top row and bottom row of cells in the map are logically adjacent and this applies also to the left and right columns. Perhaps this adjacency is most readily apparent in the 4-variable map. For grouping, the map, as set out, may be regarded as being rolled up, top joining bottom and left side joining right, into the shape of a motor tire inner tube. This means, for example, that the cells marked "1" on the 4-variable map are adjacent and can therefore form a group of two (group *b'c'd*), and the cells marked 2 are also groupable into a group of 4 (group *bd'*). Grouping is often indicated by enclosing the cells grouped with broken or continuous lines as shown in Figure 3.25.

Grouping on a Karnaugh map

We have already seen how terms can be mapped and once mapped a process of grouping can take place.

The rules for the grouping of the cells (normally "1" entries) is as follows:

1. Form groups of 2^n adjacent cells where n is an integer ≥ 0. Groups must be symmetrical and diagonal grouping is not allowed.
2. Form largest groups first then progressively the smaller groups.
3. All "1" entries must be included in a least one group and any "1" entry may be used in more than one group if convenient.
4. On completion, check for redundant groups.

The process will be illustrated using a four-variable environment.

Example 1

$$X = f(pqrs)$$

where $X = \Sigma m\,(0, 1, 2, 3, 7, 8, 9, 12, 14)$.

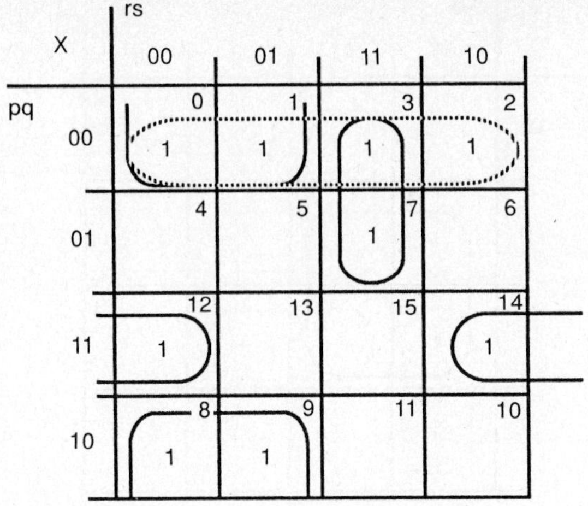

Figure 3.26 Map for example 1

The function maps as shown in Figure 3.26. From grouping shown on the map it will be seen that there are two groups of four and two groups of two. Reading off the groups from the map we have:

$$X = p'q' + q'r' + p'rs + pqs'$$

a simplified form of the function X.

It will be clearly seen that each "1" is included in at least one group and that three of the "1"s are used in more than one group.

Example 2

In order to illustrate the possibility of generating redundant groups lets us consider the second example.

$$T = a'bc' + a'b'c'd + abc' + abcd + ab'c'd' + a'bcd'$$

Mapping the terms gives the 1 (and the consequent 0) entries on the map (Figure 3.27). Grouping will give one group of four and four groups of two as shown giving:

$$T = bc' + a'bd' + abd + ac'd' + a'c'd$$

However, a more careful check reveals that the largest group—the group of four— is *redundant* since *all* the "1s" comprising that group are also included in other groups and, although these groups are smaller, each of them is *essential*. Essential means that there is no way in which an essential group can be omitted in a valid solution.
The correct solution to this particular simplification example is

$$T = a'bd' + abd + ac'd' + a'c'd.$$

Using the map of Figure 3.27 we could also find the simplest solution for the function T'. (Group the 0's on the map.)

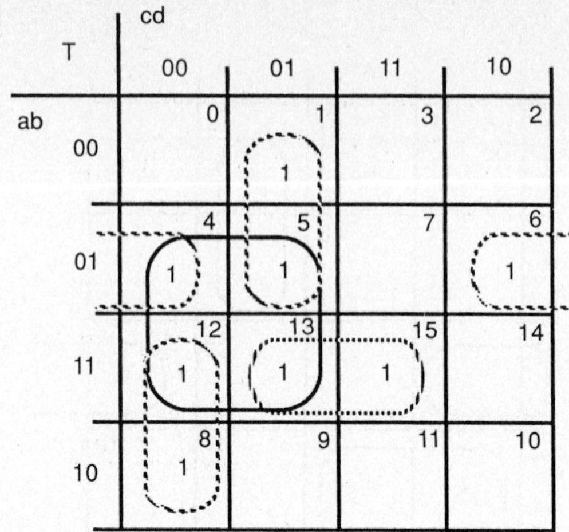

Figure 3.27 Map for example 2

A further point to remember is that a simplified solution taken from a Karnaugh map is not necessarily unique. This is illustrated in the following example.

Example 3

Consider $X = f(abcd)$

where: $X = \Sigma m\,(1, 3, 4, 5, 6, 7, 8, 9, 11, 12, 14)$

Mapping, as in Figure 3.28 we can see that there are two essential groups of four, a choice between two further groups of four and then a choice between two groups of two as follows:

$$X = bd' + b'd + \begin{Bmatrix} a'b \\ a'd \end{Bmatrix} + \begin{Bmatrix} ab'c' \\ ac'd' \end{Bmatrix}$$

where $\{\ \}$ indicates a choice of grouping. Thus, four equally good and valid solutions can be obtained.

The first two terms are *essential*, each of the latter two terms has an alternative.

Prime implicants
Valid groups formed are known as prime implicants (PIs) and it will be seen that the map yields six PIs in this case. Four PIs are sufficient for the specified relationship and a choice exists as indicated. However it will be seen that the groups *bd'* and *b'd* represent *essential prime implicants* both of which *must* appear in any valid solution.

Implications of simplification on hardware realizations
The engineer is mostly concerned with processes which yield positive cost or performance benefits in the system being designed.

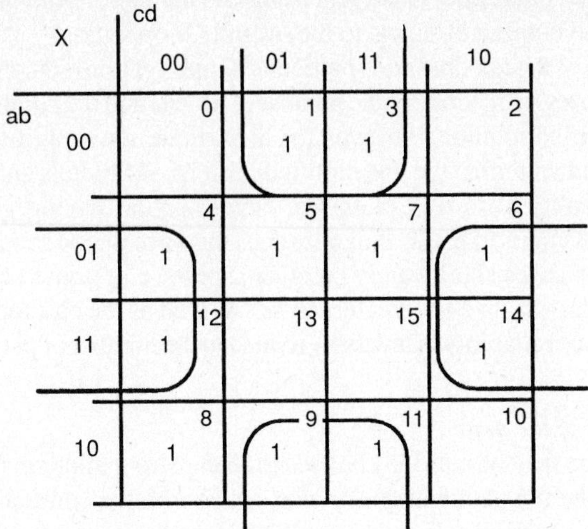

Figure 3.28 Map for example 3

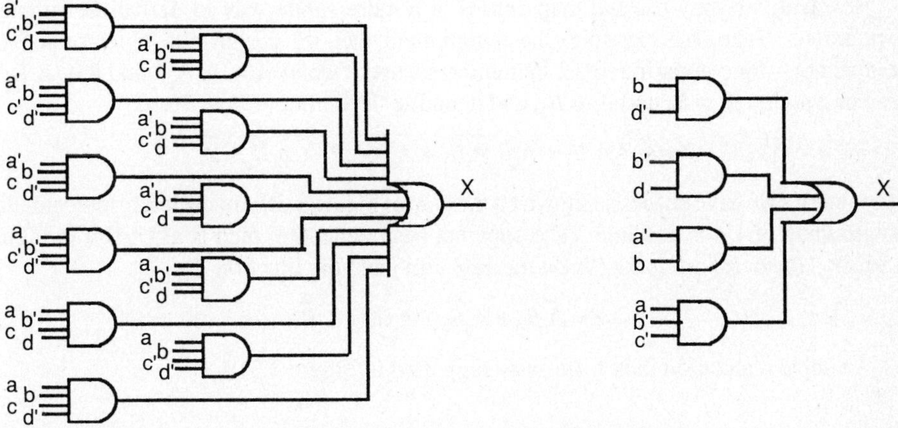

Figure 3.29 Effect of simplification on hardware - an example

Considering the previous example, we started with a function in minterm form, namely:

$$X = \sum m\,(1, 3, 4, 5, 6, 7, 8, 9, 11, 12, 14)$$

where $X = f\,(abcd)$.

Remembering that minterm form implies a sum of products form of expression and that each minterm (in this case) represents a 4-input *And* operation, we have the unsimplified realization of Figure 3.29 (a). Simplification, however, results in an expression with three 2-literal and one 3-literal terms. The hardware is considerably reduced as shown in Figure 3.29(b)—obviously a great reduction in complexity and cost.

In general each literal present in the implemented expression represents an input to an

And gate, the number of product terms represents the number of *And* gates and number of *And* gates equals the number of inputs to the overall *Or* operation.

Thus, a measure of the goodness of a particular simplification (sometimes called the cost function) can be expressed in terms of the number of literals and the number of product terms present in the simplified solution. To revert to our example, the simplified solution has nine literals and four product terms; the unsimplified form has 44 literals and 11 product terms.

Clearly the simplified form is preferred, simplification having greatly reduced the number of gates and inputs to gates. Not only does this save on the actual cost of logic gate packages, but it also saves significantly on other aspects, e.g. printed circuit board layout time and space and/or wiring costs and test costs. As well as the cost reductions, reliability is also improved since reliability is inversely related to the number of parts and connections.

Making further use of the map

We have already seen that the map may be used to change from minterm (the 1's on the map) to maxterm form (the 0's on the map) (or vice versa) with the consequent implication on hardware. Whereas one form is based on the SOP form of expression, the other is based on the POS form. Hardware realizations are *And* followed by *Or* for SOP changing to *Or* followed by *And* for POS (or equivalent logic in either case).

However, we may use the map entries in a more subtle way to derive the simplest expression. Take, for example, the requirement that we design the logic for a 2-bit comparator—for comparing two 2-bit numbers *A* and *B* where $A = (A_1A_0)_2$ and $B = (B_1B_0)_2$.

For *equality* $A_1 = B_1$ and $A_0 = B_0$, and denoting the output as *Z* we have:

$$Z = A_0B_0A_1B_1 + A_0B_0\overline{A}_1\overline{B}_1 + \overline{A}_0\overline{B}_0A_1B_1 + \overline{A}_0\overline{B}_0\overline{A}_1\overline{B}_1$$

Mapping 1's we have entries as shown on the map of Figure 3.30 which clearly indicates that no grouping of 1's is possible. The apparent best realization then is as shown in Figure 3.31(a). However, *grouping 0's on the map* will yield the function $\overline{Z}$:

$$\overline{Z} = \overline{A}_0B_0 + A_0\overline{B}_0 + \overline{A}_1B_1 + A_1\overline{B}_1.$$

A simple realization then follows as suggested in Figure 3.31(b).

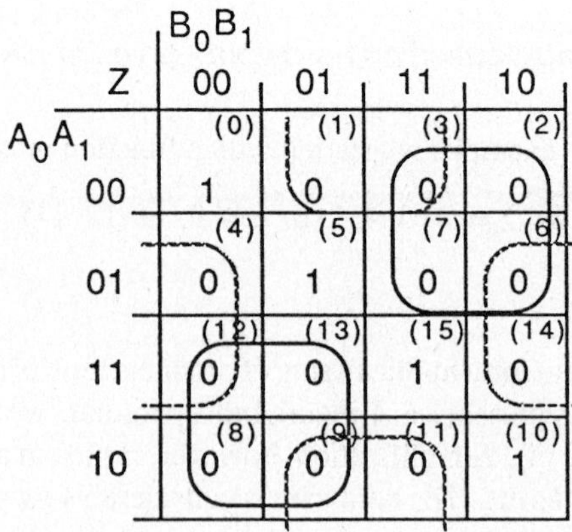

Figure 3.30 Map for 2-bit comparator

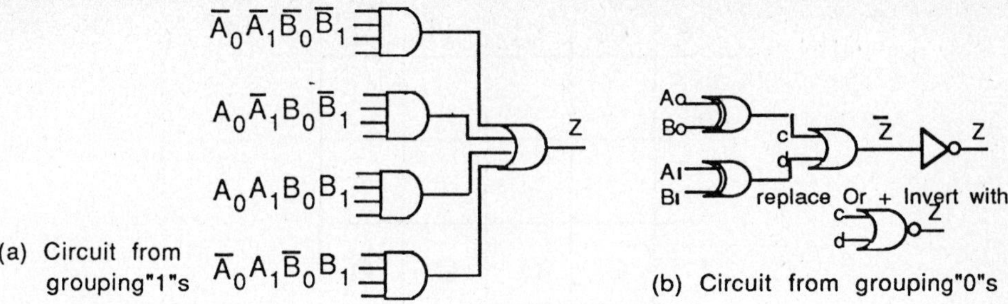

(a) Circuit from grouping"1"s

(b) Circuit from grouping"0"s

Figure 3.31 Alternative arrangements for 2-bit comparator

"Don't care" entries (denoted Ø)

"Don't cares" are unspecified conditions in a map which may thus be treated, in grouping, as either "0" or "1", and may be included in or excluded from groups as convenient to improve simplification.

An example will serve to illustrate the point.

Example 4

A 4-line binary weighted parallel input is used to convey decimal information as in Table 3.5. A logic circuit is to be designed to detect the occurrences of decimal 9 or of decimal 5 (encoded) on inputs *abcd*. The circuit is to produce an output of $P = 1$ when an occurrence is detected.

Table 3.5 Binary coded decimal

Binary inputs a	b	c	d	*Decimal equivalent*
0	0	0	0	0
0	0	0	1	1
0	0	1	0	2
0	0	1	1	3
0	1	0	0	4
0	1	0	1	5
0	1	1	0	6
0	1	1	1	7
1	0	0	0	8
1	0	0	1	9
1	0	1	0	
1	0	1	1	↑
1	1	0	0	Unused
1	1	0	1	inputs
1	1	1	0	↓
1	1	1	1	

The problem is conveniently tackled with a 4-variable map as in Figure 3.32. Note that *P* is mapped as 1's and that "don't care" entries are entered for all "unused" combinations of *abcd* in Table 3.5. Grouping for P, using don't cares as convenient gives:

P ab \ cd	00	01	11	10
00				
01		1		
11	Ø	Ø	Ø	Ø
10		1	Ø	Ø

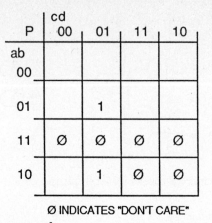

Ø INDICATES "DON'T CARE"

Figure 3.32 Map for P (example 4)

$$P = ad + bc'd$$

Note that the don't care entries do not have to be covered in the solution for *P* but are used to advantage in group formings.

Karnaugh maps for five and six variables

Five- and 6-variable maps are constructed of 4-variable maps and may be treated that way for the purposes of simplification.

It is the author's practice to set out such maps about "reflecting surfaces" and to seek mirror image groups about these surfaces after having grouped within individual 4-variable segments.

Let us consider a 6-variable situation in which the variables are *abcdef* in descending order of significance. Four-variable (*cdef*) and 5-variable (*bcdef*)situations may then be regarded, for convenience, as subsets of the 6-variable case. Four-variable maps in variables *cdef* would then be set out as in Figure 3.33 (with the minterm numbers associated with a particular segment being entered for reference).

A 5-variable map in variables *bcdef* is then constructed as in Figure 3.34. It may be seen that two 4-variable segments are put together about a vertical "reflecting surface" as shown. (Note the order of the variables *ef*.) The state of most significant variable *b* identifies the two halves of the map. Since *b* is of weight 16, corresponding cells on either side of the reflecting surface differ by 16 in minterm number.

As an example, consider the function *T* where $T = f\,(bcdef)$, and $T = T1 + T2$ given that:

$$T1 = \Sigma m\ (0, 4, 8, 15, 24, 26, 31)$$

and,

$$T2 = c'd'ef' + bc'ef' + bd'e'f' + b'cd'f'$$

The function *T* may be mapped as shown in Figure 3.35.

An orderly approach to grouping is possible by first grouping in the individual 4-variable segments of the map.

T cd \ ef	00	01	11	10
00	(0)	(1)	(3)	(2)
01	(4)	(5)	(7)	(6)
11	(12)	(13)	(15)	(14)
10	(8)	(9)	(11)	(10)

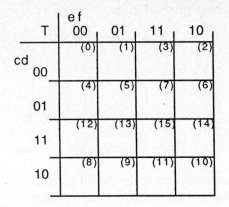

Figure 3.33 A 4-variable quadrant

	b'				b				
T cd \ ef	00	01	11	10	10	11	01	00	ef / cd
00	(0)	(1)	(3)	(2)	(18)	(19)	(17)	(16)	00
01	(4)	(5)	(7)	(6)	(22)	(23)	(21)	(20)	01
11	(12)	(13)	(15)	(14)	(30)	(31)	(29)	(28)	11
10	(8)	(9)	(11)	(10)	(26)	(27)	(25)	(24)	10

Figure 3.34 Basic 5-variable map

	b'				b				
T cd \ ef	00	01	11	10	10	11	01	00	ef / cd
00	(0) 1	(1)	(3)	(2) 1	(18) 1	(19)	(17)	(16) 1	00
01	(4) 1	(5)	(7)	(6)	(22) 1	(23)	(21)	(20)	01
11	(12)	(13)	(15) 1	(14)	(30)	(31) 1	(29)	(28)	11
10	(8) 1	(9)	(11)	(10) 1	(26) 1	(27)	(25)	(24) 1	10

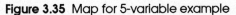

Figure 3.35 Map for 5-variable example

For the left-hand segment, grouping gives:

1 group of four (minterms 0, 2, 8, 10)	defined by term	$b'd'f'$
1 group of two (minterms 0, 4)	defined by	$b'c'e'f'$
1 ungrouped minterm (15)	defined by	$b'cdef$

Similarly, the right-hand segment yields:

1 group of four (m16, 18, 24, 26)	defined by term	$bd'f'$
1 group of two (m18, 22)	defined by	$bc'ef'$
1 ungrouped minterm (m31)	defined by	$bcdef$

The "mirror image" groups may now be sought and merged.

Clearly, the two groups of four are mirror images giving:

1 group of eight (minterms 0, 2, 8, 10, 16, 18, 24, 26) ... term $d'f'$

The groups of two remain as given above since they are not mirror images.

Finally, the two as yet ungrouped minterms will group now across the centre line to give:

1 further group of two (minterms 15, 31) defined by $cdef$

The simplified solution for function T is therefore:

$$T = d'f' + b'c'e'f' + bc'ef' + cdef$$

The perspicacious reader may, by now, have deduced the form of a 6-variable Karnaugh map (Figure 3.36).

The 6-variable map has two mutually perpendicular reflecting axes and four segments or quadrants, each comprising a 4-variable map. The quadrants are each defined by a particular combination of the two most significant variables ab as shown.

Clearly, as with all Karnaugh maps, the minterm numbers may be entered if required but are usually omitted if the problem specification is in algebraic form.

As in the 5-variable case a systematic approach to grouping is advised. A simple example will serve to carry our considerations further and to indicate a way in which a single map may be used for multiple output functions (since the actual drawing-up of 6-variable maps is not a trivial exercise).

Example 1. X, Y and Z are all functions of the same variables $abcdef$ and are defined as follows:

$$X = \Sigma m\ (0, 4, 8, 12, 16, 20, 24, 28, 42, 46, 58, 62)$$

$$Y = \Sigma m\ (7, 8, 9, 10, 11, 15, 23, 24, 25, 26, 27, 31, 40, 41, 42, 43, 56, 57, 58, 59)$$

$$Z = \Sigma m\ (0, 2, 8, 10, 16, 18, 26, 32, 34, 40, 42, 48, 50, 56, 58)$$

Find the simplest form for each output X, Y and Z.

Entries for each function are made as X, Y, Z rather than as 1s (Figure 3.37).

It is now a simple process to group the X's for X, Y's for Y, etc. giving:

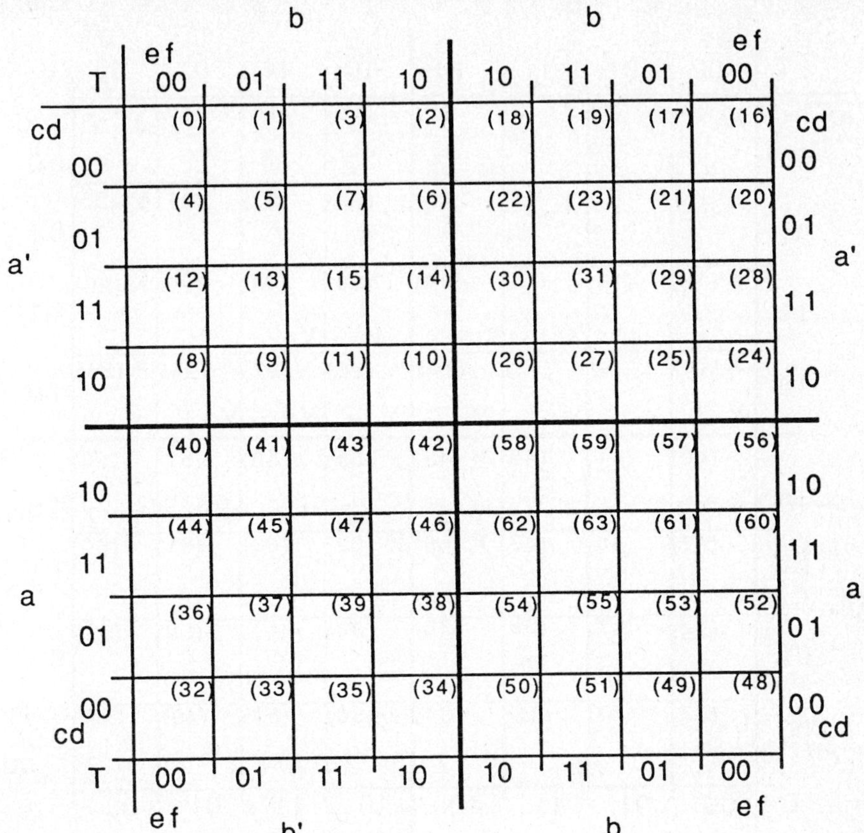

Figure 3.36 Basic 6-variable map

$$X = a'e'f' + acef'$$
$$Y = cd' + a'def$$
$$Z = b'd'f' + ad'f' + d'ef' + c'd'f'$$

Note that a group of two eliminates one variable, a group of four eliminates two variables, a group of eight eliminates three variables, etc.

Thus, for example, the expression arrived at for Z comprises four groups of eight cells, each group thus representing a 3-literal term, since each grouping eliminates three of the six variables.

Some guidelines for 6-variable map grouping

1. The map can be set out in many ways depending on the order in which the variables are represented. Assuming that the quadrants are to be defined by the two most significant variables, then there are two possible arrangements as in Figure 3.38.
2. Form groups within each quadrant treating each as a 4-variable map.
3. Look for matching groups about either the horizontal or vertical centre-line (mirror images). *No diagonal grouping is allowed.*

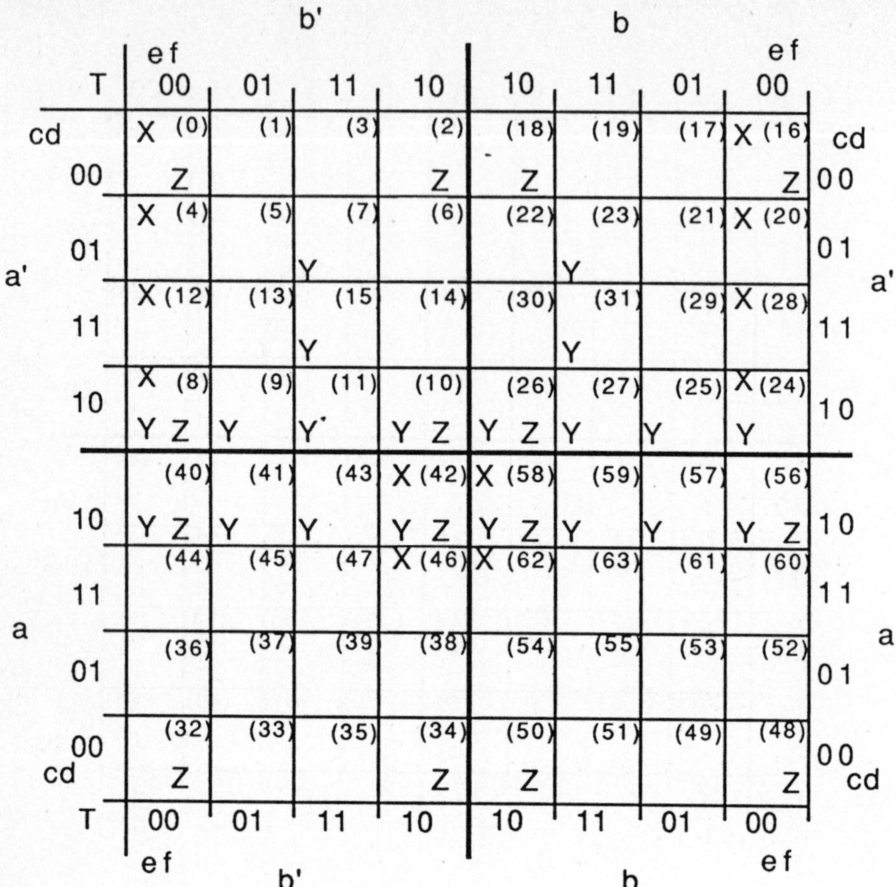

Figure 3.37 Variable map for multiple output example

4. Having paired groups about, say, the horizontal axis, seek similarly paired groups for the next level of grouping across the other (e.g. vertical) axis.
5. Form the largest possible groups and then choose a set of groups to cover all minterms at least once with as few groups as possible. Remember that "don't cares" may be used at all stages of group forming.
6. Check for and discard redundant groups, that is, groups in which *all* minterms have been covered more than once or are all "don't cares". The remaining groups form an SOP expression as the simplified solution.
7. Check the goodness of the solution in terms of:

(the number of product terms) + (the total number of literals)

in the solution. The number of product terms determines the number of *And* gates and the number of literals determines the total number of inputs to those *And* gates in an *And -Or* hardware realization. The fewer the better in both cases. The number of product

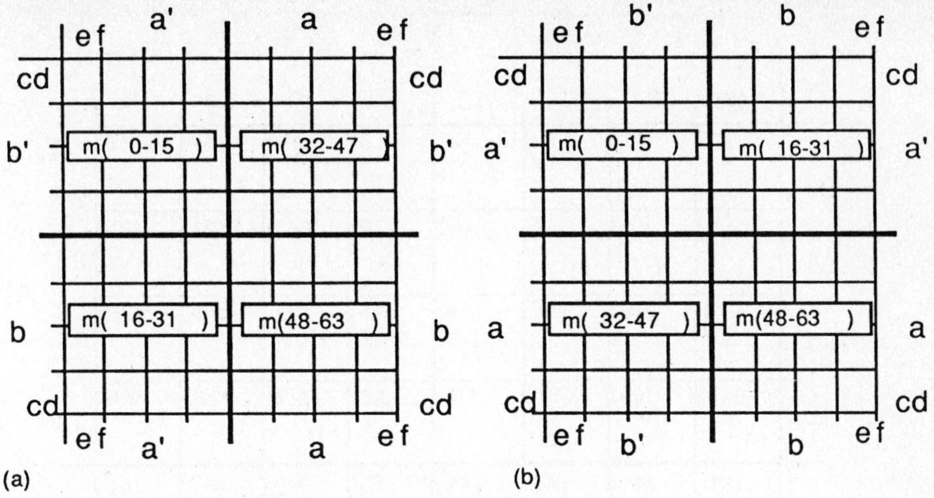

Figure 3.38 Two possible arrangements for 6-variable map
Minterm numbers in any quadrant are determined by *cdef*
(The author prefers arrangement (b) because it is the logical extension of the form of 5-variable map dealt with earlier.)

terms also determines the number of inputs to the overall *Or* operation. An example follows to illustrate some of these considerations.

Example 2 Find the best solution for $T = f(ABCDEF)$ where:

$$T = \Sigma m\,(1, 5, 6, 7, 8, 9, 11, 13, 15, 16, 17, 18, 20, 21, 24, 27, 29, 30, 32, 33, 34, 35, 36, 40, 42, 46, 49, 50, 51, 52, 53, 54, 55, 59, 63)$$

and:

$$D = \Sigma m\,(14, 22, 23, 47, 56, 58, 60)$$

where D denotes "don't care".

The map for this example is shown in Figure 3.39.

Grouping has not been shown on the map and the reader is invited to try out the processes of finding the best solution (fewest terms and fewest literals) for this example.

In order to avoid too much frustration a good solution (according to the author) comprises 13 terms with 54 literals as follows:

$$T = B\bar{C}D + \left\{\begin{array}{l}\bar{A}\bar{C}DE + \bar{A}\bar{B}CF + A\bar{C}\bar{D}E + \bar{C}\bar{D}\bar{E}F + ABEF + C\bar{D}\bar{E}\bar{F} + \bar{A}B\bar{C}\bar{F} + \\ \bar{A}\bar{B}DF\end{array}\right.$$

$$\bar{A}DE\bar{F} + \bar{A}D\bar{E}F + A\bar{B}CE\bar{F} + A\bar{B}\bar{C}\bar{E}\bar{F} + \left\{\begin{array}{l}\bar{A}C\bar{D}EF \\ BC\bar{D}EF\end{array}\right.$$

Can you improve on this solution?

The hardware realization (*And - Or*) is represented by the logic diagram of Figure 3.40.

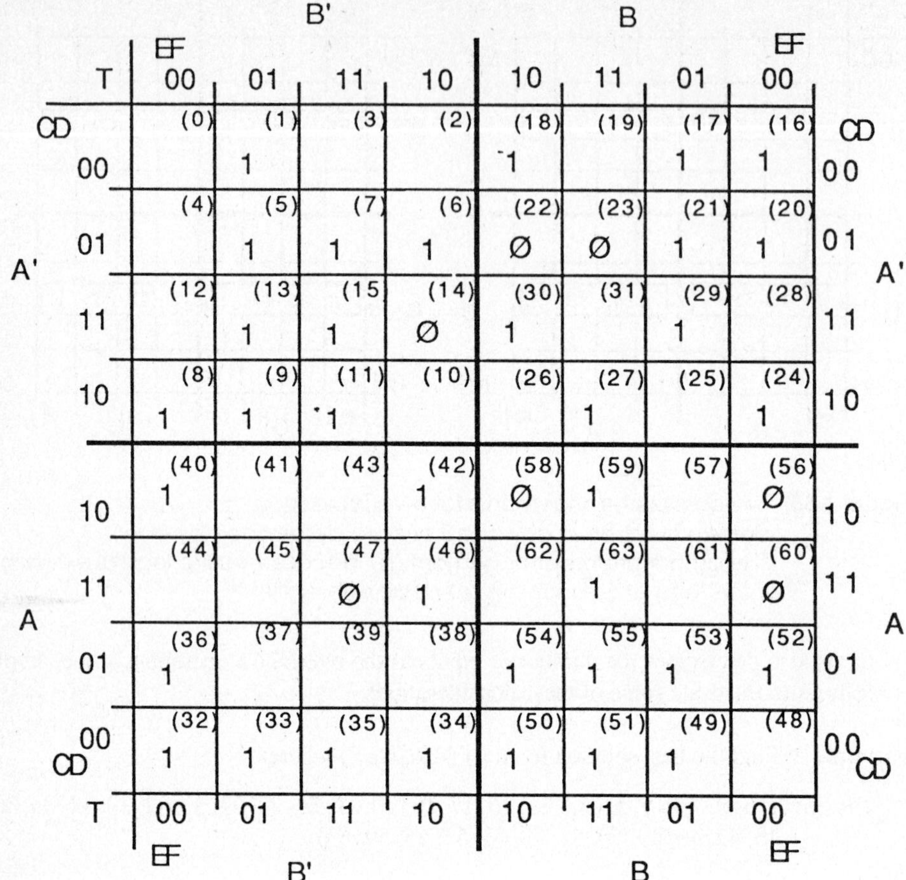

Figure 3.39 Map for 6-variable example 2

3.4.4 A tabular approach to simplification: The Quine-McCluskey method

It will have become apparent that any straightforward application of the Karnaugh map is limited to six variables or less; indeed you may well argue that 6-variable maps are hardly straightforward. Obviously problems cannot always be conveniently limited or partitioned to six variables or less and, bearing in mind the uncertainties of algebraic methods, we see that other means of simplification are needed. Furthermore, simplification by Karnaugh map is essentially a pattern recognition problem which is not well suited for adaptation to implementation by computer. Thus tabular methods are popularly used as the basis of minimization algorithms. A particular approach, the Quine-McCluskey (or Q-M) algorithm, is well known and has been widely used.

Basic procedure

1. Obtain all prime implicants (valid groups).
2. Choose the optimum set of prime implicants to satisfy the problem requirements.

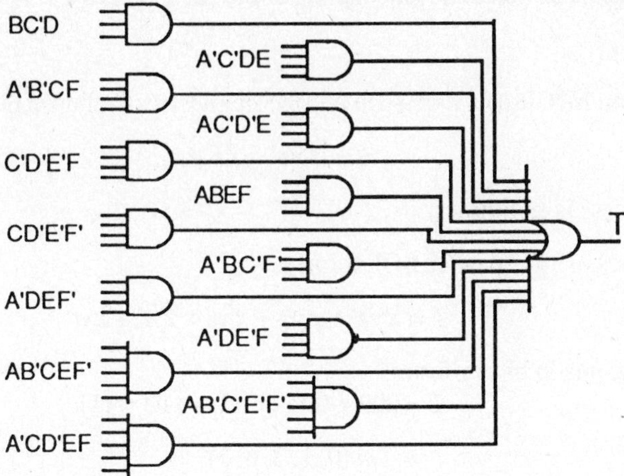

Figure 3.40 Logic for 6-variable example 2

The objective of obtaining the set of prime implicants is achieved through the following steps.

(a) Express the function in expanded SOP form or as a sum of minterms.
(b) Represent each term by its minterm number in *binary form* (i.e. write a 1 for the appearance of a variable and 0 for the appearance of its complement). Minterms readily converted, for example, $m_{10} = 1010$, $m_{14} = 1110$, etc. Note that variables must be ordered in accordance with weighting.
(c) Make a table of terms, ordering them according to *index* (index = number of 1's in the binary representation, e.g. 0100, 0001, are of index 1, 1001 is of index 2, etc.).
(d) By making groups, wherever possible across boundaries between terms which differ by 1 in index (e.g. index 0 with index 1; index 1 with index 2; index *n*-1 with index *n*), form a second table of grouped terms ordered again by index and replacing grouped variables by a dash (-), (i.e. where a 0 and 1 have been grouped). Terms will group when they differ in only one literal, e.g. 001 will group with 101 to form -01; 100 groups with 110 to give 1-0, etc.
(e) Tick all terms in table 1 which have been grouped. Unticked terms must appear in the final PI set.
(f) Now repeat the grouping exercise across the index boundaries in table 2 with the constraint that dashes must also line up for grouping to be possible.
(g) Form table 3 of terms grouped in table 2. Tick table 2 terms as they are grouped. Unticked terms must appear in the final PI set.
(h Continue this grouping procedure, forming subsequent tables until no further grouping is possible.
(j) *The terms in the final table together with all unticked terms in all previous tables now constitute the PI set.*

Having obtained the PI set the second objective of the basic procedure, namely choosing the optimum set of PIs, is then achieved. One way of so doing is to use a PI chart.

The best way of learning the approach is to use it and to this end we will now tackle some simple problems to illustrate both aspects of the basic process.

Example 1

Use the Quine-McCluskey method to find the simplest realization of the function of T where:

$$T = x'y'z' + x'z + xy.$$

1. The process is as follows:

(a) Express T in expanded SOP form:

$$T = x'y'z' + x'y'z + x'yz + xyz' + xyz$$

(b) Write this in binary form:

$$T = 000 + 001 + 011 + 110 + 111$$

$$= \Sigma m\,(0, 1, 3, 6, 7)$$

(c) - (f) Set out table 1 according to index as follows and carry out the grouping processes.

Table 1

Index	*"m"*	*term*
Zero	0	✔ 000
One	1	✔ 001
Two	3	✔ 011
	6	✔ 110
Three	7	✔ 111

Boundary (Zero/One): ← Group to form 00- [m(0,1)]

Boundary (One/Two): ← Group to form 0-1 [m(1,3)]
No group between m1 and m6

Boundary (Two/Three): ← Group to form -11 [m(3,7)]
Group to form 11- [m6,7)]

End of table 1 — no unticked terms

Table 2

Index	*m*	*term*
Zero	0,1	(A) 00-
One	1,3	(B) 0-1
Two	3,7	(C) -11
	6,7	(D) 11-

Boundary / Boundary: No grouping is possible ... dashes do not line up across any boundary. Therefore, identify all (unticked) terms (A), (B) etc. to form PI set.

No further tables in this case.

(g) to (h) No further grouping is possible thus, no terms are ticked in Table 2 and there can be no Table 3.

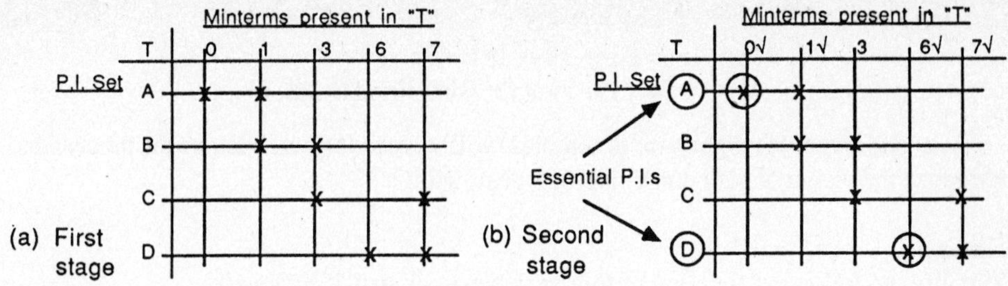

Figure 3.41 PI chart for *T* Example 1

(j) The PI set is thus all the terms in Table 2 (all terms in Table 1 having been ticked, i.e. covered in Table 2).

For convenience, identify the PIs as A to D as shown above.

The PI set for this problem is thus:

$$T = A + B + C + D$$

where *A* to *D* can be evaluated later in terms of the variables *xyz* if desired.

2. (a) We now set out a PI chart to facilitate the choice of an optimum set of PIs to satisfy the function *T*.

   From our previous work ((b) above), we see that, in minterm form $T = \Sigma m$ (0, 1, 3, 6, 7).

   The PI chart is set out as follows (Figure 3.41):

   (b) From our records (Table 2 in this case) we see that PI (*A*) covers minterms 0 and 1, PI (*D*) covers minterms 6 and 7, etc. Enter *X* in chart to indicate minterms covered by each PI as shown in Figure 3.41(a).
   (c) Now scan each minterm column in turn and identify those columns in which only *one* *X* appears. Circle the *X* entries in these columns as shown in Figure 3.41(b). Circle also, the PI which covers each of the circled *X*s as shown.
   (d) Circled PIs are *essential PIs* since they *cover minterms which are not covered by any other PI*. Thus the "essential PIs" *must* be present in any solution for *T*.
   (e) Now, tick all minterms which are covered by the essential PIs (0, 1, 6 and 7 in this case). Note all minterms which are yet to be covered (*m*3 in this case).
   (f) Choose PIs as economically as possible to cover the remaining minterm(s). [In this case *either* PI *B* or *C* will do equally well since they are equal size groups].
   (g) Write out the final solutions as a set of PIs:

$$T = A + D + \begin{cases} B \\ C \end{cases}$$

   and translate to algebra again by noting the binary representation of the PIs from tables 1 to *N* (2 in this case).

e.g. $T = A + D + B$
$= 00\text{-} + 11\text{-} + 0\text{-}1$
$= x'y' + xy + x'z$ (simplified solution)

A second example (slightly more complex) will serve to further illustrate the process and introduce the way in which don't cares are dealt with.

Example 2
Simplify the following function of four variables such that $T = f(abcd)$

where $T = \Sigma m\,(0, 1, 2, 3, 8, 9, 11, 13, 15)$

$D = \Sigma m\,(10, 14)$

where D = don't cares.

Clearly, this is a 4-variable environment since the highest minterm is 15.

The procedure is as before with "don't cares" being treated as minterms for grouping *but don't cares are omitted from the PI chart in procedure part* 2 for choosing the PI set.

Table 1

Index	*m*	*term*
Zero	0	✔ 0000
One	1	✔ 0001
	2	✔ 0010
	8	✔ 1000
Two	3	✔ 0011
	9	✔ 1001
	10	✔ 1010
Three	11	✔ 1011
	13	✔ 1101
	14	✔ 1110
Four	15	✔ 1111

Table 2

Index	*m*	*term*
Zero	0,1	✔ 000-
	0,2	✔ 00-0
	0,8	✔ -000
One	1,3	✔ 00-1
	1,9	✔ -001
	2,3	✔ 001-
	2,10	✔ -010
	8,9	✔ 100-
	8,10	✔ 10-0
Two	3,11	✔ -011
	9,11	✔ 10-1
	9,13	✔ 1-01
	10,11	✔ 101-
	10,14	✔ 1-10
Three	11,15	✔ 1-11
	13,15	✔ 11-1
	14,15	✔ 111-

(Groups of 2)

Table 3

Index	*m*	*term*
Zero	0,1	✔ 0 0 - -
	2,3	
	0,1	
	8,9	✔ - 0 0 -
	0,2	
	8,10	✔ - 0 - 0
One	1,3	
	9,11	✔ - 0 - 1
	2,3	
	10,11	✔ - 0 1 -
	8,9	
	10,11	✔ 1 0 - -
Two	9,11	
	13,15	(B) 1 - - 1
	10,11	
	14,15	(C) 1 - 1 -

(Groups of 4)

Table 4

Index	*m*	*term*
Zero	0,1	
	2,3	(A) - 0 - -
	8,9	
	10,11	

(Groups of 8)

No further grouping is possible. Note the two unticked terms in table 3.

From the tables, the PI set is $(A) + (B) + (C)$.

Now set out a PI chart as in Figure 3.42.

Solution:

$$\begin{aligned} T &= (A) + (B) \\ &= \text{-}0\text{-}\text{-} + 1\text{-}\text{-}1 \\ &= b' + ad \end{aligned}$$

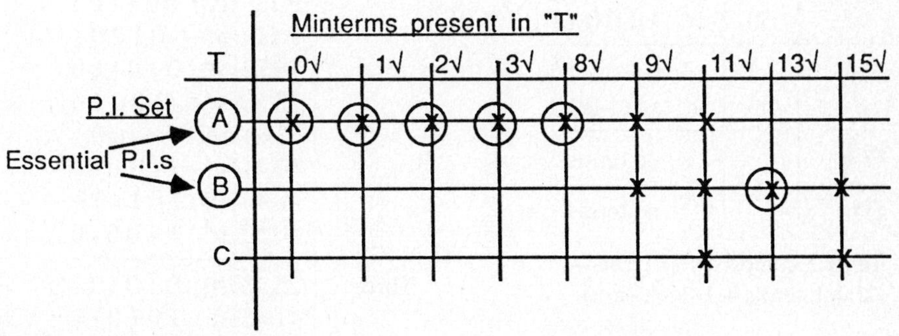

Figure 3.42 PI chart for example 2

You may well contend that this method of simplification is time-consuming and prone to error, and has little to offer in comparison with other methods. For example, the two examples dealt with are very easily tackled with Karnaugh maps. However, we have already agreed that 6-variable Karnaugh maps are not the ultimate in "light entertainment" so, for comparison, let us now set out a 6-variable problem to tackle by the Q-M algorithm.

Example 3

$$F = \sum m\,(0, 2, 3, 6, 8, 9, 13, 14, 16, 18, 22, 23, 24, 32, 40, 41, 43, 44, 48, 50)$$

$$D = m\,(33)$$

Table 1

Index	*m*	*term*
Zero	0	✔ 000000
One	2	✔ 000010
	8	✔ 001000
	16	✔ 010000
	32	✔ 100000
Two	3	✔ 000011
	6	✔ 000110
	9	✔ 001001
	18	✔ 010010
	24	✔ 011000
	33	✔ 100001
	40	✔ 101000
	48	✔ 110000
Three	13	✔ 001101
	14	✔ 001110
	22	✔ 010110
	41	✔ 101001
	44	✔ 101100
	50	✔ 110010
Four	23	✔ 010111
	43	✔ 101011
Five		no terms
Six		no terms

Table 1 comprises 21 terms—
(20 minterms + 1 don't care).

Table 2

Index	*m*	*term*
Zero	0,2	✔ 0000-0
	0,8	✔ 00-000
	0,16	✔ 0-0000
	0,32	✔ -00000
One	2,3	(A) 00001-
	2,6	✔ 000-10
	2,18	✔ 0-0010
	8,9	✔ 00100-
	8,24	✔ 0-1000
	8,40	✔ -01000
	16,18	✔ 0100-0
	16,24	✔ 01-000
	16,48	✔ -10000
	32,33	✔ 10000-
	32,40	✔ 10-000
	32,48	✔ 1-0000
Two	6,14	(B) 00-110
	6,22	✔ 0-0110
	9,13	(C) 001-01
	9,41	✔ -01001
	18,22	✔ 010-10
	18,50	✔ -10010
	33,41	✔ 10-001
	40,41	✔ 10100-
	40,44	(D) 101-00
	40,50	✔ 1100-0
Three	22,23	(E) 01011-
	41,43	(F) 1010-1

Table 2 comprises 28 terms —
(i.e. 28 groups of two)

Table 3

Table 3

Index	*m*	*term*
Zero	0,2 16,18	(G) 0 - 0 0 - 0
	0,8 16,24	(H) 0 - - 0 0 0
	0,16 32,48	(J) - - 0 0 0 0
	0,8 32,40	(K) - 0 - 0 0 0
One	2,6 18,22	(L) 0 - 0 - 1 0
	8,9 40,41	(M) - 0 1 0 0 -
	32,33 40,41	(N) 1 0 - 0 0 -
	16,18 48,50	(P) - 1 0 0 - 0

No further grouping. Table 3 comprises 8 terms (i.e. 8 groups of 4).

From the tables, PI set comprises 14 terms. Now set out PI chart as in Figure 3.43.

$$F = A + B + C + D + E + F + H + P + \begin{cases} J \\ K \\ N \end{cases}$$

The solution comprises nine terms of 42 literals in total and, if $F = f(abcdef)$:

$$F = a'b'c'd'e + a'b'def' + a'b'ce'f + ab'ce'f' + a'bc'de + ab'cd'f + a'd'e'f' + bc'd'f'$$

$$+ \begin{cases} c'd'e'f' \\ b'd'e'f' \\ ab'd'e' \end{cases}$$

The reader may wish to attempt this same problem using a 6-variable Karnaugh map in order to see how easy (or otherwise) it is to come up with an optimum solution by that means. Note that some "obvious"groups of four do not appear in the best solution.

You may well conclude that neither method is particularly easy but when it comes to computer-based tools for doing such tasks then the Q-M method translates into software quite readily.

3.5 Summing up the introductory chapters

The first three chapters have been devoted to fundamental aspects with which the digital designer should be familiar since they come into the category of "tools of the trade". The material presented also forms a platform on which the rest of this text is built.

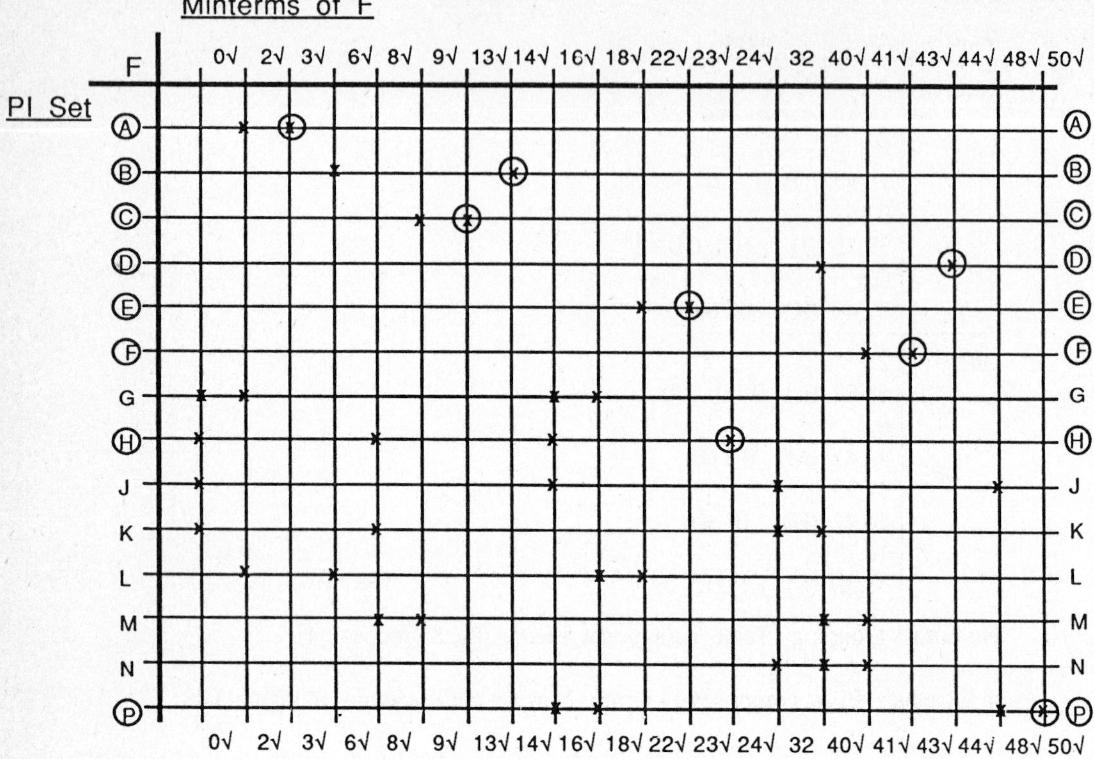

Figure 3.43 PI chart for *F*-example 3

Two main topic areas have been dealt with, each of which rates at least a full text in its own right if all relevant and related matters are to be covered. However, as far as semiconductor/MOS technology is concerned the material essential to this text and to the practicing electronic engineer has hopefully been put in place and further aspects of the realization of circuits in silicon will emerge throughout the rest of this text.

As far as the second topic area, the design of logic circuits, is concerned, the following chapters will set out some effective methods for designing combinational and sequential logic circuitry and present appropriate realizations in silicon MOS technology.

3.6 Worked examples

1. Draw up a map and write the equation for output *X* in each of the five combinational logic circuits in Figure 3.44.

   Solution :

   The equations and maps are:

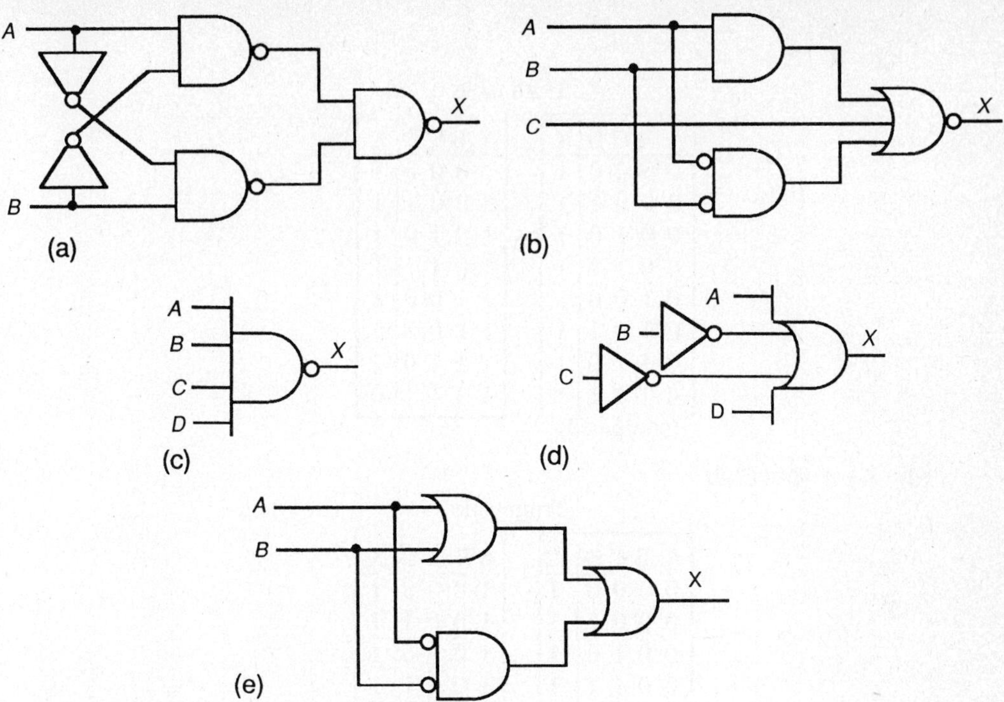

Figure 3.44 Some combinational circuits

(a) $X = AB' + A'B$

Truth table

A	B	X
0	0	0
0	1	1
1	1	0
1	0	1

(b) $X' = AB + A'B' + C$
$X = A'BC' + AB'C'$

Truth table

A	B	C	X
0	0	0	0
0	0	1	0
0	1	0	1
0	1	1	0
1	0	0	1
1	0	1	0
1	1	0	0
1	1	1	0

(c) $X' = ABCD$

$X = A' + B' + C' + D'$

Truth table

A B C D	X
0 0 0 0	1
0 0 0 1	1
0 0 1 0	1
0 0 1 1	1
0 1 0 0	1
0 1 0 1	1
0 1 1 0	1
0 1 1 1	1

(continued)

A B C D	X
1 0 0 0	1
1 0 0 1	1
1 0 1 0	1
1 0 1 1	1
1 1 0 0	1
1 1 0 1	1
1 1 1 0	1
1 1 1 1	0

(d) $X = A + B' + C' + D$

Truth table

A B C D	X
0 0 0 0	1
0 0 0 1	1
0 0 1 0	1
0 0 1 1	1
0 1 0 0	1
0 1 0 1	1
0 1 1 0	0
0 1 1 1	1

(continued)

A B C D	X
1 0 0 0	1
1 0 0 1	1
1 0 1 0	1
1 0 1 1	1
1 1 0 0	1
1 1 0 1	1
1 1 1 0	1
1 1 1 1	1

(e) $X = A + B + A'B'$

Truth table

A B	X
0 0	1
0 1	1
1 1	1
1 0	1

<u>X = 1</u>

2. Using DeMorgan's theorem, derive and set out the logic circuits for the complementary functions for each figure in Figure 3.45.

Solution:

(a) From figure: $F = PR' + P'R$

By Demorgan's theorem: $F' = (P' + R).(P + R')$

$= P'R' + PR$

The logic circuit is given as Figure 3.46(a).

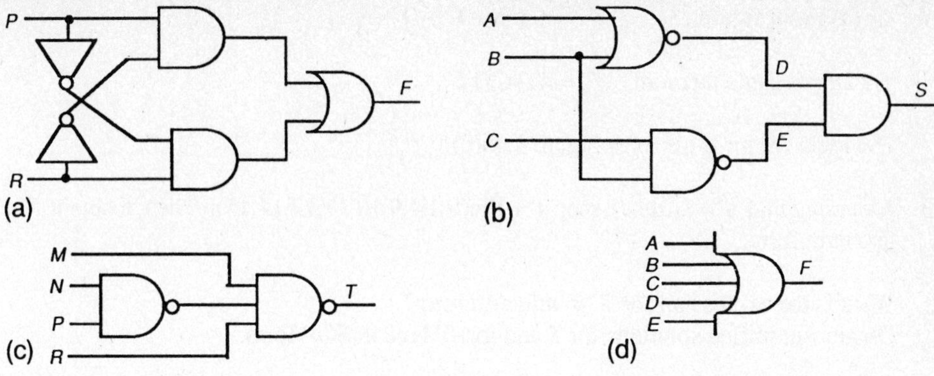

Figure 3.45 Circuits for which the complementary functions are required

(b) From figure: $S = D.E$

where $D = A'B' \quad E = B'+C'$

$$S = A'B'(B'+C')$$
$$= A'B' + A'B'C'$$
$$= A'B'(1 + C') = A'B'$$

By Demorgan's theorem: $S' = A + B$

The logic circuit is given as Figure 3.46(b).

(c) From figure: $T = M' + R' + NP$

By Demorgan's theorem: $T' = MR(N'+P')$

$$= MRN'+MRP'$$

The logic circuit is given as Figure 3.46(c).

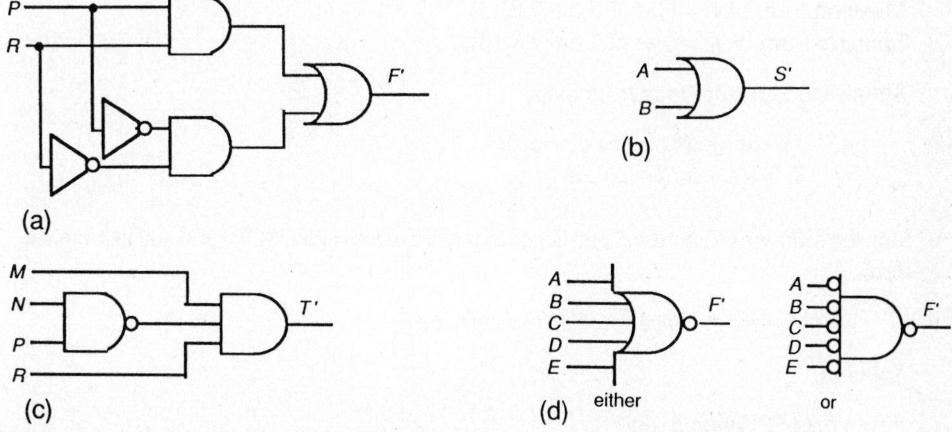

Figures 3.46 Circuits which form the complement of those in Figure 3.45

(d) From figure, $F = A + B + C + D + E$

By Demorgan's theorem, $F' = A'B'C'D'E'$

The logic circuit is given as Figure 3.46(d).

3. Assuming that $T = f(abcd)$, map $T = \Sigma m(0,1,4,9,10,11,12,14,15)$. Thence obtain T in maxterm form.

What is the expression for T' in minterm form?
Obtain simplified solutions for T and for T' both in SOP form.

Solution :

The map for T is given as Figure 3.47.

T ab \ cd	00	01	11	10
00	1	1	0	0
01	1	0	0	0
11	1	0	1	1
10	0	1	1	1

Figure 3.47 Map for question 3

Maxterm form of $T = \Pi M\ (2,3,5,6,7,8,13)$.
Minterm form of $T' = \Sigma m\ (2,3,5,6,7,8,13)$.

Simplified SOP solutions from map:

$$T = ac + a'b'c' + bc'd' + ab'd$$
$$T' = a'c + bc'd + ab'c'd'$$

4. Map the following function F and hence express it in expanded SOP form and in minterm form.

$$F = b'c'e' + b'ce'f + d'ef' + bcef + be'f.$$

Solution:

From map of Figure 3.48:
Expanded SOP form:

	b'				b				
F cd \ ef	00	01	11	10	10	11	01	00	F cd
00	1	1	0	1	1	0	1	0	00
01	1	1	0	0	0	0	1	0	01
11	0	1	0	0	0	1	1	0	11
10	0	1	0	1	1	1	1	0	10

Simplest SOP form... F = e'f + b'c'e' + bcf + d'ef'

Figure 3.48 Map for question 4

$$T=b'c'd'e'f' + b'c'd'e'f + b'c'd'ef' + b'c'de'f' + b'c'de'f + b'cd'e'f + b'cd'ef' + b'cde'f + bc'd'e'f + bc'd'ef' + bc'de'f + bcd'e'f + bcd'ef' + bcd'ef + bcde'f + bcdef$$

Minterm form:

$$T = \Sigma m\ (0,1,2,4,5,9,10,13,17,18,21,25,26,27,29,31)$$

5. Simplify the following expression:

$$T = \Sigma m\ (0,2,4,5,6,7,9,10,12,13,18,19,22,28,30)$$

Solution:

From map of Figure 3.49 (assuming $T=f\ (bcdef)$:
Simplified SOP form is:

$$T = b'c'd + c'ef' + b'de' + b'c'f' + b'ce'f + b'd'ef' + bc'd'e + bcdf'.$$

	b'				b				
T cd \ ef	00	01	11	10	10	11	01	00	T cd
00	1	0	0	1	1	1	0	0	00
01	1	1	1	1	1	0	0	0	01
11	1	1	0	0	1	0	0	1	11
10	0	1	0	1	0	0	0	0	10

Figure 3.49 Map for question 5

6. Simplify the following expression:

$F = f(abcd)$ $\quad\quad$ $F = \Sigma m\ (0,4,5,7,13,)$
$D = \Sigma m\ (1,6,8,)$

Solution:

From map of Figure 3.50:
Simplified SOP form is:

$$F = a'b + a'c' + bc'd$$

F ab \ cd	00	01	11	10
00	1	Ø	0	0
01	1	1	1	Ø
11	0	1	0	0
10	Ø	0	0	0

Figure 3.50 Map for question 6

T pq \ rs	00	01	11	10
00	0	1	1	0
01	0	1	1	0
11	1	0	0	1
10	0	1	1	0

Figure 3.51 Map for question 7

7. Simplify the following expression:

$T = f(pqrs)$ $\quad\quad$ $T = \Pi M\ (0,2,4,6,8,10,13,15)$

Solution:

From map of Figure 3.51:
Simplified SOP form is:

$$T = p's + q's + pqs'$$
$$T' = p's' + q's' + pqs$$

Note that variable *r* is redundant.

8. Find the best (simplified) SOP solution to:

$$S = a'b'c'd'f' + a'b'cd'f + a'cef' + a'c'd'f' + a'bcdf + a'bc'e'f' + ab'cef' + ab'cdf + ac'd'f' + ab'c'd'e'f' + bcef' + abcd'f + abc'e'f'$$

Solution:

From map of Figure 3.52:
Simplified SOP form is:

		b'				b				
	cd \ ef	00	01	11	10	10	11	01	00	cd
a'	00	1	0	0	1	1	0	0	1	00
a'	01	0	0	0	0	0	0	0	1	01
a'	11	0	0	0	1	1	1	1	0	11
a'	10	0	1	1	1	1	0	0	0	10
a	10	0	0	0	1	1	1	1	0	10
a	11	0	1	1	1	1	0	0	0	11
a	01	0	0	0	0	0	0	0	1	01
a	00	1	0	0	1	1	0	0	1	00
	cd \ ef	00	01	11	10	10	11	01	00	cd
		b'				b				

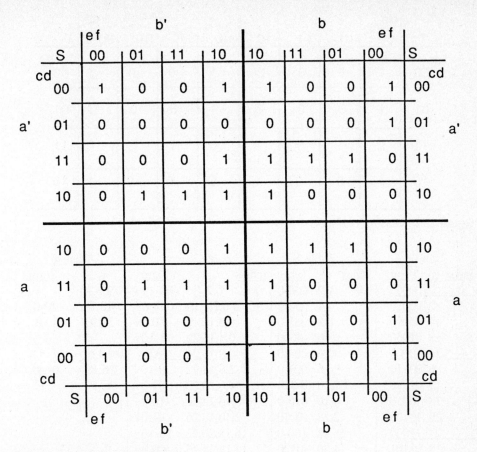

Figure 3.52 Map for question 8

$$S = cef' + c'd'f' + a'b'cd'f + ab'cdf + a'bcdf + abcd'f + bc'e'f'$$

9. (a) Using a Karnaugh map find the simplest SOP solution to:

$$T = f(abcde) \quad T = \Sigma m\,(0,4,8,10,13,15,16,18,22,24,31)$$
$$D = \Sigma m\,(2,6,20,26,29)$$

(b) Repeat the simplification using the Quine-McCluskey method.

Solution (a):

From map of Figure 3.53:

$T = c'e' + b'e' + bce.$

(b):

Using Quine-McCluskey method:

	a'				a				
T (bc \ de)	00	01	11	10	10	11	01	00	T (bc)
00	1	0	0	Ø	1	0	0	1	00
01	1	0	0	Ø	1	0	0	Ø	01
11	0	1	1	0	0	1	Ø	0	11
10	1	0	0	1	Ø	0	0	1	10

Figure 3.53 Map for question 9

Table1

Index	m	Term
0	0	00000
1	2	00010
	4	00100
	8	01000
	16	10000
2	6	00110
	10	01010
	18	10010
	20	10100
	24	11000
3	13	01101
	22	10110
	26	11010
4	15	01111
	29	11101
5	31	11111
	all ticked√	

Table 2

Index	m	Term
0	0,2	000-0
	0,4	00-00
	0,8	0-000
	0,16	-0000
1	2,6	00-10
	4,6	001-0
	2,10	0-010
	8,10	010-0
	2,18	-0010
	16,18	100-0
	4,20	-0100
	16,20	10-00
	8,24	-1000
	16,24	1-000
2	6,22	-0110
	10,26	-1010
	18,22	10-10
	20,22	101-0
	24,26	110-0
3	13,15	011-1
	13,29	-1101
4	15,31	-1111
	29,31	111-1
	all ticked√	

Table 3

Index	m	Term
0	0,2,4,6	00- -0√
	0,2,8,10	0-0- 0√
	0,2,16,18	-0 0-0√
	0,4,16,20	-0-0 0√
	0,8,16,24	- -000√
1	2,6,18,22	-0-10√
	4,6,20,22	-01-0√
	8,10,24,26	-10-0√
	2,10,18,26	- -010√
	16,18,20,22	10- -0√
	16,18,24,26	1-0-0√
2	13,15,29,31	-11-1 A

Table 4

m	Term
0,2,8,10 16,18, 24,26	- -0-0 B
0,2,4,6 16,18 20,22	-0- -0 C

PI map

m	0	4	8	10	13	15	16	18	22	24	31
PIs											
A					(X)	(X)					(X)
B	X		(X)	(X)			X	X		(X)	
C	X	(X)					X	X	(X)		

All PIs are Essential.
Solution is $T = A + B + C$
$= bce + c'e' + b'e'$
as before.

3.7 Tutorial 3

1. Using a Karnaugh map in each case, show that:

(a) $a + a'.b = a + b$
(b) $(p.q' + p'.q)' = p.q + p'.q'$
(c) If $F = A'.B'.C'$ then $F' = A + B + C$

2. (a) How many minterms may be generated in (i) a 4-variable situation, and (ii) a 5-variable situation?
   (b) Express the following minterms algebraically:
      (i) $m(0,7,13)$ when $F = f(pqrs)$
      (ii) $m(1,5,20,30)$ when $F = f(abcde)$.
   (c) Map $F = f(pqrs)$ where $F = \Sigma m(0,1,3,8,9,11)$. Thence derive the maxterm form of F.
   Express F in simplified and then fully expanded SOP form.

3. Map the following function:

$$T' = (a + c).(b + d)$$

   Hence, assuming a is weighted 8, b is weighted 4, c is weighted 2, d is weighted 1, express T in minterm and in maxterm forms.
   Write the fully expanded algebraic SOP form of T.

4. Set out a Karnaugh map for the following expression:

$$T = \Sigma m(0, 1, 2, 5, 8, 9, 11, 12).$$

   Draw a logic circuit for T assuming $T = f(abcd)$.
   Write T in maxterm form and find the simplest expression for T' in *SOP form.*

5. Set out a 6 variable Karnaugh map for T $= f(abcdef)$ where:

$$T = \Sigma m(0, 2, 5, 7, 8, 9, 11, 12, 14, 15, 18, 19, 21, 23, 31, 36, 38, 39, 40, 41, 48, 51, 53, 58, 61, 63)$$

   and $D = \Sigma m(3, 6, 17, 26, 45, 50, 60)$
   where D indicates "don't care".
   Find the simplest SOP solution for T and draw a logic circuit.

6. Using the Quine-McCluskey tabular procedure, simplify the following expression to give the best solution (i.e. fewest terms and fewest literals):

$$T = \Sigma m(0, 4, 5, 7, 9, 12, 13, 14).$$
$$D = \Sigma m(2,10)$$

   where D indicates " don't care".
7. Using the Quine-McCluskey tabular procedure, simplify the following expression for $F = f(pqrs)$, where:

$$F = qr's' + pq'r's' + p'q'r's + p'qrs'$$

   and where there are don't care minterms 2,5,10,14.

4 The design of combinational logic

4.1 Introduction

Combinational logic forms an important and major part of digital systems engineering. It is important, not only in its own right, but also as the basis of sequential logic, since it is only the manner in which logic circuits are connected which distinguishes one from the other, as we shall see. However, we shall be treating combinational and sequential logic as separate classes due to the different manner in which we analyze or synthesize the behavior. Having said that, we will nevertheless establish the close link between the two and observe that there is much in common in the processes used to approach the business of design in each class.

Combinational (or combinatorial) logic may be envisaged within the general model set out in Figure 4.1(a). In essence, a combinational logic circuit is a circuit in which every possible steady state of the output(s) at any time is determined wholly and solely by the combination of input conditions at that time. Other significant factors are propagation delay, and rise and fall times as defined in Figure 4.1(b).

(a)

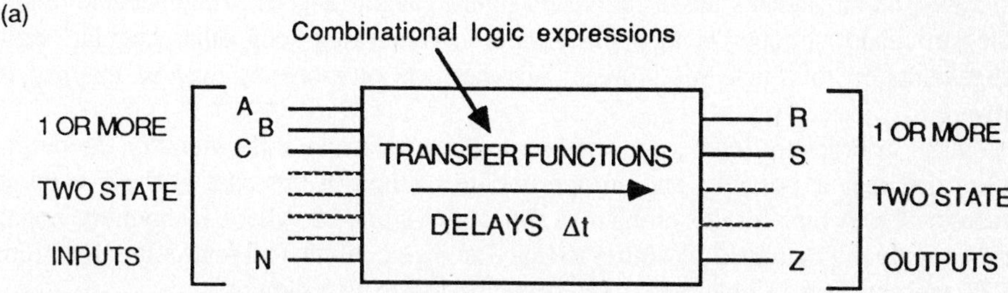

A different Transfer function will apply for each output.
Similarly, the delay Δt may be different for various input-output combinations.

(b)

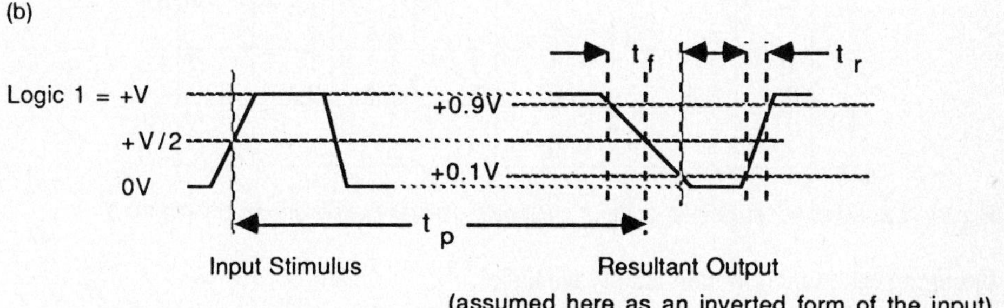

(assumed here as an inverted form of the input).

t_p = propagation delay, input to output (leading edge delay illustrated here).
t_r = rise time. t_f = fall time (output used by way of example)

Figure 4.1 Characteristics of combinational logic

Combinational logic and the desired logical performance is often specified using Boolean algebra and it is a suitable starting point from which to approach design methodology and examine some of the alternative hardware realizations.

In order to pursue these matters , it is useful to take a problem as an example through which we may examine alternative approaches to the design of combinational logic.

Let us consider a multiple output function of five variables which are specified as *abcde* and for which there are four output functions, Z_1, Z_2, Z_3 and Z_4, such that:

$$Z_1 = b'c'de + ab'd'e + bcd' + bcde' + bde + b'cde + a'b'c'd'e'$$
$$Z_2 = a'b'c'd' + a'c'de + a'bc'e$$
$$Z_3 = bc + b'de + bc'd + b'c'de + bc'd'e' + ac'd'e' + a'b'c'd'e'$$
$$Z_4 = a'b'e + ace + a'be$$

4.2 "Random" logic

Random logic is not really random but is taken to imply a solution using the minimum set (preferably) of *And* and *Or* gates or equivalent simple gates arranged in a manner determined by the particular function(s) being realized. This is in contrast to some other "regular" logic arrangements on to which this problem or other sets of problems may be mapped in hardware.

Clearly, in order to find a good random logic solution we must simplify the output functions, if that is possible, and an appropriate method in this case is the 5-variable Karnaugh map. A map for this problem is shown in Figure 4.2, where, it should be noted, 1 entries apply to function Z_1; 2 entries to function Z_2, etc., so that all four output functions may be presented on a single map (minterm numbers are also shown).

	a'				*a*				
bc \ de	00	01	11	10	10	11	01	00	bc
00	12 3 (0)	2 4 (1)	12 34 (3)	(2)	(18)	1 3 (19)	1 (17)	3 (16)	00
01	(4)	4 (5)	1 34 (7)	(6)	(22)	1 34 (23)	1 4 (21)	(20)	01
11	1 3 (12)	1 34 (13)	1 34 (15)	1 3 (14)	1 3 (30)	1 34 (31)	1 34 (29)	1 3 (28)	11
10	3 (8)	2 4 (9)	12 34 (11)	3 (10)	3 (26)	1 3 (27)	(25)	3 (24)	10

Figure 4.2 Map for multiple output problem (small numbers=minterm no.)

Grouping each function individually we have:

$$Z_1 = de + bc + ab'e + a'b'c'd'e'$$
$$Z_2 = a'c'e + a'b'c'd'$$
$$Z_3 = de + bc + bd + c'd'e'$$
$$Z_4 = a'e + ce$$

This solution would imply a need for ten *And* gates since the functions involve ten different product terms, two being common to two functions. However a better solution is obtained if the second term of Z_2 is "ungrouped" to $a'b'c'd'e'$ the other minterm being already included in another group and this product term being already required for Z_1. Similarly, the first term of Z_4 may be ungrouped to $a'c'e$ this already having been formed for Z_2. A better solution then is:

$$Z_1 = de + bc + ab'e + a'b'c'd'e'$$
$$Z_2 = a'c'e + a'b'c'd'e'$$
$$Z_3 = de + bc + bd + c'd'e'$$
$$Z_4 = a'c'e + ce$$

Note that the product terms *de, bc, a'c'e* and *a'b'c'd'e'* are each common to 2-output functions. This solution thus requires eight *And* gates (with 22 literals) and, of course, four *Or* gates in the most straightforward implementation as shown in Figure 4.3. The solution obtained by simplifying output functions individually would require ten *And* gates (with 28 literals).

Thus, multiple output functions require special attention in simplification since the simplification of individual output functions may well yield a more complex solution than is necessary. In general, a better solution is obtained if the number of common terms can be

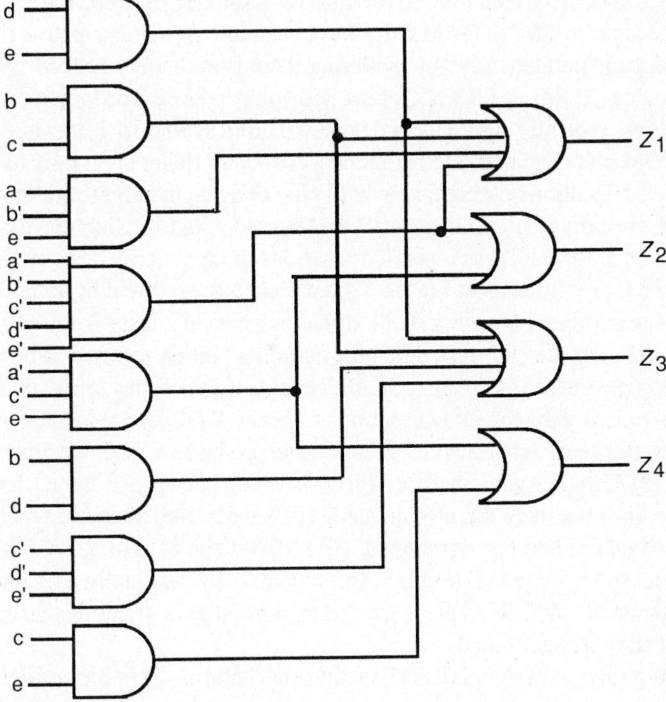

Figure 4.3 Random logic (*And*/*Or*) realization of problem

increased. Any "don't care" conditions are utilized as convenient to achieve this objective and, of course, to reduce overall and individual term complexity.

Although the example we have taken is a straightforward one, in earlier work we have seen how difficult and tedious the simplification processes for combinational logic may become. However, if we wish to take advantage of modern technology there are other ways of realizing complex combinational logic functions, some of which will not involve the same degree of simplification or may not require any at all. The first method to be discussed falls into this latter catagory.

4.3. Using read only memories (ROM) (or non-volatile random access memories (RAM)) to realize combinational logic

4.3.1 General memory configurations

A ROM or RAM chip provides the user with the capability of addressing individual locations, a "write" capability through which data in the form of one or more bits may be stored in each individual location (or address), and a "read" capability which enables the data in any location to be addressed and read out again.

In the case of a ROM, the data is written in once only and is then permanently stored, but for the RAM chip the data may be rewritten or altered as desired. RAMs are mentioned here because modern CMOS RAM chips have such low power dissipation that they can be used to hold data "permanently" by powering them from battery backed-up sources.

In both cases (ROMs and RAMs) chips with many addressable locations, and with each location holding typically eight bits of data, are readily available. For example, commonly available ROM chips are marketed at modest cost comprising up to 64K locations (where K = 1024), each location holding one byte (8 bits) of data. In order to address, for example, a 16K ROM, the individual locations will be addressed via a 14-bit address input vector (16K = 2^{14}) supplied at the "address input" connections to the chip. A basic chip configuration might then be represented as in Figure 4.4 and data accessed will be available on the data output lines when the chip select (CS') signal is asserted*. Data is initially read into the locations by putting the chip into the "programming" mode using facilities not described here. For completeness, the reader should be aware that some forms of ROM are programmed to order by the manufacturer and are generally unalterable. However, ROMs of certain types can be programmed by the user and are known as "programmable read only memories" (PROMs). Again, there are PROMs which once programmed cannot be altered (e.g. fusible link) but there are also erasable (EPROMs) and alterable types (EAPROMs) which can be erased and reprogrammed. EAPROMs can be reprogrammed in circuit but EPROMs must be removed to be erased (for example, by irradiating with ultraviolet light). Once programmed, such chips are regarded as non-volatile memories since data remains until erased and reprogrammed.

The random access memory (RAM) on the other hand has a similar configuration (Figure

* "Asserted" means the active state of the signal line in question. i.e. a signal line S is active in the Hi or Logic 1 state while a signal line S' or $\overline{S}$ is active in the Lo or Logic 0 state.

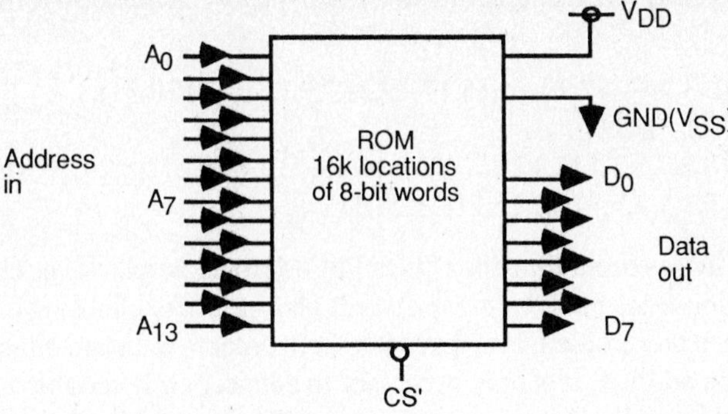

Figure 4.4 Typical 16K ROM arrangement

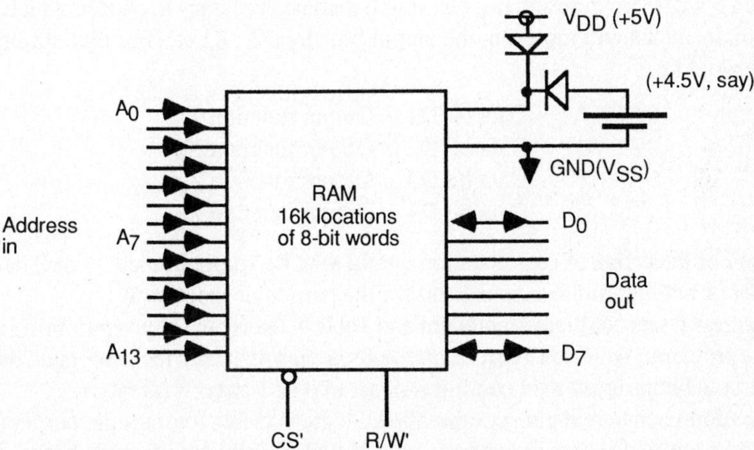

Figure 4.5 Typical 16k battery backed RAM arrangement

4.5) but has read/write control facilities so that the memory contents can be dynamically changed. The contents of such a memory is volatile and will be lost when the power is removed from the chip, but, as stated earlier, very low dissipation CMOS memories are available which can have long life on relatively small back-up batteries. An arrangement is suggested in Figure 4.5.

4.3.2 Mapping combinational logic expressions into a ROM or PROM chip

As an example, let us take the previously considered multiple output combinational function of five variables (*abcde*) for which the four output functions (Z_1 to Z_4) are defined as follows:

$$Z_1 = b'c'de + ab'd'e + bcd' + bcde' + bde + b'cde + a'b'c'd'e'$$
$$Z_2 = a'b'c'd' + a'c'de + a'bc'e$$
$$Z_3 = bc + b'de + bc'd + b'c'de + bc'd'e' + ac'd'e' + a'b'c'd'e'$$
$$Z_4 = a'b'e + ace + a'be$$

First **expand** (*not simplify*) each output function and express it in minterm form (see section 3.3.14):

$$Z_1 = m(0,3,7,11,12,13,14,15,17,19,21,23,27,28,29,30,31)$$
$$Z_2 = m(0,1,3,9,11)$$
$$Z_3 = m(0,3,7,8,10,11,12,13,14,15,16,19,23,24,26,27,28,29,30,31)$$
$$Z_4 = m(1,3,5,7,9,11,13,15,21,23,29,31)$$

In order to map these output functions into a PROM, for example, let us choose a small PROM of 256 locations each holding an 8-bit word. The reason for choosing a small PROM will be apparent when this problem is mapped into it. In order to translate minterm numbers directly into location address, it is only necessary to connect each variable to the address input line of equivalent weight (in this case, least significant variable *e* to address line A0, most significant variable *a* to address line A4, etc). Location 0 will then represent m_0, location 5 will then represent m_5, etc. It will then be necessary to choose which data bits of the 8-bit locations will represent the output functions Z_1, Z_2, etc. For this example a logical choice would be:

Data bit D1 ... Output function Z_1;
Data bit D2 ... Output function Z_2;
Data bit D3 ... Output function Z_3;
Data bit D4 ... Output function Z_4.

Each of these bits of each location would then be "programmed" 1 or 0 depending on whether or not that minterm is present for the particular function Z.

Figure 4.6 sets out the arrangement and Table 4.1 sets out the programming of locations for this problem. Any "don't care" conditions are just that, and the associated data bits may be left in the unprogrammed condition or set to 0 or 1 as convenient.

This approach to realizing combinational logic is viable for multiple output functions in up to (currently) 18 variables and is well suited to "slower" designs based on standard packaged ICs.

However, this method is wasteful of space in a VLSI context and is also slow compared with the *And/Or* logic equivalent. The reason why it is wasteful of space is that all possible minterms are covered and 8-output functions per chip can be accommodated although less may be used. Response time can be more than 10 times that of the equivalent *And/Or* arrangement.

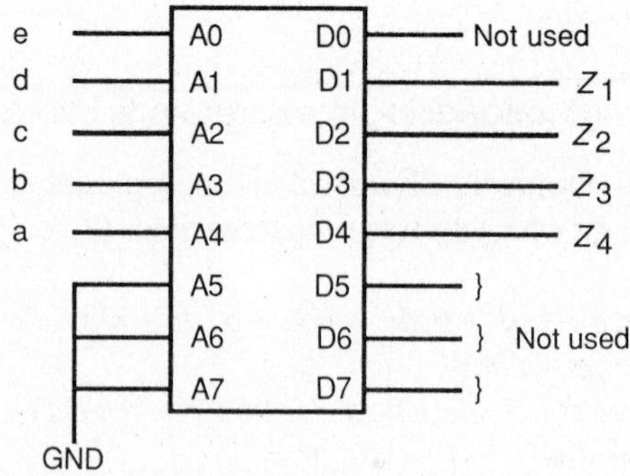

Figure 4.6 Connections to PROM for the example

Table 4.1 for Figure 4.6

Input variables	Location (address)	Data stored D1 D2 D3 D4
a'b'c'd'e'	0	1 1 1 0
a'b'c'd'e	1	0 1 0 1
a'b'c'd e'	2	0 0 0 0
a'b'c'd e	3	1 1 1 1
a'b'c d'e'	4	0 0 0 0
a'b'c d'e	5	0 0 0 1
a'b'c d e'	6	0 0 0 0
a'b'c d e	7	1 0 1 1
a'b c'd'e'	8	0 0 1 0
a'b c'd'e	9	0 1 0 1
a'b c'd e'	10	0 0 1 0
a'b c'd e	11	1 1 1 1
a'b c d'e'	12	1 0 1 0
a'b c d'e	13	1 0 1 1
a'b c d e'	14	1 0 1 0
a'b c d e	15	1 0 1 1
a b'c'd'e'	16	0 0 1 0
a b'c'd'e	17	1 0 0 0
a b'c'd e'	18	0 0 0 0
a b'c'd e	19	1 0 1 0
a b'c d'e'	20	0 0 0 0
a b'c d'e	21	1 0 0 1
a b'c d e'	22	0 0 0 0
a b'c d e	23	1 0 1 1
a b c' d'e'	24	0 0 1 0
a b c'd'e	25	0 0 0 0
a b c'd e'	26	0 0 1 0
a b c'd e*	27	1 0 1 0
a b c d'e'	28	1 0 1 0
a b c d'e	29	1 0 1 1
a b c d e'	30	1 0 1 0
a b c d e	31	1 0 1 1

4.4 Using multiplexers (MUXs) to realize combinational logic

A multiplexer (MUX) is a standard circuit arrangement, available in all forms of packaged integrated circuits and readily configured in custom VLSI designs, which acts as a multiposition switch and can be set to any position by the "SELECT" input lines. A basic 8-way multiplexer is illustrated in Figure 4.7 together with a truth table and an equivalent circuit. (For the "purists", many multiplexers also have a complementary output and some forms have a "strobe" or an "output enable" input).

Multiplexers may be used individually or in arrays to "factor out" some or all variables in combinational logic expressions or problems.

4.4.1 Problems with four variables or less

Take, by way of example, a 4-variable problem defined as:

$$T = f(abcd) = m(0,1,2,4,5,6,7,10,11,13,14)$$

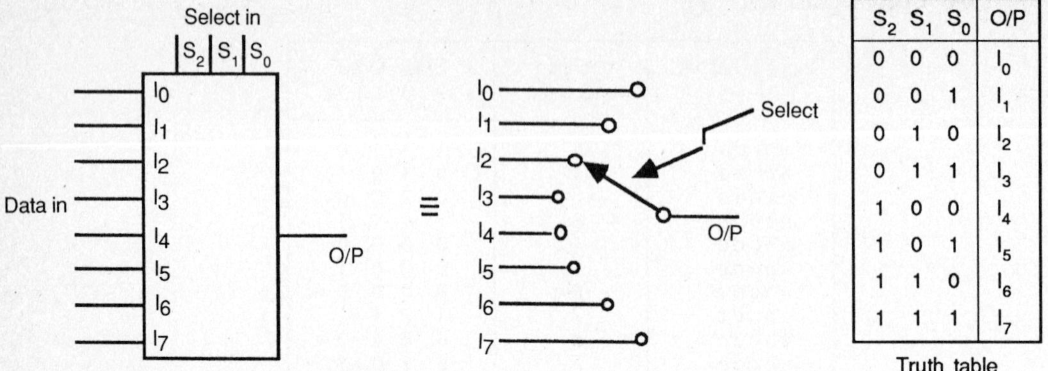

S_2	S_1	S_0	O/P
0	0	0	I_0
0	0	1	I_1
0	1	0	I_2
0	1	1	I_3
1	0	0	I_4
1	0	1	I_5
1	1	0	I_6
1	1	1	I_7

Truth table

Figure 4.7 Eight-way multiplexers

Consider two possible MUX-based implementations of this expression:

(a) based on an 8-way MUX, or (b) based on a 16-way MUX.

Using an 8-way multiplexer (as in Figure 4.8)
An 8-way MUX has three select input lines (2^3=8) which may be used to "factor out" the three most significant variables of the problem, namely *a,b,c*, by connecting them in descending order of significance to select lines S_2, S_1, S_0, respectively.

Clearly now, input I_0 will be selected whenever $a=b=c=0$ ($a'b'c'$) so that both minterms 0 ($a'b'c'd'$) and 1 ($a'b'c'd$) will be covered by input I_0. In this particular case both m_0 and m_1 are present in the expression so that I_0 must be connected to logic 1. Now it follows that input I_1 will be selected whenever $a=b=0, c=1$ ($a'b'c$). This input therefore covers both m_2 ($a'b'c$ d') and m_3 ($a'b'c\ d$). In this problem only m_2 ($a'b'c\ d'$) is present so that input I_1 must be connected to d'. By similar reasoning, I_2, I_3 and I_5 must be connected to logic 1, I_4 to logic 0, I_6 to d, and I_7 to d'. The overall arrangement is then as shown in Figure 4.8 and it will be seen that a very simple implementation is achieved. The circuit comprises one 8-way MUX with, at most, one inverter.

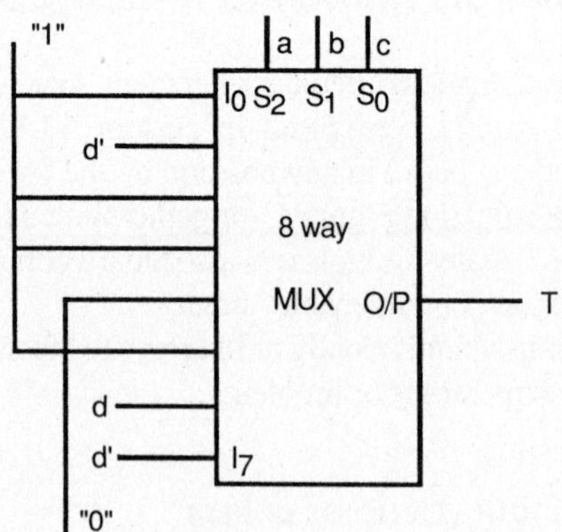

Figure 4.8 Connection to 8-way multiplexer

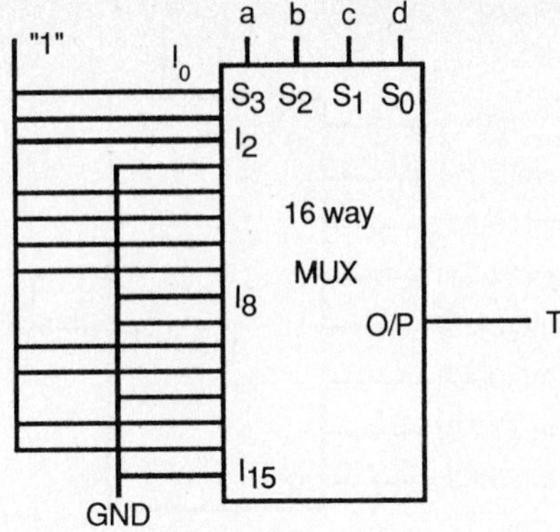

Figure 4.9 Connection to 16-way multiplexer

Using a 16-way multiplexer (as in Figure 4.9)
It stands to reason that a 16- way MUX (having four select inputs) can be used to "factor out" all four variables of this problem. A little thought will reveal that, for the connections given in Figure 4.9, each input I_0 to I_{15} represents a minterm (m_0 to m_{15} respectively). Thus the connections to the inputs are logic 1 or logic 0 as determined directly from the expression for *T*. For this reason, the MUX is sometimes referred to as a "universal logic module" (ULM).

The implementation is illustrated in Figure 4.9 and will be seen to consist of a single 16-way MUX only with inputs connected to 1, where minterms are present.

Exercise:
The reader may wish to consider implementing this same problem with a 4-way multiplexer (two select inputs) and compare the resulting arrangement with those discussed above.

4.4.2 Using MUXs for problems with larger numbers of variables

The approach is best illustrated by means of an example. In this case we will take a 6-variable problem (already presented in Chapter 3, section 3.4.3, as 6-variable example 2) and for which a Karnaugh map simplification and consequent *And/Or* realization were presented as Figures 3.39 and 3.40. For convenience, the problem is repeated here rephrased to require a multiplexer solution.

Example: Find a solution for $T = f\,(ABCDEF)$ using an 8-way MUX plus any necessary random logic where:

$$T = m\,(1,5,6,7,8,9,11,13,15,16,17,18,20,21,24,27,29,30,32,33,34,35,36,40,\\ 42,46,49,50,51,52,53,54,55,59,63), \text{ and,}$$

$$D = m\,(14,22,23,47,56,58,60) \quad [D = \text{don't care}]$$

This is clearly a 6-variable problem which would require a 2^6 = 64-way multiplexer for

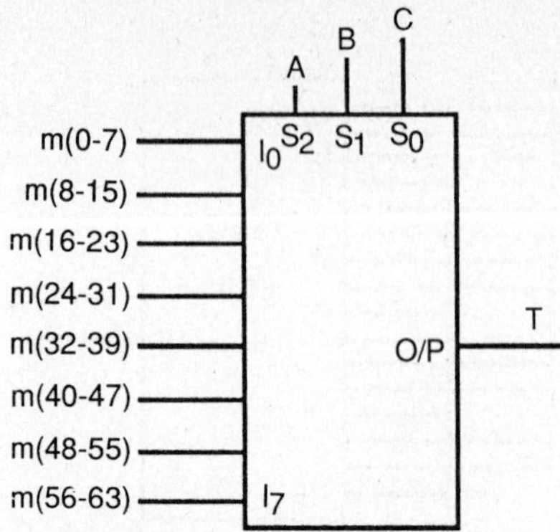

Figure 4.10 Problem partitioning with an 8-way MUX

complete "factoring out" of all variables. Such an arrangement, using a single multiplexer, is impractical for both standard IC and VLSI implementations.

However, the 8-way MUX specified in the question will allow for the "factoring out" of three variables which will partition the 6-variable problem into eight simple 3-variable problems, known as residual functions, one at each of the eight MUX inputs.

Let us consider using the three most significant variables *ABC* to connect to the three select inputs as shown in Figure 4.10.

Now the problem is partitioned since, for example, when $A = B = C = 0$ then input I_0 is selected and connections to I_0 must thus cover all minterms with $A = B = C = 0$ $(A'B'C')$, namely m_0 to m_7 inclusive.

Similarly, input I_1 must cover all minterms for which $A = B = 0, C = 1$ $(A'B'C)$, namely m_8 to m_{15} inclusive. The minterms to be covered by each input are indicated in Figure 4.10. A simple 3-variable problem has to be solved for each input to minimize the necessary logic circuits (Table 4.2).

"Don't cares" are used in the simplification process wherever it is advantageous to do so.

A possible realisation for T is presented as Figure 4.11.

It should be noted that, in general, the random logic (illustrated in Figure 4.11) could be replaced by an 8-way multiplexer at each input.

Exercise: Evolve an arrangement, using four 8-way multiplexers plus the necesary random logic, to realize the multiple output function set out earlier (end of section 4.1 and again at the beginning of section 4.3.2).

4.5. Programmable logic arrays(PLAs) for combinational logic

Programmable logic arrays are basically combinational logic structures for the direct implementation of SOP (*And/Or*) logic expressions.

Basically, the expressions to be realised in the PLA determine its dimensions (or for packaged logic, the dimensions of the PLA determine the size of expression which will fit the PLA).

Table 4.2 Multiplexer input connections for the example

Input	A B C	Minterms covered	Maps for residual functions	Connections to input	Minterms in *T*
I_0	0 0 0	m_0 to m_7	EF: 00, 01, 11, 10 D 0: ^{0}, 1^{1}, ^{3}, ^{2} D 1: ^{4}, 1^{5}, 1^{7}, 1^{6}	DE+E'F	$m_{1,5,6,7}$
I_1	0 0 1	m_8 to m_{15}	EF: 00, 01, 11, 10 D 0: 1^{8}, 1^{9}, 1^{11}, ^{10} D 1: ^{12}, 1^{13}, 1^{15}, $Ø^{14}$	D'E'+F	$m_{8,9,11}$ $m_{13,15}$ d_{14}
I_2	0 1 0	m_{16} to m_{23}	EF: 00, 01, 11, 10 D 0: 1^{16}, 1^{17}, ^{19}, 1^{18} D 1: 1^{20}, 1^{21}, $Ø^{23}$, $Ø^{22}$	E'+F'	$m_{16,17,18}$ $m_{20,21}$ $d_{22,23}$
I_3	0 1 1	m_{24} to m_{31}	EF: 00, 01, 11, 10 D 0: 1^{24}, ^{25}, 1^{27}, ^{26} D 1: ^{28}, 1^{29}, ^{31}, 1^{30}	D'(E'F'+EF) +D(E'F+EF')	$m_{24,27,29}$ m_{30}
I_4	1 0 0	m_{32} to m_{39}	EF: 00, 01, 11, 10 D 0: 1^{32}, 1^{33}, 1^{35}, 1^{34} D 1: 1^{36}, ^{37}, ^{39}, ^{38}	D'+E'F'	$m_{32,33,34}$ $m_{35,36}$
I_5	1 0 1	m_{40} to m_{47}	EF: 00, 01, 11, 10 D 0: 1^{40}, ^{41}, ^{43}, 1^{42} D 1: ^{44}, ^{45}, $Ø^{47}$, 1^{46}	D'F'+ { EF' DE	$m_{40,42,46}$ d_{47}
I_6	1 1 0	m_{48} to m_{55}	EF: 00, 01, 11, 10 D 0: ^{48}, 1^{49}, 1^{51}, 1^{50} D 1: 1^{52}, 1^{53}, 1^{55}, 1^{54}	D+E+F	$m_{49,50,51}$ m_{52-55}
I_7	1 1 1	m_{56} to m_{63}	EF: 00, 01, 11, 10 D 0: $Ø^{56}$, ^{57}, 1^{59}, $Ø^{58}$ D 1: $Ø^{60}$, ^{61}, 1^{63}, ^{62}	EF	$m_{59,63}$ $d_{56,58,60}$

d_{nn} = don't care minterm

A PLA has the dimensions: $v \times p \times z$ (see Figure 4.12) where,

v = number of input variables
p = number of product terms
z = number of outputs.

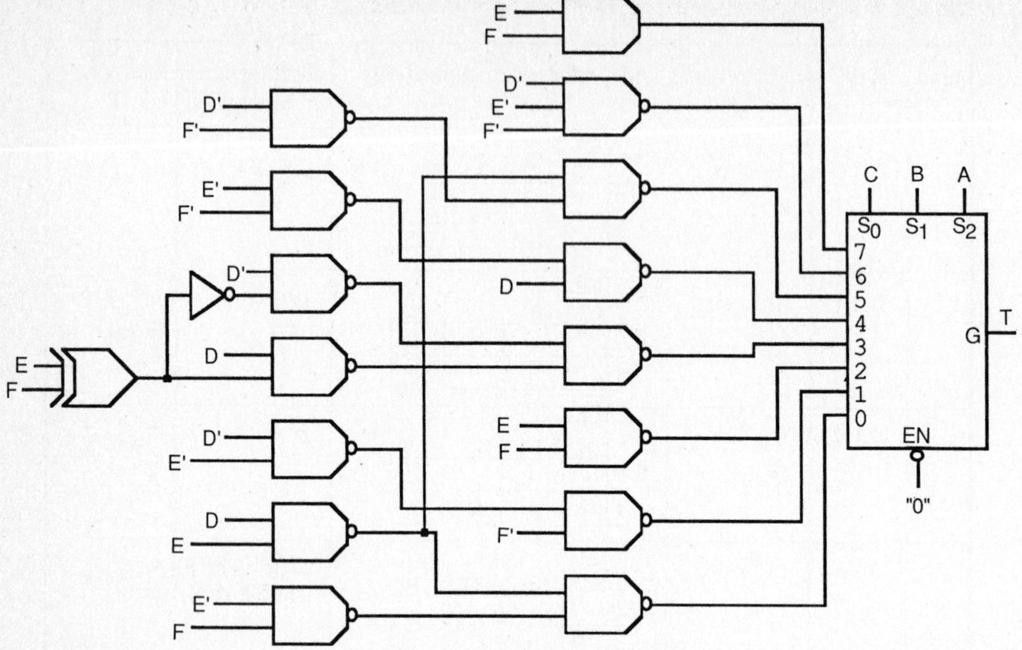

Figure 4.11 Connections of logic circuits for the example*

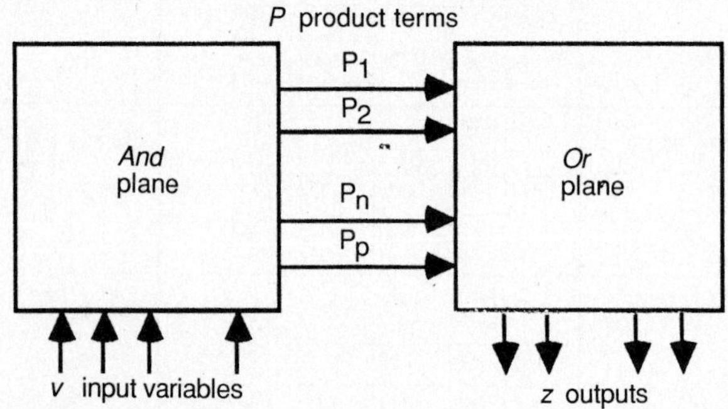

Figure 4.12 Basic PLA structure

Let us first examine the general PLA logic with the aid of a schematic diagram (Figure 4.13). Connections are for the problem specified in this figure. The problem in this case is the previously considered multiple output function of five variables, the simplified form of which involves eight separate product terms from which the four output functions are formed (see page 151). The PLA then has the dimensions 5 X 8 X 4.

* This diagram has been drawn, and the circuit stimulated, using the facilities of "LogicWorks"™ (Capilano Computing) used with a Macintosh Plus™ (Apple Computer, Inc.) personal computer system.

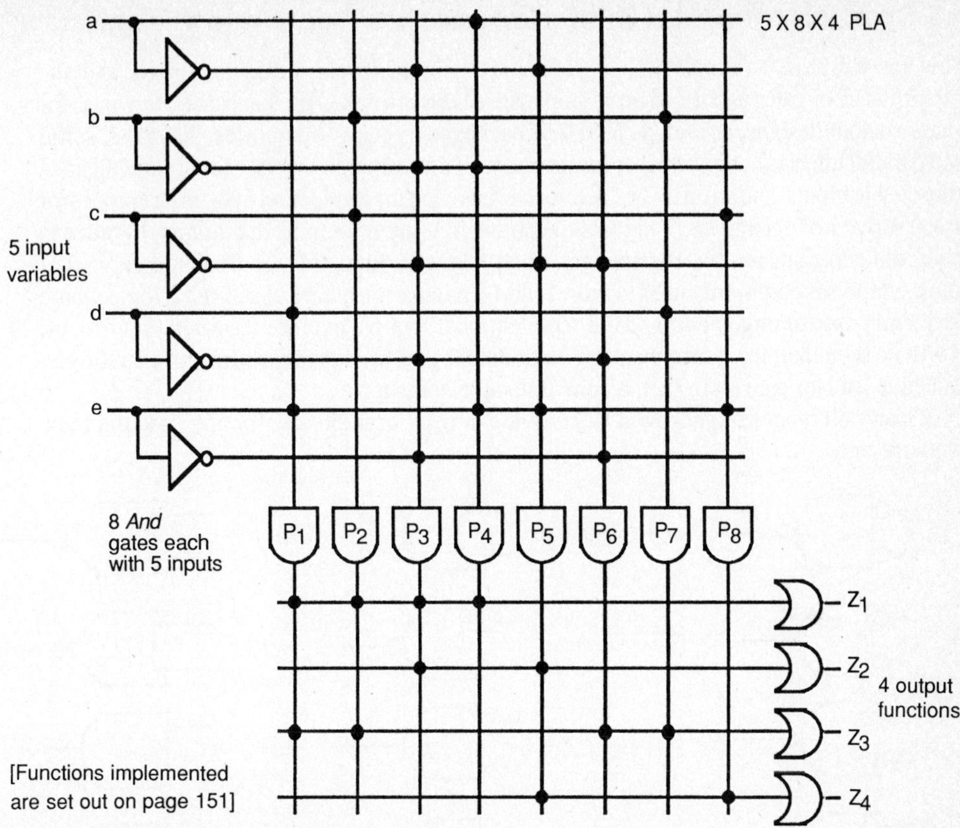

Figure 4.13 PLA for multiple O/P function

Figure 4.13 shows that the five input variables and complements are made available for connection as desired to eight *And* gates (*p*) , each having five inputs (*v*) any one or more of which may be connected so that all five variables may be accommodated. In Figure 4.13, however, only one input is shown in the interests of clarity and up to five connections will be shown to this single input to represent the actual connections.

The output from each product term (*And* gate) is propagated across the inputs to the *Or* gates forming the four (*z*) output functions. Again, for clarity, only one input per gate is shown but this is taken as representing the eight (*p*) inputs, one per product term, which are actually present.

More will be said about PLAs in later chapters of this text but suffice it to say that PLAs are commonly used both as packaged logic and for custom designs of integrated circuits. In the latter case, PLAs are readily designed for both nMOS and CMOS technologies and represent the "regular" type of structure preferred for VLSI designs. When custom design is undertaken, all simplification should aim to reduce the PLA dimensions, i.e. by reducing the number of input variables (*v*), and/or reducing the number of product terms (*p*).(as in section 4.2) and /or by reducing the number of output functions (*z*).

4.6 Symbols for and a brief discussion of "active low" logic

This far we have assumed that logic 1 and Hi logic levels are synonymous and the performance of gates and the logic symbols have been discussed on that basis. However, for many custom designs, as well as MSI/LSI packaged circuitry (memories, decoders, serial and parallel interface chips, microprocessors, etc.) it is common practice for control signals, chip enable inputs, and so on, to be "active Lo"*. For example, the Intel 8080 microprocessor uses "active Lo" IN and OUT signals to control I/O (input/ output) operations. In order to save confusion, and remove the need for much "mental arithmetic", a complementary set of logic symbols is commonly used to more readily indicate the nature of active Lo logic. Some frequently used arrangements and the "conventional" equivalents are shown in Figure 4.14. It will be seen that, for example, a conventional Or gate is an And gate in active Lo Logic; an active Lo Nor gate is, in fact, a conventional Nand, etc.

Clearly all ranges of packaged ICs provide directly or indirectly for any function to be implemented.

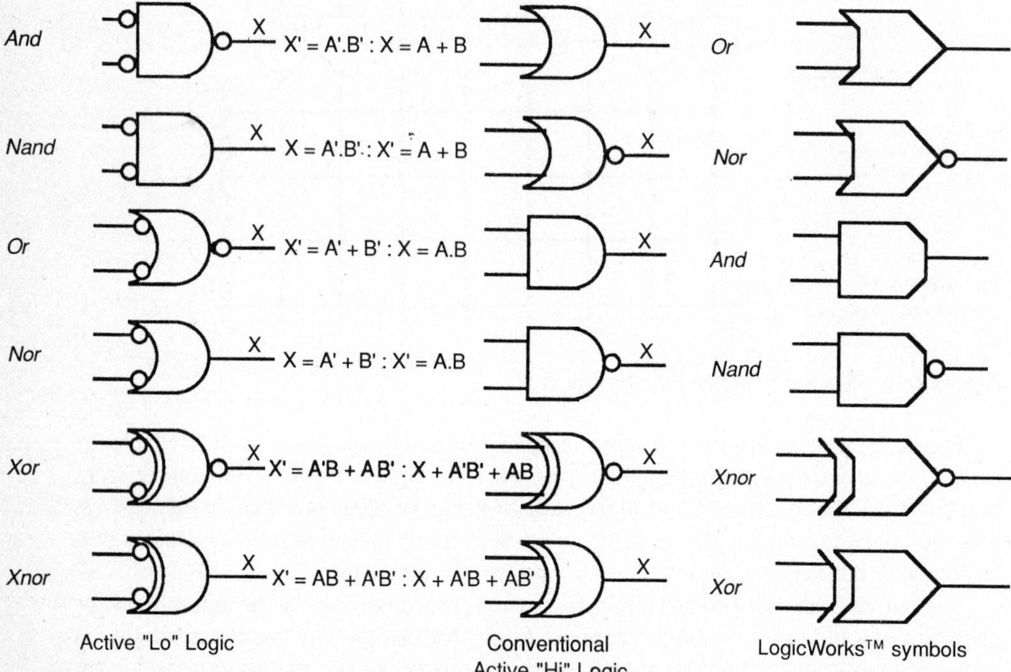

Figure 4.14 Alternative logic symbols

4.7 Summary

We have now established the basic procedures in logic design and used them in this chapter to illustrate possible realisations for combinational logic and the design processes that are relevant. An examination of alternative realisations is usually worthwhile and the most

* For active Hi(conventional) logic logic 1 = +V ; logic 0 = 0V
For active Lo logic logic 1 = 0V ; logic 0 = +V

appropriate choice depends on specification detail and on available technologies. For example, if the utmost in speed is required then random logic (e.g., *And*, *Or*, *Nand*, *Nor* gates) is the most appropriate realization.

In VLSI designs there is a further choice of technology, nMOS or CMOS, and a choice in CMOS of various types of logic circuit.

The next chapter examines some possibilities and puts the design of combinational logic in silicon into perspective through various examples.

4.8 Worked examples

1. A 7-segment display, with segments numbered as follows, is to be used with an 8421 BCD counter and decoder designed to display the numerals 0 to 9 as shown. Design the decoder logic using a random logic approach.

```
   S2
S1| S7| S3   Segment numbering
S6| S5| S4
```

Display 0 1 2 3 4 etc.

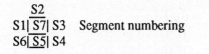

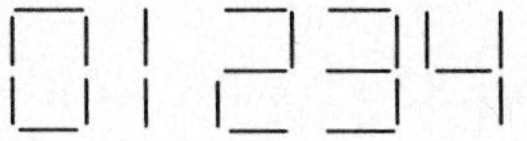

Solution:

First allocate, say, DCBA to the counter output and then draw up a transition table for the count sequence as shown:

D C B A	Count
0 0 0 0	0
0 0 0 1	1
0 0 1 0	2
0 0 1 1	3
0 1 0 0	4
0 1 0 1	5
0 1 1 0	6
0 1 1 1	7
1 0 0 0	8
1 0 0 1	9
0 0 0 0	0 Repeat

Then S1 is on for display digits 0,4,5,6,8,9 mapping gives:

S1 (DC \ BA)	00	01	11	10
00	1			
01	1	1		1
11	Ø	Ø	Ø	Ø
10	1	1	Ø	Ø

From the map S1 = B'A' + CA' + CB' + D

S2 is on for display 0,2,3,5,6,7,8,9

S2 (DC \ BA)	00	01	11	10
00	1		1	1
01		1	1	1
11	Ø	Ø	Ø	Ø
10	1	1	Ø	Ø

From map S2 = B + D+ CA + C'A'

S3 is on for display 0,1,2,3,4,7,8,9

S3 (DC \ BA)	00	01	11	10
00	1	1	1	1
01	1		1	
11	Ø	Ø	Ø	Ø
10	1	1	Ø	Ø

From map S3 = C'+ B'A'+ BA

S4 is on for display 0,1,3,4,5,6,7,8,9

	S4 \ BA	00	01	11	10
DC	00	1	1	1	
	01	1	1	1	1
	11	Ø	Ø	Ø	Ø
	10	1	1	Ø	Ø

From map S4 = B'+ A + C

S5 is on for display 0,2,3,5,6,8

	S5 \ BA	00	01	11	10
DC	00	1		1	1
	01		1		1
	11	Ø	Ø	Ø	Ø
	10	1		Ø	Ø

From map S5 = BA' + C'B + C'A' + CB'A

S6 is on for display 0,2,6,8

	S6 \ BA	00	01	11	10
DC	00	1			1
	01				1
	11	Ø	Ø	Ø	Ø
	10	1		Ø	Ø

From map S6 = BA' + C'A'

S7 is on for display 2,3,4,5,6,8,9

	S7 \ BA	00	01	11	10
DC	00			1	1
	01	1	1		1
	11	Ø	Ø	Ø	Ø
	10	1	1	Ø	Ø

From map S7 = D + BA' + CB' + C'B

Summarizing results:

S1 = B'A' + CA' + CB' + D;
S2 = B + D + CA + C'A';
S3 = C' + B'A' + BA;
S4 = B' + A + C;
S5 = BA' + C'B + C'A' + CB'A;
S6 = BA' + C'A';
S7 = D + BA' + CB' + C'B.

A random logic realization is given as Figure 4.15.

2. For the 7 segment decoder arrangement specified in question 1, design a decoder using a 256 location X 8-bit word ROM.

Solution:

In this case, each of the functions S1–S7 is required in fully expanded (or sum of minterms) form.

In the weighting of inputs DCBA we have the digit being displayed being the same number as the minterm number of the particular combination of DCBA representing that digit. Thus by considering the digits for which each segment must be on (as in Q1) we may write:

S1 = Σm(0,4,5,6,8,9);
S2 = Σm(0,2,3,5,6,7,8,9);
S3 = Σm(0,1,2,3,4,7,8,9);

S4 =Σm(0,1,3,4,5,6,7,8,9);
S5 = Σm(0,2,3,5,6,8);
S6 = Σm(0,2,6,8);
S7 = Σm(2,3,4,5,6,8,9)
[and of course D = Σm(10,11,12,13,14,15)]. (D = Don't care)

Connect the ROM as in Figure 4.16, with DCBA connected to address inputs $A3, A2, A1, A0$ respectively and data bits D1 to D7 providing the decoded outputs to drive segments S1 to S7 respectively.

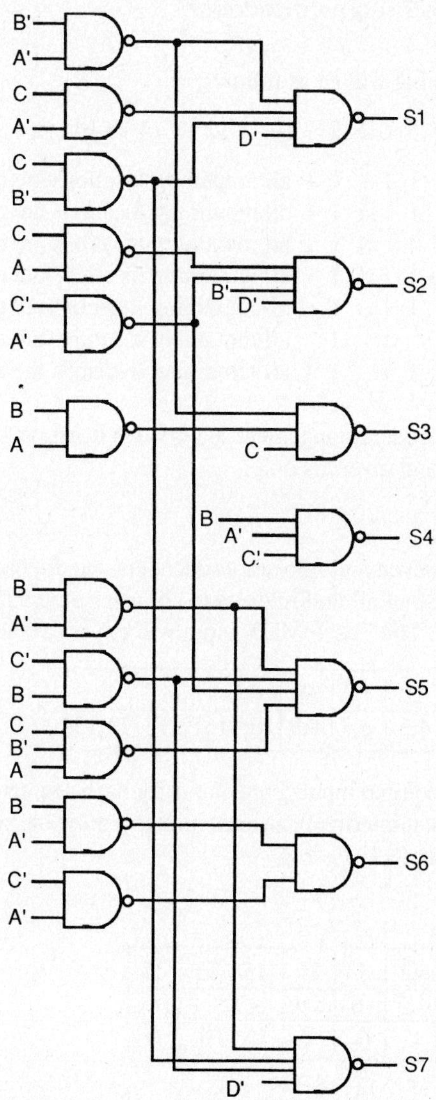

Figure 4.15 Random logic 7-segment decoder

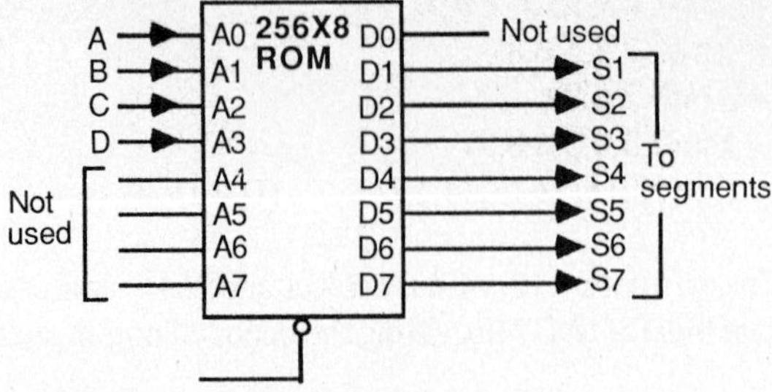

Figure 4.16 ROM based 7-segment decoder

The ROM programming will be as follows:

Location	0	1	2	3	4	5	6	7	8	9	10 11 12 13 14 15 16 etc.
D1	1				1	1	1		1	1	all remaining locations are not used.
D2	1		1	1		1	1	1	1	1	all remaining locations are not used.
D3	1	1	1	1	1			1	1	1	all remaining locations are not used.
D4	1	1		1	1	1	1	1	1	1	all remaining locations are not used.
D5	1		1	1		1	1		1		all remaining locations are not used.
D6	1		1				1		1		all remaining locations are not used.
D7			1	1	1	1	1		1	1	all remaining locations are not used.

3. For the 7-segment decoder arrangement specified in question 1, design a decoder using 8-way multiplexers and inverters only.

Solution:

In this case, we will use seven multiplexers as decoders, one for each segment, and connect the select lines S2, S1, S0 of all multiplexers to counter outputs DCB respectively so that DCB are "factored out". Thus, each MUX input will cover two minterms as follows:

Input	I0	I1	I2	I3	I4	I5	I6	I7
minterms	0,1	2,3	4,5	6,7	8,9	10,11	12,13	14,15

We may now tabulate required input connections for each segment decoder MUX [noting that for any input "In" no minterm =0; an even minterm only =A'; an odd minterm only =A; and both minterms present =1].

S1 MUX

Input	I0	I1	I2	I3	I4	I5	I6	I7
minterms	0	-	4,5	6	8,9	-	-	-
connection	A'	0	1	A'	1	0	0	0

S2 MUX

Input	I0	I1	I2	I3	I4	I5	I6	I7
minterms	0	2,3	5	6,7	8,9	-	-	-
connection	A'	1	A	1	1	0	0	0

S3 MUX

Input	I0	I1	I2	I3	I4	I5	I6	I7
minterms	0,1	2,3	4	7	8,9	-	-	-
connection	1	1	A'	A	1	0	0	0

S4 MUX

Input	I0	I1	I2	I3	I4	I5	I6	I7
minterms	0,1	3	4,5	6,7	8,9	-	-	-
connection	1	A	1	1	1	0	0	0

S5 MUX

Input	I0	I1	I2	I3	I4	I5	I6	I7
minterms	0	2,3	5	6	8	-	-	-
connection	A'	1	A	A'	A'	0	0	0

S6 MUX

Input	I0	I1	I2	I3	I4	I5	I6	I7
minterms	0	2	-	6	8	-	-	-
connection	A'	A'	0	A'	A'	0	0	0

S7 MUX

Input	I0	I1	I2	I3	I4	I5	I6	I7
minterms	-	2,3	4,5	6	8,9	-	-	-
connection	0	1	1	A'	1	0	0	0

The solution is now complete and the circuit is given as Figure 4.17. Note that an inverter is not required if both A and A' are available.

4.9 Tutorial 4

1. Use a 16-way multiplexer (and random logic for input functions) to realize the following six-variable problem.

$$T = c'd'f + ae'f' + b'cf + bce.$$

2. For the problem specified in question 1, indicate the connections to and programming of a ROM to realize the function T. You may assume a 256 location X 8-bit ROM.

3. A 7-segment display, with segments numbered as follows, is to be used with a 4-bit binary counter. Design a decoder to display all the hex numerals 0 to F as shown, using a random logic approach.

```
   S2
S1|S7| S3   Segment numbering
S6|S5| S4
```

0 1 2 3 4 ... etc ... A b C d E F

4. (a) For the decoding requirements specified in question 3, design a decoder using a PLA based approach.
   You may assume that a custom PLA design is possible. You are to complete your design by setting out the circuit diagram for the PLA for this problem.

   (b) Draw a stick diagram for the PLA designed in (a).

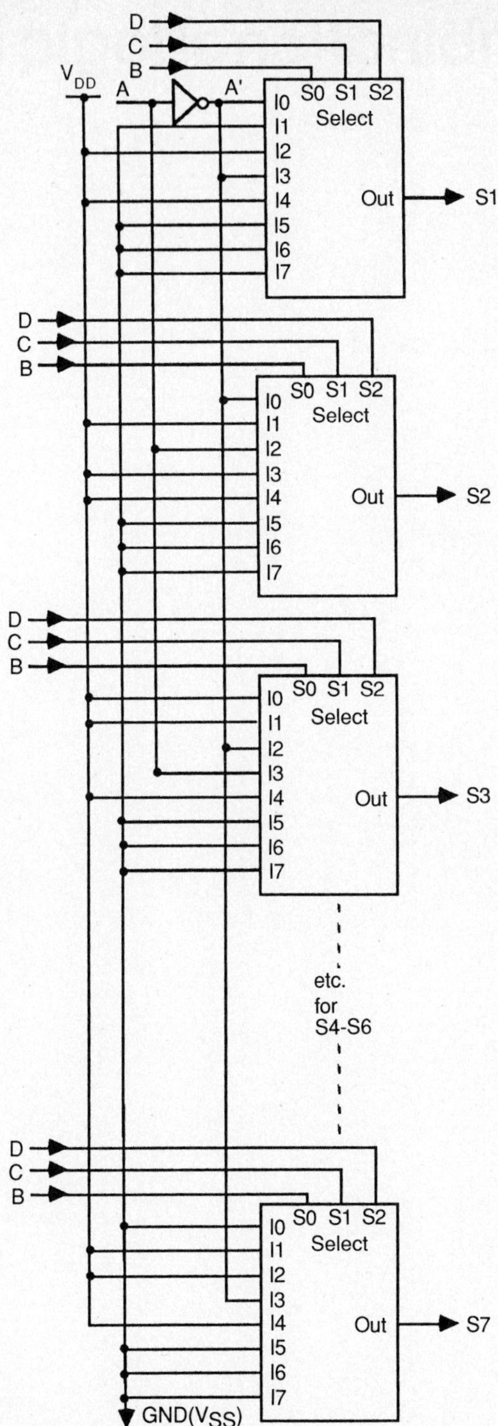

Figure 4.17 MUX based 7-segment decoder

5 Combinational logic in silicon

5.1 Implementing combinational logic functions in silicon

Now that the capability to design circuits in silicon is no longer the exclusive province of the "in-house" semiconductor specialist, it is appropriate to examine some of the relevant aspects in an area in which the designer is offered increasing scope .

5.2 nMOS and CMOS custom design of logic circuits

It is appropriate to examine some aspects of the design of common logic circuits in the two most popular technologies.

In both cases there are two general classes of logic circuit:

1. switch-based logic using pass transistors and/or transmission gates, and
2. inverter based (or restoring) logic.

5.2.1 Switch-based logic

In nMOS and in pMOS technology (since nMOS and pMOS devices are available in a CMOS design) the basic enhancement mode transistor may be used as a series or parallel switching element controlled by an electrically isolated gate. In such applications the transistor is referred to as a "pass transistor" and then, of course, its switching characteristics are of interest in designing logic circuits.

Switching times

A simple model of a pass transistor is set out in Figure 5.1. The time required to "turn on" the gate (G) is determined mainly by the gate-to- channel capacitance C_g and by the output resistance of the input signal source. For example, for a 5-micron feature size nMOS or pMOS transistor of minimum size (as is most often the case) the value of C_g is about .01pF* typically driven from a source of resistance of between 10 kΩ and 80 kΩ. [* See Chapter 1 (section 1.7.3 and 4) for typical parameter values.]

Once the voltage V_{gs} applied between gate and source exceeds the threshold voltage V_t, then conduction may take place between source and drain, and any current flow will be

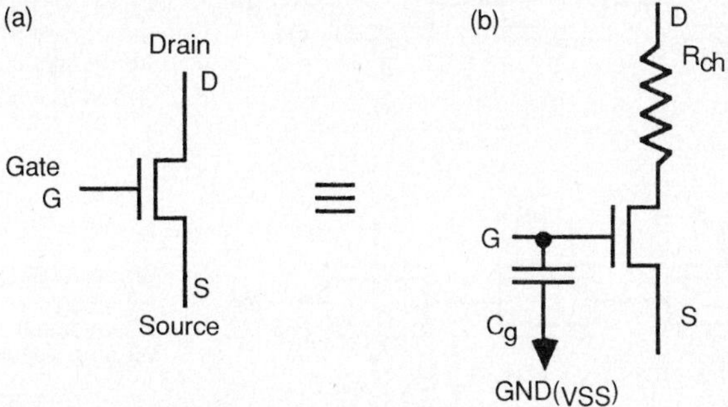

Figure 5.1 Simple model of nMOS pass transistor

through the channel resistance R_{ch}, which for a 5 μm nMOS minimum size transistor (channel length L = channel width W = 5 μm) is typically 10 kΩ. For a similar pMOS transistor, the value is typically 25 kΩ. If such transistors are used as switches, then any capacitive load must charge/discharge through R_{ch}. The resistance of the pMOS device can be reduced to the same value as nMOS by increasing the channel width by a factor of 2.5 (often, 2 is used to maintain whole lambda boundaries) but this will result in increasing C_g by the same factor. Thus, if the capacitive load on the transistor is contributed by the input of another similar device, there will be no net gain in speed (this applies to increasing the width of nMOS transistors also).

Logic level transmission

We have already discussed (section 1.7.4) the characteristics of nMOS and pMOS pass transistors as switches in a "positive active Hi logic" environment, that is:

$$\text{Logic 1} = \text{Hi} = +V; \quad \text{Logic 0} = \text{Lo} = 0V$$

and considered the output voltages obtained through pass transistor switch(es) when the input is a good logic1 and when it is a good logic 0 as in Figure 5.2. We have noted that threshold voltages are such that logic level degradation can typically be in the region of 20 percent and care must be taken in using pass transistor switches in consequence.

In CMOS circuits, we can eliminate these problems by making use of the transmission gate (complementary switch) as a switching element. The properties in this case are also summarized in Figure 5.2.

Clearly this type of switch is superior to either the nMOS or the pMOS pass transistor but, as is always the case, there is a "cost" penalty somewhere, and in this case it is in the area occupied and in the complementary input switching signals required. Because of these factors, CMOS designers may use simple nMOS pass transistor switches when degraded logic 1 and pMOS when poor logic 0 levels are acceptable.

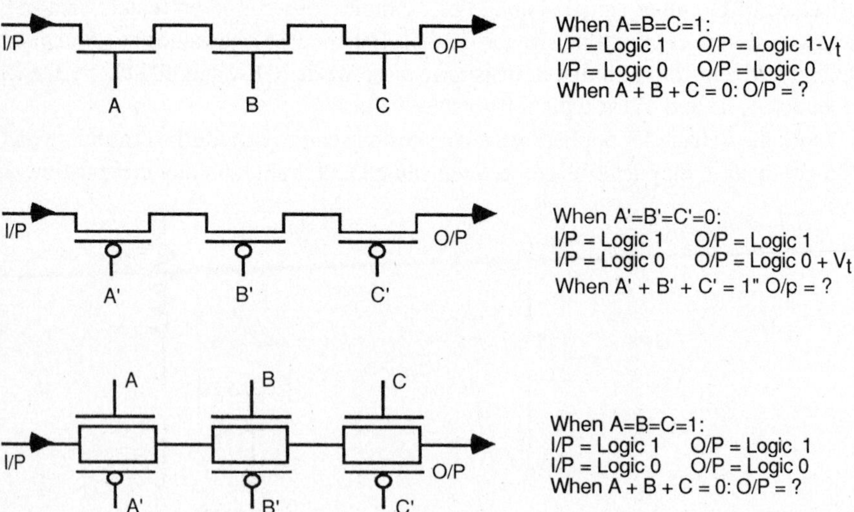

Figure 5.2 Logic level switching by pass transistors and transmission gates

Some common logical arrangements of MOS switches

Switches may be connected in series or parallel so that basic *And* and *Or* arrangements are readily configured. Basic *And* circuits and their properties have already been introduced in Chapters 1 and 3 and in Figure 5.2. The *And* connection in Figure 5.2 illustrates a feature of switch-based logic which must be taken account of in logic design. Note that when the inputs *A*, *B* and *C* are not all equal to logic 1 then there is no connection between I/P and O/P so that the output is left floating. The output will thus retain the value it had when the condition *ABC*=1 was last met, degraded of course by whatever amount the charge on the output capacitance has changed. This is in contrast to the conditions normally experienced in, say, TTL designs in which, if the conditions for a logic 1 output are not met, then the output will be logic 0 and vice versa

Thus, generally, when designing switch-based logic both the logic 1 and the logic 0 output conditions must be designed for.

5.2.2 Complementary switch based logic

To function correctly as an *And* gate, for example, pass transistors or transmission gates would comprise complementary logic circuitry as shown in Figure 5.3 for a 3 I/P *And* gate. *Note also the effect of the choice of type of switch on output logic levels and the fact that both the true and complementary input logic signals (e.g both A and A') are needed.*

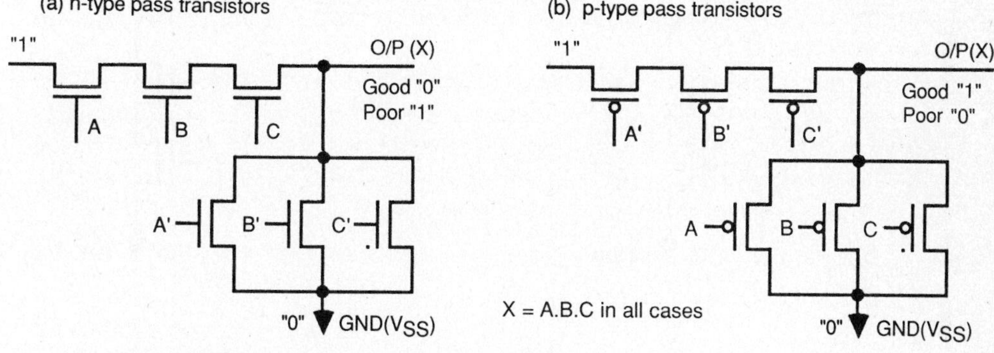

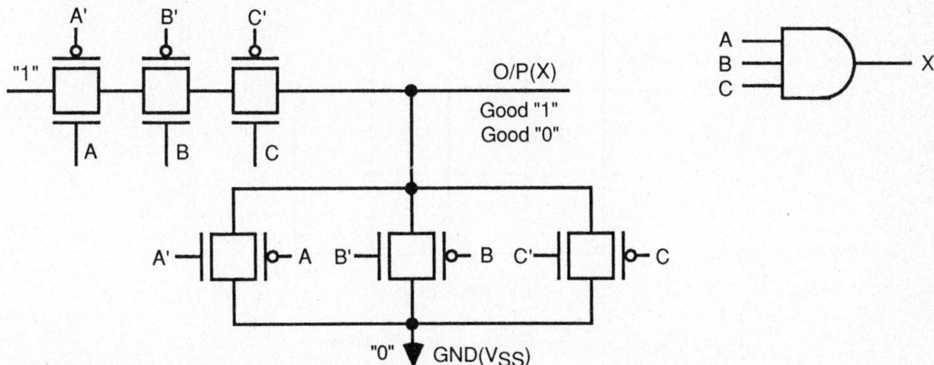

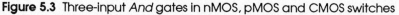

Figure 5.3 Three-input *And* gates in nMOS, pMOS and CMOS switches

An *Or* gate can also be readily configured. It is possible to *eliminate the need for complementary input signals* in a CMOS environment by utilizing both nMOS and pMOS pass transistors. For example, 3-input pass transistor based *Or* gates together with the relevant Karnaugh map are set out in Figure 5.4 (a) and (b). Note the effect of the choice of configuration on output logic levels and that only (a) *A, B* and *C* or, (b) *A', B'* and *C'*, are needed as inputs. Clearly we can also configure the arrangement *using transmission gates as in Figure 5.5*, but in this case, *complementary input signals will be needed.* The logic level outputs will be "fully restored" (i.e. good 1's and 0's) which is also the case for Figure 5.4 (b) but not for 5.4 (a). However, the arrangement of Figure 5.4 (b) comprises half the number of transistors used in the transmission gate based circuit of Figure 5.5 with a consequent saving in area occupied.

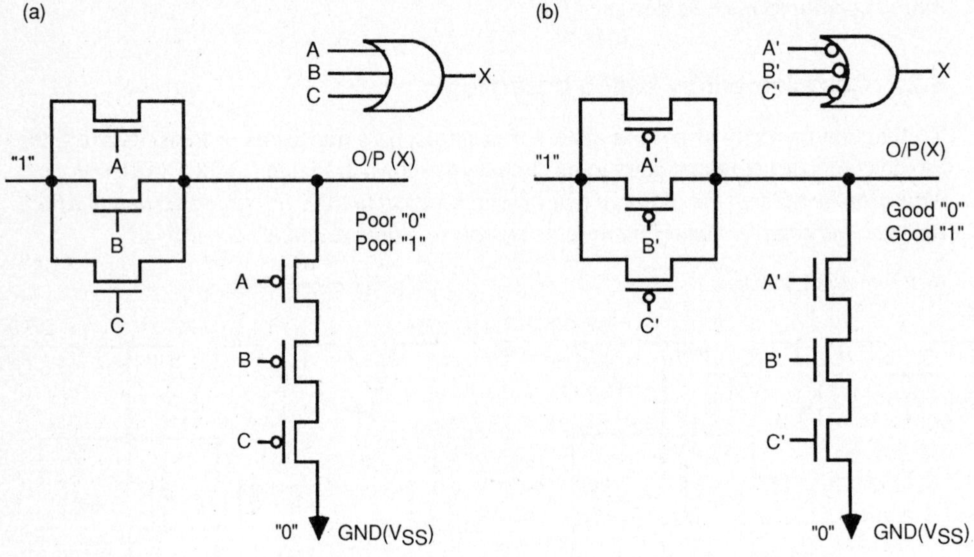

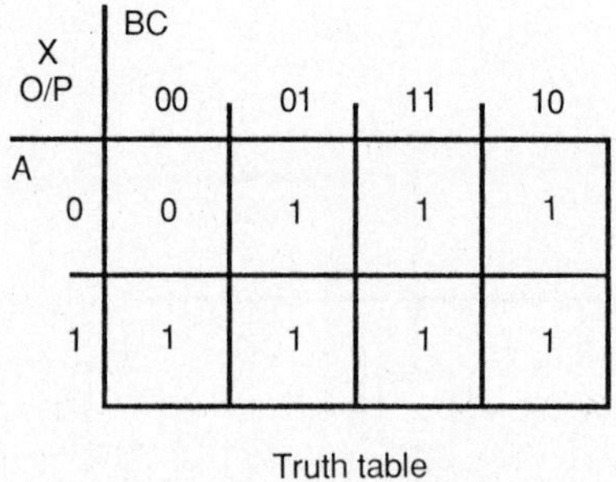

X O/P, A \ BC	00	01	11	10
0	0	1	1	1
1	1	1	1	1

Truth table

X = A + B + C in both cases

Figure 5.4 Complementary pass transistor logic 3-input *Or* gates

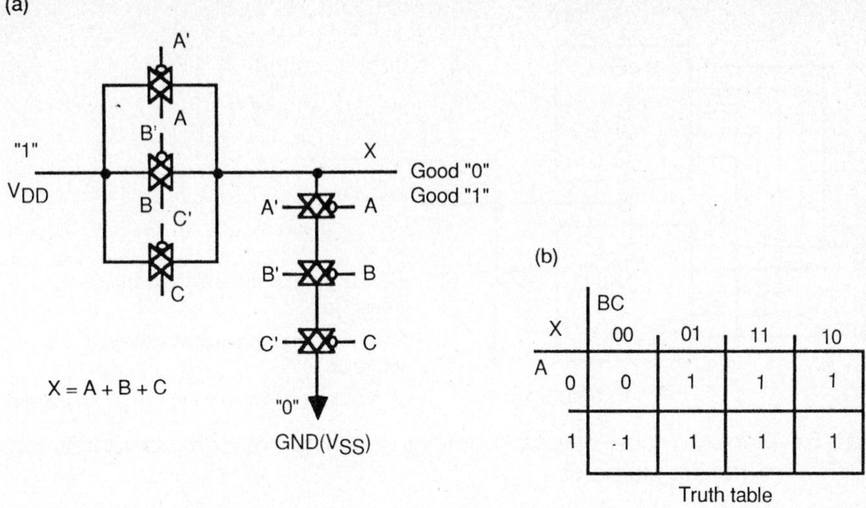

Figure 5.5 (a) Transmission gate 3-input or; (b) Truth table

Now note the correlation between the map and the switch configuration. The 0 entries (one only in this case) define the conditions under which O/P = 0 and translates here to an *And* term $A'B'C'$, which in turn directly correlates with a series connection of switches to logic 0 (GND or V_{SS}). Similarly, the 1s in the map group to $A + B + C$ which clearly defines a parallel connection of switches to logic 1(V_{DD}) in the Figures. Note however that wherever pMOS switches are used each input variable to a switch is complemented since pMOS switches operate on complementary inputs, so that here only A' B' and C' inputs are needed. Note that for transmission gate circuits, both true and complementary signals are needed, thus implying a need for inverters unless both are available.

Before going on to the general design processes for switch-based logic, let us examine switching performance of the alternative *Or* arrangements set out in Figures 5.4 and 5.5 (noting that further arrangements using all nMOS or all pMOS switches are also possible). A general model for complementary logic as in Figure 5.6 helps in this process. The pull-down resistance to V_{SS} (logic 0) in each case is contributed by three switches in series while the pull up resistance to V_{DD} (logic 1) depends on how many switches are on. Since the capacitive load C_L on the circuit will charge/discharge through these resistances then the response time of the logic circuit depends on them. For the circuits of Figure 5.4, for example, it is possible to map the various time constants as suggested in Figure 5.7. Remembering that p-channel devices have 2.5 times the resistance of similarly sized n-channel devices, then anywhere that pMOS switches are in series we will have larger resistances and consequently larger time constants(i.e rise Δ and/or fall ∇ time and propagation delay). It will be seen that the arrangement of Figure 5.4(b) gives better performance; the transmission gate version (Figure 5.5) will have a similar performance but will be generally faster because of the parallel n and p type switches in each transmission gate.

Note that in both cases, there is no static current flow between the V_{DD} and V_{SS} rails since the pull-up and pull-down logic cannot be "on" together. Hence there is no static power dissipation (except for leakage).

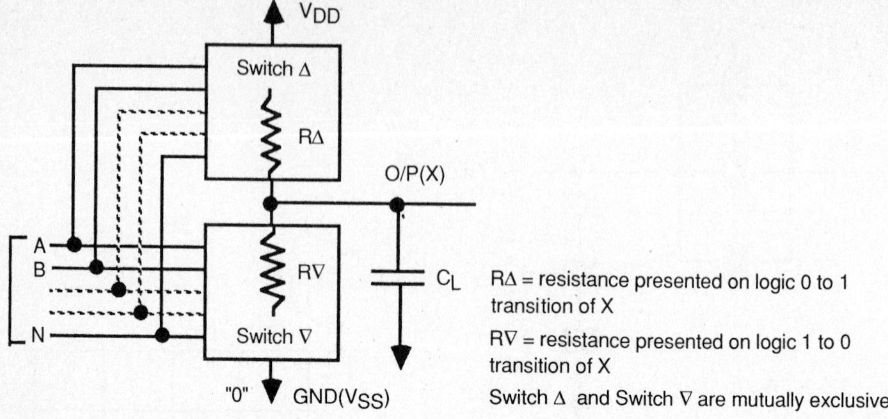

Figure 5.6 General model of logic functions generated by complementary logic switching elements

(a)

O/P X — A \ BC	00	01	11	10
0	∇ = 7.5T	Δ = T	Δ = T/2	Δ = T
1	Δ = T	Δ = T/2	Δ = T/3	Δ = T/2

(b)

O/P X — A \ BC	00	01	11	10
0	∇ =3T	Δ = 2.5T	Δ = 1.25T	Δ = 2.5T
1	Δ = 2.5T	Δ = 1.25T	Δ = .83T	Δ = 1.25T

Where: $T = RC_L$; Δ or ∇ indicates direction of O/P transition

R = "On" resistance of min. size nMOS transistor.

All transistors assumed min. size.

Figure 5.7 Switching time for complementary logic 3-input *Or* gates: (a) for Figure 5.4(a), (b) for Figure 5.4 (b)

5.2.3 A general procedure for the design of CMOS complementary logic

The general model for such circuits has already been presented as Figure 5.6 and an alternative form of the model relevant to CMOS can be set out as in Figure 5.8. In order to avoid the degradation of both logic levels (as in Figure 5.4(a)), pull-up logic must be p-type and pull-down logic n-type (assuming +ve V_{DD}). In designing the pull-up and pull-down logic, the safest method to approach all but the simplest functions is to set out a Karnaugh map. An example helps to illustrate aspects of a design process.

Example 1: Design of a CMOS complementary logic Xor gate

1. Map the function, for this example, the exclusive *Or* function of two variables *P Q* as in Figure 5.9. From the map we may deduce the possible circuit configurations. For

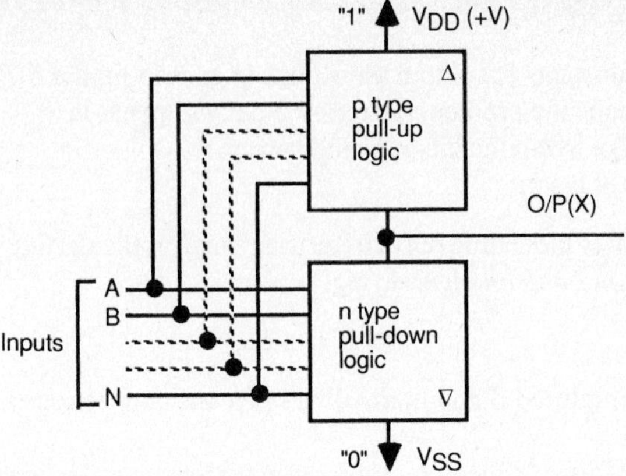

Figure 5.8 Complementary logic model

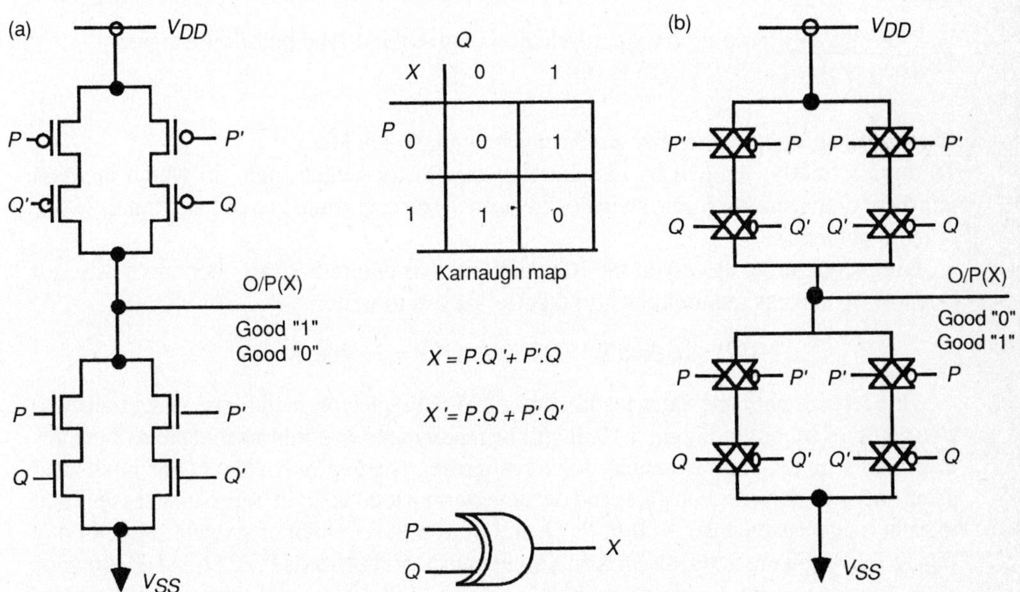

Figure 5.9 CMOS complementary logic *Xor* gate

complementary logic, the 1's in the map give the required connection from output X to logical 1 (i.e.the p-type pull-up logic, remembering that each literal must be complemented for connection to a p-transistor) and the 0's in the map give the connections between the output X and logical 0 (i.e. the n-type pull-down logic, the expressions obtained being directly implemented) as shown (a) for pass transistors only and (b) using transmission gates.

Consider alternative designs (if any).

2. Consider and identify any logic level degradation. In the case of nMOS only and pMOS only, due to the degradation of one or other logic level, the complementary logic

structures, as in Figure 5.3(a) and (b), cannot be used if fully restored logic levels are needed.

3. Estimate time constants for output transitions (e.g. as in Figure 5.7).
4 Partition or reshape the problem if excessive delays appear in 3.
5. Draw up a stick or symbolic diagram and hence
6. Develop the mask layout.

As *an alternative* to the setting out of a Karnaugh map as the starting point in (1) we may *set out the function in Boolean algebra*, e.g:

$$X = P'.Q + P.Q'$$

This (suitably simplified if appropriate) will give the connections to logic 1, i.e. the p-type pull-up logic.

Next, using DeMorgan's theorem, form the complementary expression:

$$X' = P.Q + P'.Q'$$

This (suitably simplified if appropriate) will give the n-type pull-down logic.

Then proceed as before, (2) to (6).

Alternative approach to an Xor gate using selector switch logic

To digress briefly, we will be looking later at selector switch logic, in which the pass transistor or transmission gate switches are used to select or route two or more inputs to one or more output(s).

Let us look more closely at the Karnaugh map in Figure 5.9. An alternative way, for example, to express requirements for the *Xor* gate is to write:

$$\text{If } P = 0\text{, then, } X = Q; \quad \text{else, if } P = 1\text{, then, } X = Q'$$

This arrangement translates readily into a two-path selector, in this case using transmission gates, as set out in Figure 5.10. It will be readily seen that this arrangement compares favorably with the complementary logic realization, needing only P and Q as inputs and using only two transmission gates and two inverters (a total of eight transistors) as opposed to eight transmission gates with P, P', Q, and Q' as inputs (a total of sixteen transistors) in Figure 5.9(b). Figure 5.9(a) also has only eight transistors but needs P, P', Q, and Q' as inputs.

We must be alert to such possibilities in custom VLSI design as the area implications of alternative structures are most important.

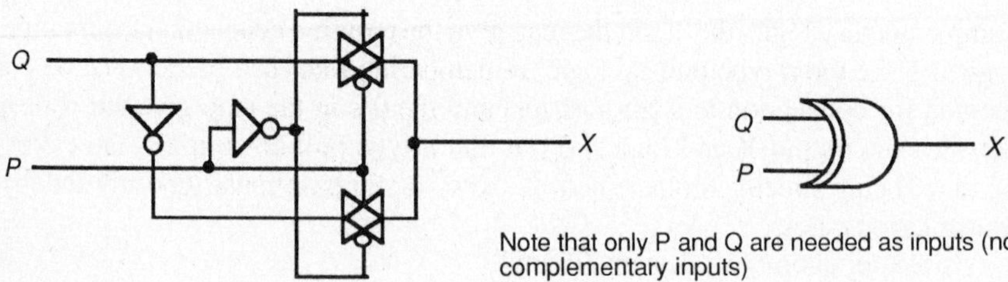
Q
P
X
Q
P
X
Note that only P and Q are needed as inputs (no complementary inputs)

Figure 5.10 Selector logic based CMOS *Xor* gate

Example 2: A logic function in five variables

This example is intended to illustrate aspects of the procedure set out earlier up to and including step (4). The problem may be stated thus:

$$F = A'B + A'E + AB'E' + BC'D'E'$$

1. A map is set out as Figure 5.11

From the map, by grouping the 0's, we may obtain the following expression for the pull-down logic:

$$F' = ABC + ABD + AE + A'B'E'$$

	A'				A				
BC \ DE	00	01	11	10	10	11	01	00	BC
00	0	1	1	0	1	0	0	1	00
01	0	1	1	0	1	0	0	1	01
11		1	1	1	0	0	0	0	11
10	1	1	1	1	0	0	0	1	10

Figure 5.11 Map for example 2

In logic, this would imply four paths in parallel with 3, 3, 2 and 3 switches in series respectively (a total of 11 n-type transistors).

However, a little algebra will reshape the expression as follows:

$$F' = A(B(C + D) + E) + A'B'E'$$

a total of eight n-type switches connected as in Figure 5.12(a).

From the map, by grouping the 1's, we may obtain the following expressions for the pull-up logic:

$$F = A'B + A'E + AB'E' + BC'D'E'$$

or an alternative grouping gives:

$$F = A'B + A'E + AB'E' + AC'D'E'$$

The second form is preferred since it allows reshaping as follows:

$$F = A'(B + E) + AE'(B' + C'D')$$

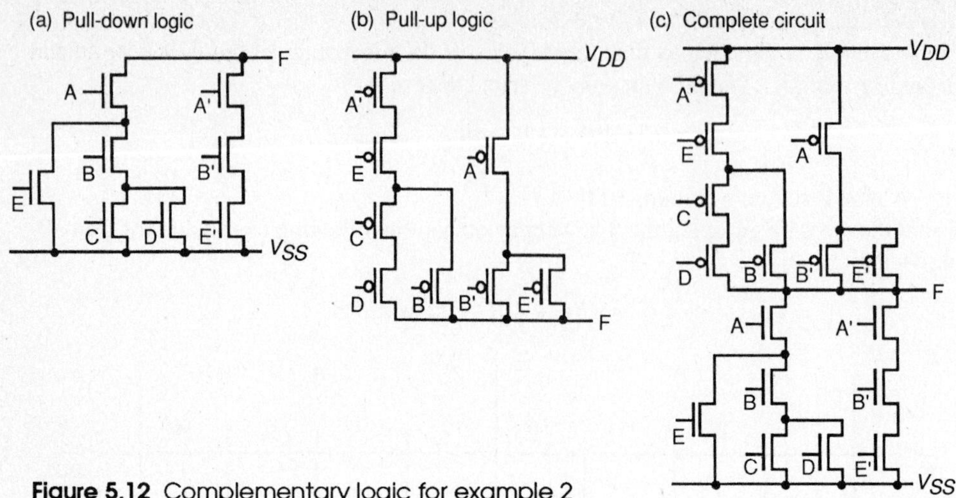

Figure 5.12 Complementary logic for example 2

a total of eight p-type switches connected as in Figure 5.12(b), remembering that literals are complemented for connection to p-type switches. Putting pull-down and pull-up logic together we have the final circuit of Figure 5.12(c).

2. *Consider and identify any logic level degradation.*
There will be none since all pull-up switches are p-type and all pull-down switches are n-type.

3. *Estimate time constants for output transitions.*
In this case it is readily seen that the worst pull-up condition will have four p-type switches in series, and the worst pull-down condition will have three n-type switches in series. Assuming all minimum size transistors, the worst pull-up and pull-down time constants (TC) are thus $10T$ and $3T$ respectively, where $T=R.C_L$ and R is the "on" resistance of a minimum size, $W = L$, nMOS transistor.

4. *Partition or reshape the problem if excessive delays appear in (3).*
For this design the long pull-up time (TC = $10T$) predicted in (3) might well be unacceptable. Consider the implemented expression:

$$F = A'(B + E) + AE'(B' + C'D')$$

We may write:

$$F = A'F_1 + AF_2$$

where $F_1 = B + E$ and $F_2 = E'(B' + C'D')$.

The function $F = A'F_1 + AF_2$ is readily implemented as in Figure 5.13(a) and will be seen to need $F_1{}'$ and $F_2{}'$ as inputs. Using algebra or by mapping we may obtain expressions for $F_1{}'$ and $F_2{}'$ as follows:

$$F_1{}' = B'E' \quad \text{and} \quad F_2{}' = E + B(C + D)$$

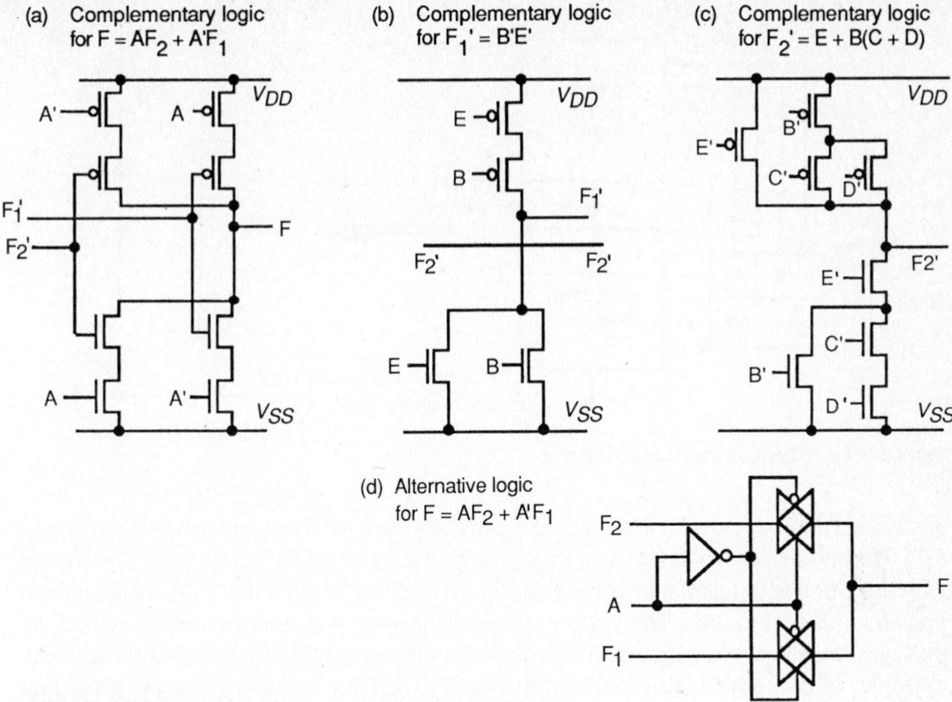

Figure 5.13 Alternative realization of example 2

These expressions are implemented in Figures 5.13(b) and (c) respectively.

A further possibility is to implement the expressions for F_1 and F_2 (not their complements) and then implement the function $F = A'F_1 + AF_2$ using transmission gates as in Figure 5.13(d). Again, the configuration used is that of a selector switch (controlled by the variable *A*).

The mask layouts of color plate 5 illustrate the area implications of Figures 5.13(a), (b) and (c) in this case.

Exercise

Draw up circuit and stick diagrams for the expressions for F_1 and F_2 to complete the arrangement of Figure 5.13(d) for this example.

5.2.4 The complementary (CMOS) inverter and inverter-based logic

The simplest form of the general arrangement of Figure 5.8 comprises a single p-transistor for the pull-up and a single n-transistor for the pull-down transistor, both switched by a single input *A* (say) as in Figure 5.14.

Clearly the arrangement is that of an Inverter as already discussed in Chapter 1. For this reason, logic based on the general arrangement of Figure 5.8 is often referred to as CMOS Inverter-based logic.

Logic based on an inverter leads naturally to *Nand* and *Nor* gates. The circuit arrange

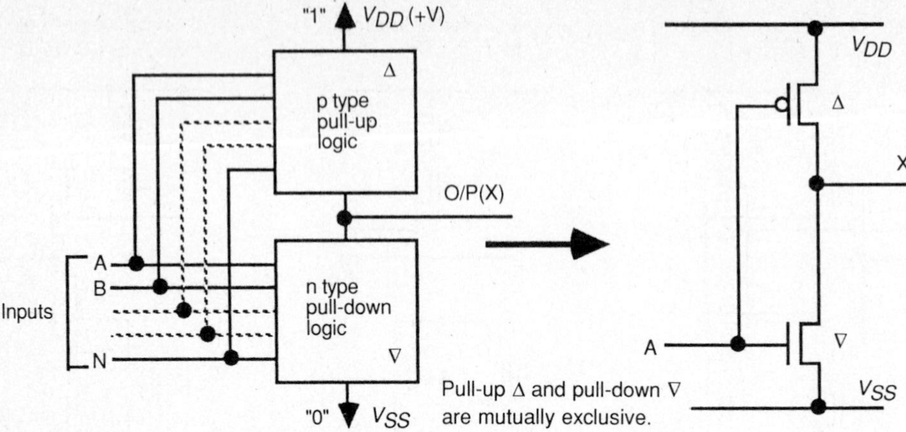

Figure 5.14 Complementary inverter

ments and stick diagrams for a two I/P *Nor* gate and a three I/P *Nand* gate are given in Figure 5.15. Possible mask layouts for these widely used circuits are given in color plate 6, and gates with more or less inputs can be readily configured bearing in mind that a practical guide to preventing excessive time delays is *to avoid more than four transistors in series in either the pull-up or in pull-down structures* (particularly for the p-type pull-up). We have already seen examples of the general design of complementary (inverter-based) logic in the preceding section.

5.2.5 nMOS and pseudo-nMOS inverter based logic (ratio logic)

We have already looked at the arrangement of ratio rules for the nMOS and pseudo-nMOS inverters in Chapter 1 (section 1.7.5) and it is clear that a family of logic circuits may be evolved from each. In doing so it is useful to recognize a general model to which they both conform as set out in Figure 5.16. The essential feature of the model is that there is a permanently "turned on" pull-up which realizes a logic 1 at the output unless it is pulled down by the n-type logic forming the pull-down structure. Clearly, when the pull-down is conducting, a potential divider effect determines the output logic 0 level (which can never be perfect—hence the ratio rules). The output logic 1 level will be good if the pull-up arrangement is chosen to avoid threshold voltage effects. The pull-up for the nMOS logic family (and inverters) is a depletion mode transistor. Depletion mode transistors are normally implanted to have a negative threshold voltage of typically $-0.6V_{DD}$. The gate and source are connected together so that the transistor is fully turned "on". For pseudo nMOS logic (and inverters), the pull-up is a p-type transistor with its source connected to V_{DD} and its gate connected to V_{SS} so that it is fully turned "on". Design then consists of mapping the 0 entries in a Karnaugh map (the complement of the desired logic function) into the n-logic and observing the Z_{pu}/Z_{pd} ratio requirements as illustrated in the examples in Figures 5.17 and 5.18. [When calculating effective pull-up resistances it must be remembered that R_{CHp} (for the pseudo nMOS p-type pull-up) is 2.5 X R_{CHn} (R_{CHn} applies for the nMOS n-type depletion pull-up)]. Note the need to *observe overall* Z_{pu}/Z_{pd} *ratios at each input of parallel*

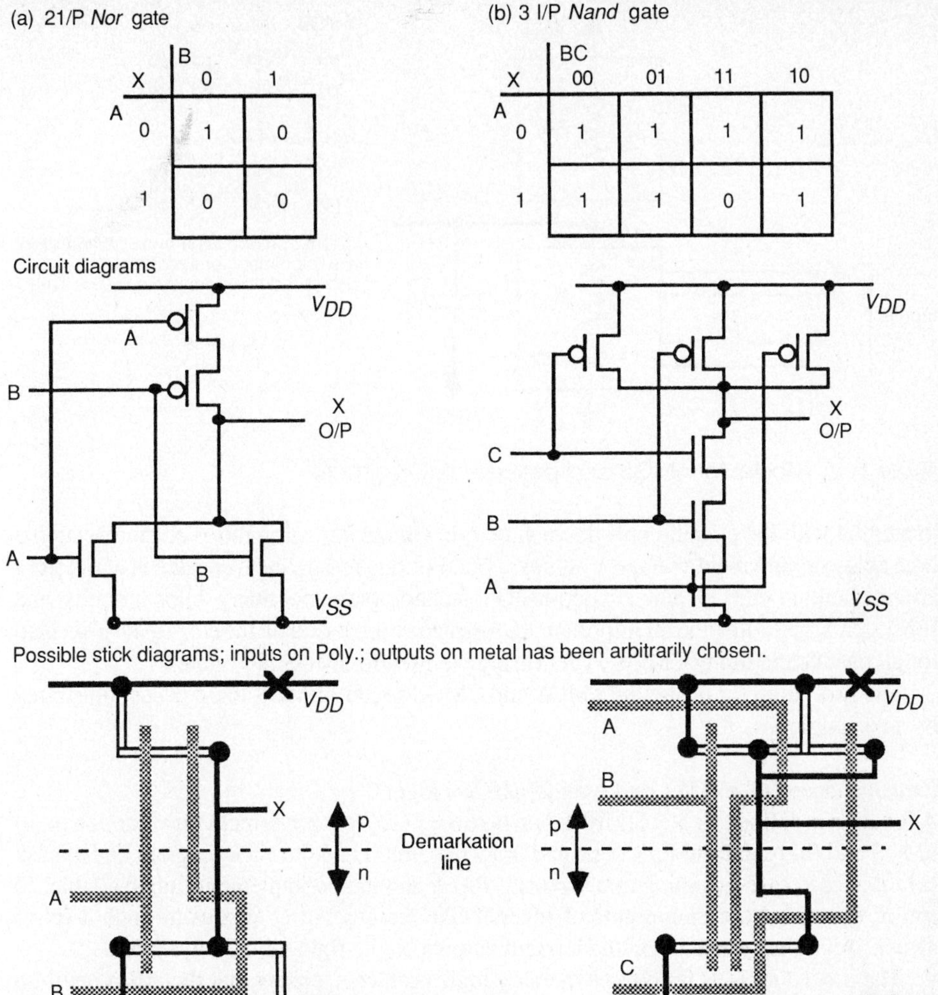

Figure 5.15 Complementary logic *Nor* and *Nand* gates

inputs or for the sum of all inputs in series taking account of the source of the input signal for nMOS logic. Color plate 7 gives possible mask layouts for the *Nor* gates of Figure 5.17.

In assessing the time delays it will be noted that an *n* input *Nand* arrangement (Figure 5.18) implies *n* pull-down transistors in series and consequent slow response. As well as this, in order to maintain the overall nMOS 4:1 or 8:1 or pseudo-nMOS 3:1 ratios, the geometry (usually *W* for the pull-down transistors) increases in size with *n* and the *Nand* arrangement in consequence is *area hungry* as is apparent from the mask layouts of Figures 5.19. Finally, the *threshold voltage* of the top input transistor (Tr3 in Figures 5.18) will be higher than that of the bottom (Tr1) when current flows through the pull-down logic due to the IR drop through the transistors in series and the consequent effect of higher substrate to channel bias. The *Nor* gate on the other hand does not suffer from these problems. Inputs may be added

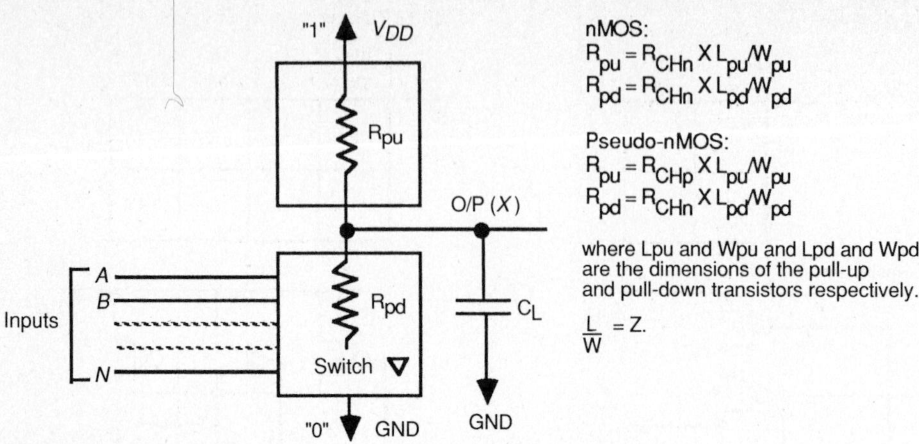

Figure 5.16 Model for nMOS and pseudo-nMOS circuits

in parallel with the existing pull-down structure without adjusting ratios or increasing the time delay or threshold voltage V_t at any input. For this reason the *Nor* gate is a preferred arrangement in nMOS and pseudo-nMOS technologies and many logic circuits and subsystems, including the all important PLA structure are based on the *Nor* gate. This is so for all nMOS and many CMOS PLAs (using the pseudo-nMOS *Nor* gate).

The procedure for designing nMOS and CMOS pseudo-nMOS logic is best illustrated by an example.

Example design of nMOS (and pseudo-nMOS) logic

A function is defined as $S = \prod M(0,1,2)$ where $S = f(PQR)$. Implement this function in (a) nMOS and (b) pseudo-nMOS logic and draw the circuit and the stick diagram and a mask layout in each case assuming that variables *P Q R* and their complements are available. *P'* and *R'* emanate from the outputs of inverter-like circuits but *Q'* comes through a series switch. All inputs are to be on the left and outputs on the right all on polysilicon.

The procedure is as for complementary logic problems, except that there is no pull-up logic to design; but we must note and take account of the required ratios and corresponding W and L dimensions at the appropriate stages of the design).

1. *A map is set out* and the problem entered (maxterms in this case). Design the n-type pull-down structure and the single pull-up, *noting ratios*. Set out a circuit diagram.
2. *Consider and identify any logic level degradation.*
3. *Estimate time constants for output transitions and consider power dissipation.* The worst conditions are usually readily deduced.
4. *Partition or reshape the problem if excessive delays appear.*
5. *Draw up a stick or symbolic diagram.* This may be done by hand or by using a graphics editor or other CAD software.
6. *Develop a mask layout.* The appropriate design rules must be applied and correctness verified (e.g. by close and careful visual checking or by using design rule checking software).

(a) nMOS 3 I/P *Nor* gate

X (A \ BC)	00	01	11	10
0	1	0	0	0
1	0	0	0	0

(b) CMOS pseudo-nMOS 3I/P *Nor* gate

X (A \ BC)	00	01	11	10
0	1	0	0	0
1	0	0	0	0

(i) Circuit diagrams: assume inputs A and B direct from another inverter with input C* taken from a series pass transistor (nMOS) or transmission gate (CMOS).

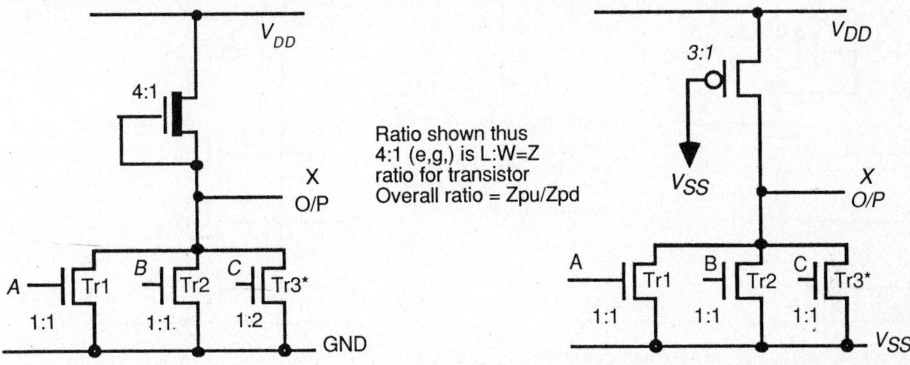

(ii) Possible stick diagrams. Inputs on Poly.; output on metal has been arbitrarily chosen.

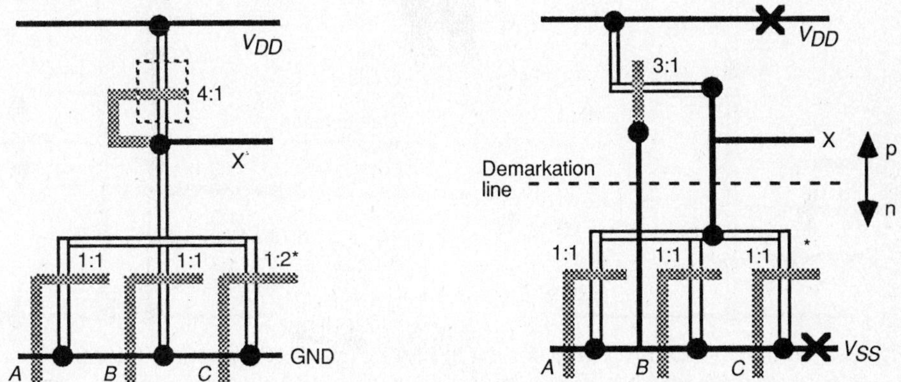

Figure 5.17 nMOS and CMOS pseudo-nMOS *Nor* gate configuration

A solution for the example:

1. *The map for this problem* is set out in Figure 5.20. From the map we may see that:

$$S' = P'Q' + P'R' = P'(Q' + R')$$

The pull-down structure thus consists of three n-type transistors: one in series with two in parallel as shown in the Figures 5.20 (i). This is common to both nMOS and CMOS realizations. The appropriate pull-up structure and the required L:W ratios are readily added in each case as shown.

(a) nMOS 3 I/P *Nand* gate

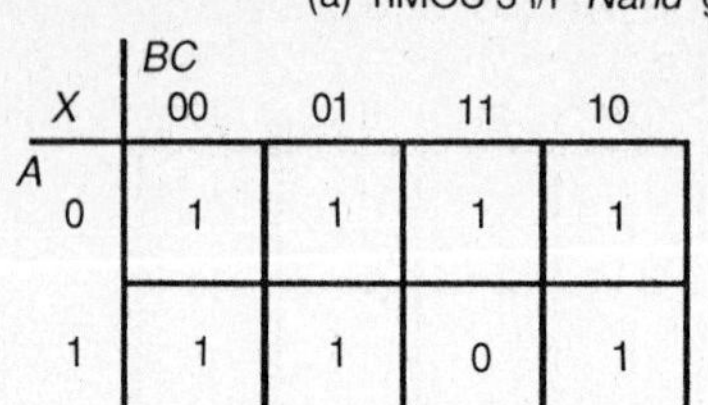

BC
X
00
01
11
10
A
0
1
1
1
1
1
1
1
0
1

(b) CMOS pseudo-nMOS 3 I/P *Nand* gate

A \ BC	00	01	11	10
0	1	1	1	1
1	1	1	0	1

(i) Circuit diagrams. All input signals assumed to come directly from inverter-like outputs.

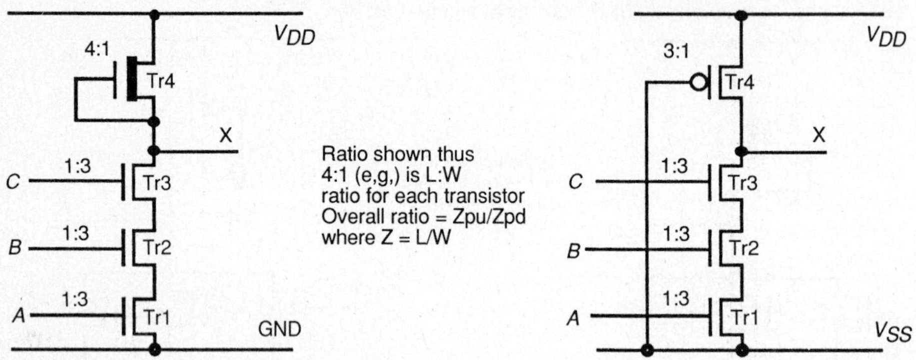

(ii) Stick diagrams. Simple layouts are shown for clarity.

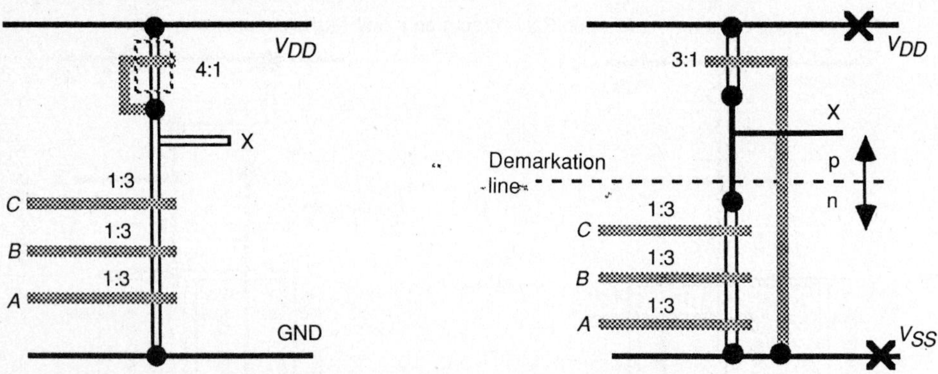

Figure 5.18 Three-input *Nand* gates in (a) nMOS, (b) CMOS pseudo-nMOS configurations

2. *Check for logic level degradation.* There will be degradation of the logic 0 output level in each case but the ratio rules which we are obeying make this acceptable. Good logic 1 levels are produced.
3. *Estimate the worst case time constants.* The pull-up delays into load C_L are (a) $8R_{CHn}.C_L$ for nMOS and (b) $3R_{CHp}\ C_L(=7.5R_{CHn}.C_L)$ for CMOS. The longest pull-down delay is (a) $2R_{CHn}\ C_L$, (b) $2R_{CHn}\ C_L$ (2 n-type in series). The maximum power dissipation is (a) $V_{DD}^2/8.8R_{CHn}$ or (b) $V_{DD}^2/8.25R_{CHn}$ in the logic 0 output state with all three n-transistors "on".
4. There is no way to reduce these by *reconfiguring the circuit* but resistances could be reduced, to reduce delays, by decreasing the *L:W* ratio of all transistors by the same factor, thus maintaining overall ratios. However, this will be at the expense of power dis-

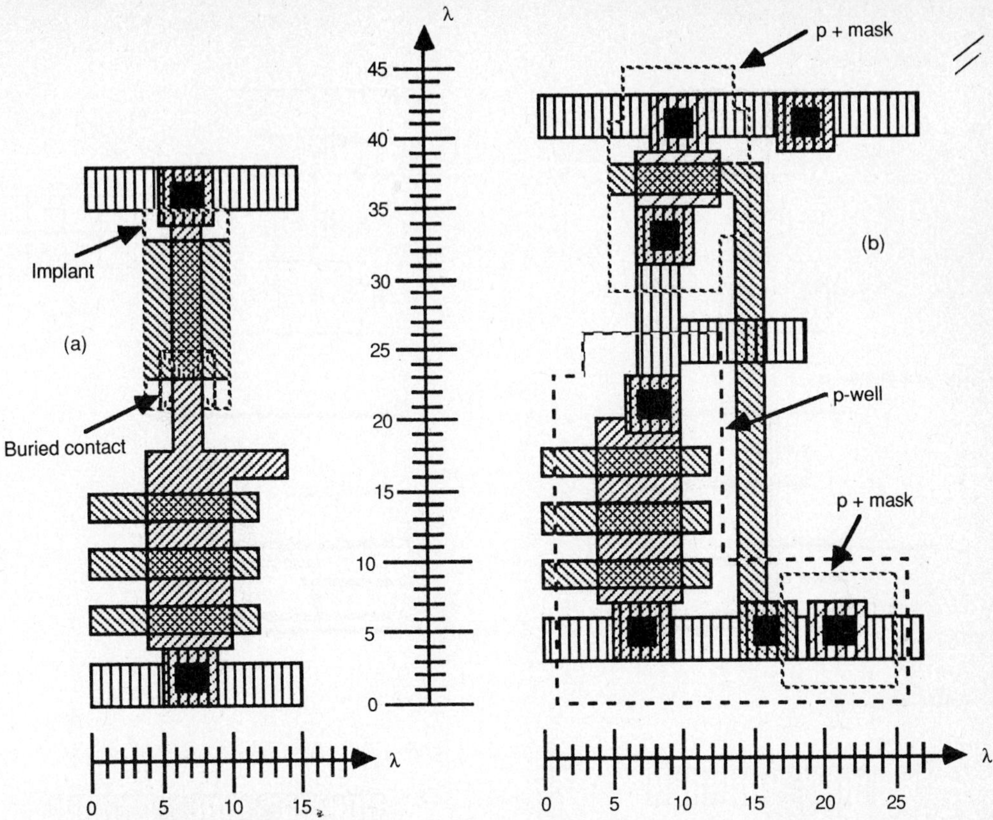

Figure 5.19 Three-input *Nand* gate mask layout: (a) nMOS, (b) CMOS pseudo-nMOS

sipation in the 0 output state and may achieve only marginal speed up if C_L is contributed mainly by the input of similar stages.

5. and 6. *Stick diagrams and mask layouts* are presented as Figure 5.20 (ii) and (iii) respectively. It should be noted that they represent only one of several possible arrangements for this circuit.

5.2.6 The use of bridging switches

It is not intended to treat this topic formally but merely introduce it through an example so that the reader is aware of the possibility.

Example: Consider the 5-variable problem, $F = f(PQRST)$, where:

$$F = \Sigma\, m\,(10,11,12,14,15,18,20,21,22,23,26,27,28,29,30,31)$$

This may be mapped, as in Figure 5.21, and a grouped solution obtained for F' to give the pull-down logic, as follows:

$$F' = P'Q' + R'S' + P'S'T + Q'R'T$$

(a) nMOS

(b) CMOS psuedo-nMOS

(i) Circuit diagrams

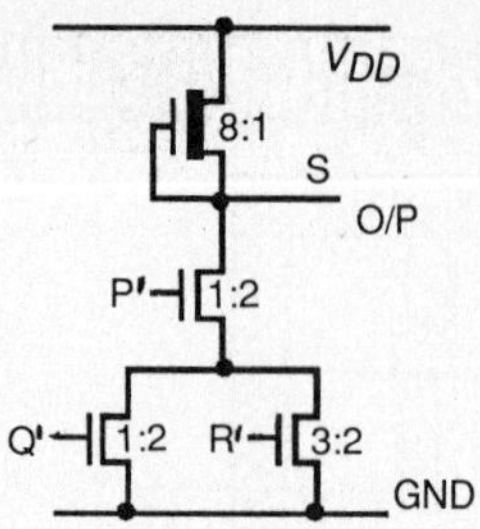

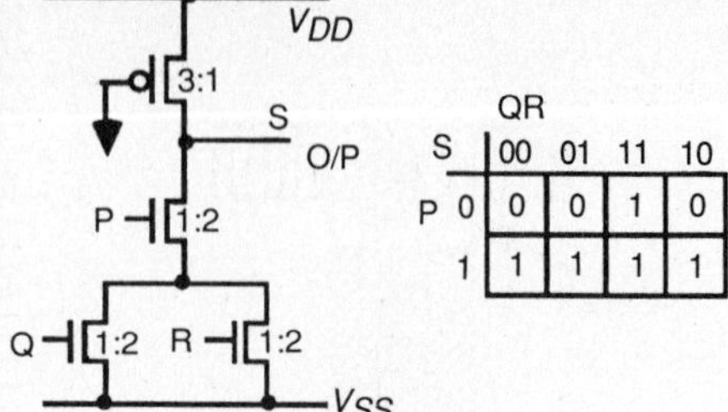

(ii) Stick diagrams

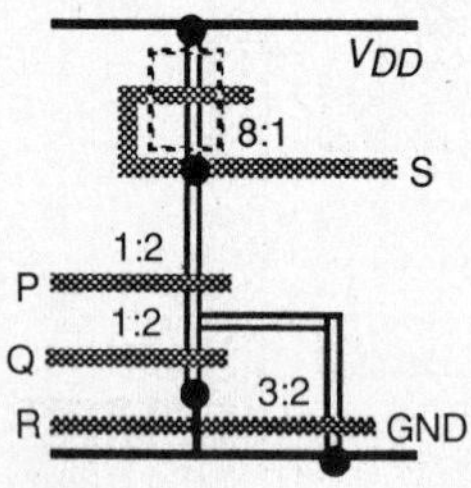

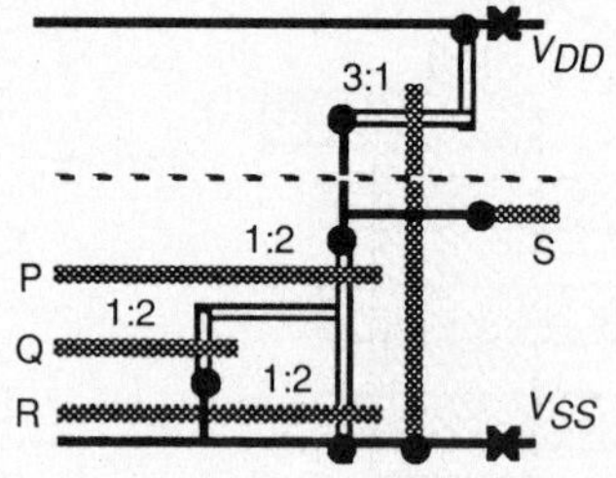

(iii) Mask layouts

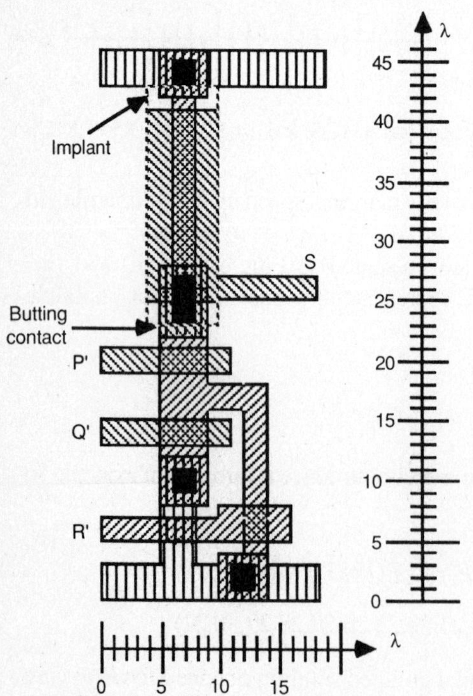

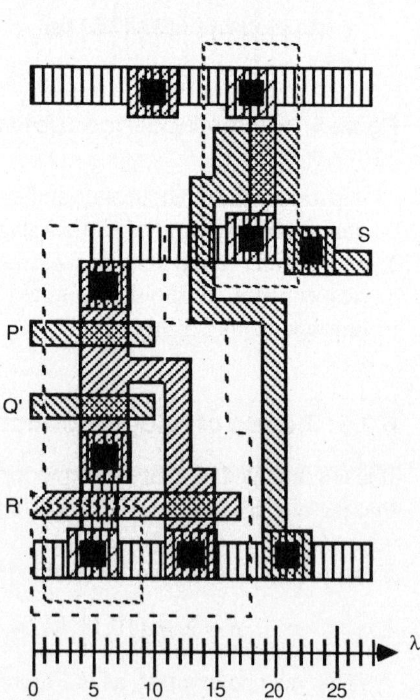

Figure 5.20 Example of nMOS and CMOS psuedo-nMOS logic design

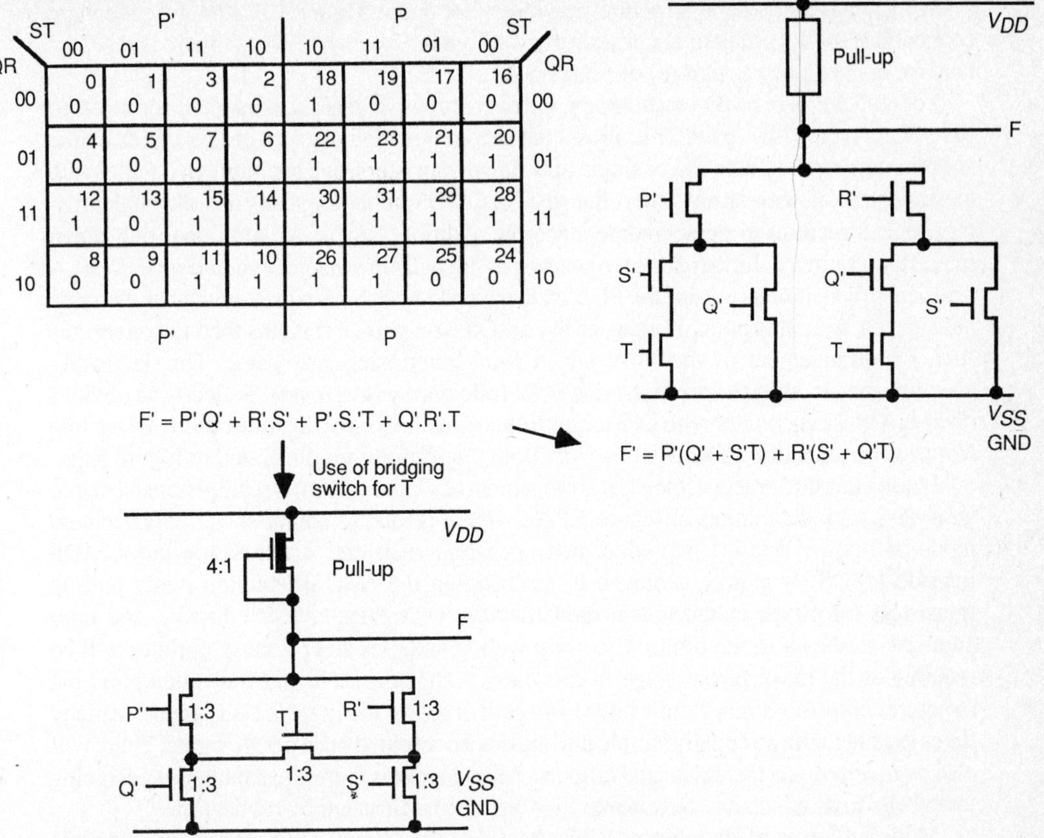

Figure 5.21 Map and logic for bridging switch example

This could be rearranged for example, to reduce the number of switches as follows:

$$F = P'(Q' + S'T) + R'(S' + Q'T)$$

a total of eight switches. However, the use of a bridging switch reduces this to five as shown in the figure. In implementing this circuit in ratioed logic, a compromise must be sought in achieving an acceptable ratio since the overall ratio will vary with the path through the pull-down. However the derivation of the required ratio is such that the operation will be safe if the desired overall ratio or greater is achieved.

An acceptable set of *L*:*W* ratios in this case for a 4:1 nMOS situation is given in Figure 5.21.

5.2.7 PLAs in silicon

The PLA has already been discussed in section 4.5 and its' use, in principle, has been demonstrated with the multiple output function used as an example through various sections of the previous chapter.

A PLA has the general structure previously set out as Figure 4.12 and the schematic connections for the problem in question were set out in Figure 4.13. This Figure is repeated here for convenience as part (a) of Figure 5.22.

For realization in nMOS technology, the only attractive, readily expandable option is to base PLAs on the *Nor* gate. This allows the accommodation of more or less inputs by the simple addition or subtraction of single pull-down transistors, one per input, in parallel with existing input(s), without any other changes. For CMOS technology, any use of complementary logic structures is cumbersome since the addition, say, of an additional pull-down transistor requires a further series transistor in the pull-up arrangement and vice versa. A practical proposition is to base the PLA on the pseudo-nMOS *Nor* gate which has the same advantages, from this point of view, as the nMOS *Nor* gate. It remains then to convert the *And, Or* arrangement of the PLA into a form comprising *Nor* gates. This is readily accomplished for the *Or* operations since *Nor* followed by *Not (Invert)* is clearly an obvious choice. A close encounter with DeMorgan's theorem allows a ready translation of *And* into *Not (Invert)* operations followed by *Nor*. Both translations are illustrated in Figure 5.22.

Putting this into practice, the PLA arrangement of Figure 5.22(a) is readily translated into an nMOS stick diagram as in Figure 5.22(c) which is seen to comprise 1:1 enhancement mode pull-down and 4:1 depletion mode pull-up transistors. A translation into CMOS (pseudo-nMOS) is readily achieved by exchanging the nMOS depletion mode pull-up transistors for p-type enhancement mode devices with gates connected to V_{SS} and ratio adjusted, as shown in the figure. Clearly p-well boundaries and p+ mask outlines will be required at the mask layout stage in this case. Although the layout looks complex, the structure comprises a very simple pull-down cell, as shown in Figure 5.22(d), repeated many times together with an equally simple pull-up device repeated once per *Nor* gate. There will also be *Inverters* on the inputs and outputs. Mask layout is therefore a matter of designing three cells from which any reasonable PLA may be constructed by replication.

Modified forms of PLA are possible, based on the "clocked" or on the "precharged" concepts, and possible arrangements are readily inferred from the following discussions of precharged and clocked logic.

5.2.8 Precharged and clocked logic

Both CMOS and nMOS logic suffer from relatively slow output rise time due in the first case to the high resistance of the p-type pull-up structure, and, in the case of nMOS, due to the ratio requirements which necessitate a relatively high resistance pull-up transistor. In both cases, time is taken to charge the output circuit capacitance which, for example, in the case of circuits driving a number of other inputs or a bus structure may be very significant comprising input and/or wiring and stray capacitances. In order to minimize these delays we might consider the following:

1. *CMOS complementary logic:* Increase the size of the p-type transistors (i.e. decrease the L:W ratio) to reduce resistance values, but, this can only be done at the expense of increased input capacitances for the p-type transistors and there may be little if any net gain. Secondly, increasing the width W of the pull up transistors affects the β_n/β_p ratio and the logic threshold voltage (thus noise immunity).

(a) General arrangement

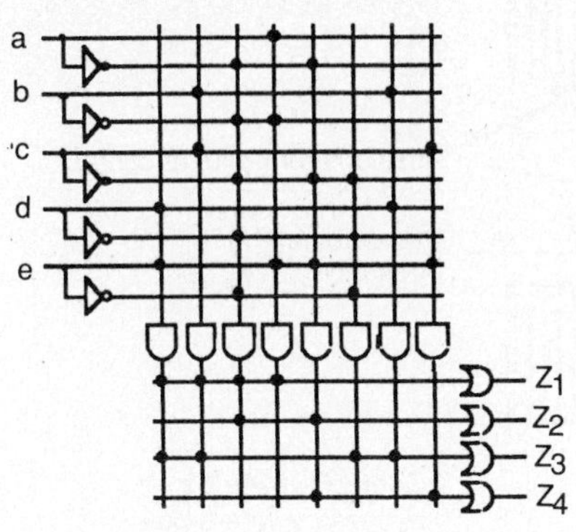

(b) Translation into *Nor* and inverter

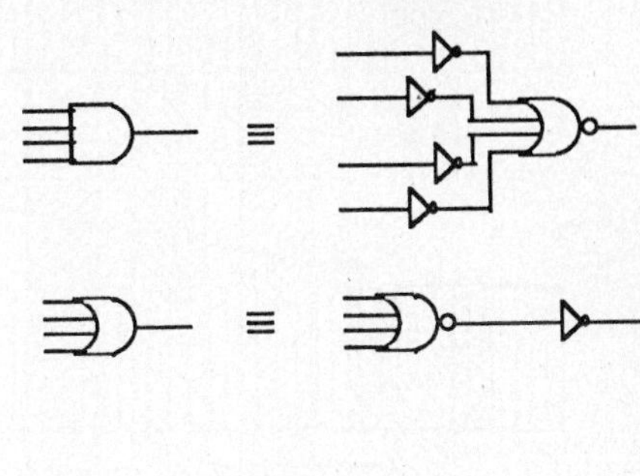

(c) A stick diagram

nMOS structure shown

V_{DD}

3:1

Alternative pull-up for CMOS

V_{SS}

4:1 4:1 4:1 4:1

All pull-down transistors 1:1

a a' b b' c c' d d' e e'

4:1 4:1 4:1 4:1 4:1 4:1 4:1 4:1

$Ø_2$

V_{DD}

$Ø_1$

a b c d e

Z_1 Z_2 Z_3 Z_4

Outputs

GND

V_{SS}

Note: Inputs and outputs may (as shown) or may not be clocked.

Figure 5.22 PLA for multiple output function

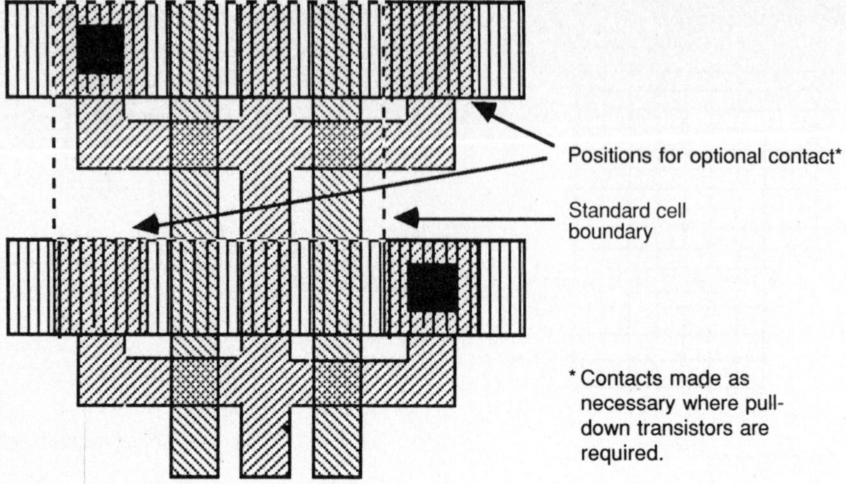

Figure 5.22(d) Possible PLA cell mask layout

A further ploy is to choose those logic structures which have predominantly parallel connections in the p-type pull-up logic, e.g. *Nand* and *Or* rather than *Nor* or *And* type logic.

2. *nMOS or CMOS pseudo-nMOS logic:* Increase the width W of the pull-up transistor by some factor to reduce resistance, but in order to maintain the desired overall Z_{pu}/Z_{pd} ratio the width of each pull-down transistor and, in consequence, the input capacitance must also increase by the same factor. Again there may be no net gain in performance *and the static power dissipation will also rise* due to the decreased total resistance from rail to rail.

In the case of (2) it is possible to reduce the resistance of the pull-up transistor without affecting the size of the pull-down transistors *if we can arrange that pull-up and pull-down are not turned on at the same time when the logic output of the circuit is being read.*

We turn then to the concept of clocked logic and to start our considerations, and with regard to Figure 5.23, consider a logic circuit of the nMOS or pseudo-nMOS type interconnected with a clock signal Ø as shown. When Ø is Lo the pull-down logic is isolated but the pull-up transistor is on and the output capacitance is charged up and, if the pull-up is of low resistance (low $L{:}W$ ratio), the output voltage will quickly reach V_{DD} (less V_t for nMOS arrangement shown).

Ø now goes Hi and the pull-up is turned off while the pull-down logic is evaluated, i.e. connected between output Z and logic 0 [GND(V_{SS})]. The output point either remains at logic 1 due to the precharged voltage at the output or it drops to logic 0 if the pull-down (n-type) logic is conducting due to the combination of its input signals.

Clearly, for a single gate using a single clock signal there is a possibility of having the pull-up transistor initially conducting when the pull-down logic evaluates. This may give rise to a momentarily incorrect evaluation and a consequent transiently incorrect output voltage. However, a far more significant problem arises when we attempt to connect several

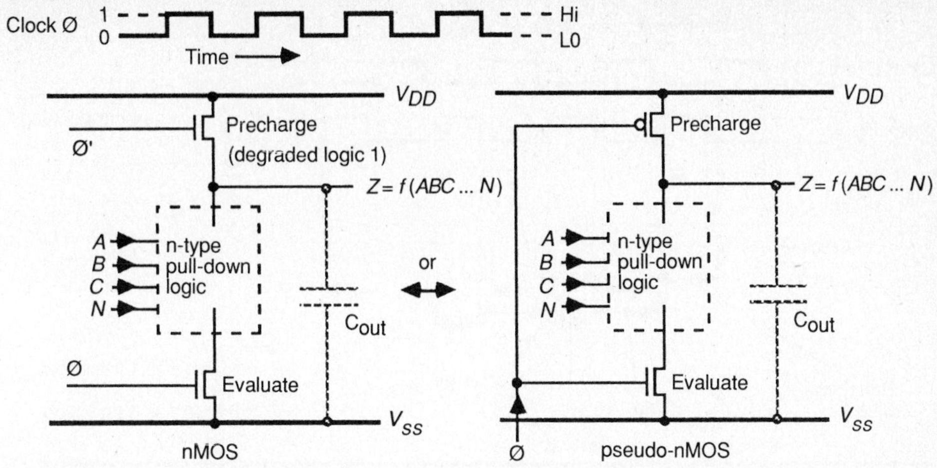

Figure 5.23 Use of clock signal to precharge the output to logic 1

logic gates in series. Clearly, if all series connected gates are evaluating simultaneously, the inputs to a gate are likely to be changing while evaluation is taking place. In consequence, the evaluation may well be incorrect which will also affect further stages, and so on.

We might anticipate an improvement if a 2-phase clock is used as illustrated in Figure 5.24. Here we consider interconnected gates using a 2-phase non-overlapping clock but it may be seen that this arrangement will not work, due to the fact that the precharge period for gate1 overlaps the evaluation period for gate 2. Thus the evaluated output of gate 1is lost before gate 2 can use it as input, and so on for gate 3 from gate 2, etc. Using a transition equation form of notation (see Appendix 2), we may identify the transitions of $Ø_1$ and $Ø_2$ as shown in the figure. A theoretically workable scheme would be to precharge gate 1 in the short period between $\nabla Ø_2$ and $\Delta Ø_1$ and evaluate gate 1 during the Hi state of $Ø_1$. Similarly we would then precharge gate 2 in the short period between $\nabla Ø_1$ and $\Delta Ø_2$ and evaluate gate 2 during the Hi state of $Ø_2$. A third gate would then repeat the conditions for gate 1, etc. However, the periods available for precharge are very short and simple combinational logic gates will not sort out the two separate precharge signals from the 2-phase clock waveforms.

In practice we usually resort to making use of a 4-phase clock and of dynamic storage elements on the output of each logic gate as follows.

A 4-phase clocking scheme for dynamic CMOS logic

The basic pseudo-nMOS gate structure is modified by the addition of an n-type evaluation transistor and a transmission gate at the output which isolates and stores the gate output to following stages. The derived clocks from a parent 4-phase clock are used in such a way that four types of gate structure are possible. They are summarized in Table 5.1., where together with Figure 5.25, the nature of each type should be apparent. The way in which the clock periods are used may be described with reference to the type 3 gate and clock waveforms shown in the figure. The precharging operation commences when $Ø_{34}$ goes Lo, i.e. in coincidence with $\Delta Ø_1$. C_{out1} is precharged and then the transmission gate opens in coincidence with $\Delta Ø_2$ (i.e. when $Ø_{23}$ goes Hi). C_{out2} now also precharges through the transmission gate and the pull-up transistor. Precharging finishes in coincidence with $\Delta Ø_3$

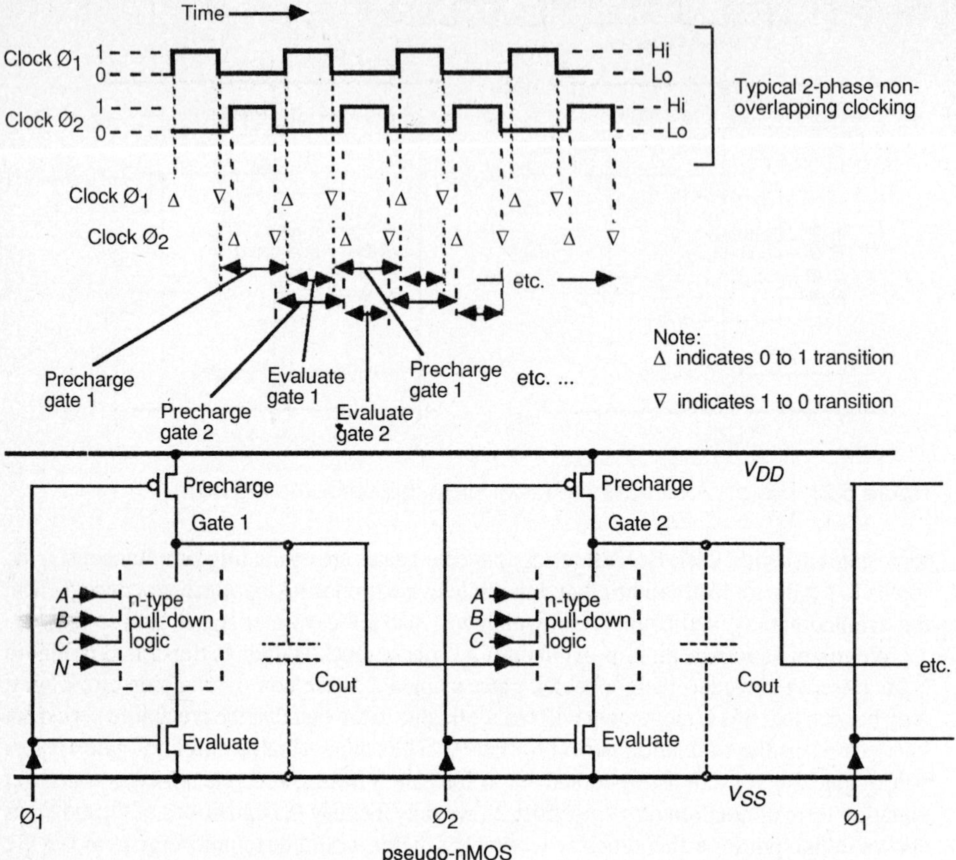

Figure 5.24 Impractical 2-phase clocking scheme for precharged logic

Table 5.1 Phase dynamic logic gate structure ... clocking and interconnection rules

Gate type	Precharge period	Evaluate period	Transmission gate on	O/P data valid	Clock 1	Clock 2	Next type
1	$Ø_{34}=\overline{Ø_{12}}$	$Ø_{12}$	$Ø_{41}$	$Ø_{23}$	$Ø_{12}$	$Ø_{41}$	2 or 3
2	$Ø_{41}=\overline{Ø_{23}}$	$Ø_{23}$	$Ø_{12}$	$Ø_{34}$	$Ø_{23}$	$Ø_{12}$	3 or 4
3	$Ø_{12}=\overline{Ø_{34}}$	$Ø_{34}$	$Ø_{23}$	$Ø_{41}$	$Ø_{34}$	$Ø_{23}$	4 or 1
4	$Ø_{23}=\overline{Ø_{41}}$	$Ø_{41}$	$Ø_{34}$	$Ø_{12}$	$Ø_{41}$	$Ø_{34}$	1 or 2

(i.e. when $Ø_{34}$ goes Hi). The correct output has been evaluated and stored at the ouput by the time $\Delta Ø_4$ occurs. Correct data is therefore available from type 3 gates during $Ø_4$ and $Ø_1$ (i.e. during the entire period of $Ø_{41}$). It will be noted that the output of any one type of gate is thus compatible with the input timing requirements of two other types of gate, this giving the designer greater freedom. Note however that the pull-down time is increased due to the

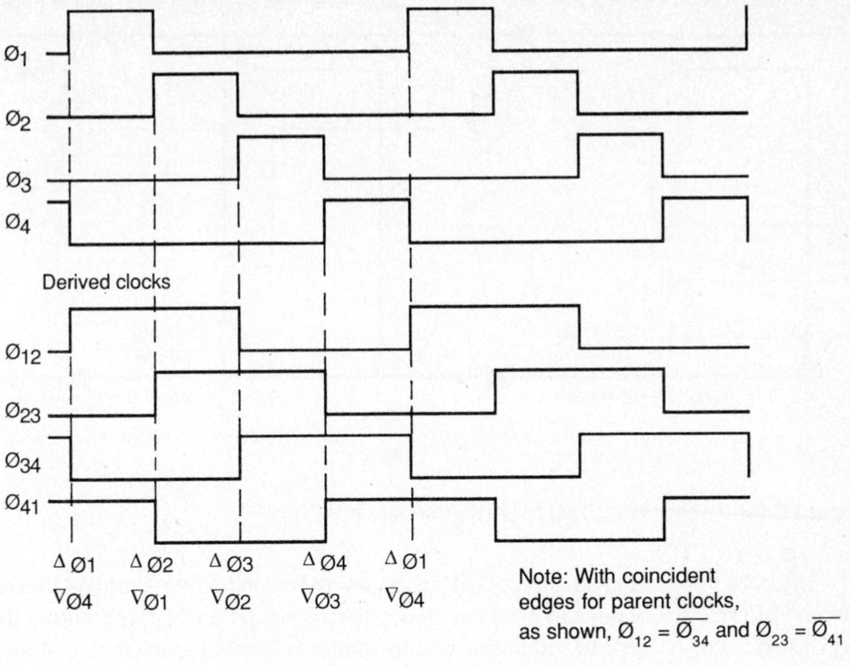

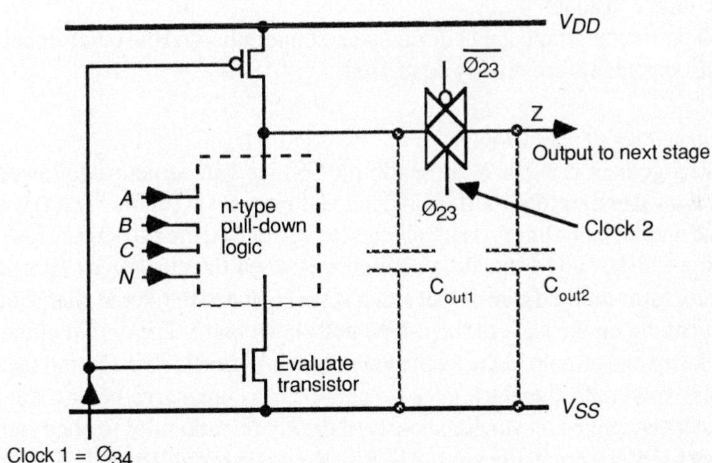

Figure 5.25 Four-phase dynamic logic gate structure (Type 3 for example) (Modified pseudo-nMOS gate)

presence of the extra (evaluation) transistor in series and the propagation delay through the gate will also increase through the presence of the transmission gate at the output. Clearly, compatible inverter circuits are readily configured and it will be realized that an inverter is formed if the n-type pull-down logic of Figure 5.25 becomes a single n-type transistor as in Figure 5.26.

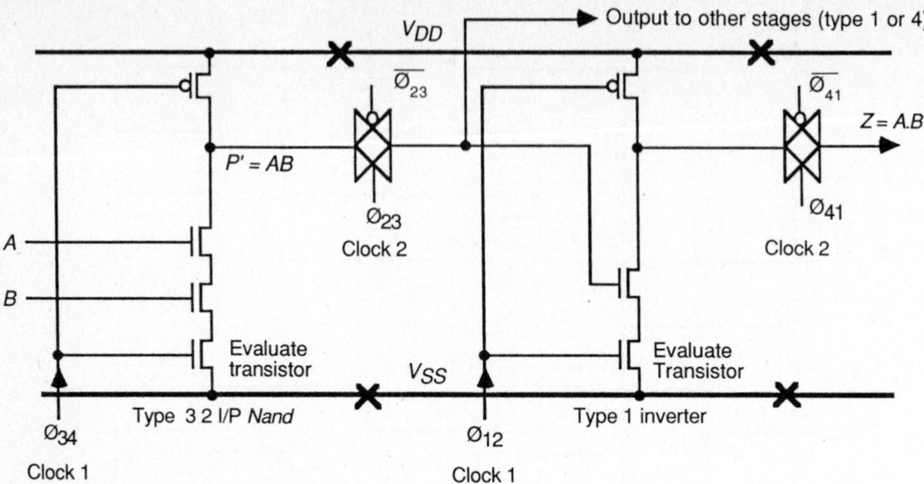

Figure 5.26 Interconnected four-phase dynamic logic

Since the n-logic and the p-type pull-up can never be conducting simultaneously, there are no ratio rule restrictions to obey and generally the n-type transistors forming the pull-down logic will be kept to minimum size to minimize input capacitance and area. The precharge and evaluate transistors, however, can be designed to have lower resistances to reduce time constants.

The particular arrangement discussed is commonly used but other clocking schemes and other arrangements are readily conceived.

Precharged single clock logic

This arrangement consists of a pseudo-nMOS-like gate structure followed by an inverter which is used as a storage /buffer element. With respect to Figure 5.27(a) it may be seen that the gate output, including C_g is precharged to V_{DD} during the Lo period of the clock (Ø'). Precharging finishes and evaluation commences when the clock goes Hi and the evaluation transistor turns on. The gate output either stays Hi due to the stored charge on C_g, or is pulled Lo depending on the state of the n-type pull-down logic. The output of the inverter/buffer will take up and remain in the evaluated state while the clock is Hi, and the output from the buffer is valid until the clock goes Lo and the next precharge begins. Since the p- and n-transistors cannot be on simultaneously, there are no ratio rules to obey and the pull-up can be wide to reduce precharge time. The n-type transistors will usually be of minimum size to reduce input capacitances. Clearly, since cascaded stages all evaluate in the same Hi period of the clock there will be a "ripple through" effect, sometimes known as the "domino" effect as the stages "fall" in sequence, and the clock Hi period must be long enough to allow for this. If logic of this type is to be clocked at low speeds, the inverter /buffer may be readily converted to a latch by the addition of a further p-type transistor, as in Figure 5.27(b), which may be of high resistance since its only purpose is to maintain the stored state of the charge on C_g. A problem to be considered in this and in other types of logic involving charge storage is that of charge sharing. With regard to Figure 5.27(a), note the presence of parasitic capacitance C_p. When the pull-down logic is in the non-conducting state and the gate

(a) Dynamic buffer

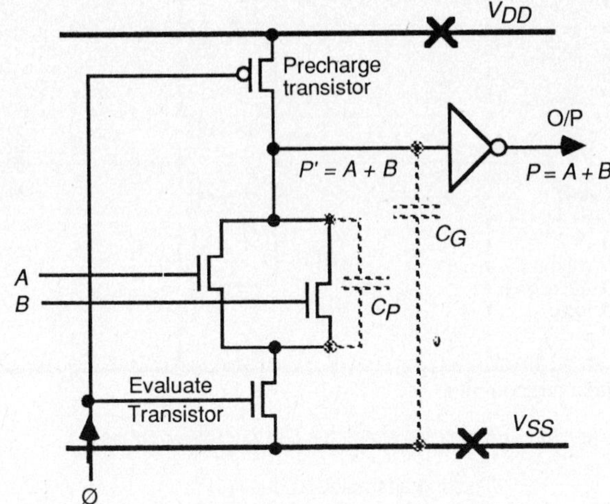

(b) Latched version

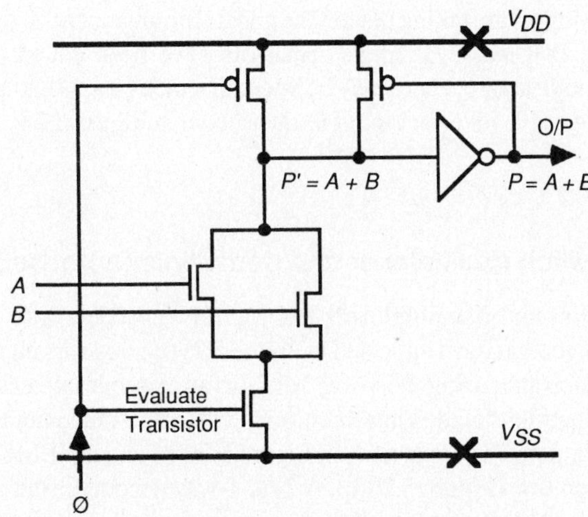

Figure 5.27 CMOS precharged single clock logic 2-input *Or*

evaluates, the stored charge on C_g will be shared between C_g and C_p and if C_p is significant with respect to C_g the precharged Hi level may be destroyed.

Note that the presence of the single inverter means that only *And* and *Or* type logic can be directly implemented.

Clocked complementary MOS logic (C²MOS)

Different clocking schemes may be applied to complementary logic which then becomes clocked complementary MOS logic, i.e. C^2MOS logic. One purpose of introducing clock signals into a logic circuit is to allow the output to be isolated, retaining its current value,

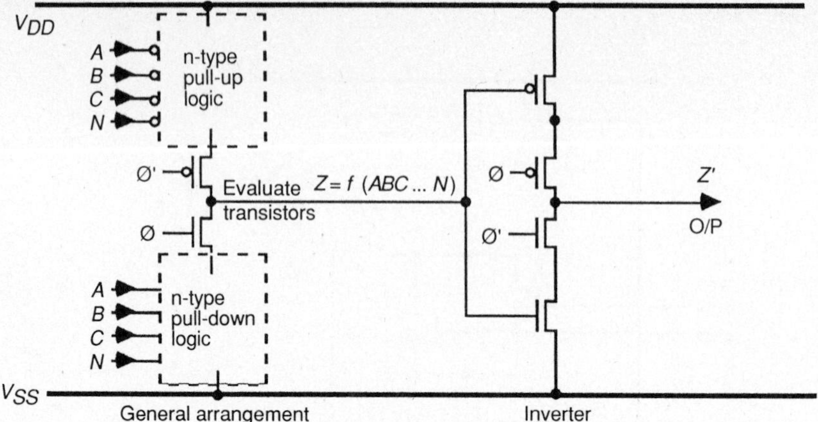

Figure 5.28 C²MOS logic and inverter

while changes to its inputs are taking place. The general arrangement is given in Figure 5.28 and it will be seen that two evaluation transistors have been added to a conventional complementary logic structure. Gates may be interconnected by alternating Ø and Ø' signals as indicated. A compatible inverter circuit is also shown in Figure 5.28. Two-phase clocks may also be used.

5.2.9 Selector switch (multiplexer and demultiplexer) based logic

Multiplexers (MUXs) and demultiplexers (DEMUXs) (decoders) are forms of selector switches having the general form indicated in Figure 5.29(a). Such arrangements are readily implemented in silicon and, using the 4-way MUX as an example, are easily translated into regular (meaning repetitions of the same circuit elements and arrangement) structures. Some possible circuit arrangements are set out in Figure 5.29(b) and a DEMUX is very closely related as can be seen from Figure 5.29(c). A 2- to 4-way decoder is the same arrangement with the input returned to a fixed logic 0 or 1 .

A practical limitation on the number of ways which can be accommodated in a single MUX (orDEMUX) in MOS technology is imposed by the need to limit the number n of pass transistors or transmission gates in series to four or fewer. As n increases so the delay increases with n^2 and acceptable limits for fast operation are exceeded when $n>4$ as we have seen earlier (in Chapter 1).

Stick diagrams and mask layout detail for 4-way multiplexers in both nMOS and CMOS technologies have been included in Figures 5.30 and 5.31 and, although the overall arrangements look a bit formidable, it will be seen that complete layouts are regular structures built up from very simple cells repeated many times so that the amount of detailed design work at the mask level is relatively small. In the case of the n-type circuits, it will be seen that 2-way, 4-way, 8-way and 16-way MUXs can be formed from very simple cells.

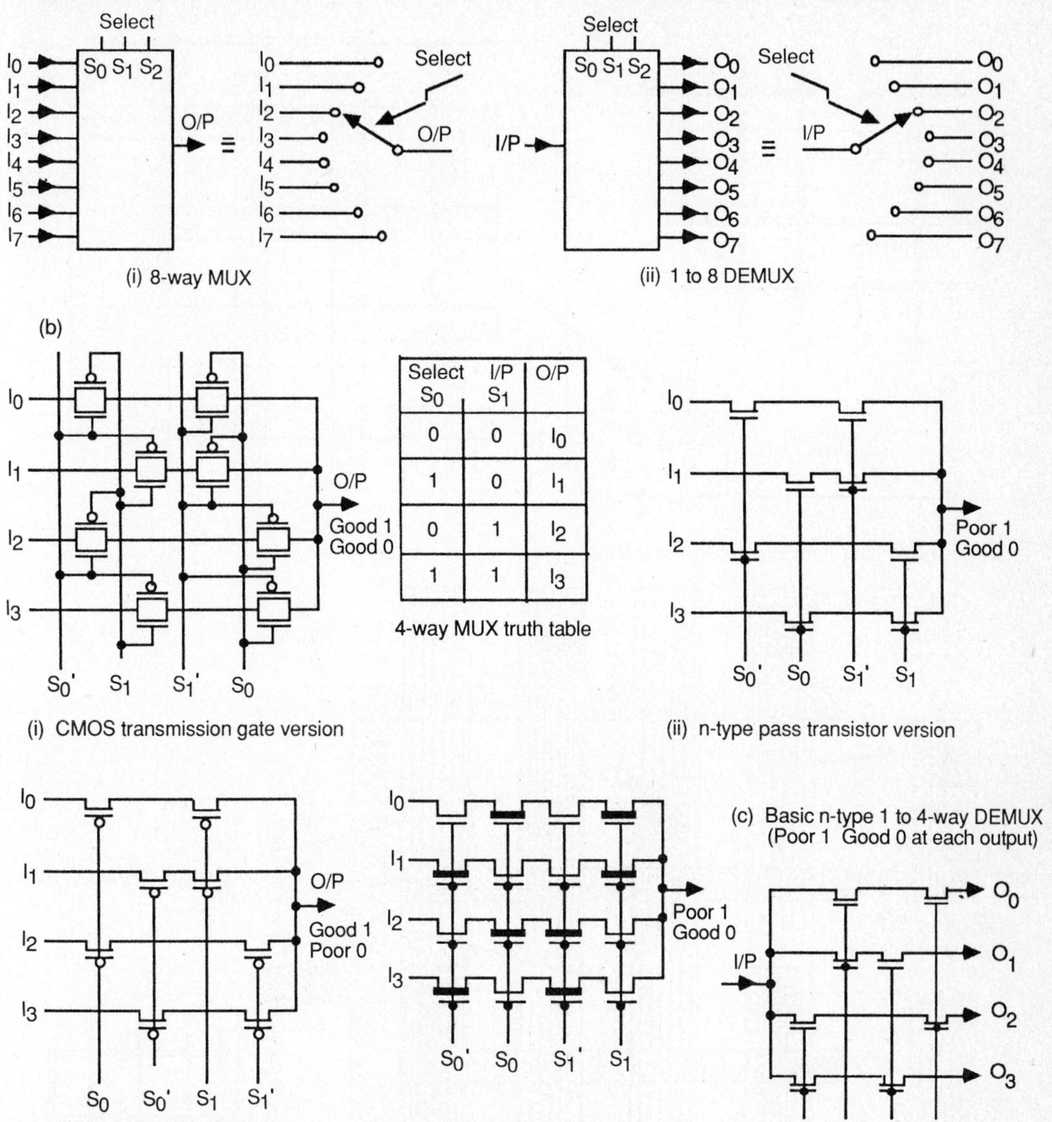

Select S_0	I/P S_1	O/P
0	0	I_0
1	0	I_1
0	1	I_2
1	1	I_3

Figure 5.29 Multiplexer (MUX) and demultiplexer (DEMUX) arrangement

The use of multiplexers allows the generation of all possible simple combinational logical functions of a number of variables (up to four for a single MUX) at a single output. A 2-variable case is shown in Figure 5.32. We can extend this up to functions of three or four variables using an 8-way or 16-way MUX respectively Beyond this, arrays of MUXs can be formed. Used this way, the MUX is sometimes referred to as a universal logic module (ULM).

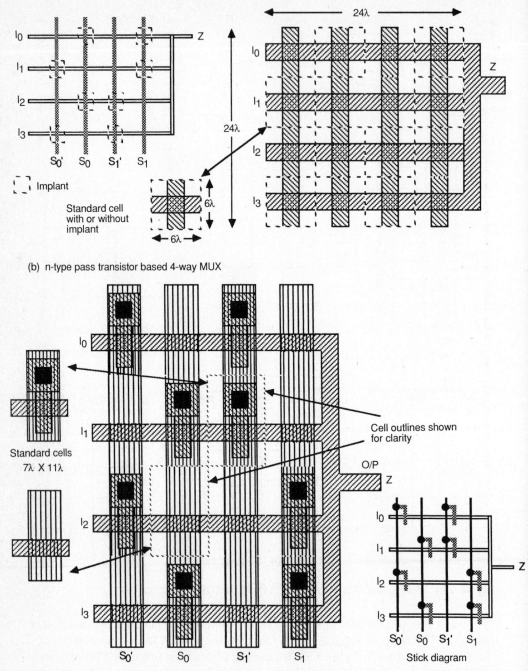

(a) Stick and mask layouts for n-type red/green based 4-way MUX

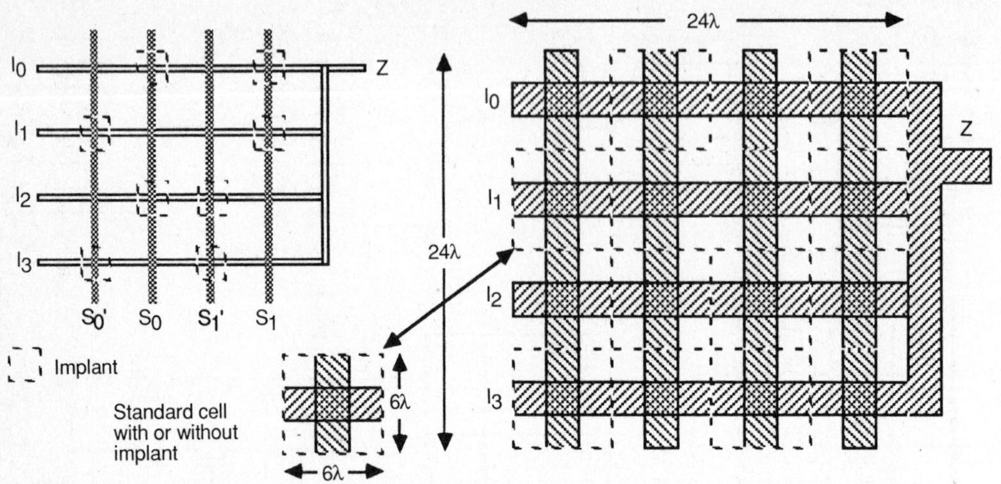

(b) n-type pass transistor based 4-way MUX

I_0
I_1
I_2
I_3
Standard cells 7λ X 11λ
Cell outlines shown for clarity
O/P
Z
S_0' S_0 S_1' S_1
Stick diagram

Figure 5.30(a) Stick and mask layouts for n-type red/green block based 4-way MUX
(b) n-type pass transistor based 4-way MUX

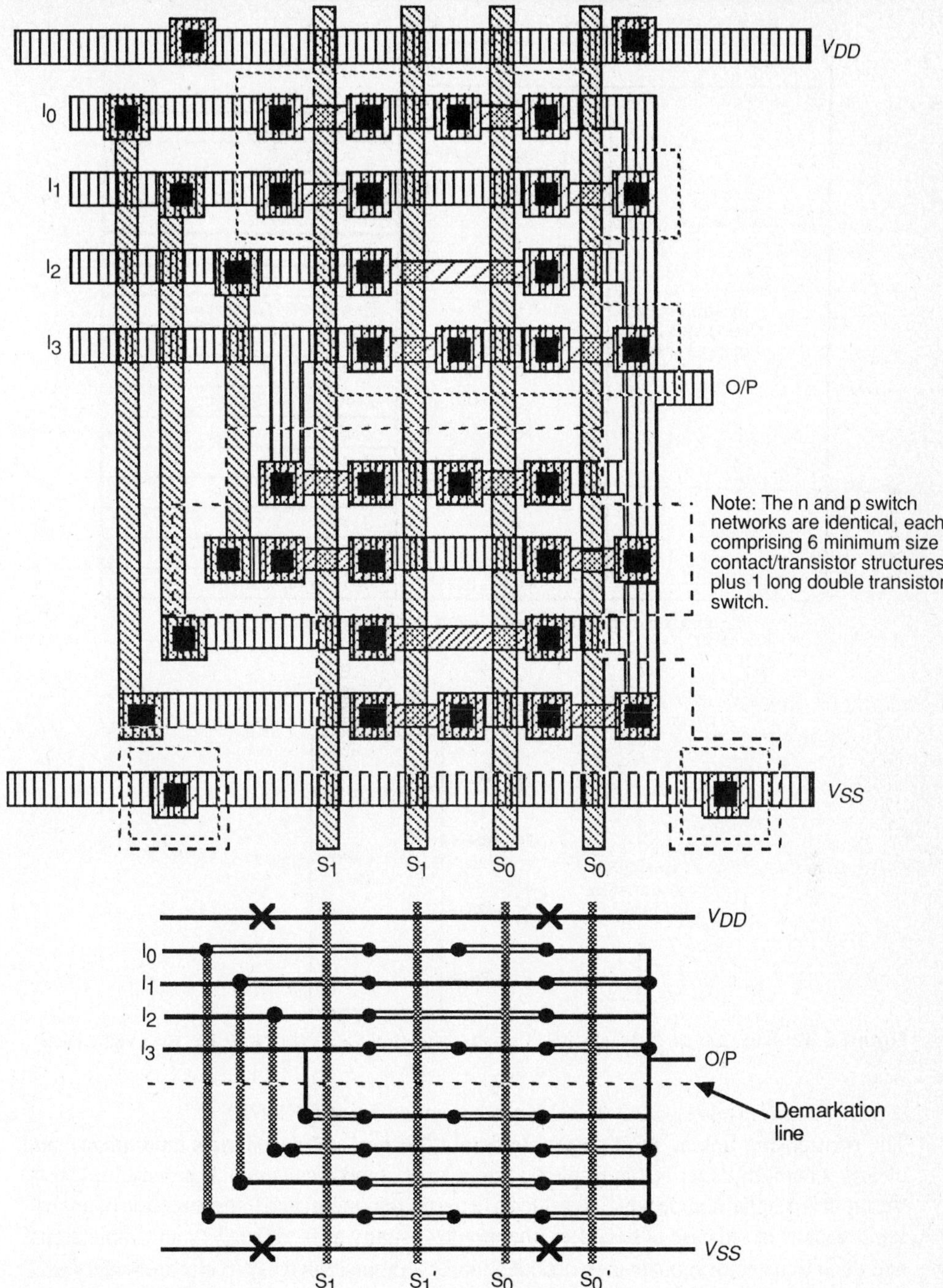

Figure 5.31 CMOS transmission gate based 4-way MUX

Input variables	AB	AB'	A'B	A'B'		
Input selected	I_3	I_2	I_1	I_0	Z =	Function
Programming i.e. connection of input lines	0	0	0	0	Z=0	Inhibit
	0	0	0	1	Z=A'.B'	*Nor*
	0	0	1	0	Z=A'.B	
	0	0	1	1	Z=A'	*Not* A
	0	1	0	0	Z=A.B'	
	0	1	0	1	Z=B'	*Not* B
	0	1	1	0	Z=AB'+A'B	*Exclusive Or*
	0	1	1	1	Z=A'+B'	*Nand*
	1	0	0	0	Z=A.B	*And*
	1	0	0	1	Z=AB.+A'B'	*Equality*
	1	0	1	0	Z=B	
	1	0	1	1	Z=A'+B	
	1	1	0	0	Z=A	
	1	1	0	1	Z=A+B'	
	1	1	1	0	Z=A+B	*Or*
	1	1	1	1	Z=1	*Enable*

Table showing connection of inputs and generated functions at "Z"

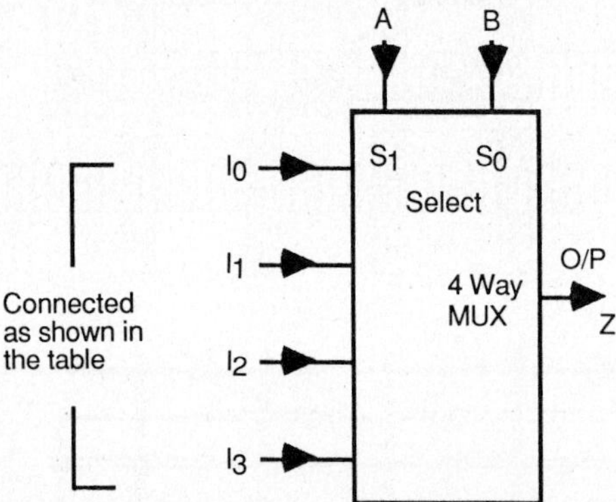

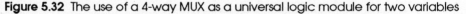

Figure 5.32 The use of a 4-way MUX as a universal logic module for two variables

A CMOS transmission gate 2- to 4-Line decoder (or 1 to 4 DEMUX)

The relationship linking multiplexers to demultiplexers and decoders is quite plain, and clearly a demultiplexer is a multiplexer "in reverse" (and vice versa). For demultiplexers, the input is a signal source which is routed to a particular output under the direction of a select input vector. In the case of decoders, the input is usually a fixed 0 or 1 or an enable signal and the *n* line vector input is then decoded into 2^n outputs. Such as circuits are widely used and the latter appear frequently, for example, in memory address decoding operations. In particular, row and column decoders make use of decoders. To put matters in perspective it is useful to examine at least one form and to this end Figure 5.33 sets out a design for a 2-line (address) to 4-line (column select) CMOS transmission gate based decoder.

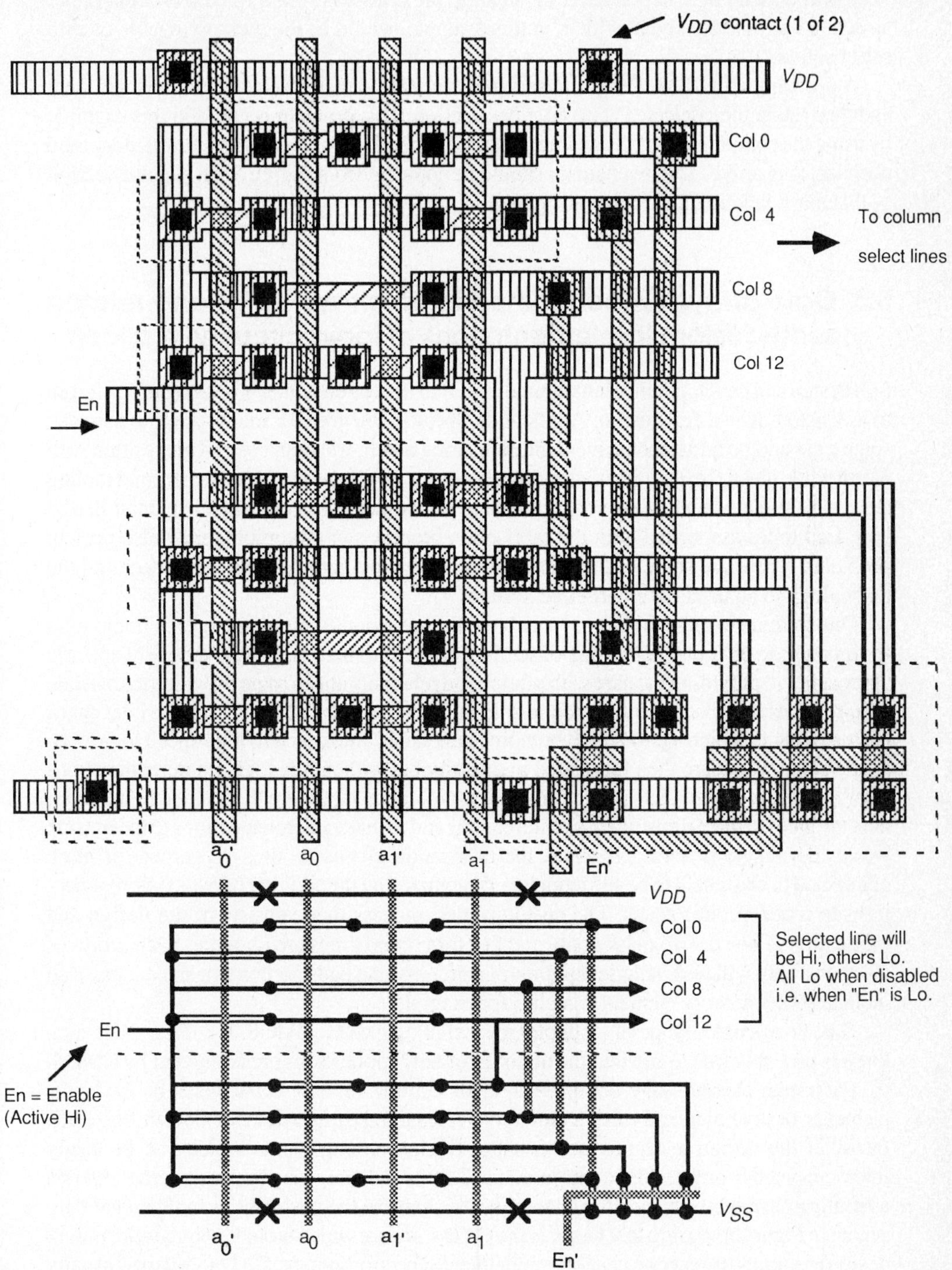

Figure 5.33 2- to 4-line decoder as column select circuitry for 16-bit RAM

It will be seen that, in this case, the input to the decoder is not a fixed 0 or 1 but rather an active Hi "enable" signal which, in this example, would be the memory enable or chip select signal.

A problem with MOS demultiplexers and decoders is that unselected outputs will float and thus retain their selected state after deselection. This problem is cured in this example by using the complement of the enable signal to "ground" all outputs when the memory (and decoder) is inactive. The mechanism should be apparent from Figure 5.33. A color version of this mask layout appears as color plate 8.

5.3 Gate arrays (uncommitted logic arrays-ULAs) and related semi-custom implementations of logic circuitry in silicon

Full design and development of Integrated circuits to meet a customer's needs (often referred to as Custom ICs or full custom Application Specific Integrated Circuits-ASICs) can take some time and be quite expensive in mask making and fabrication costs. This is often well worth while when the quantity of circuits to be produced allows the development and tooling costs to be spread, or when the very ultimate in performance is necessary.There may also be long lead times associated with the fabrication process and prototyping may also present problems. These are some of the factors which are taken into account when deciding the technology to be used for a particular system.

One solution to these problems, which retains some of the key advantages of integration in silicon, is to be found in the area of semi-custom ICs. There is, now, a variety of specific approaches to providing the user with a quick and relatively cheap route by which to translate requirements into circuits integrated in silicon. However they all stem from the concept of *Gate Arrays* in which up to 20000 or more gates can be integrated in high speed logic onto single chips which are then fabricated in quantity but without the logical interconnections between gate elements. Single or double (or more) metal layers are then added to interconnect the gates to meet a customer's requirements and to make interconnections to the outside world. For single metal approaches, the fabrication process is thus 90 percent or more completed in chips held to be subsequently customized by the addition of metal interconnections to meet specific needs. The customization process then consists of the design and fabrication of one or two masks only and fast turn round times are possible. Such arrays of gates on a chip will also include input/output circuitry and pads for bonding to and may also include some memory elements, latches for example.

Gate arrays use a range of technologies including nMOS, CMOS, I^2L, ECL,TTL, etc., but it is probably true to say that the majority of new applications are being met by CMOS.

Each gate element may comprise a small number of, say, *Nand* gates or *And - Or* elements, or several mixed partially interconnected n and p transistors for CMOS, and a key factor in the design is to produce fast but flexible arrangements which can be easily interconnected in any desired configuration. To this end there may be special areas, known as routing channels, reserved for interconnects. Some of the possibilities in this regard are set out in Figure 5.34 and a key factor is the CAD software which includes automatic routing design programs to generate near optimum interconnection layouts.CAD software is usually developed and made available by semiconductor houses marketing gate arrays, to allow speedy and efficient translation of specification into silicon for a given design. In fact, some software is commonly efficient enough to allow utilization factors of 80 percent or more of the gates on a chip.

Interconnection routing is clearly a major factor in determining the ease with which a design can be completed and the efficiency of utilization of the resources on the chip. Figure 5.34(a) illustrates the basic gate array floorplan in which it would be assumed that the layout of the individual elements is such as to allow for connections to be made to the cell and to allow for the connections to the cell and other signal lines to run right through the cell. A comparison of the three arrangements of Figure 5.34 seems to indicate that the greatest density of elements would be achieved by this basic layout due to the fact that cells are directly abutted in rows and columns. However, the fact that interconnections may run through each cell means that cells need to be larger than layouts where interconnections are run in channels, therefore the gain in density may be offset. Again, it would appear that this layout might yield the shortest interconnects and thus the fastest logic, but this may not be the case since crowding of interconnects through particular cells may require some circuitous routes to be taken.

The type of floorplan introduced in Figure 5.34(b) allows connection routing channels across either the rows or columns of the cells. Clearly this will relieve the pressure on layout detail which, for (a), requires allowance for connections through the cell. However, this is at the expense of the density of cells (or elements) per chip and of making many connections longer than they would need to be if interconnects were not constrained to the channels.

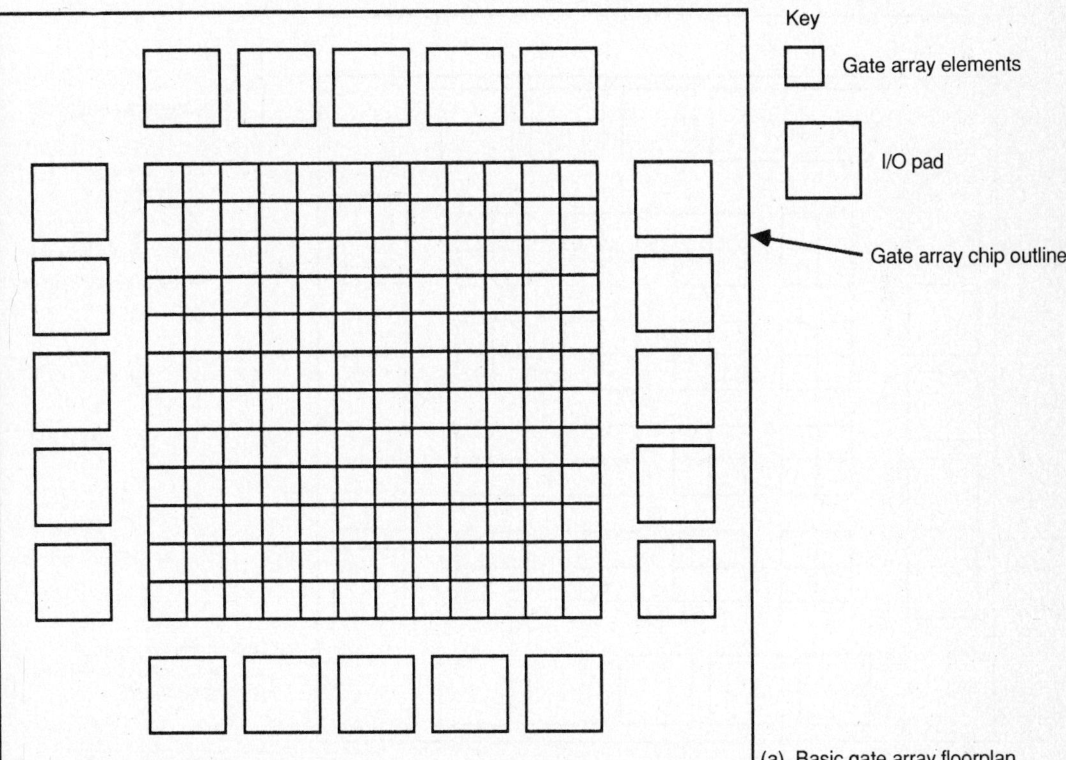

Figure 5.34(a) Basic gate array floorplan

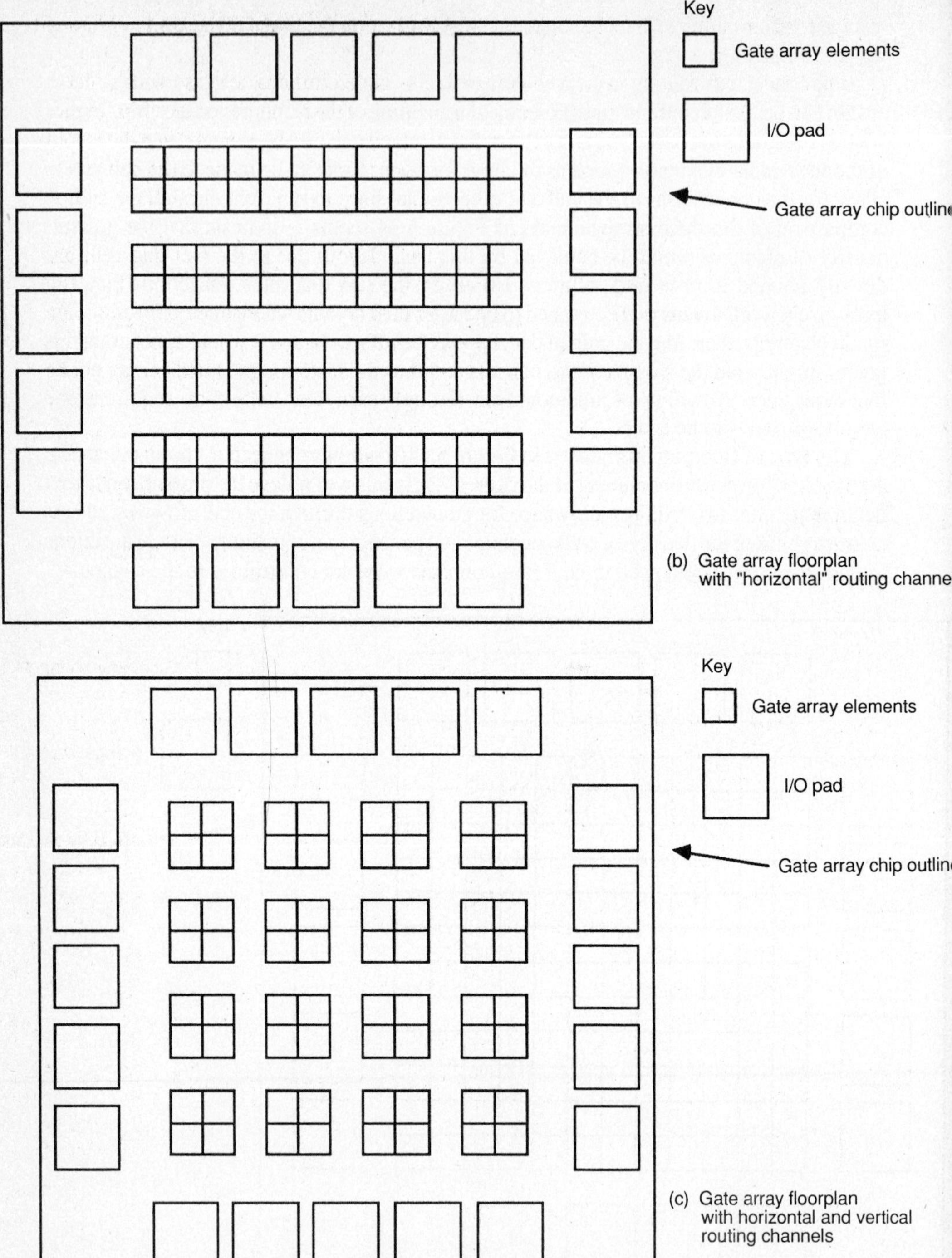

Figure 5.34(b) Gate array floorplan with "horizontal" routing channel
(c) Gate array floorplan with horizontal and vertical routing channels

To avoid overlong runs, an alternative floorplan layout is suggested in Figure 5.34(c). In this case routing channels replace selected rows and columns, as shown, with the idea of facilitating short interconnection paths. Again, this will be at the expense of density.

However, with careful, well-thought-out layouts, it is possible to produce fast gate arrays which provide customized ICs which achieve component densities which are a good proportion of those achieved with full custom design and which approach the speed of full custom circuits.

5.3.1 User programmable logic arrays

Many applications require more modest numbers of logic gates and also benefit from quick turn around and low fabrication costs. A number of families of devices are now available on the market which allow the potential user to hold stocks of finished gate array like chips typically comprising up to 2000 or so equivalent gates, finished that is except for the programming of the programmable interconnection pattern.

Into this category come programmable logic devices (PLDs) which include the PLA (already discussed) and PAL (programmable array logic) devices, the latter being an optimized version of the PLA but with less generality in terms of the output *Or* functions which may be accommodated. Both are well suited to designs requiring the equivalent of a few hundred gates. As well as being user programmable, some versions allow the current programming to be erased and the device reprogrammed while others, for example, fusible link technology, may be programmed once and once only.

Larger designs may be accommodated in other forms of user configurable integrated circuits (UCICs) which are variously known, for example, as EPLDs (erasable programmable logic device). EPLDs from Altera combine CMOS gate with EPROM technology to achieve user-configurability and accommodate circuits with complexities of a few hundred to more than 2000 equivalent gates. The general arrangement of an Altera EPLD is set out in Figure 5.35 (reproduced here by courtesy of Altera).

Another approach, due to Xilinx, is to market a range of programmable gate arrays based on the concept of logic cell arrays (LCA)™ and an "on board" memory which is usually loaded on power up from a user-programmed EPROM or from a processor, this specifying the interconnection pattern. Clearly, such devices erase on power down and may therefore be reprogrammed at any time. Again equivalent gate complexities of 2000 or more can be accommodated. The general arrangement of a programmable LCA is set out in Figure 5.36 (reproduced here by courtesy of Xilinx).

These particular references to proprietary items have been included as illustrations of approaches to user-programmable devices which give quick turn round, low development cost, and are well suited to the design of small to medium complexity systems which would otherwise require significant numbers of, say, TTL or CMOS SSI and MSI packaged logic devices.

5.4 The adder

The adder is a commonly used digital subsystem and is the main "number-crunching" element in any digital computer.It is therefore quite clearly a subsystem which deserves

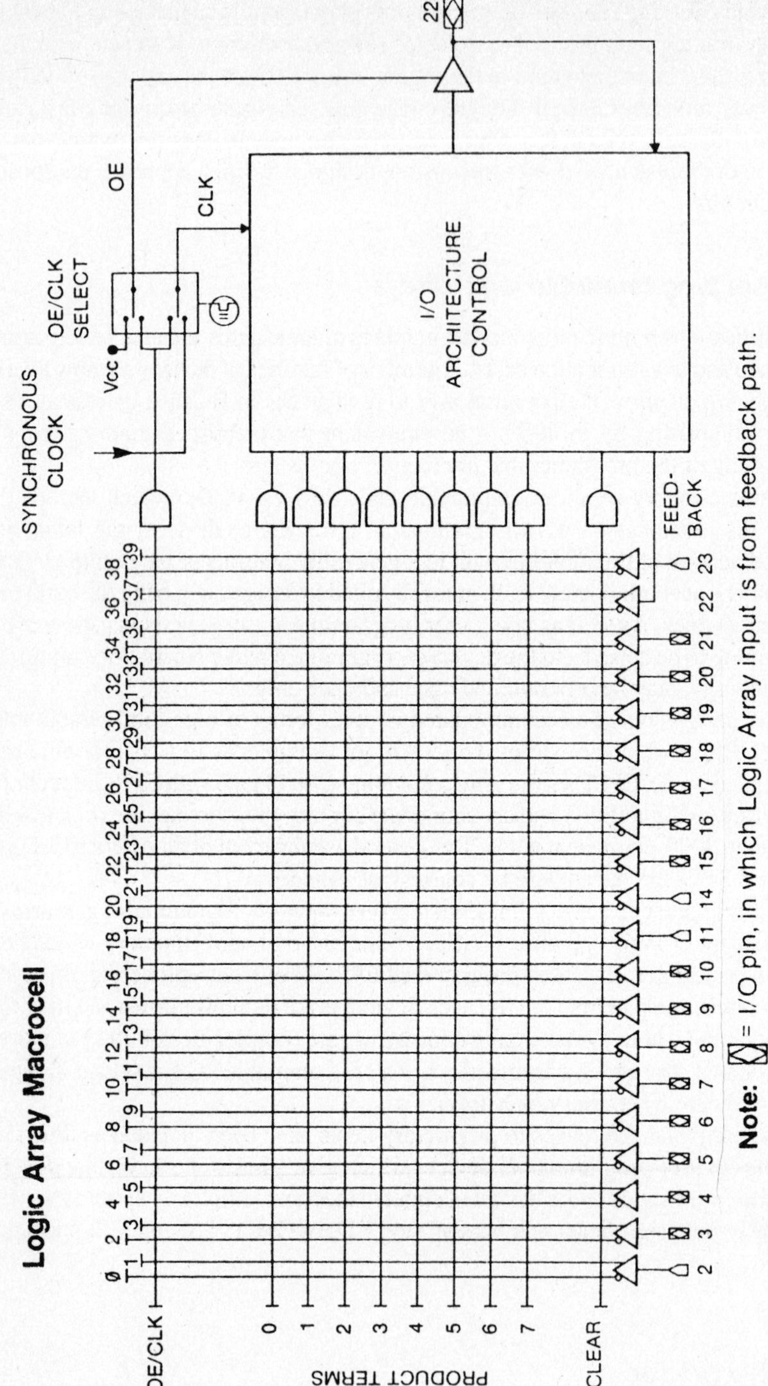

Figure 5.35(a) General EPLD architecture
(reproduced through the courtesy of Altera Corporation)

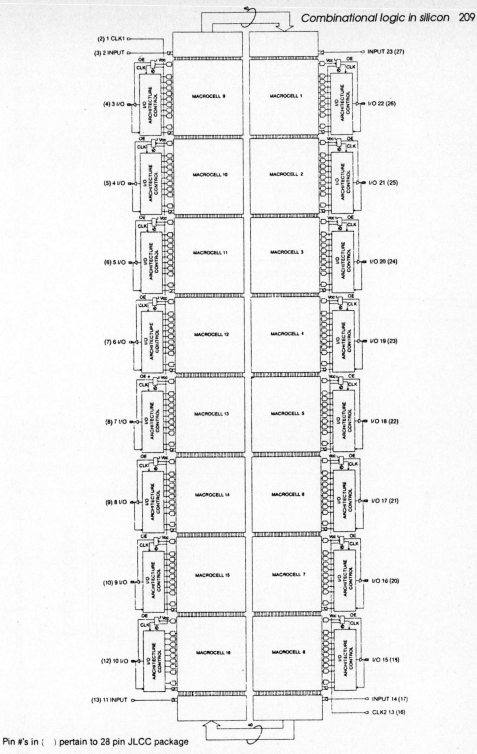

Pin #'s in ( ) pertain to 28 pin JLCC package

Figure 5.35(b) A particular EPLD arrangement
(reproduced through the courtesy of Altera Corporation)

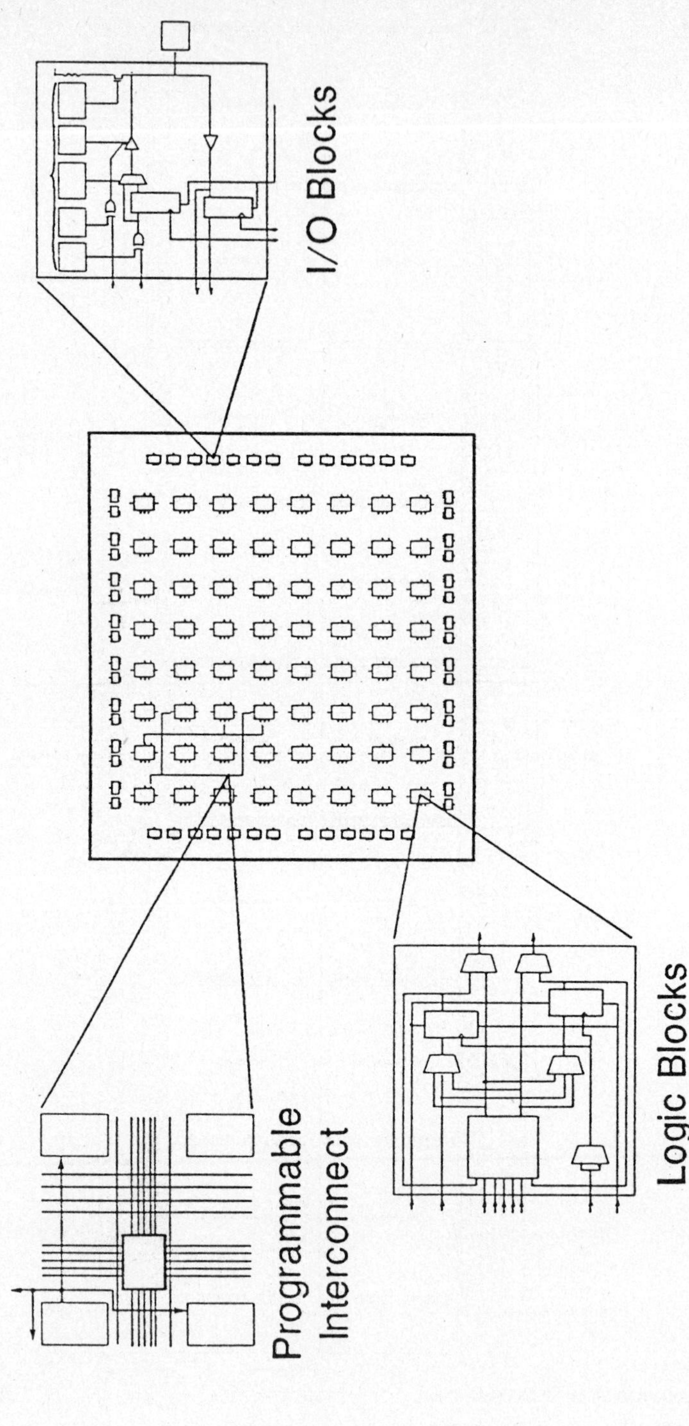

Figure 5.36 Programmable gate array architecture (reproduced through the courtesy of Xilinx)

some attention. Various approaches can be taken to the design of multi-bit adders and the correct choice can be an important one on which, say, the speed limitations of fast processors may well depend.

Thus the adder is important enough to warrant special treatment and although our treatment cannot be exhaustive, we can nevertheless establish the principles on which designs are based and show, later, that the adder can be applied to other arithmetic and to logical operations. Indeed, the adder is the heart of the arithmetic and logic unit (ALU) of digital computers.

5.4.1 The adder equations

In order to establish the adder equations we must first consider the general requirements for the addition of two multi-bit numbers. Let those numbers be A and B, each assumed to be of, say, 8-bit word length and let us further assume that both numbers are positive and in "twos complement" form*, although the adder we are to consider will perform additions equally well on unsigned binary numbers.

$$\begin{array}{rl|c|ll} A & 010 & 1 & 1011 & = A_7A_6A_5 \ldots A_1A_0 \\ +\ B & 000 & 1 & 1101 & = B_7B_6 \quad \ldots B_1B_0 \\ \hline \text{Sum } S & 011 & 1 & 1000 & \leftarrow\leftarrow\leftarrow \text{ carry} \\ & & k & & \end{array}$$

In any one column "k" there are three inputs: A_k, B_k, and C_{k-1} (the carry from the previous column), and there are two outputs: S_k (the sum), and C_k (the new carry).

Thus, it is possible to set out a truth table for a 1-bit slice (column "k") of a binary adder as in Table 5.2.

Table 5.2 Truth table for 1-bit slice of an adder

Inputs			*Outputs*	
A_k	B_k	C_{k-1}(Prev. carry)	S_k (Sum)	C_k (New carry)
0	0	0	0	0
0	1	0	1	0
1	1	0	0	1
1	0	0	1	0
0	0	1	1	0
0	1	1	0	1
1	1	1	1	1
1	0	1	0	1

From the table we may deduce the adder equations in several different but obviously interrelated forms. A convenient form on which designs may be readily based is as follows:

* Number representation and arithmetic are covered in Chapter 2.

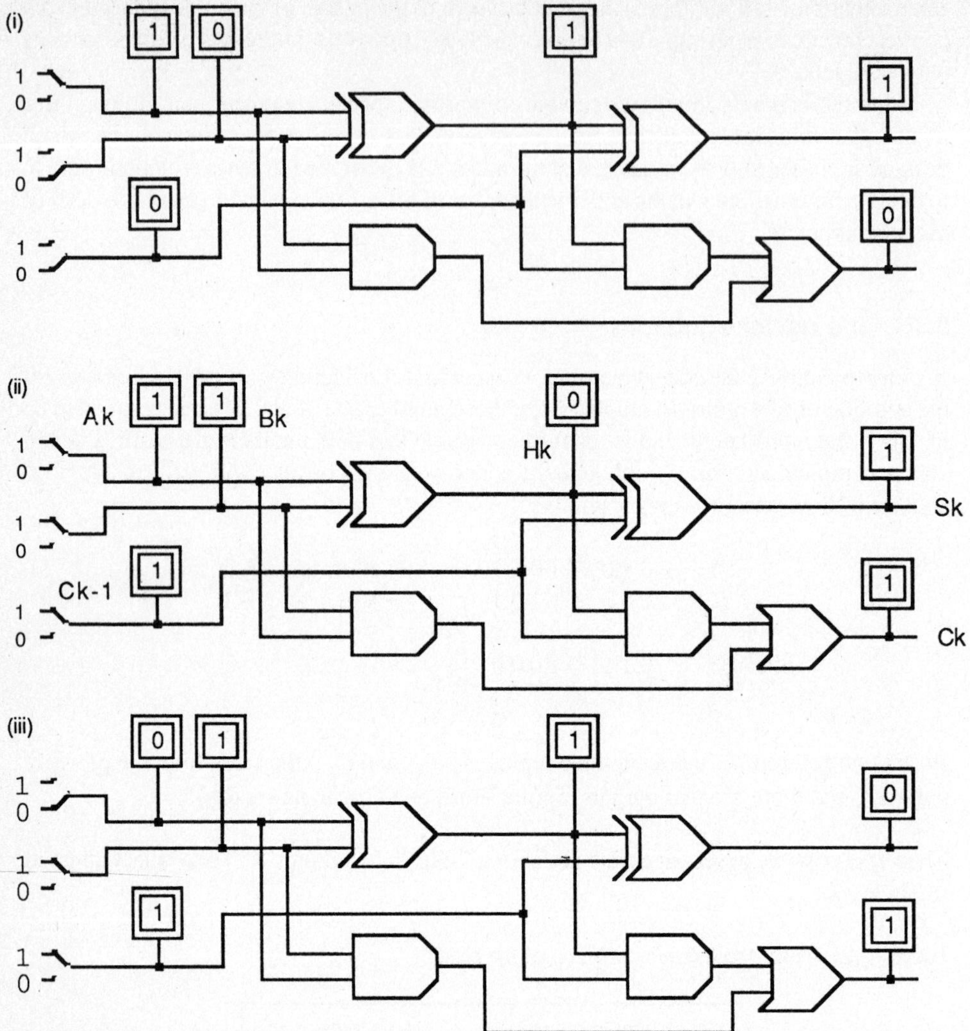

Figure 5.37(a) Adder bit-slice with simulations

$$\text{Sum} \quad S_k = H_k . C'_{k-1} + H'_k . C_{k-1}$$

where:

$$H_k = A_k . B'_k + A'_k . B_k$$

$$\text{New carry} \quad C_k = A_k . B_k + H_k . C_{k-1}$$

Note that two of the above expressions are *Exclusive Or* operations so that a readily implemented form of a 1-bit slice is as shown in Figure 5.37.

The figure shows the logic circuitry with "LogicWorks™" simulations for three sets of input conditions by way of illustration.

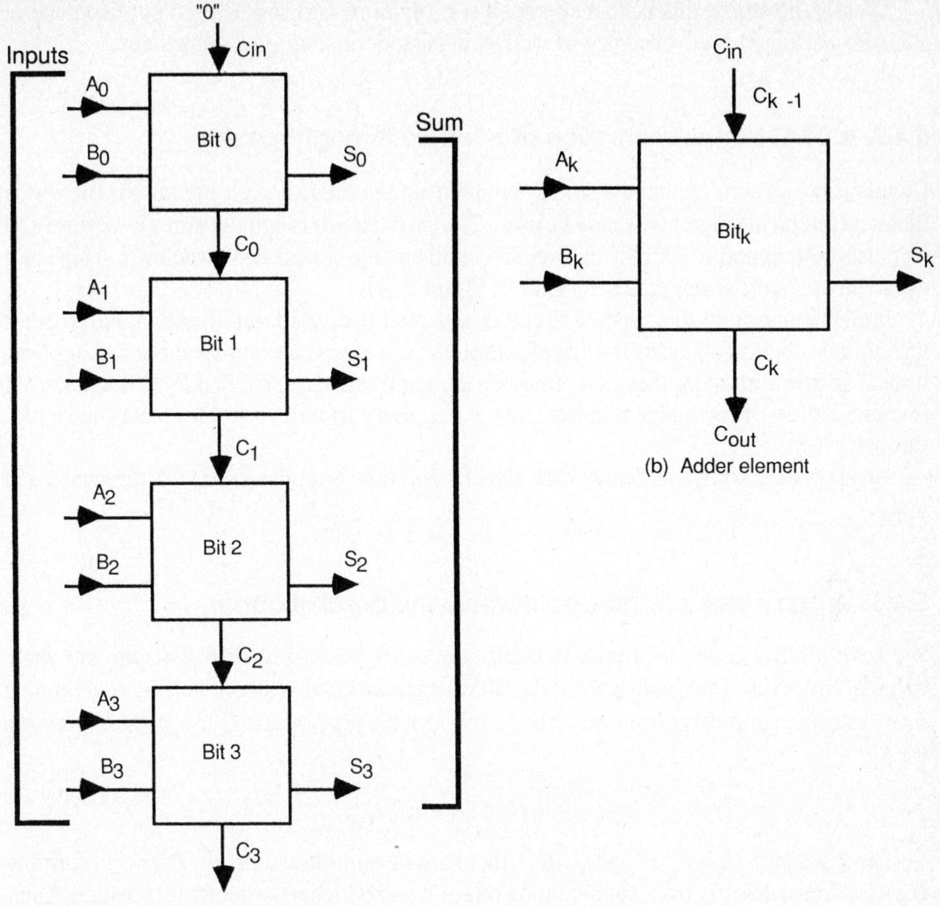

Figure 5.38(a) Four-bit parallel adder arrangement
(b) Adder element (bit-slice)

We may now configure an *n*-bit, say 4-bit, adder as in Figure 5.38 using the symbolic representation set out in Figure 5.38(b).

This particular form of adder is first of all a *parallel adder* since all bits of both numbers to be added are presented simultaneously in parallel to the inputs of the adder, and the sum is produced at the outputs after some short delay due to propagation delay effects through the adder circuits. Secondly, this particular approach to design yields what is called *a ripple through adder* since, for example, in order to form carry C_3 and sum S_3 carry C_2 must have been formed which, in turn, required the prior formation of carry C_1 which, again, could not be generated until carry C_0 was present. This effect causes a cumulative delay (or ripple through) in forming the sum and carry bits which may be unacceptable for the high order bits of multi-bit adders and limits the speed with which additions can be carried out.

Clearly, however, this design approach is straightforward and is often implemented in standard packaged logic circuitry as well as in custom or semi-custom designs.

5.4.2 A CMOS implementation of a "ripple through" adder

Clearly, the half sum H_k and the sum S_k are both simple *Exclusive Or* operations for which identical circuit arrangements may be used. The carry circuit is equally straightforward and can also be mapped to extract the required pull-up and pull-down structures. Maps and consequent circuit arrangement follows as Figure 5.39.

Implementation of this type of circuit is also readily carried out in nMOS and in other technologies but, in all cases the "ripple-through" effect will be significant in fast applications if adders with more than, say, four bits are required. Thus, for fast multi-bit adders, a re-examination of the adder requirements is necessary to reduce or eliminate the ripple-through effects.

An approach, known as "carry look ahead", will now be considered with these needs in mind.

5.4.3 A carry look ahead approach to the parallel adder

We have already seen that ripple-through effects are present in both the sum and carry outputs. However, if we look again at the adder expressions we will see that the sum outputs S_k are subject to this effect because of the presence of the previous carry C_{k-1} in the expression for S_k:

$$S_k = H_k.C'_{k-1} + H'_k.C_{k-1}$$

Clearly, since $H_k = A_k.B'_k + A'_k.B_k$, there are no cumulative delay effects in forming H_k; it is formed directly from the incoming bits of A and B in that particular bit position. Thus, if we can eliminate ripple-through from the carry generation, all will be well. Let us look then once again at the expression for carry C_k:

$$C_k = A_k.B_k + H_k.C_{k-1} \qquad (1)$$

Consider now the formation of carry C_0. Since there can be no previous carry then:

$$C_0 = A_0.B_0 \qquad (2)$$

Note: no ripple through effects.
Now consider carry C_1:

$$C_1 = A_1.B_1 + H_1.C_0$$

and substituting for H_1 we have:

$$C_1 = A_1.B_1 + A_1.B'_1.C_0 + A'_1.B_1.C_0$$

whence:

$$C_1 = A_1.B_1 + (A_1 + B_1)C_0 \qquad (3)$$

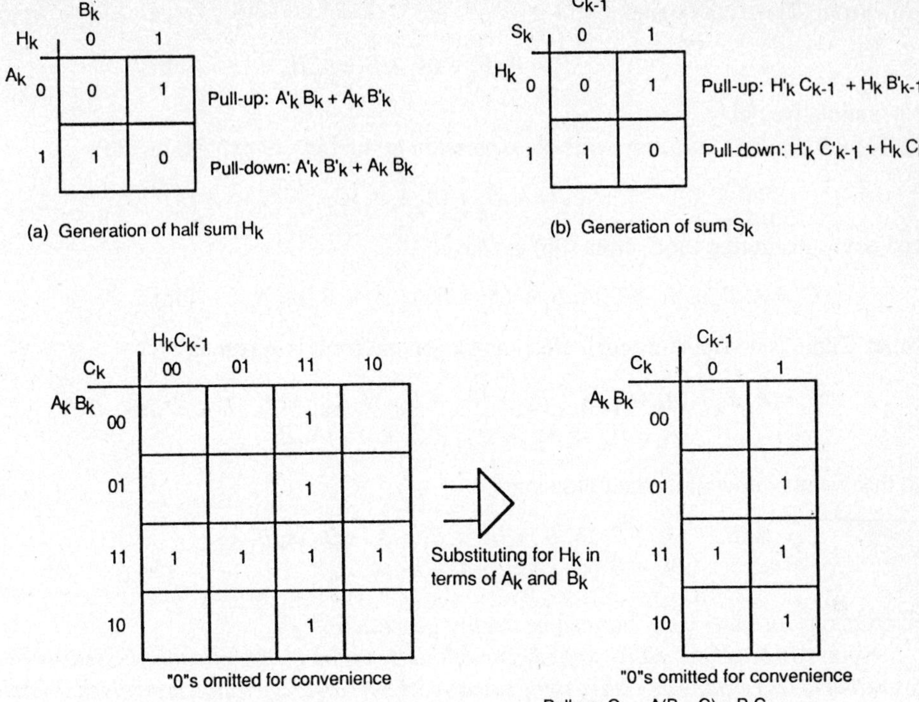

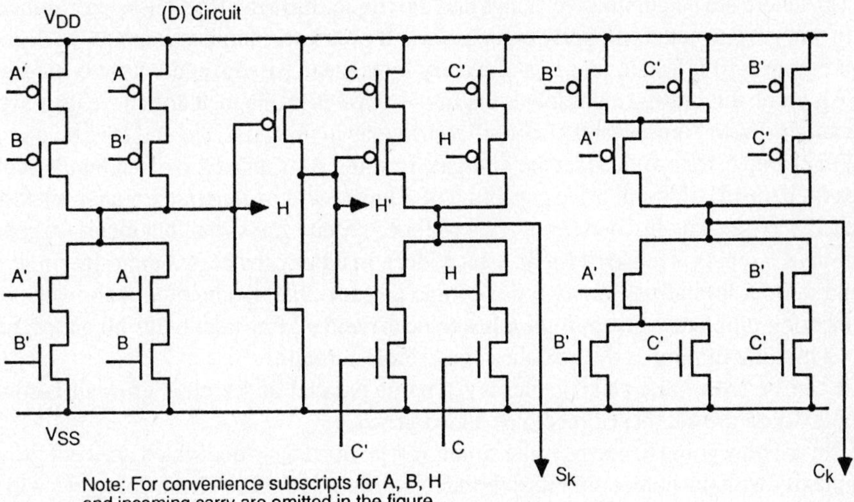

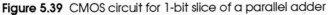

Figure 5.39 CMOS circuit for 1-bit slice of a parallel adder

Note the general form of (3) which may be written:

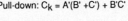

$$C_k = A_k B_k + (A_k + B_k) C_{k\text{-}1} \quad \text{(3a)}$$

Now from (2) we may write:

$$C_1 = A_1.B_1 + (A_1 + B_1).A_0.B_0 \qquad (3b).$$

No cumulative delay.

Now for carry C_2 we may write an expression in the form of expression (3a).

$$C_2 = A_2.B_2 + (A_2 + B_2)C_1 \qquad (4)$$

and now substituting for C_1 from (3b) we have:

$$C_2 = A_2.B_2 + (A_2 + B_2)A_1.B_1 + (A_2 + B_2).(A_1 + B_1)A_0.B_0 \qquad (4a)$$

Note: There is no ripple-through effect and a general form is apparent,

$$C_k = A_k.B_k + (A_k + B_k)A_{k-1}.B_{k-1} + (A_k + B_k).(A_{k-1} + B_{k-1})A_{k-2}.B_{k-2} + \ldots + (A_k + B_k).(A_{k-1} + B_{k-1})(\ldots)(\quad)A_0.B_0$$

so that we may now write the following:

$$C_3 = A_3.B_3 + (A_3 + B_3)A_2.B_2 + (A_3 + B_3)(A_2 + B_2)A_1.B_1 + (A_3 + B_3)(A_2 + B_2)(A_1 + B_1)A_0.B_0 \qquad (5)$$

Expressions for other carry bits can be readily generated.

Note: *Although the expressions become longer as the bit significance increases the overall delay in forming any carry is the same. A 3-Level logic operation is involved in each case*, (namely, individual *And* and *Or* operations on A_k and B_k etc. followed by *And* operations to form the product terms of the expression, then an overall *Or* operation).

Thus there are no cumulative delays and the ripple-through effect has been eliminated.

In order to put matters properly in perspective the hardware implications may be deduced from Figure 5.40 in which four bits of a carry look ahead parallel adder have been drawn.

Although this arrangement solves the ripple-through problem, it does have its practical limitations due to "fan-in" and "fan-out" restrictions in practical *And* and *Or* gate circuits.

For example, for a 4-bit adder the *And* gate forming $A_0.B_0$ must drive four inputs of other gates (in general, a fan-out drive capability of *n* inputs will be required for an *n*-bit adder). You will also see that *And* and *Or* operations for C_3 require gates with four inputs (in general a fan-in of *n* inputs is required for an *n*-bit adder). In all practical cases there are finite fan-in and fan-out limitations in any type of logic gates so that, in common with many other engineering problems, a compromise has to be arrived at. Practical multi-bit adders often have a mixture of pure carry look ahead and ripple-through.

It can be seen quite clearly that any type of parallel adder must grow in hardware complexity as the number of bits to be added grows.

We are now going to consider the simplest type of adder—one which does not grow in complexity with the number of bits to be added—*the serial adder*. In so doing, we will use combinational logic but also basic storage circuits which will be formally discussed in later chapters.

5.4.4 The serial adder

Instead of having as many sections to the adder as there are bits to be added, the serial adder

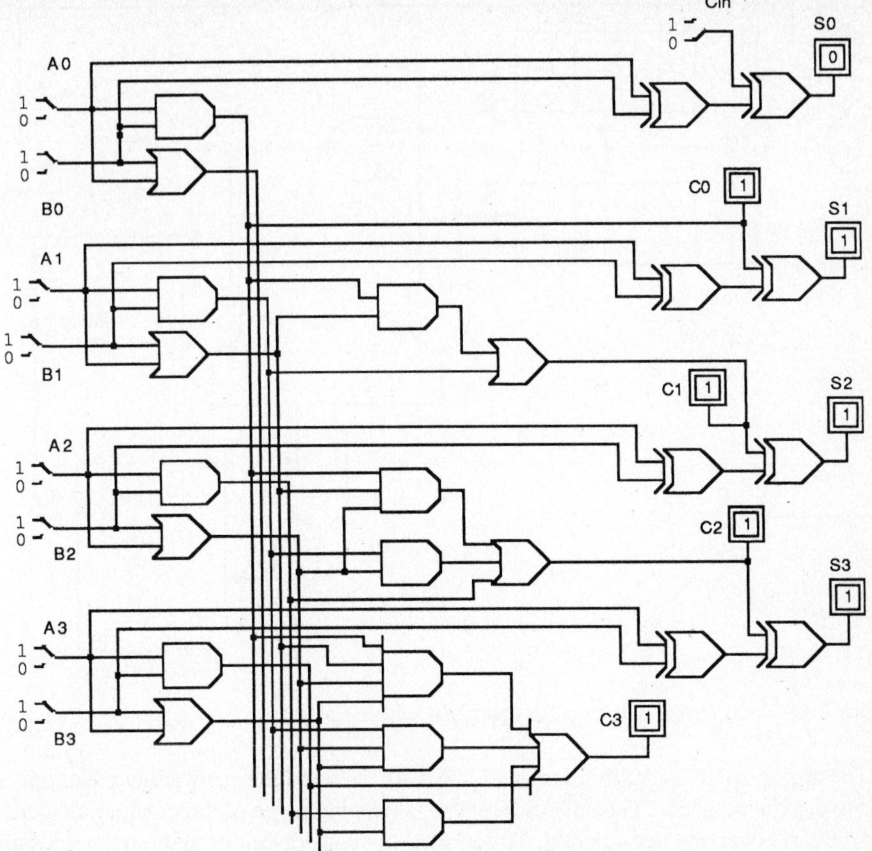

Figure 5.40 4-bit carry look-ahead adder logic (with LogicWorks simulation)

is in fact a single bit slice or adder element through which bit pairs to be added are processed *serially with respect to time*. Bits A_0 and B_0 are added first and a sum and carry generated. The sum is stored in a register outside the adder but the carry is stored inside the adder so that it can be used to form the next sum and new carry with the next bit pair A_1 and B_1 and so on.

Conveniently, each of the two numbers to be added may be held in *shift registers*. For example, if we wish to add, say, two four-bit numbers then we would need two 4-bit shift registers with the least significant bit in each case in the bit position at the serial output end of each register, these outputs providing the inputs to the adder. Thus the two least significant bits are added first. The registers are then clocked and all bits move one place right in significance so that the next two (2nd least significant) bits now occupy the least significant bit positions in each register and are thus added, the previous bit pair "falling off" the end of the registers during the clocking right operation . As this happens the most significant bit position of both registers is vacated so that the sum (just formed) may be clocked into the input end of one of the registers. At the end of four clock pulses (in this case) the sum will be complete in that register. The same clock pulse in each case clocks the newly generated carry into the temporary store or memory.

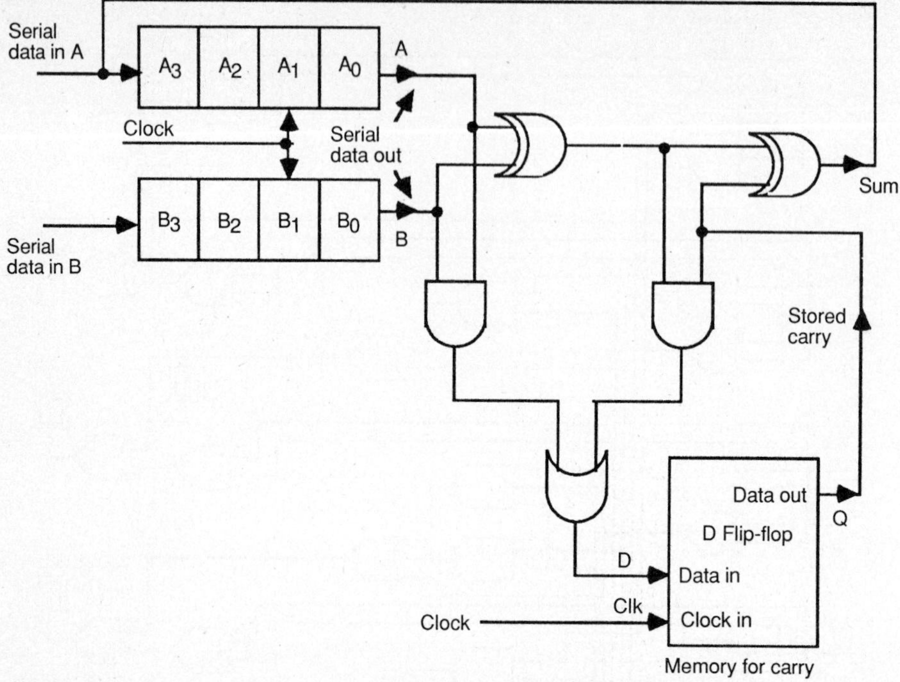

Figure 5.41 4-bit serial adder arrangement with registers

The arrangement is clearly illustrated in Figure 5.41 and the same adder equations are used to design the adder. The only difference is in the inclusion of the memory element to hold the carry over one clock period. To complete our considerations at this point a possible selector switch (MUX) implementation, using *n*-type pass transistor switches is given, in stick diagram form, as Figure 5.42.

Whatever form the implementation takes, there will be a need for a storage, or memory, element and this brings out the need for a second class of logic circuits, *sequential* or *memory circuits*. Such circuits are discussed in the following chapters.

5.5 Summary

We have now covered the basic techniques for designing combinational logic and, having earlier taken an overview of MOS VLSI design, this chapter has been largely concerned with implementation of combinational logic in silicon. It will be seen that CMOS technology in particular offers considerable scope for imaginative design work and that MOS technology generally provides a powerful approach to custom design.

The adder has been used both as an important arrangement in its own right and to illustrate subsystem design in combinational logic.

The serial adder has brought to light the need for a storage or memory element and this falls in the class of sequential circuits. We are now about to embark on an examination of this class of circuit and on the associated design and analysis techniques.

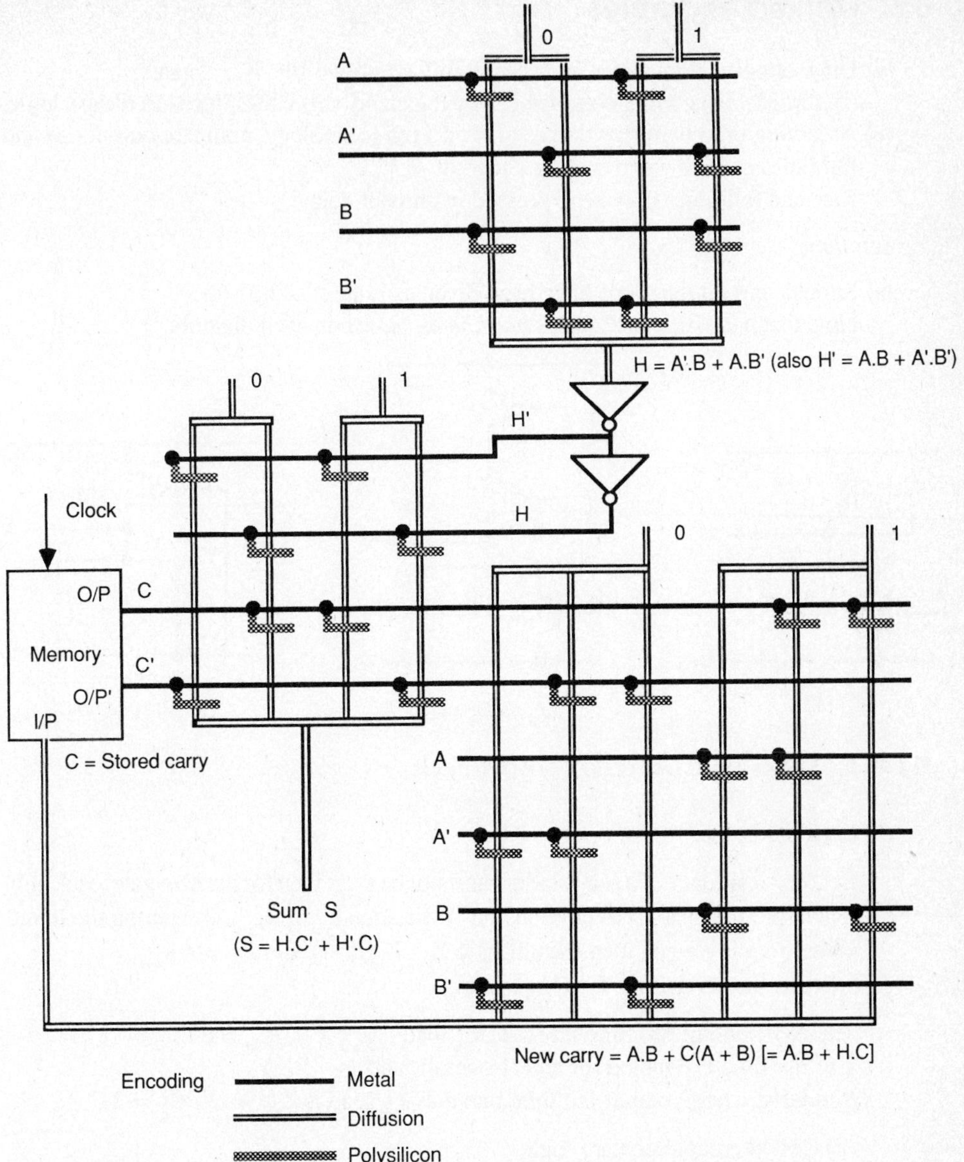

Figure 5.42 Stick diagram for possible selector switch based serial adder

Much of the work will use familiar techniques and we will see that sequential circuits are formed from combinational logic, but it is convenient to recognize certain standard configurations, like the JK flip-flop, and characterize the performance as a sequential circuit element without regard to the actual logic from which each is formed.

We will also see that combinational logic is also required when designing sequential subsystems, such as counters and registers.

5.6 Worked examples

1. (a) Draw circuit diagrams for each of the following two I/P "*Nor* " gates:
   (i) nMOS, (ii) CMOS complementary logic and (iii) CMOS pseudo nMOS logic.
   (b) Assuming minimum size transistors and 5 μm technology, compare output rise and fall times of each when driving a load of $2\square C_g$.
   Rise and fall times may be expressed in units of delay.

Solution:

(a) Suitable circuit diagrams have been given as Figure 5.43(a)-(c).
Note that *transistor L:W ratios must be as shown* on each diagram.

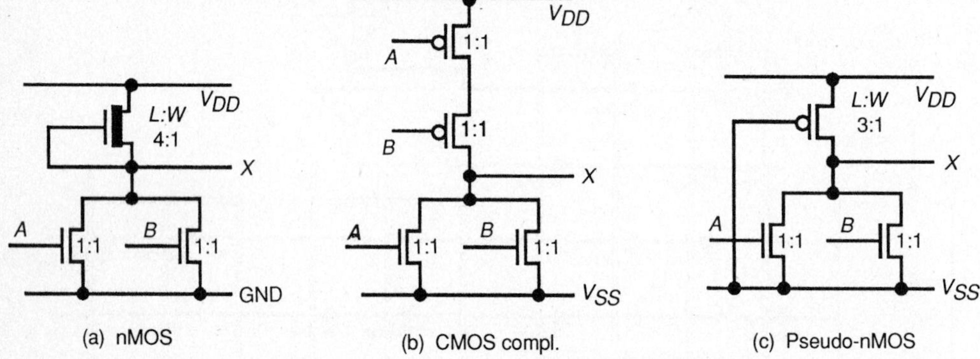

Figure 5.43 Two-input *Nor* gate configurations

(b) (i) nMOS case:

Let "on" resistance of a min. size n-transistor be R_{CHn}, then for the *Nor* gate, each pull-down (pd) transistor will have an "on" resistance Z_{pd} of R_{CHn}. Assuming the inputs come from inverters, then overall ratio Z_{pu}: Z_{pd} is 4:1 and $Z_{pu} = 4\,R_{CHn}$.
Let us further define $T = R_{CHn} \cdot 1\square C_g$.

Therefore output rise time into a $2\square C_g$ load is Z_{pu} X $2\square C_g = 8T$.
The rise time is doubled for an 8:1 overall ratio.
Similarly, worst* output fall time into a $2\square C_g$ load is Z_{pd} X $2\square C_g = 2T$.

(ii) CMOS complementary logic.

For a minimum size pull-up p-type transistor, $R_{CHp} = 2.5\,R_{CHn}$, thus $Z_{pu} = 5\,R_{CHn}$.
Therefore output rise time into a $2\square C_g$ load is Z_{pu} X $2\square C_g = 10T$.
Similarly, worst* output fall time into a $2\square C_g$ load is Z_{pd} X $2\square C_g = 2T$.

(iii) CMOS pseudo nMOS logic:

Here we must obey an overall ratio Z_{pu}:Z_{pd} of 3:1 and noting that for a p-type transistor, $R_{CHp} = 2.5\,R_{CHn}$, the figure for Z_{pu} is 3 X 2.5 $R_{CHn} = 7.5\,R_{CHn}$.

* Halved if both inputs are 1.

Therefore output rise time into a $2\square C_g$ load is $Z_{pu} \times 2\square C_g = 15T$.
Similarly, worst* output fall time into a $2\square C_g$ load is $Z_{pd} \times 2\square C_g = 2T$.

2. Given that:

$$F' = PQS' + PQRST + P'RS + P'S'T \text{ and } D = PQRST'$$

Form the simplest circuit realization for nMOS and for CMOS pseudo-nMOS technology. Your answer should include circuit diagrams.

Solution:

First, resist the temptation to form the expression for F rather than F'. This form is well suited to *Nor* realization which is an easy arrangement to implement in these two technologies.

A further temptation would be to simplify and factorize the expression as follows:

$$F' = PQS' + QRT + P'RS + P'S'T$$

leading to:

$$F' = Q(PS' + RT) + P'(RS + S'T)$$

implying a pull-down structure of 10 transistors as shown in Figure 5.44(a). However, leaving the expression in the original form will allow for the use of a bridging switch and a simpler realisation as shown in Figure 5.44(b) and(c).

The pull-down structure in each case is now seven transistors.

3. For the problem stated in question 2, try to achieve the simplest CMOS complementary logic form of circuit.

Solution:

The design of complementary CMOS logic consists of finding a pull-down expression (i.e. for F') for the n-type logic and the complementary expression for F for the pull-up p-type structure.

We have already arrived at the simplest pull-down arrangement in solution 2 and there is no reason to depart from that.

In order to achieve an equally simple pull-up structure we may map, the function F and group as shown in Figure 5.45.

From the map we may see that:

$$F = PQ' + R'S + P'S'T'.$$

This will give seven transistors in the pull-up structure which matches well with the pull-down as shown in the figure.

4. Draw a circuit diagram for a 4-phase clocked logic circuit to realise the logic circuit given as Figure 5.46(a).

* Halved if both inputs are 1.

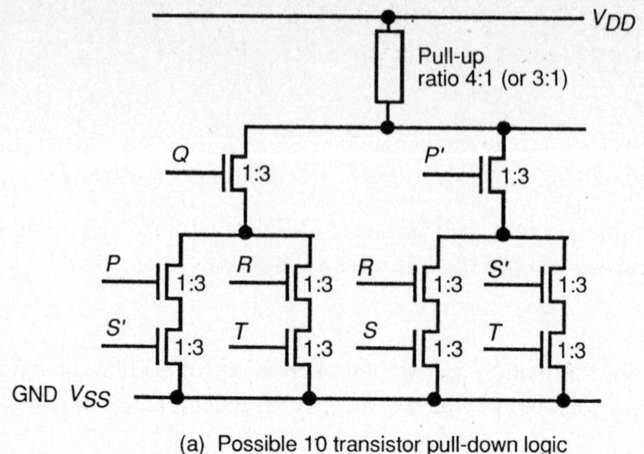

(a) Possible 10 transistor pull-down logic

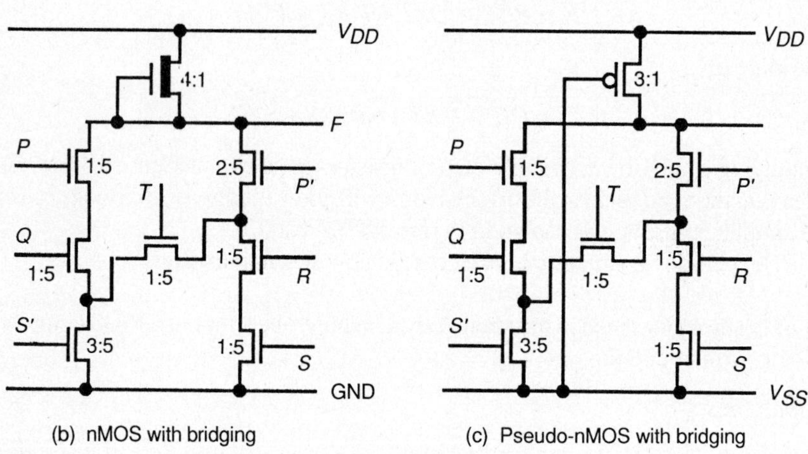

(b) nMOS with bridging

(c) Pseudo-nMOS with bridging

Figure 5.44 Possible nMOS and pseudo-nMOS logic for Question 2

Solution:

Remembering the allowable sequences for interconnecting 4-phase logic (as in Table 5.1 and Figure 5.25) we may draw up a suitable arrangement as in Figure 5.46(b).

5. An 8-way multiplexer can be connected as a universal logic module in three variables. How many functions of three variables are possible? Indicate how the MUX would be so connected and *indicate* how a table of all possible functions would be set out. Do *not* attempt to fill out the complete table.

Solution:

It may be readily shown that for n variables, there are $2^{(2n)}$ possible functions.
Thus, for $n = 3$ there will be $2^{(8)} = 256$ functions.

In order to list them all, we may set out a table, having a form similar to the table in Figure 5.32, as follows:

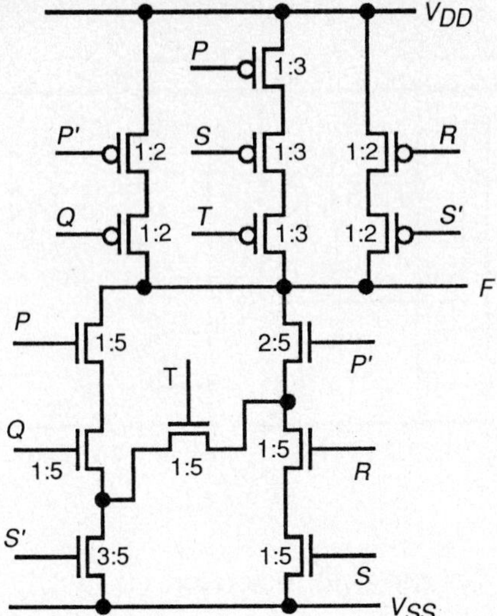

L:W ratios shown are suggested to achieve reasonably fast rise and fall times.

Figure 5.45 Possible CMOS complementary logic for Question 2

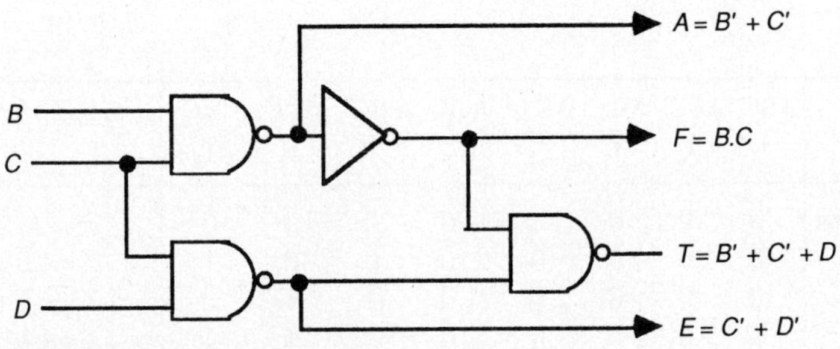

(a) Logic circuit for 4-phase logic realization

Figure 5.46(a) Logic circuit for 4-phase logic realization

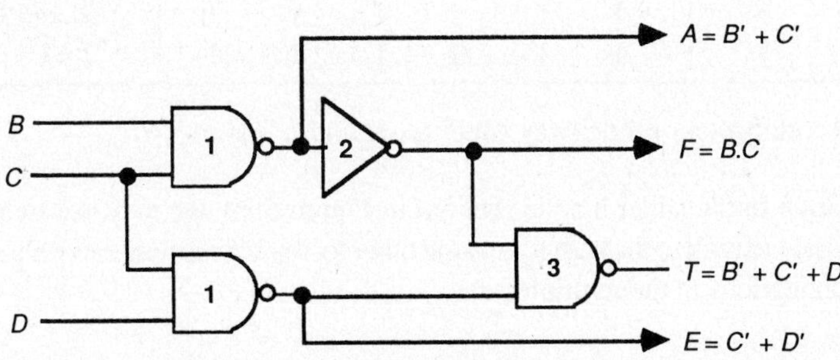

(b) (i) Possible gate types (bold) for 4-phase logic realization

Figure 5.46(b) (i) Possible gate types (bold) for 4-phase logic realization

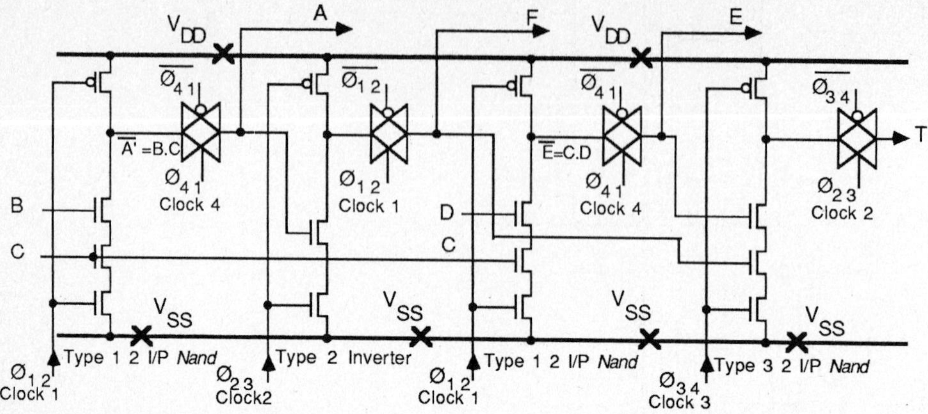

Figure 5.46(b) (ii) Circuit arrangement for 4-phase logic realization

Input variables Input selected	ABC I_7	ABC' I_6	AB'C I_5	AB'C' I_4	A'BC I_3	A'BC' I_2	A'B'C I_1	A'B'C' I_0	O/P Z =f(ABC)
Input connections	0	0	0	0	0	0	0	0	$Z=0$
	0	0	0	0	0	0	0	1	$Z=A'B'C'$
	0	0	0	0	0	0	1	0	$Z=A'B'C$
	0	0	0	0	0	0	1	1	$Z=A'B'$
					etc...				
	1	0	0	0	0	0	0	0	$Z=ABC$
					etc...				
	1	1	1	1	1	1	1	0	$Z=A+B+C$
	1	1	1	1	1	1	1	1	$Z=1$

The required connection of the 8-way MUX is set out in Figure 5.47.

6. As an alternative to the adder bit slice set out in Figure 5.39, we may use two 8-way multiplexers, one to realize the sum S_k and the other to realise the new carry C_k. Set out the circuit connections to the multiplexers.

Solution:

Let us first set out a map of requirements in a convenient form.

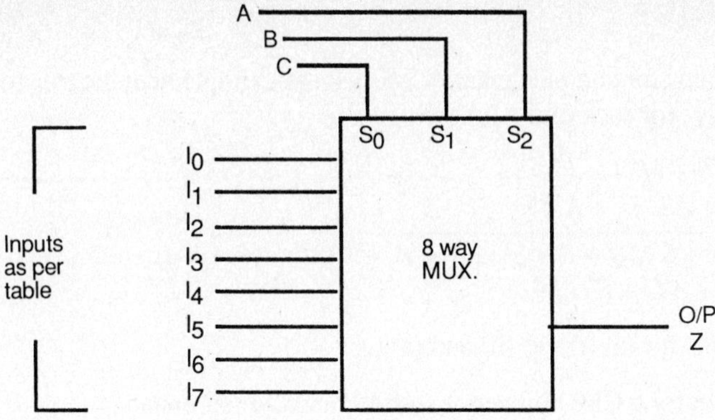

Figure 5.47 Connections to 8-way MUX as a 3-variable ULM

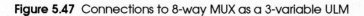

	S_k/C_k	$(AB)_k$ 00	01	11	10
C_{k-1}	0	**0/0** *0*	**1/0** *1*	**0/1** *3*	**1/0** *2*
	1	**1/0** *4*	**0/1** *5*	**1/1** *7*	**0/1** *6*

Weighting $C_{k-1} = 4$; $A_k = 2$; $B_k = 1$ we make the connections shown in Figure 5.48, connecting 0s and 1s to the MUX inputs in accordance with the minterm numbers (*in italics*) in the map as shown.

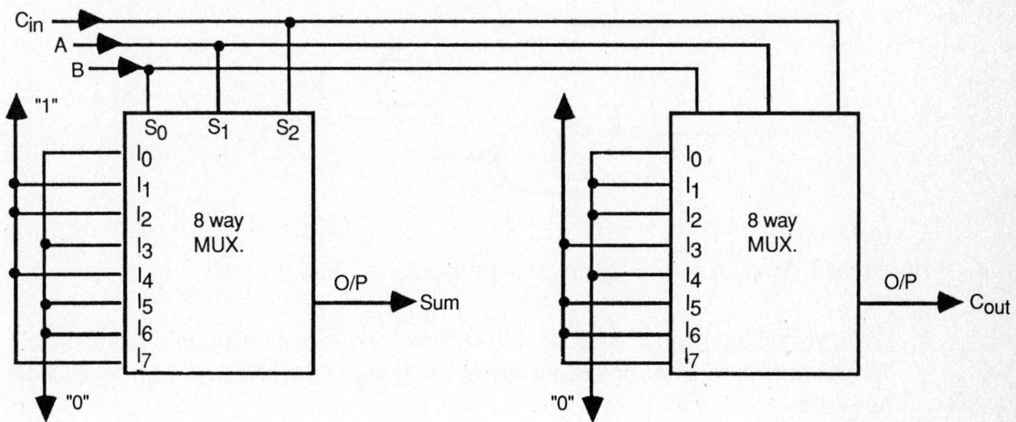

Figure 5.48 Multiplexer based adder

5.7 Tutorial 5

1. (a) Draw circuit diagrams for complementary CMOS logic to implement the true form (Q, F, P, T respectively) for each of the following:

   (i) $Q' = M + M'P + M'N$
   (ii) $F = A'C'D' + A'C.D' + A.B'D'$
   (iii) $P = \overline{(Q + R' + S).(Q + R + S').(Q' + R' + S).(Q' + R + S')}$
   (iv) $T' = A'C'D' + A'D + B'C.D'$

   (b) Draw stick diagrams for (a) (i), (a) (ii) and (a) (iv).

   (c) Draw a mask layout for a CMOS inverter with I/P and O/P on metal.

2. (a) Draw a circuit diagram for a 4-bit CMOS dynamic shift register (right shift only) and thence determine a "leaf cell" from which registers of any length can be formed by replication of the cell.

   (b) Draw the stick diagram and the mask layout for the leaf cell.

   (c) From (b) determine the bounding box dimensions and connection detail for the leaf cell. Thence determine the overall dimensions and connecting points for a 4-bit register.

3. (a) The logic for a form of *Exclusive or* gate is given as Figure 5.49. Draw a stick diagram and the mask layout for this arrangement.

   (b) Devise a circuit arrangement to act as a 1-bit CMOS static data latch. Draw a stick diagram and mask Layout for this cell.

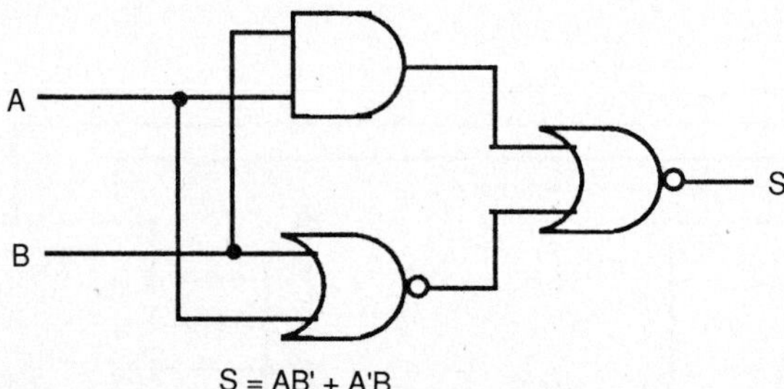

Figure 5.49 An exclusive Or gate arrangement

4. Project/laboratory work. Using the shift register, *Exclusive or* gate and static data latch designs developed in the preceding questions (2) and (3), configure a 4-bit serial adder subsystem.

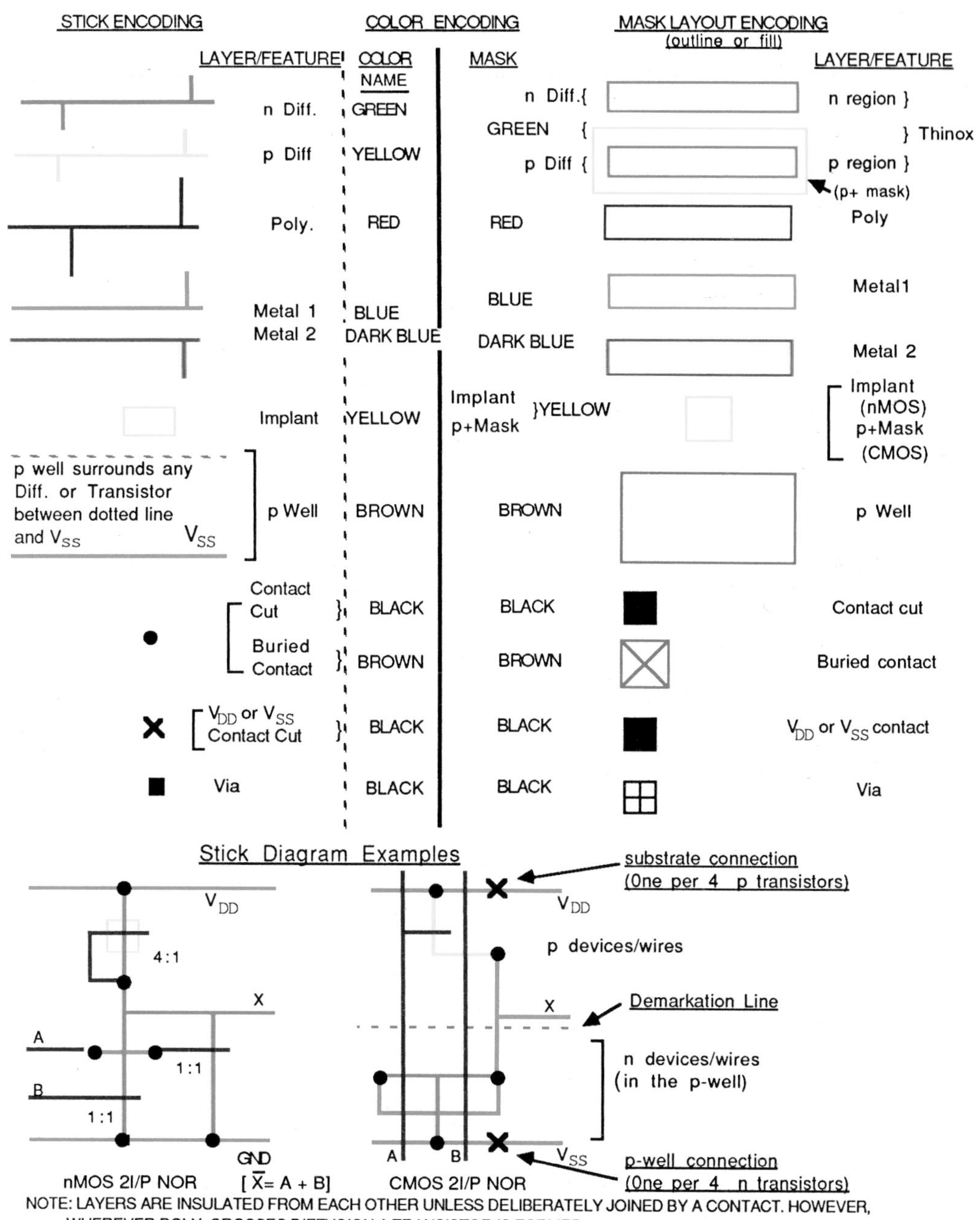

Color Plate 1 Color Encoding Schemes

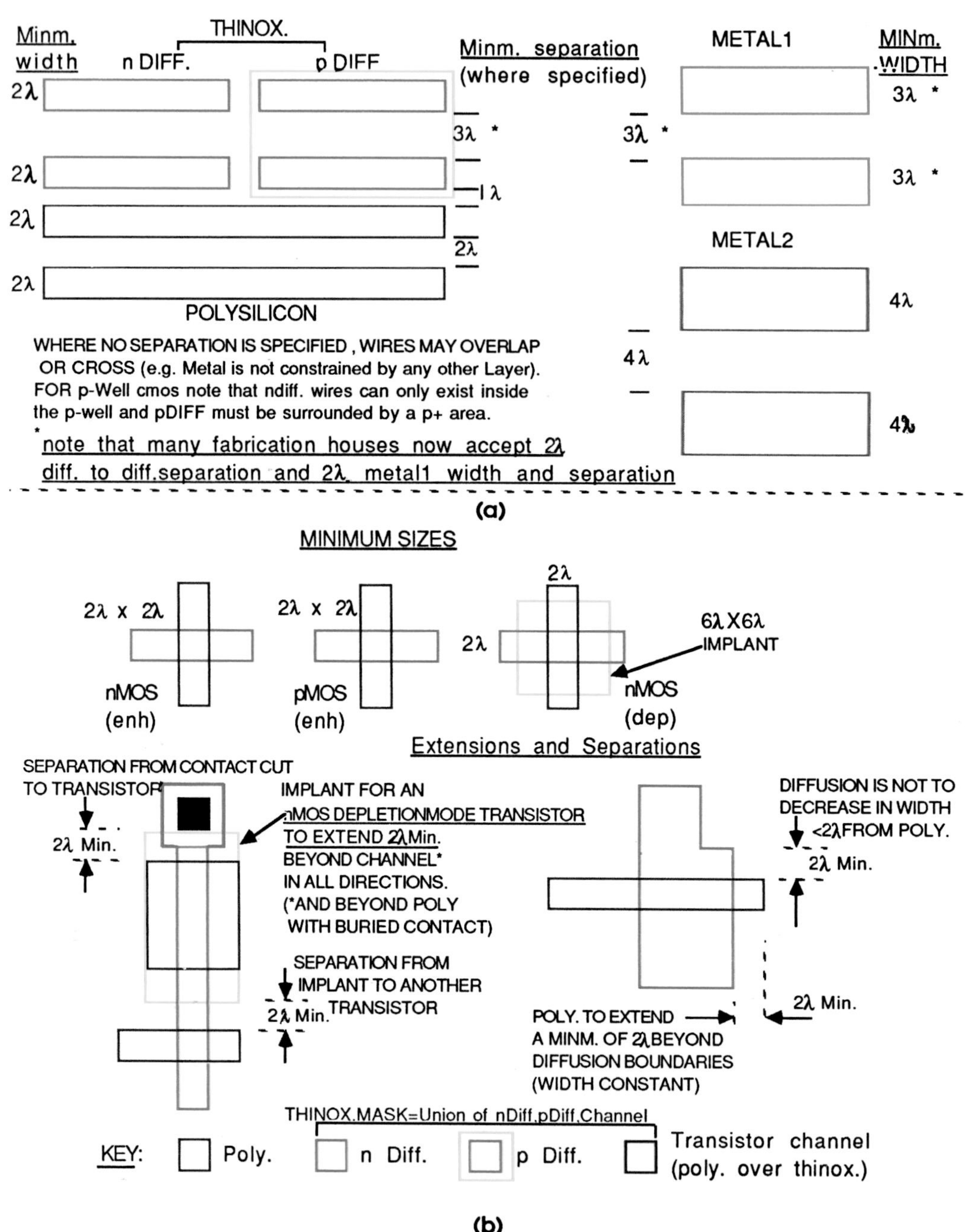

Minm. width
THINOX.
n DIFF.
p DIFF
Minm. separation (where specified)
METAL1
MINm. WIDTH
2λ
3λ *
3λ *
3λ *
1 λ
2λ
METAL2
4λ
4λ
POLYSILICON
WHERE NO SEPARATION IS SPECIFIED, WIRES MAY OVERLAP OR CROSS (e.g. Metal is not constrained by any other Layer).
FOR p-Well cmos note that ndiff. wires can only exist inside the p-well and pDIFF must be surrounded by a p+ area.
*note that many fabrication houses now accept 2λ diff. to diff.separation and 2λ metal1 width and separation
(a)
MINIMUM SIZES
2λ x 2λ
nMOS (enh)
pMOS (enh)
6λ X6λ IMPLANT
nMOS (dep)
Extensions and Separations
SEPARATION FROM CONTACT CUT TO TRANSISTOR
2λ Min.
IMPLANT FOR AN nMOS DEPLETIONMODE TRANSISTOR TO EXTEND 2λ Min. BEYOND CHANNEL* IN ALL DIRECTIONS. (*AND BEYOND POLY WITH BURIED CONTACT)
SEPARATION FROM IMPLANT TO ANOTHER TRANSISTOR
DIFFUSION IS NOT TO DECREASE IN WIDTH <2λ FROM POLY.
POLY. TO EXTEND A MINM. OF 2λ BEYOND DIFFUSION BOUNDARIES (WIDTH CONSTANT)
THINOX.MASK=Union of nDiff,pDiff,Channel
KEY:
Poly.
n Diff.
p Diff.
Transistor channel (poly. over thinox.)
(b)

Color Plate 2(a) Design rules for width and separation of wires (nMOS and CMOS)

(b) Transistor design rules (nMOS, pMOS and CMOS)

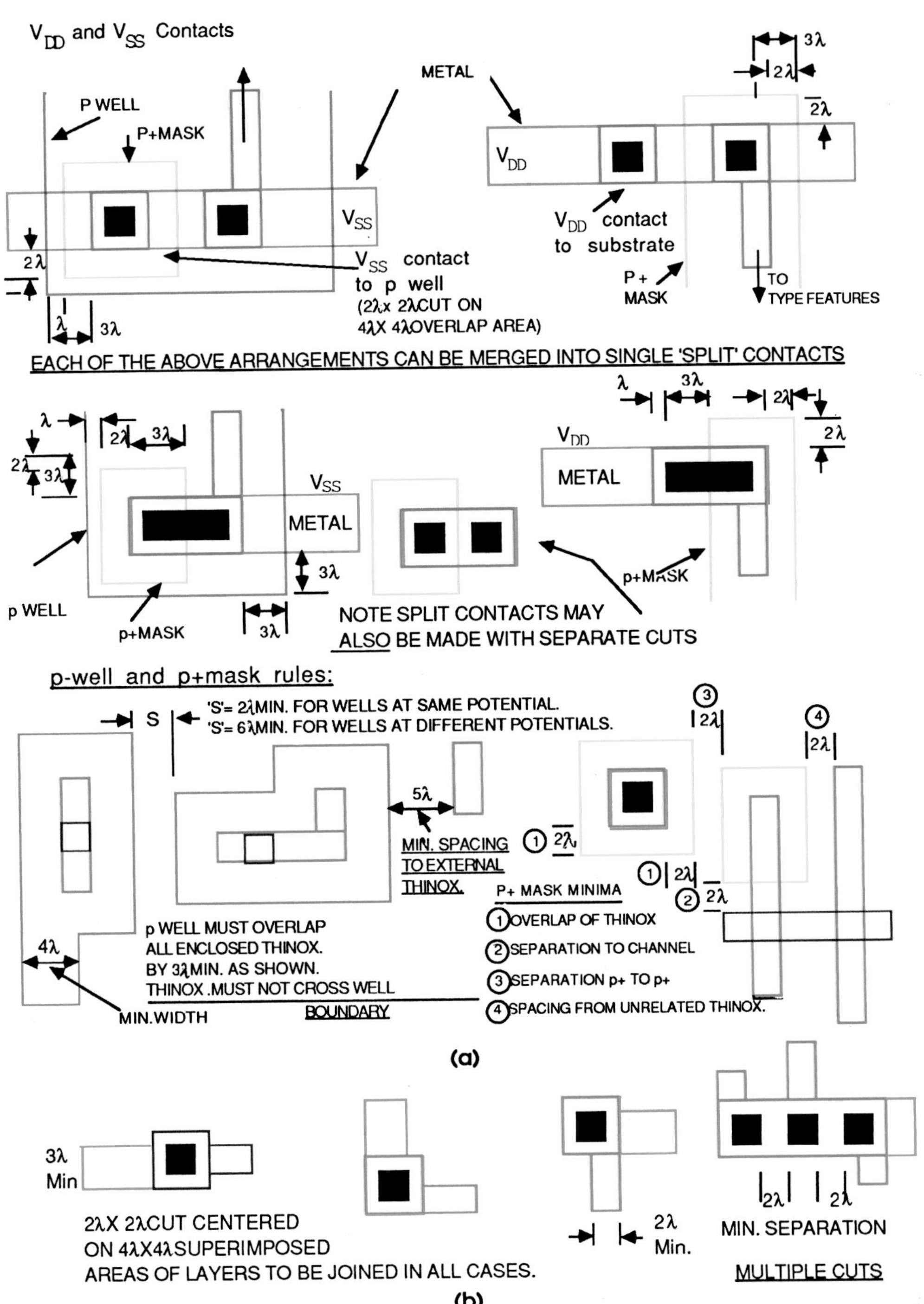

Color Plate 3(a) Particular rules for p-well CMOS process

(b) Simple contacts (nMOS and CMOS)

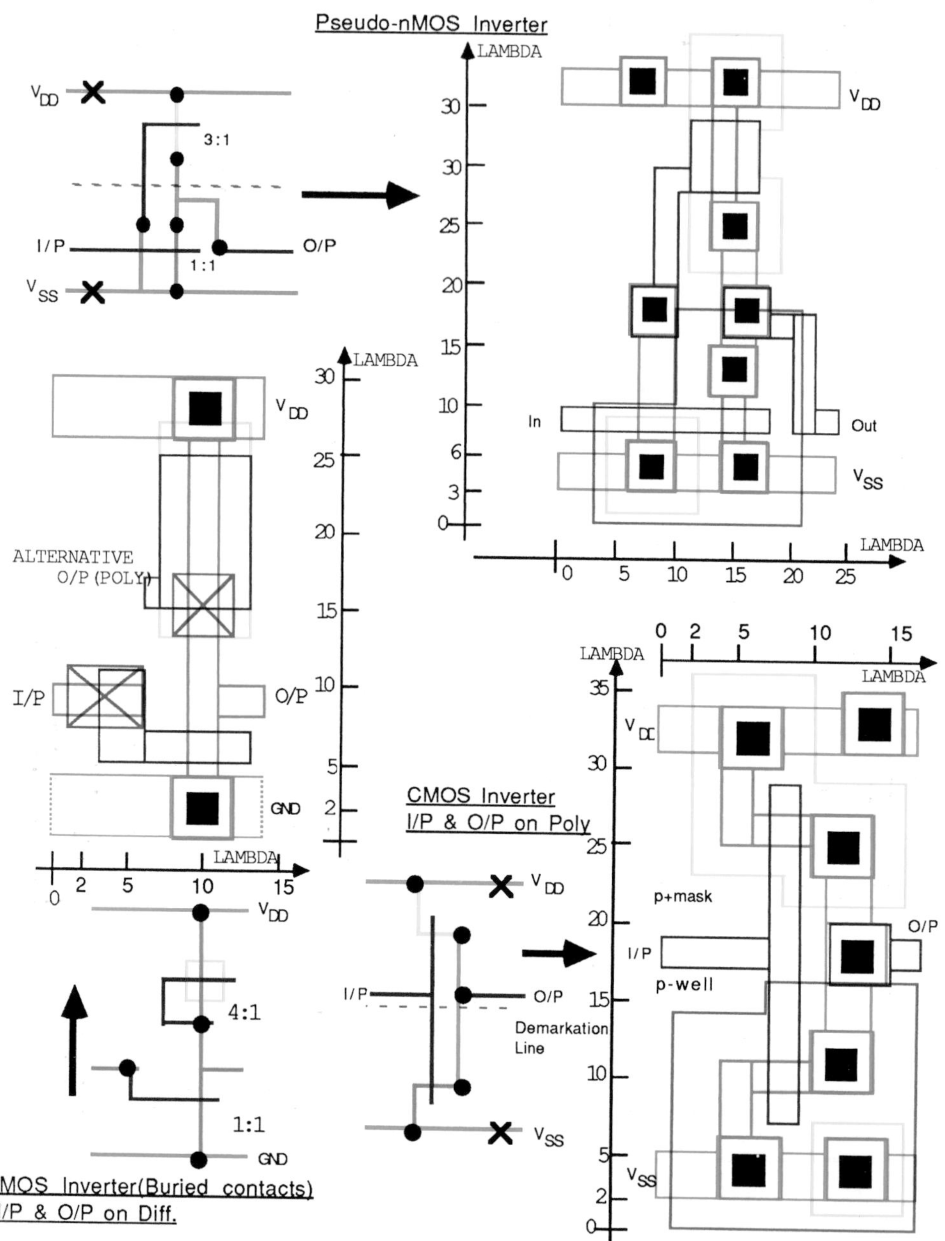

Color Plate 4 Example Layouts

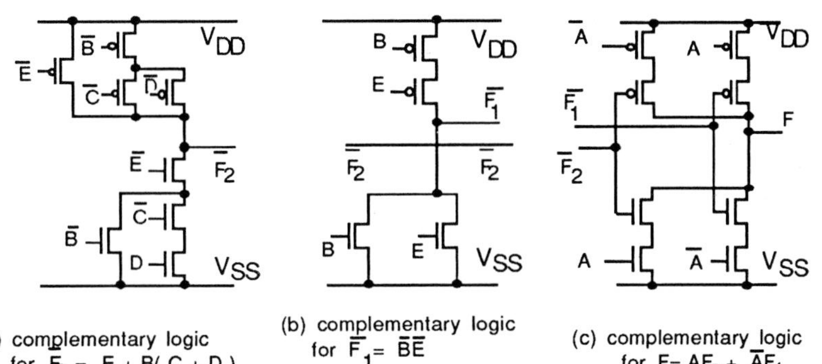

(a) complementary logic for $\bar{F}_2 = E + B(C + D)$

(b) complementary logic for $\bar{F}_1 = \bar{B}\bar{E}$

(c) complementary logic for $F = AF_2 + \bar{A}F_1$

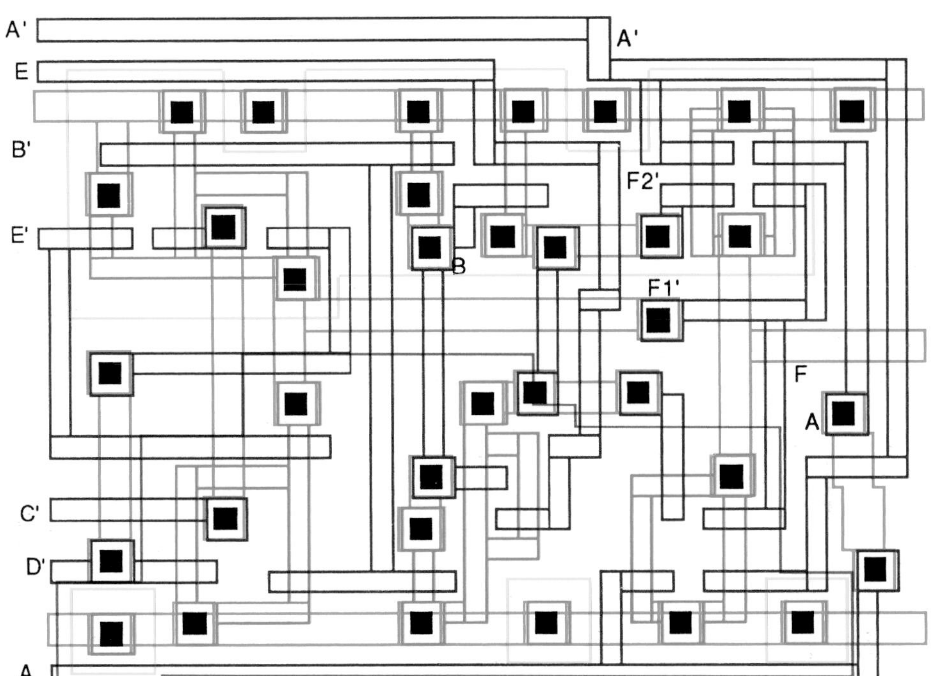

Color Plate 5 One possible mask layout for Example 2

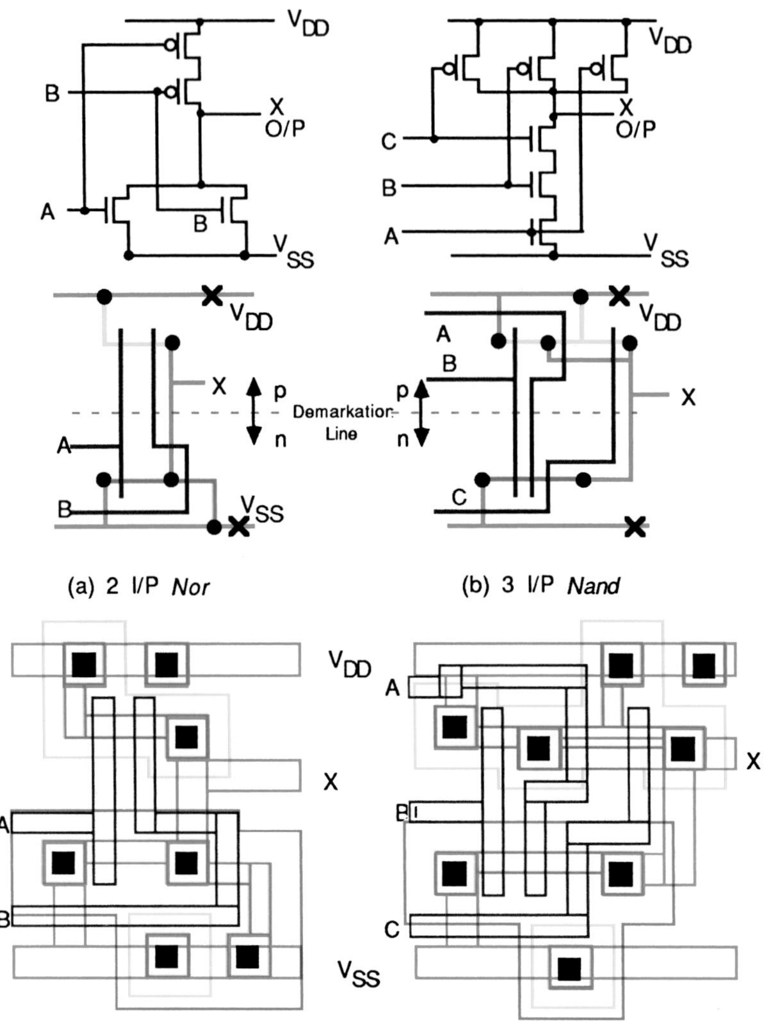

Color Plate 6 Possible Stick diagrams and Mask Layouts for typical CMOS gates

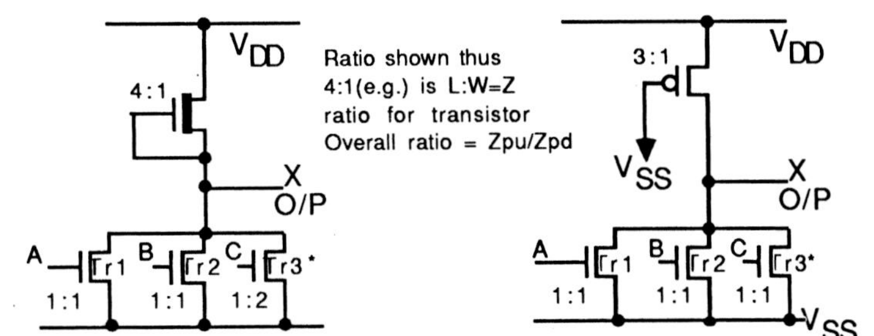

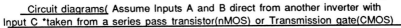

Circuit diagrams(Assume Inputs A and B direct from another inverter with Input C *taken from a series pass transistor(nMOS) or Transmission gate(CMOS)

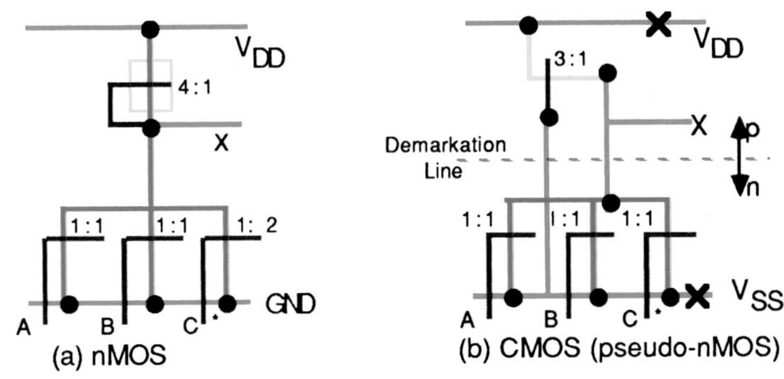

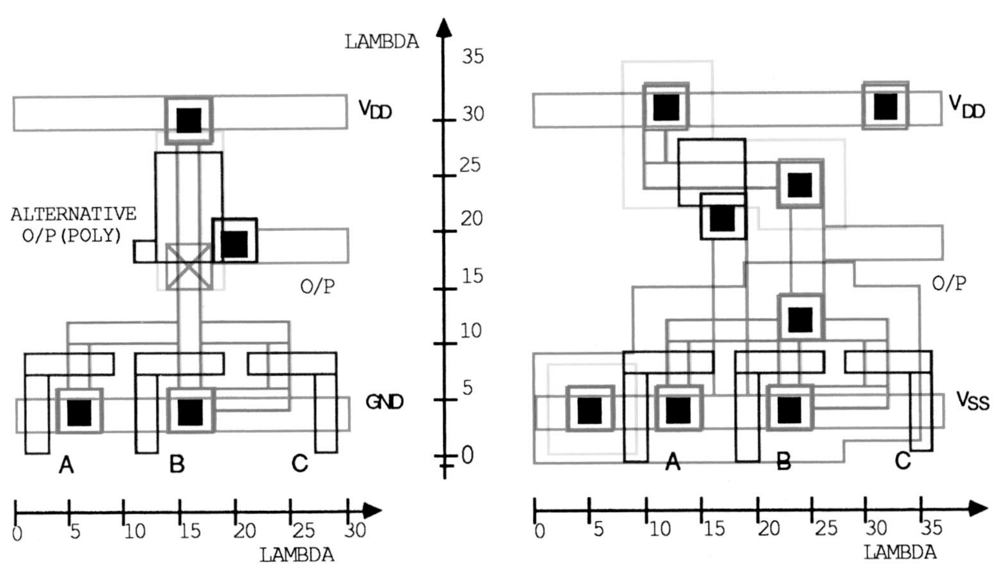

Color Plate 7 Possible Stick and Mask Layout for 3 I/P *Nor*

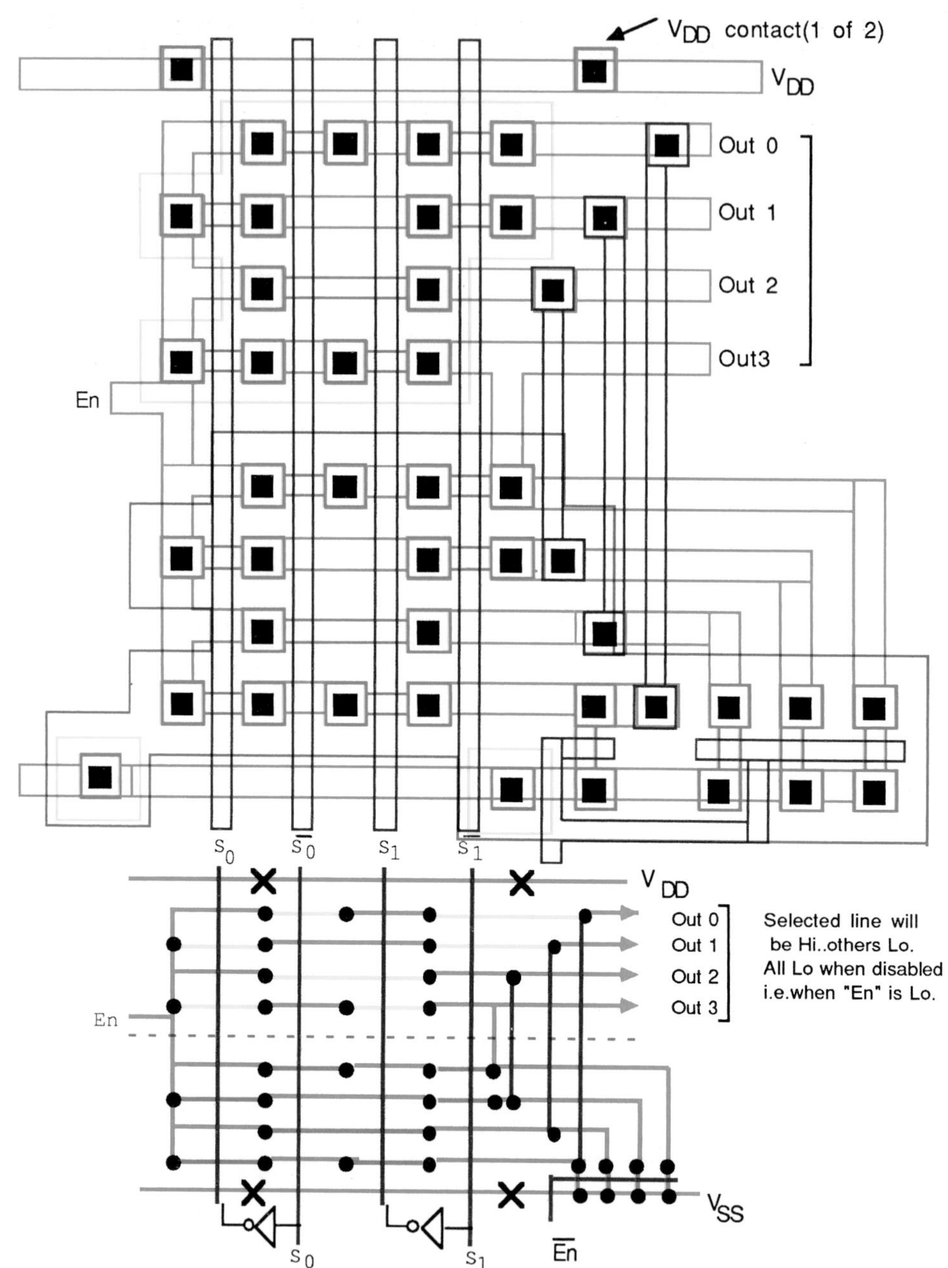

Color Plate 8 2 to 4 line Decoder

6 Asynchronous (fundamental mode) sequential logic

We have already demonstrated (Chapter 5) a need for circuits to act as registers (temporary data storage elements) and to act as memories for individual logic signals (for example, the carry information in the adder).

Clearly, the logic circuits we have dealt with so far are entirely combinational in nature and do not exhibit a memory property.

As well as memory elements, we shall also need to develop circuits which are *sequence dependent* in order to design counters, code generators etc.

These requirements are met by sequential logic circuits of which there are two major types: *asynchronous* and, *synchronous* or *clocked* sequential circuits.

6.1 A basic sequential logic circuit

Let us start with a circuit composed of *Nand* gates the behaviour of which is by now well known to us. A simple arrangement, with two inputs A and B and a single output Y is given in Figure 6.1(a).

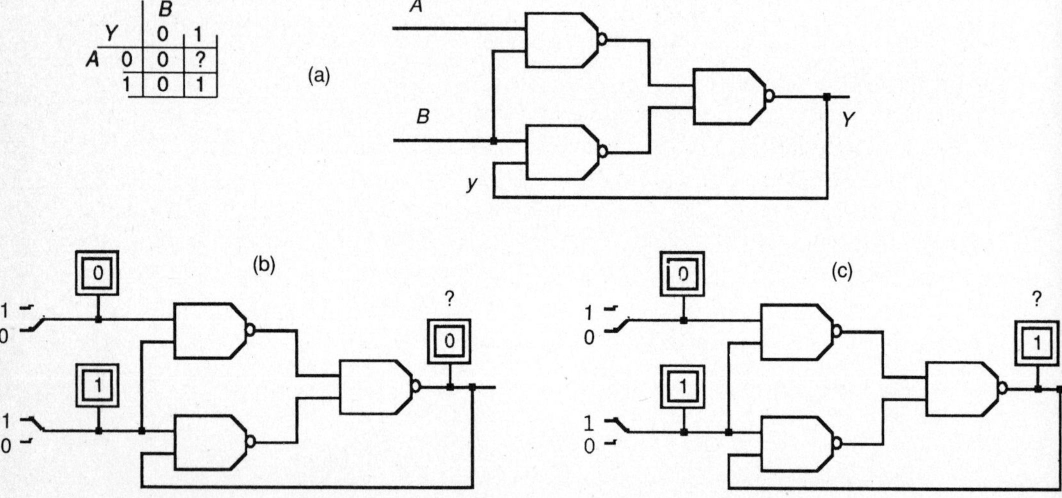

Figure 6.1 Simple circuit for consideration

In simulating (as in Figure 6.1(b) (c)), or in wiring up and testing this circuit, you would find that it is impossible to draw up a simple truth table to fully define the circuit operation. In particular, the top right-hand cell (truth table of Figure 6.1) cannot be filled in with a single logic level because, for these input conditions, output Y can be either 0 or 1 depending on the state of the circuit prior to the inputs taking on the value $AB = 01$. In other words, the behavior is sequence dependent. If we constrain the inputs so that only A or B can change at any instant (simultaneous changes are prohibited) then, $AB = 01$ can follow $AB = 00$, in which case Y will be 0; or, $AB = 01$ can follow $AB = 11$, in which case Y will be 1 in the cell in question.

The reason for the sequence sensitivity is that we now have feedback from output to input in the form of a direct connection between output Y and input y. Thus, when input changes to A or B occur, the circuit takes account of the present state of the output Y in determining the next state of the output Y. *Thus we have created a sequential logic circuit.* The circuit belongs to the asynchronous class of circuits because changes of state are determined by circuit delays only and are not dependent on clock activation.

A final comment on Figure 6.1 is that some readers may be mystified by the fact that a single signal line is labelled Y at one end and y at the other. Although these points are one and the same, we must consider the case when an input change occurs. The new state of Y takes some small time to propagate through the circuit and during that time new Y and present y may well differ.

Once we have established this then there is no conceptual difficulty in treating y as an input to the circuit and Y separately as the output. We may now draw up a general model for such circuits.

6.2 A general model for sequential logic circuits

A general model, applicable to both classes of sequential circuits, is given as Figure 6.2 and will be seen to comprise combinational logic having some inputs which are output derived. There may also be independent inputs as shown. All or some of the outputs are fed back as in the figure. The information fed back is referred to as the *state vector* and the number of lines comprising the state vector determines the number of states which can be assumed by the sequential circuit (or finite state machine) such that the maximum number of states = 2^v where v is the number of state vector bits.

The general model shows delay elements of some sort connected in series in the feedback path (through which signals may be "clocked" in synchronous systems) to introduce delay t_d, which may not be the same for each path.

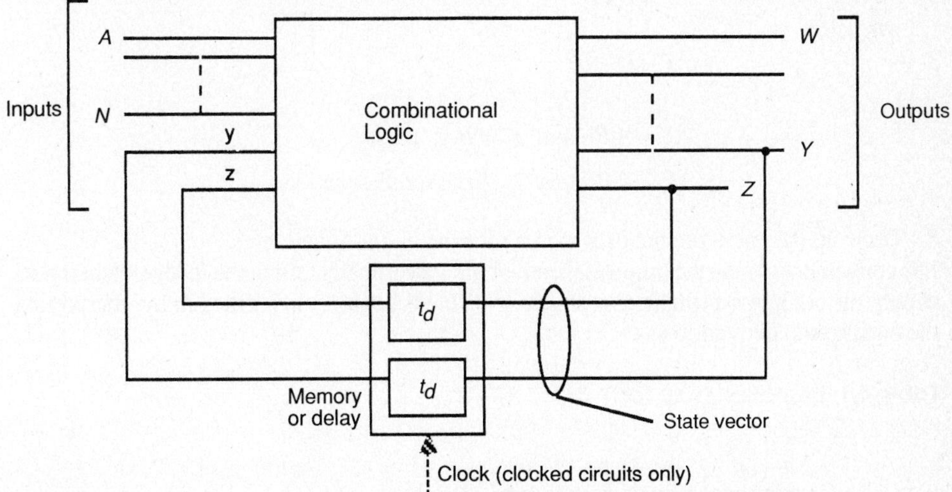

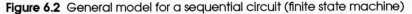

Figure 6.2 General model for a sequential circuit (finite state machine)

We must also recognize that delays t_d are not necessarily confined to the feedback path but may instead (or as well) be present in the forward path through the combinational logic.

The action implied by this arrangement is such that, when input changes occur, the next state to be assumed is dependent not only on the independent inputs A, B etc., but also on the present state of the state vector. The circuit is thus sequence dependent, or, to put it another way, the circuit exhibits a memory property. The first task we will undertake will be to examine the analysis of the behavior of a simple asynchronous logic circuit.

6.3 Simple analysis of asynchronous sequential logic circuitry

Let us start by analysing the simple circuit discussed earlier and reproduced here for convenience as Figure 6.3. The process of analysis comprises the following steps:

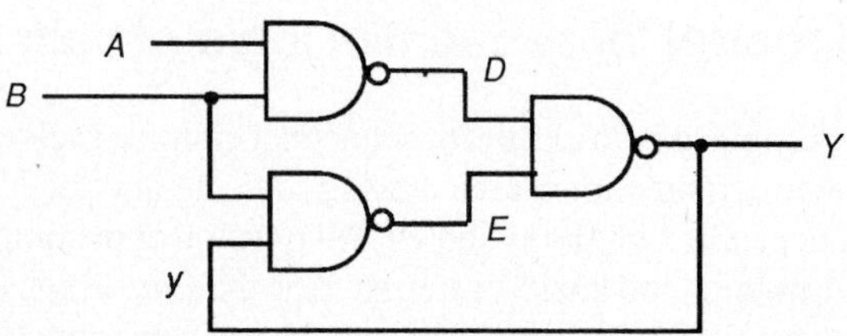

Figure 6.3 Circuit for analysis

1. Write the combinational logic expressions for the entire circuit.
   In this case:

$$Y' = D.E$$

whence:

$$Y = D' + E'$$

but,

$$D' = A.B \quad \text{and} \quad E' = B.y; \text{ thus:}$$

$$Y = A.B + B.y \qquad \text{the overall equation.}$$

2. Draw up the excitation map(or matrix) for the output function(s).

It is convenient to map the output dependent inputs vertically letting the independent inputs supply the column information as shown in Table 6.1. The map is filled in by interpreting the equation(s) derived in (1).

Table 6.1 Excitation map for *Y*

y \ *AB* (*Y*)	00	01	11	10
0	0	0	1	0
1	0	1	1	0

3. Determine the stable states of the circuit

These are the states in the map for which the entry for Y and the condition of y in that row are in agreement. Clearly, any cells in which Y differs from y must be unstable and can only persist for a short time t_d. The stable states are circled in Table 6.2.

Table 6.2 Excitation map with stable states circled

Y \ AB	00	01	11	10
y 0	(0)	(0)	1	(0)
1	0	(1)	(1)	0

4. Set out the circuit behavior in the form of a flow matrix

Clearly, there are five stable states. Two different ways of indicating this are illustrated in Table 6.3

Table 6.3 Forms of flow matrix

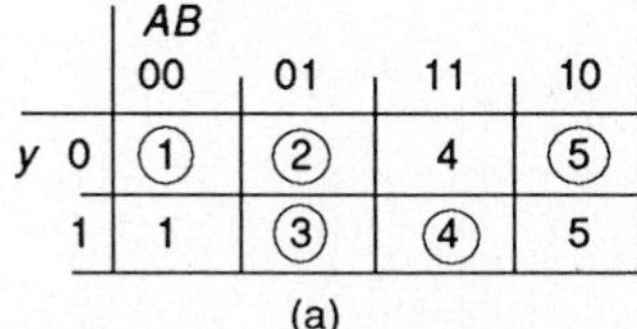

AB	00	01	11	10
y 0	(1)	(2)	4	(5)
1	1	(3)	(4)	5

(a)

AB	00	01	11	10
y 0	0↑	0		↑0
1		0	0↓	

(b)

In (a) the stable states are circled and numbered. The unstable states are also numbered (but not circled) with the number of the stable state to which each leads in the same column. In (b) the stable states are shown by Os with · for unstable states. Arrows are used to link unstable to corresponding stable states.

Clearly, from Table 6.3, when $A.B = 0.0$ the circuit will take up stable state 1 (Y=0). For $A.B$ =1.1, the stable state is 4 (Y=1), and for $A.B$ =1.0 the stable state is 5 (Y=0). However when $A.B = 0.1$ there is a choice of stable states 2 or 3 and correspondingly complementary values for Y. State 2 will be assumed when the circuit was previously in state 1 or 5; state 3 will result when the change is from state 4.

6.3.1 A further, more complex, example of analysis

Having ventured into the analysis of a simple circuit with a single variable state vector, let us now analyze a more complex, but nevertheless quite straightforward, circuit with a 2-variable state vector (two output variables fed back).

The circuit to be considered is given as Figure 6.4 with inputs and outputs identified as shown.

The analysis proceeds through the steps already set out as follows:

1. Write the combinational logic expressions for the complete circuit.

$$X' = R + y \quad X = R'.y'$$
$$Y' = S + x \quad Y = S'.x'$$

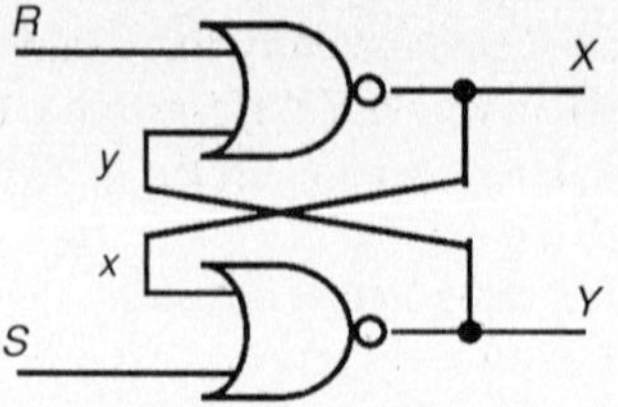

Figure 6.4 Circuit with two feedback paths

2. Draw up the excitation map for the output functions (Table 6.4).

Table 6.4 Excitation map

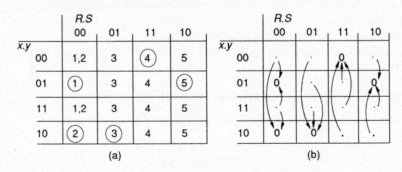

xy \ R.S (X.Y)	00	01	11	10
00	11	10	(00)	01
01	(01)	00	00	(01)
11	00	00	00	00
10	(10)	(10)	00	00

For convenience, *X* and *Y* may be mapped side by side in each cell of the map as shown in Table 6.4. Also, it will be noted that the output dependent inputs *x* and *y* are, again, mapped vertically to define the rows of the map. The independent inputs define the columns of the map.

3. Determine the stable states of the circuit.
The stable states are entries for *X* and *Y* which match *x* and *y* for the row in which they reside. The five stable states in this case are circled in Table 6.4.

4. Set out the circuit behavior in the form of a flow matrix.
The appropriate flow matrix is given as Table 6.5(a) with an alternative form in (b).

Table 6.5 Flow matrix for the example

(a)

x.y \ R.S	00	01	11	10
00	1,2	3	(4)	5
01	(1)	3	4	(5)
11	1,2	3	4	5
10	(2)	(3)	4	5

(b)

x.y \ R.S	00	01	11	10
00	.	.	0	.
01	0		.	0
11	.	.	.	.
10	0	0	.	.

It will be seen that the circuit has five stable states, one in each column for $R.S = 0.1$, $R.S = 1.1$, $R.S = 1.0$ and two in the $R.S = 0.0$ column. Clearly, when $RS = 10$ the circuit will take up stable state 5 [$XY = 01$] since all cells in this column lead to this state. Similarly, when $RS = 01$ the circuit will stabilize in state 3 [$XY = 10$]. If RS changes to 00 from either of these conditions then the circuit transfers across columns to stable state 1 [$XY = 01$], or to stable state 2 [$XY = 10$] respectively.

This behavior demonstrates the memory property since the input conditions on RS are "remembered" when the inputs RS are returned to 00 (the "no input" condition).

However, if both R and S are connected to 1 simultaneously, the circuit will assume state 4[$XY = 00$] and if we then return to the "no input" condition, the circuit conditions will be those represented by the top left hand cell of Table 6.5(a) which represents an unstable condition from whence the circuit may assume either stable state 1 or 2. The actual state assumed is determined by what is known as a "critical race" in which the final state depends on the relative propagation delays of the two *Nor* gates forming the circuit. This arrangement is, in fact, the basic RS flip-flop* which we may now further examine for its properties.

6.3.2 The RS flip-flop (reset-set flip-flop)

Consider the flow matrix and the repeated circuit in Figure 6.5.

xy / X.Y \ R.S	00	01	11	10
00	1,2	3	(4)	5
01	(1)	3	4	(5)
11	1,2	3	4	5
10	(2)	(3)	4	5

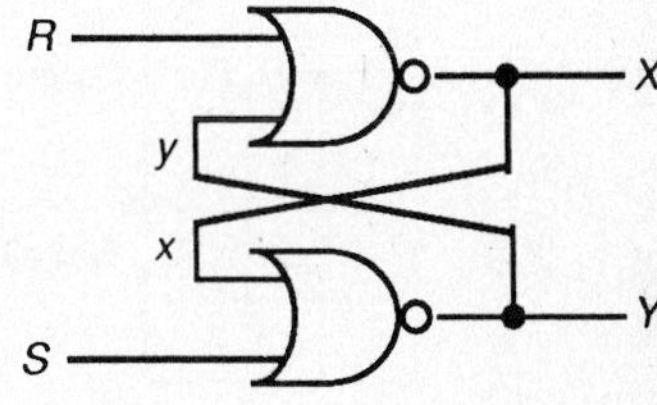

Figure 6.5 The RS flip-flop

We may analyze the behaviour of this circuit by first specifying that:

(a) A "no input" condition exists when $R = S = 0$.
(b) Simultaneous changes of R and S are not allowed to occur.

A reasoned analysis:

1. From the no input condition, let R become 1 (i.e. ΔR, then $RS = 10$). Thus the circuit goes to stable state 5 [$XY = 01$] (possibly via unstable states 5 in the $RS = 10$ column of the flow matrix).

   Now let R revert to 0 (i.e. ∇R, then $RS = 00$, the no input condition). The circuit conditions will move across columns to stable state 1 and XY will remain = 01.

   Thus the condition induced by the original ΔR input (i.e. 1 to R) is remembered at the outputs.

* Note: the name "flip-flop" is believed to have originated from the noise made by electro-magnetic relays originally used in switching applications.

2. From the no input condition, let S become 1 (i.e. ΔS, then $RS = 01$). Thus the circuit takes up stable state 3[$XY = 10$] (possibly via unstable states 3 in the $RS = 01$ column).
   Now let S revert to 0 (∇S, back to the no input condition). The circuit conditions will move across one column to stable state 2 and XY will remain =10.
   Thus the condition induced by the original ΔS input (i.e. 1 to S) is remembered at the outputs.

3. In the event of both R and S becoming 1 following (1) or (2), the circuit takes up stable state 4 and the output of the circuit becomes $XY = 00$.
   Now, we have constrained the input conditions such that R and S cannot change simultaneously, so that the next input condition must be either $RS = 10$ or $RS = 01$ and the contentions in (1) or (2) will then hold.

It is important that the stipulation that simultaneous changes of R and S be not allowed is adhered to, otherwise a return to "no input" from $RS = 11$ will result in an uncertain outcome which is no good for a memory element

Flip-flop characteristics are conveniently set out as a table and those of an *RS* flip-flop are set out in Table 6.6.

Table 6.6 RS flip-flop characteristics

R_t	S_t	Q_{t+1}	
0	0	Q_t	Output unchanged
1	0	0	
1	1	N.A.	Not allowed
0	1	1	

This table assumes the constraints already set out for inputs R and S cannot be imposed in practice but we now specify that R and S can never be 1 simultaneously (i.e. $RS = 11$ is not allowed).

The t subscript indicates the time at which we assume an input to R or S occurs.

The t+1 subscript indicates the time at which the consequent output condition occurs and the following time period when the no input condition is resumed.

Since the condition $R = S = 1$ is not allowed, then, in all the allowable stable states, X and Y are complementary, so that we could replace the letter X by Q and write Q' in place of Y. Table 6.6 has been drawn up on that basis and a symbolic representation of the basic RS flip-flop follows as Figure 6.6. A CMOS stick diagram and Mask layout for an *RS* flip-flop are set out in Figure 6.7.

The *RS* flip-flop is an important memory/sequential circuit element and we will consider it further in Chapter 7.

Exercise: Using two 2 I/P *Nor* gates, or a hardware facility, such as a Logic Trainer, or using a software simulator, such as LogicWorks™, wire up or plug up or set up the arrangement of Figure 6.5 and test out its operation, checking with the contentions in the text.

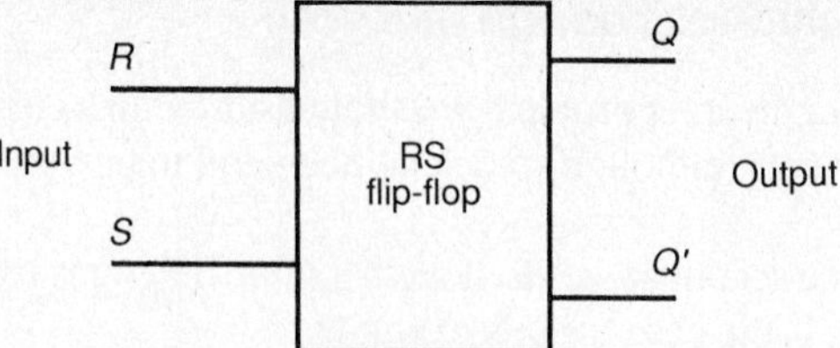

Figure 6.6 RS flip-flop symbol

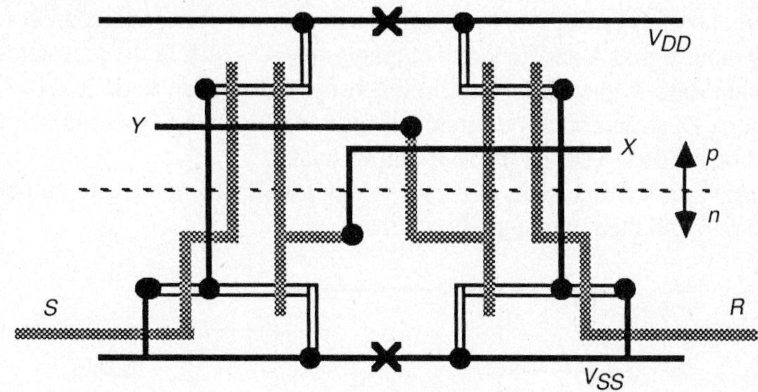

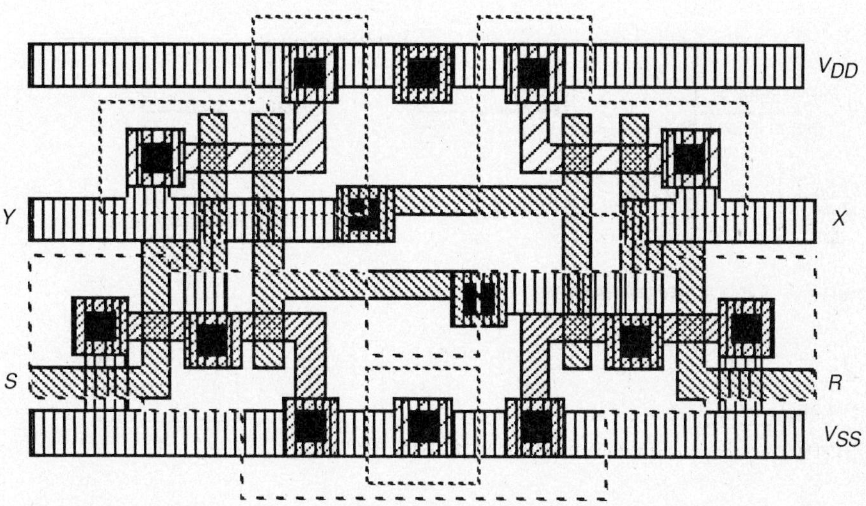

Figure 6.7 Stick diagram and mask layout for asynchronous RS flip-flop

6.4 Synthesis of simple asynchronous sequential circuits

Synthesis is in effect the process of analysis in reverse.

The general approach is best illustrated by examples and we will start off by taking a simple case in which a sequential circuit has to be designed.

6.4.1 A simple asynchronous design problem

As is the case with any engineering problem we should start with a clear and unambiguous specification of requirements (although some engineers fail to pay proper attention to this important matter).

Sometimes this can be set out in words alone but often diagrams may be useful or even essential for clarity. This is the case for this example.

1. *Specification*
   An asynchronous circuit is to be designed to respond to changes in logic level on a single binary input signal A and generate output sequences of four distinct states, the change from one state to the next being initiated by a change in logic level of *A* (in either direction). Diagrams of the requirements follow in Figure 6.8 (two output lines, say *X* and *Y*, will be needed to define the four output states).
2. Allocate variables and states of these variables to represent the four output states (as in Figure 6.9) and then draw up a flow table.

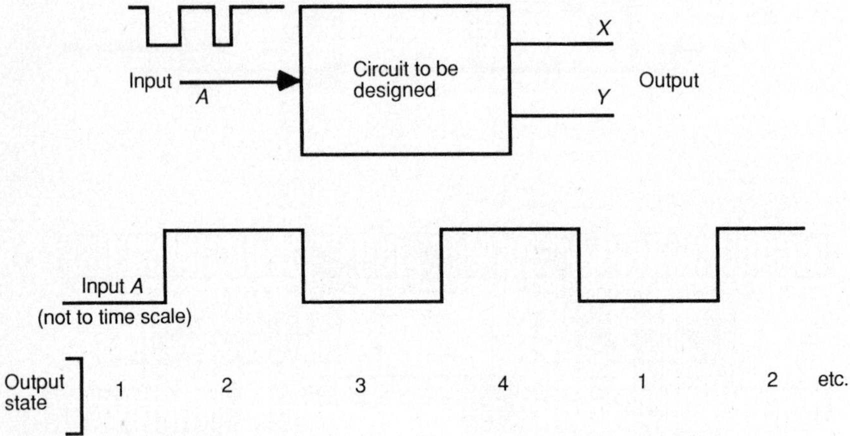

Figure 6.8 Circuit requirements

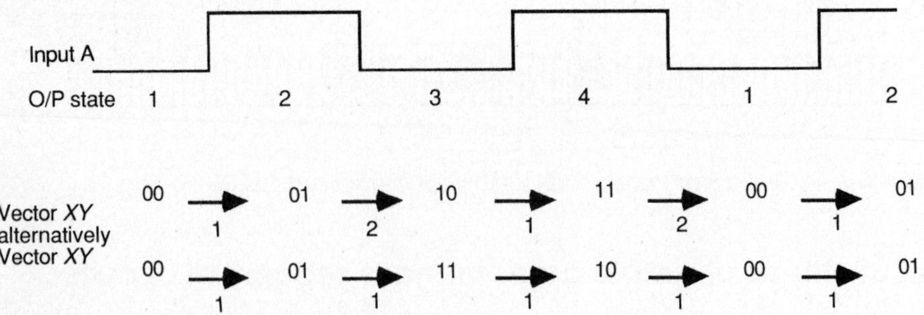

Figure 6.9 Two possible state allocations

Two possible allocations of output vector states *XY* are given in Figure 6.9. The second of the two allocations is to be preferred in this case since it minimizes the total number of transitions of the output variables and ensures that only one variable changes at each state change. (The numbers of transitions involved in each allocation are written under the arrows between states in the figure).

A flow table may now be drawn up (Table 6.7) since we now know that the required circuit has the form set out in Figure 6.10. There must be two bits for the feed-back vector in order to identify the four states required by the specification. The state vector also provides the output in this case.

Table 6.7 Flow table

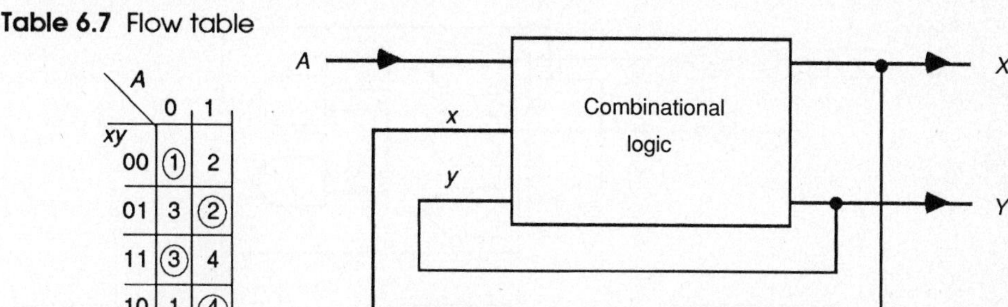

xy \ *A*	0	1
00	(1)	2
01	3	(2)
11	(3)	4
10	1	(4)

Figure 6.10 Form of circuit for the example

The entries in the flow table follow directly from the allocations of *X* and *Y* made in Figure 6.9. Having entered the stable states unstable states are then entered when *A* changes, (i.e. in the same row), and the number entered will be that of the stable state which follows next.

3. From the flow table we may now draw up the excitation map as in Table 6.8. To do this, first enter the values of *XY* which create the desired stable states (for a stable state, the entry for *XY* = the value of *xy* in that row) as in Table 6.8(a). The table is then completed by entering the values of *XY* appropriate to the number allocated to each unstable state (Table 6.8(b)).

Table 6.8 Development of the excitation map

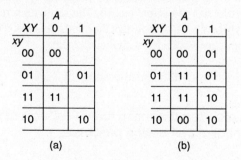

xy \ *XY*, *A*	0	1
00	00	
01		01
11	11	
10		10

(a)

xy \ *XY*, *A*	0	1
00	00	01
01	11	01
11	11	10
10	00	10

(b)

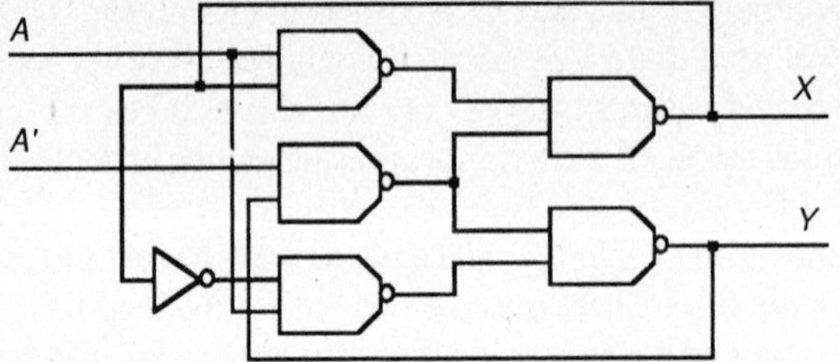

Figure 6.11(a) Required circuit in *Nand* form

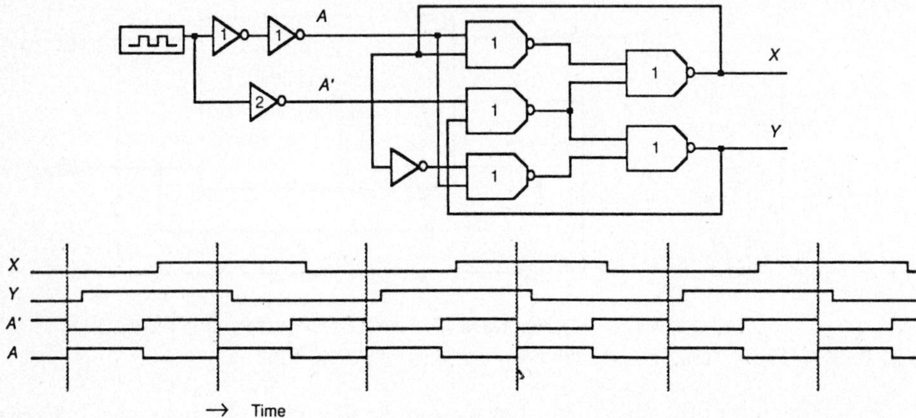

Figure 6.11(b) "LogicWorks" simulation of circuit of Figure 6.11(a) (note delay effects)

4. Derive the logic circuit equations from the excitation map (matrix). In this case, by grouping on the map we have:

$$X = A'y + Ax$$
$$Y = A'y + Ax'$$

5. Realise these equations in logic circuitry as in Figure 6.11(a), which is a *Nand* based realization.
6. Test the arrangement using a logic simulator and/or by plugging-up, e.g. on a logic trainer board, and/or wire up and test the actual circuit. The simulation results using "LogicWorks" are included as Figure 6.11(b) (note that the number enclosed in each gate symbol represents the units of delay associated with that element).

Exercise: Try re-designing the circuit with different allocations of *X* and *Y*.

6.4.2 A more difficult example, demonstrating further aspects of the asynchronous sequential circuit design processes

1. Specification

Given two input signals, A and B, which are constrained not to change state simultaneously,

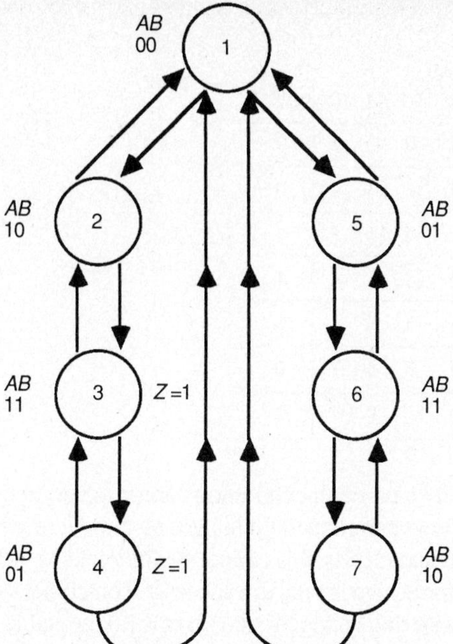

Figure 6.12 State diagram representing specified circuit

design an asynchronous circuit to present a single output Z =1, when both A and B are High but only if A went High before B and the sequence occurs after $A = B = 0$, Z is then to remain High as long as B is High. (Note: High = Logic 1.)

The interpretation of the specification can be greatly assisted by setting out a diagram, a state diagram, showing the states to be assumed by the circuit for all possible input sequences of A and B as in Figure 6.12.

The sequence commences when A = B = 0, so let us allocate state 1 to this condition as shown in the diagram. Now, concentrating on a "correct sequence" first, and recognizing that A and B cannot change together, we can set out state 2 which occurs in the sequence when A goes High. Clearly, since only one variable can change we can now either revert to state 1 if A changes, or go on to state 3 if B changes. Z will become 1 in state 3 and will remain so in state 4, which is assumed if A changes next.

The rest of the left-hand side of Figure 6.12 is then evolved by similar reasoning. Again, following state 1, B may change first in which case an "incorrect sequence" will be generated as shown on the right-hand side of the diagram and Z will not become 1 at all.

This information, or the specification directly, is now used to draw up a primitive state table as in Table 6.9.

The primitive state table is drawn up with one row for each (stable) state so that there will be seven rows in this case (states1-7 in Figure 6.12).

Rows are conveniently denoted (a), (b), etc., each row being associated with one stable state at this stage.

Unstable state entries also appear in each row indicating the next (stable) state taken up by the circuit for the particular values of AB associated with each column of the table. Dashes

Table 6.9 Primitive state transition table for the example

Row	AB 00	01	11	10	O/P Z
(a)	(1)	5	-	2	0
(b)	1	-	3	(2)	0
(c)	-	4	(3)	2	1
(d)	1	(4)	3	-	1
(e)	1	(5)	6	-	0
(f)	-	5	(6)	7	0
(g)	1	-	6	(7)	0

appear in the table where a particular condition cannot occur, in this case due to the overall restriction which disallows simultaneous changes of *A* and *B* (e.g. in row (a), the stable state is 1 for which *AB* = 00 and clearly this cannot be followed by *AB* = 11).

Since this stage of the design is vital to a successful outcome, we will dwell on the setting out of the table by considering it step by step. In row (a) the stable state 1 is for *AB* = 00 and the entry reflects that. If, however, *AB* changes to 01 or 10, then the circuit becomes unstable until it takes up either state 5 in row (e) or state 2 in row (b). The output Z is 0 for row (a) since the specified conditions have not yet been met. Similarly, in row (d), the circuit is stable in state 4 in the *AB* = 01 column, and in this case the output Z is 1 since a correct sequence leads to state 4. Further changes in *AB*, which can only be to 00 or 11, will give rise to unstable states 1 and 3 respectively.

We could now interpret this state table in logic and if we did so we would need a circuit having a state vector three bits wide in order to accomodate the seven rows. It would take the form indicated in Figure 6.13.

However, as we have seen before, an attempt to simplify the configuration might be worth while if it reduces the width of the state vector and simplifies the logic circuitry. The way of approaching this problem is to *seek state mergers.*

State merging. The number of rows in a primitive state table may often be reduced by row mergers. If this is possible, it could well reduce the circuit complexity and the size of the state vector (clearly, in this case we must reduce the final number of rows to four or less

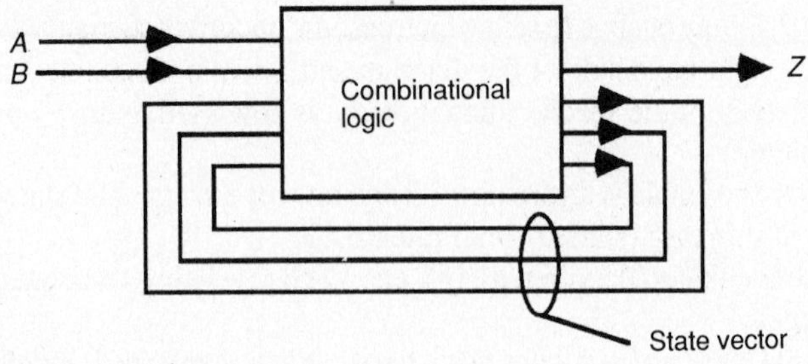

Figure 6.13 Possible arrangement

to affect the state vector since $n \leq 2^v$ where n = number of rows in the state table and v = number of bits in the state vector).

Rows may be merged if the state assignments for a particular column have the same number, treating a dash as a don't care. Unstable states thus merge into the corresponding (same number) stable states and dashes merge into any stable or unstable states. *Output(s) are ignored in merging.*

Applying these rules we may draw up a table showing all possible row mergers as in Table 6.10.

Table 6.10 Possible row mergers.

	Row entries	*Merged row*
(a)	① 5 - 2	
(b)	1 - 3 ②	① 5 3 ②
(a)	① 5 - 2	
(e)	1 ⑤ 6 -	① ⑤ 6 2
(b)	1 - 3 ②	
(c)	- 4 ③ 2	
(d)	1 ④ 3 -	1 ④ ③ ②
(e)	1 ⑤ 6 -	
(f)	- 5 ⑥ 7	
(g)	1 - 6 ⑦	1 ⑤ ⑥ ⑦

It will be seen that two 2-row and two 3-row mergers are possible. It now remains to choose a suitable set of row mergers so that the final number of rows is minimized and each original row is included in *one and one only* merged row.

To help in the choice it is convenient to set out a merger diagram as in Figure 6.14.

In this case it is clearly advantageous to choose the two 3-row mergers which then leaves unmerged row (a) to complete the coverage and results in a reduction of the state table to three rows (identified here as rows (h), (j), and (k)) as shown in Table 6.11.

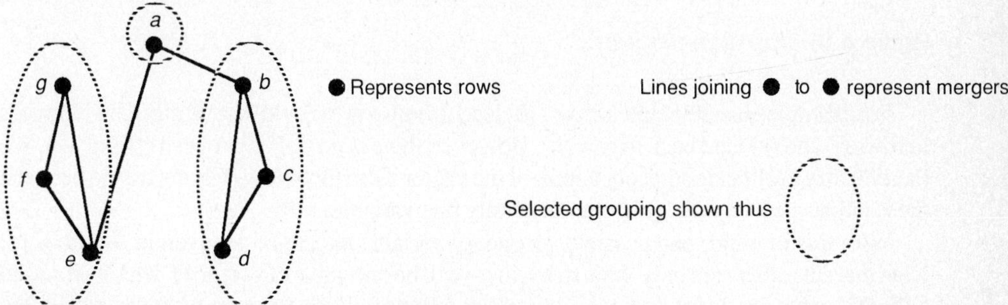

Figure 6.14 Merger diagram for the example

Table 6.11 Merged state transition table

Row	AB 00	01	11	10
(h)	(1)	5	-	2
(j)	1	(4)	(3)	(2)
(k)	1	(5)	(6)	(7)

The excitation map evolves directly from this state diagram by *allocating two secondary variables*, say *XY*, to identify the three rows and so provide the state vector in the overall arrangement which now has the form of Figure 6.15.

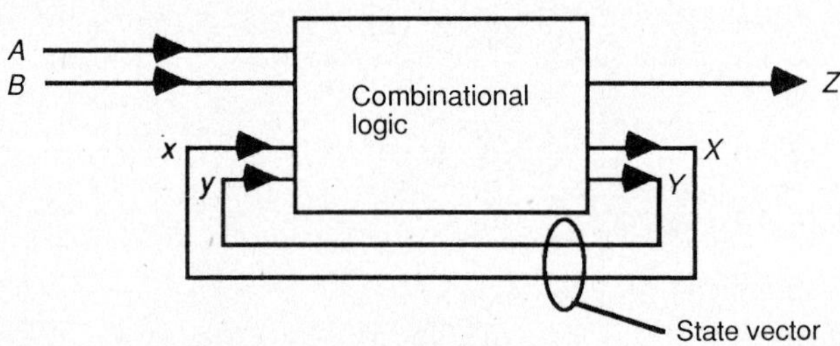

Figure 6.15 Final configuration for the example

One further step helps in making a sensible allocation of the fedback inputs, *xy*, to the rows of Table 6.11. This is to *draw a transition diagram* as in Figure 6.16. This diagram indicates the "traffic flow" between the rows ((h), (j) and (k), (in this case).

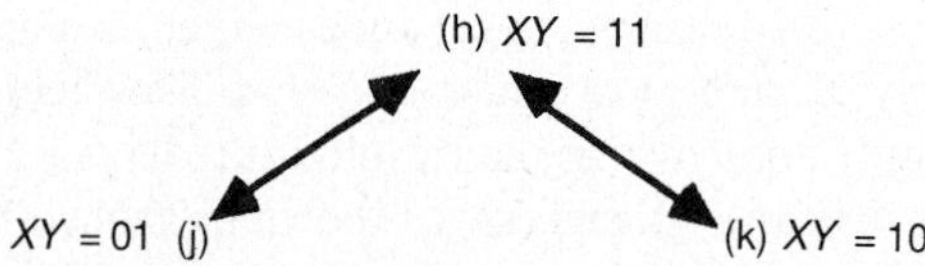

Figure 6.16 Transition diagram

For example, unstable states in row (h) lead to both rows (j) and (k) and the unstable states in rows (j) and (k) lead back to row (h). However, there is no traffic between (j) and (k). We therefore try to allocate adjacent values of the vector *XY* to those rows having traffic between them (adjacency implies a difference in only one variable).

With this in mind, one possible secondary variable allocation is given in Figure 6.16. Note that since there are only three rows, there will be one value of vector *XY* which is unused (*XY*=00 in this example) and any unused conditions will give rise to "don't cares" in the excitation map.

The flow matrix which follows is set out as Table 6.12(a) where a dash indicates "don't care" or "can't happen". In table 6.12(b) sensible state numbers have been allocated to all dash entries with an eye to final grouping in the excitation map which is to follow.

Table 6.12 Flow matrix for the example

(a)

xy \ AB	00	01	11	10	
00	-	-	-	-	
01	1	(4)	(3)	(2)	(j)
11	(1)	5	-	2	(h)
10	1	(5)	(6)	(7)	(k)

(b)

xy \ AB	00	01	11	10
00	1	4	3	7
01	1	(4)	(3)	(2)
11	(1)	5	6	2
10	1	(5)	(6)	(7)

Now draw the excitation map. This is done with reference to Table 6.12(b) by entering values for *XY* for each stable state which agree with the entry for *xy* in the particular row. Thus each stable state becomes associated with a value of *XY* which is then copied for the unstable states of the same number. In this case, the resultant excitation map is given as Table 6.13.

Table 6.13 Excitation map for the example

XY / xy \ AB	00	01	11	10
00	11	01	01	10
01	11	(01)	(01)	(01)
11	(11)	10	10	01
10	11	(10)	(10)	(10)

Next, derive the logic equations from the excitation map by grouping the entries for *X* and then, the entries for *Y*, giving:

$$X = A'.B' + B'.y' + B.x$$
$$Y = A'.B' + B'.y + B.x'$$

Do not forget the output function(s), in this case, Z.

At this point there is a temptation to admire one's own smartness in getting the secondary variable realization into logic, but this is only a means to an end, the end being to achieve an output Z as required in the specification.

We now turn back to the state diagram of Figure 6.12 or the primitive state diagram of Table 6.9, in which we set out the output conditions for Z. We see that $Z = 1$ in stable states 3 and 4 and, if it suits our convenience, we may also associate unstable states 3 and 4 with $Z = 1$. Now, from the flow matrix of Table 6.12(b), we may set out a map for output Z as in Table 6.14.

Table 6.14 Map for output Z

Z	AB 00	01	11	10
xy 00	0	1	1	0
01	0	1	1	0
11	0	0	0	0
10	0	0	0	0

Whence

$Z = Bx'$

An implementation in logic now completes the design (as in Figure 6.17) and simulation in software or testing in hardware may now take place.

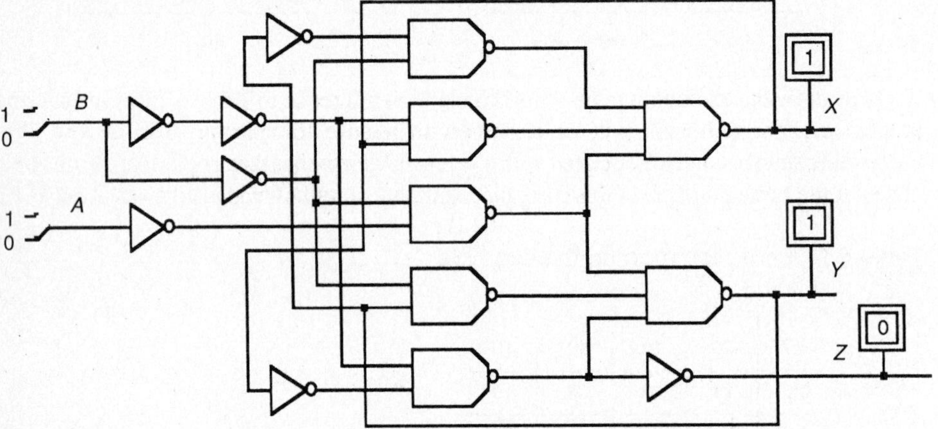

Note: The inverters used on inputs A and B might not be needed in practice, depending on the availability of B' and A'.

Figure 6.17 Complete circuit for the example (with simulation showing state ①)

Note: The inverters used in inputs A and B might not be needed in practice, depending on the availability of $\overline{B}$ and $\overline{A}$ as well as inputs B.

6.4.3 On the importance of words

In writing or interpreting a written specification, it is important that words are chosen carefully to avoid ambiguity (or worse) and that words are carefully and properly interpreted.

Take for example, the specification for the just completed example set out at the beginning of section 6.4.2.

Omission of the words "and the sequence occurs after $A = B = 0$" puts a different complexion on the problem and the state diagram now becomes that set out as Figure 6.18 (compare this with Figure 6.12). Note that the required sequential circuit has one less state than previously.

A worked example at the end of this chapter completes this design.

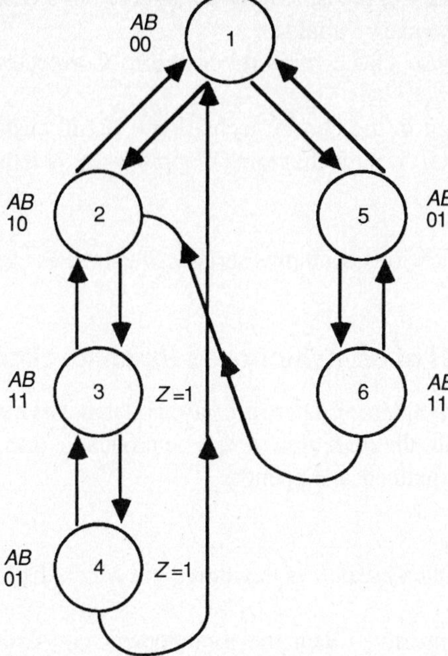

Figure 6.18 State diagram for modified specification

6.4.4 Outline of a procedure for designing asynchronous sequential logic

From the previous example we may see that the following steps constitute a sensible and orderly approach:

1. Set out the specification in full and in an unambiguous manner. It is often difficult to cover all input possibilities unambiguously and the logic signal flow graph* introduced in Section 6.4.5 and discussed in Appendix I can be used to ensure that specifications are indeed full and unambiguous.
2. (a) Set out a state diagram showing input and output conditions, and/or
   (b) Draw a primitive state transition table, one row per stable state.
3. (a) Simplify the state diagram or table by merging where possible, output conditions may be ignored.
   (b) Set out a merger diagram in order to choose an appropriate set of merged rows, and thence,
   (c) Set out the merged form of the state table.
4. Allocate secondary variables and states of same to separately identify all rows of the merged state table.
5. Draw up a flow matrix.
6. Set out the excitation map.
7. Derive the logic equations.

* The logic signal flow graph technique is due to Kikkert.

8. Map, or otherwise obtain expressions for output functions unless the outputs are obtained directly from the secondary variables.
9. Simulate, or otherwise check, circuit operation. Correct or modify the design as necessary.
10. Implement the design in the chosen technology. If full custom design is envisaged, produce mask layouts, via stick diagrams if appropriate, and then check and simulate at this level of design.

It is hoped that the previous examples serve to illustrate the general process of design.

6.4.5 Further aspects of asynchronous sequential circuits

Some discussion and examples used here are based on material kindly contributed by Prof. C.J.Kikkert. In particular, the *logic flow graph* approach is due to Prof. Kikkert and this concept is more fully explained in Appendix I.

Latches and Oscillators

We have already seen that *feedback* is the ingredient which distinguishes sequential from combinational logic.

If the feedback is *positive* then the arrangement can exhibit *latching* properties. Consider, for example, the circuit set out in Figure 6.19(a).

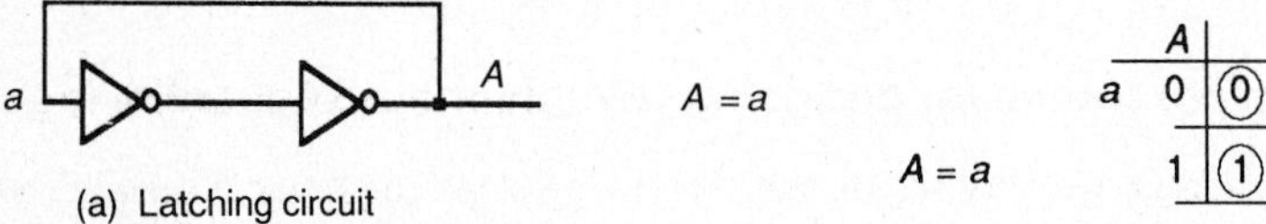

Figure 6.19(a) Latching circuit

The circuit is stable with either $A = 0$ or with $A = 1$ and the circuit will power-up into one of these two states and remain there until power-down. Since it is most likely that the circuit always powers-up into the same state, this arrangement is of little use as it stands.

To make it useful we must make the feed back *conditional* on some control signal. In so doing, we can create a data latch type storage circuit as suggested in Figure 6.19(b);other arrangements follow in later chapters.

Exercise: Analyse the behavior of the circuit of Figure 6.19(b).

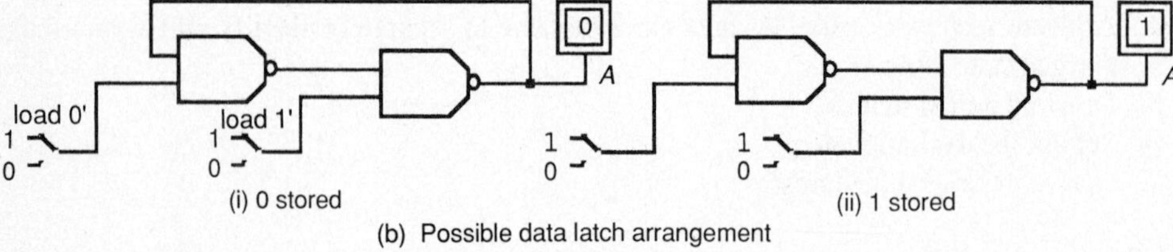

Figure 6.19(b) Possible data latch arrangement

If, on the other hand, we make the feedback *negative*, then this can form the basis of an *oscillator* as suggested in Figure 6.20. Note that this particular circuit allows the output to be "turned off" or initialized into the logic 0 state. This is reflected in the maps of Table 6.15.

Table 6.15 Maps for the circuit of Figure 6.20

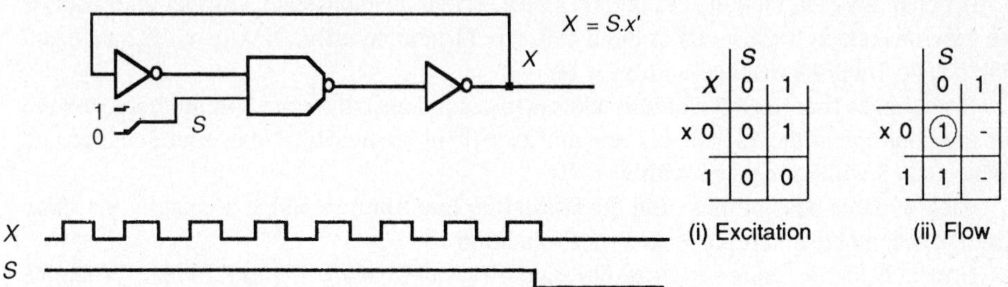

Figure 6.20 Oscillator circuit with output inhibit

When $S = 1$ neither condition in the right-hand column of the excitation map is stable, so that the circuit oscillates from one row to the other and a square wave output appears at the output X, at a frequency determined by the sum of propagation delays t_p through the circuit.

When $S = 0$ the circuit takes up stable state 1 (the only one!) and the output X is held at 0. These aspects should be apparent from the "LogicWorks" simulation waveforms with Figure 6.20.

Some "facts of life"—races and hazards

In moving from one stable state to the next, an asynchronous sequential circuit may go directly or move through unstable states in transit. If there is one and one only route, which is consequently always followed, the sequence followed is referred to as a *cycle.*

If the route can vary depending on relative delays between gates, but if the circuit always reaches the correct target state, then the sequence is referred to as a *race* .

If, on changing state, the circuit moves into a column containing more than one stable state and, depending on gate delays, the circuit may end up in any one of the stable states in that column, then this is referred to as a *critical race situation* and the outcome is uncertain.

In order to illustrate race situations, consider the flow map for the RS flip-flop given earlier as Table 6.5 and reproduced here as Table 6.16.

Table 6.16 Flow map for the RS flip-flop

x.y \ R.S	00	01	11	10
00	1,2	3	(4)	5
01	(1)	3	4	(5)
11	1,2	3	4	5
10	(2)	(3)	4	5

Consider a change of state from stable state 5 ($xy = 01$; $RS = 10$) to stable state 3 ($xy = 10$; $RS = 01$), for which both x and y must change. When RS changes to 01 the circuit will immediately take up the unstable state 3 in the $RS = 01$ column for which xy remains = 01. To get to stable state 3, both x and y must change, but due to unequal gate delays it is likely that either x or y will change first.

If x changes first, then the circuit takes up a second unstable state 3 immediately below the previous one in the $RS = 01$ column and $xy = 11$ momentarily. Next, y will change and stable state 3 will be reached with $xy = 10$.

If y changes first, then the circuit takes up a second unstable state 3 immediately above the previous one in the $RS = 01$ column and xy = 00 momentarily. Next, x will change and stable state 3 will be reached with $xy = 10$.

Each of these possibilities (and the possibility that x and y change at exactly the same time) is acceptable and represents a *race* situation.

Now consider a change from stable state 4 ($xy = 00$; $RS = 11$) to $RS = 00$. When RS changes to 00 the circuit moves into an unstable condition at the top of the $RS = 00$ column. The stable states in this column have either x or y =1 so that either x or y must change from 0. Now the changes of inputs R and S may not be coincident and/or the circuit gate delays may not be identical, so that either x or y may change first. If x changes first then the circuit will take up stable state 2; if y changes first will the circuit go to stable state 1.

This illustrates a *critical race* situation and such situations should be avoided (or excluded) in design. In this regard, care is required at the state assignment stage of design, and setting out a transition diagram (as in Figure 6.16) helps to make adjacent state assignments to states between which there is "traffic flow". The golden rule is to *try to make assignments such that all transitions between states involve changes to a single secondary variable only*, but clearly this is not always (often?) possible. One way of eliminating critical race situations, which will often work, is to try various possible permutations of allocations in a given situation. Another ploy is to allocate additional states, and thus move away from a minimal situation. In some instances, two differing combinations of secondary variables may be allocated to a single state. For further reading on this and, in particular, on the allocation of extra states, the reader may consult the original work of Huffman* and a large number of texts on switching theory, some of which are listed in the general reading bibliography in this text.

A *static hazard* is said to exist during a particular state transition if one or more secondary variable or output is required to retain its present value and if, due to circuit delay differences, it may change momentarily.

An example, set out as Table 6.17, may help to illustrate this condition.

With reference to the diagram it will be seen that three gates form this circuit and that they may be individually identified by numbering them 1,2,3 as shown. Consider the circuit when in the stable state, for which F is identified with a bold **1** in the $AB = 11$ column, and, further, let us envisage a change in AB to 10 and a consequent transition to the bold **1** in the right-hand column $AB = 10$. It would appear to be a quite straightforward matter until we look a

* D.A. Huffman "The synthesis of sequential switching circuits", *J. Franklin Institute* 257, pp. 161-90 and 275-303, Mar/Apr. 1954.
D.A. Huffman "A study of the memory requirements of sequential switching circuits" *Technical report* 293, Electronic Research Lab., M.I.T., Apr. 1955.
D.A.Huffman "The design and use of hazard free switching networks" *J. ACM*, 4, 47, 1957.

Table 6.17 Example demonstrating a static hazard

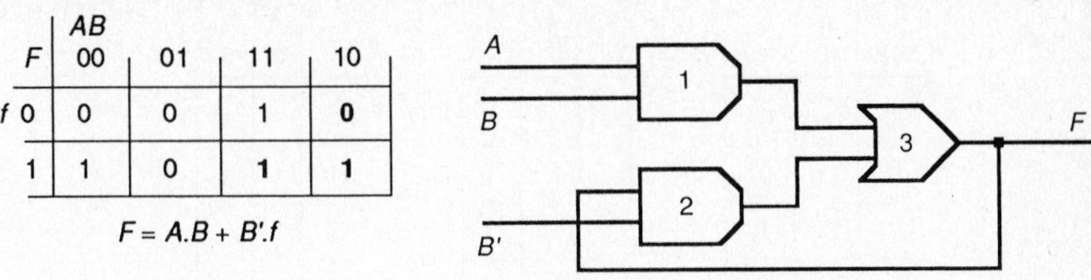

F \ AB	00	01	11	10
f 0	0	0	1	**0**
1	1	0	**1**	**1**

$F = A.B + B'.f$

Note: Numbers identify gates, not gate delays

little deeper. In the initial state, $F = 1$, the 1 being generated by *And* gate 1, gate 2 having a 0 output state. In the target state the converse applies with *And* gate 2 generating a 1 and gate 1 having a 0 output state. However, consider the condition which holds if gate 1 turns "off" before gate 2 turns "on". The circuit conditions will then be such that both gates 1 and 2 will have an output of 0 and F (and f) will also become 0, thus putting the circuit into the stable state identified by a bold $F = \mathbf{0}$ immediately above the desired state in the $AB = 10$ column. The circuit output F will thus become a stable 0 rather than the intended stable 1. This *static hazard* can be eliminated in this case by including an extra *And* term so that the expression implemented becomes:

$$F = A.B. + B'.f + A.f$$

It may be seen that this additional, apparently redundant, term serves to "bridge the gap" since it generates a 1 output in both the initial and target states, thus holding $F = 1$.

This introduction to static hazards is brief but should serve to make the designer aware of potential problems and of a possible remedy.

As well as static hazards, which cause incorrect performance, we may also encounter *dynamic hazards* which may or may not be fatal but which impair the operation of a sequential circuit. Dynamic hazards may be likened to contact bounce in mechanical switching operations, a particular variable may change three or any greater odd number of times before settling, where only one change was intended. For example, if the desired change was from 0 to 1 then a dynamic hazard exists if the actual sequence is 0 to1, then 1 to 0, then 0 to1, before the circuit settles. Such hazards should not exist if proper reduction of excitation maps is effected

Further problems which may be encountered are known as *essential hazards* which, at first glance, seem to be one and the same thing as static hazards. However, a closer look at a previous example worked through in section 6.4.1 may help in explaining the difference. This was a counter circuit for which the excitation map was given in Table 6.8, reproduced here as Table 6.18(a), and the logic circuitry was set out as Figure 6.11(a), also reproduced here with the table.

The excitation maps for X and Y have been separated in Table 6.18(b) in order to simplify our discussion and, with our newly acquired insight, we may recognize potential *static hazards* in both excitations since we have the two *And* terms forming X and those forming Y in separate columns with no bridging (when moving between the $A = 0$ and the $A = 1$ columns).

To cure this we may add an *And* gate to each, as follows in Figure 6.21:

Table 6.18 Excitation maps for the counter under consideration

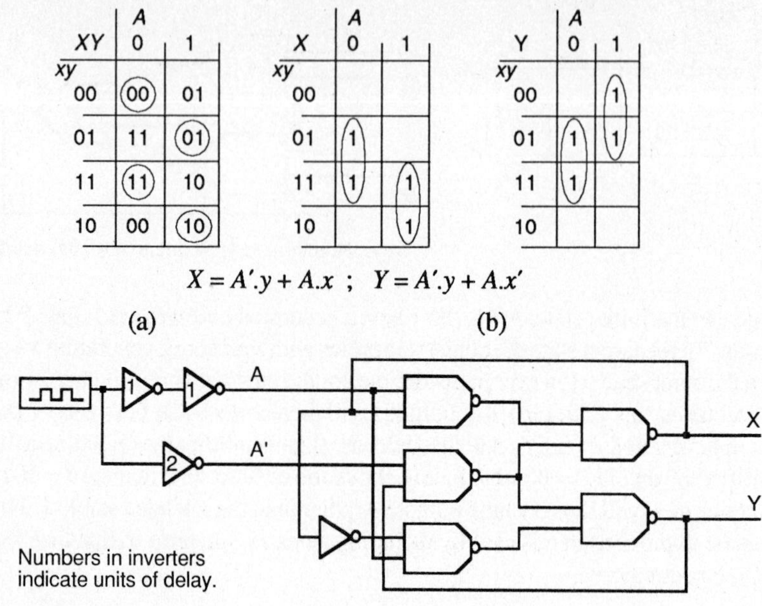

xy \ A	0	1
00	(00)	01
01	11	(01)
11	(11)	10
10	00	(10)

XY

xy \ A	0	1
00		
01	1	
11	1	1
10		1

X

xy \ A	0	1
00		1
01	1	1
11	1	
10		

Y

$$X = A'.y + A.x \;;\quad Y = A'.y + A.x'$$

(a) (b)

The circuit then becomes that of Figure 6.21 (shown here with simulator waveforms).

$$X = A'.y + A.x + x.y$$
$$Y = A'.y + A.x' + x'.y$$

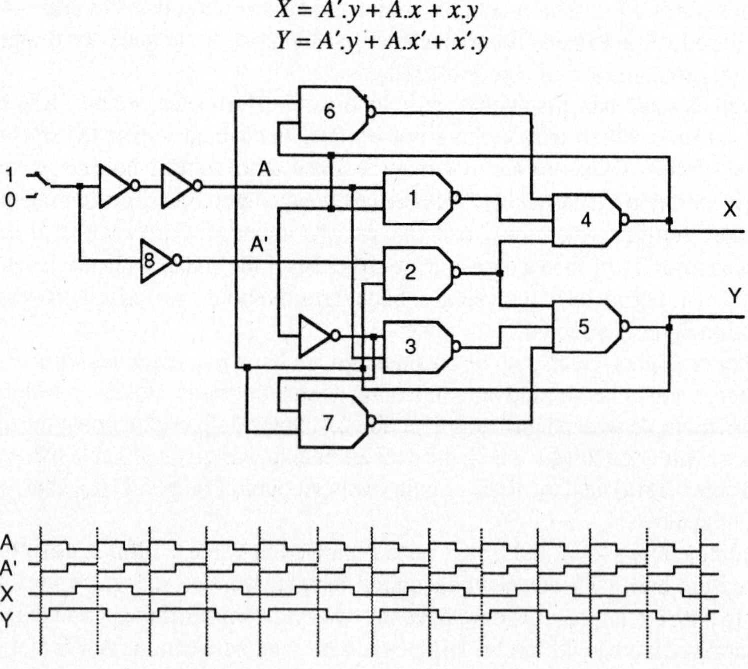

Figure 6.21 Modified arrangement to avoid static hazards (with simulation)

A further study now reveals a further potential serious hazard referred to as an *essential hazard*. With respect to Table 6.18(a) and Figure 6.21, consider the circuit in the top left-hand (in the table) stable state, $XY = 00, A=0$. Let A change to 1 so that Y should change to 1 for stable state $XY = 01, A=1$. The change in A puts all 1s on the inputs of *Nand 3* so that Y becomes 1. This is fed back so that $y = 1$ and *Nand 7* is activated, thus holding $Y = 1$. If there is a substantial delay in *Inverter 8*, the top input of *Nand 2* will remain 1 for some time but the second input will also become 1 since $y = 1$. Thus X also becomes 1 and the circuit takes up a temporary state $XY = 11$. Eventually the correct value propagates through *Inverter 8* so that the circuit conditions are now $A = 1, xy = 11$, so that the inputs to *Nand 2* are now 0 and 1 and the circuit takes up stable state $XY = 10, A = 1$ (the bottom right state in Table 6.18(a)). Thus states are skipped over and the counter will not operate correctly.

The remedy is obvious in this case and is to reduce or eliminate the delay through *Inverter 8*, that is the delay between signal A and its complement A'. This is a simple process in this circuit arrangement and you will note that Figure 6.21 has two inverters in line A and one in line A', the delays being adjusted to give coincident A and A' signals to the circuit for the purposes of simulation. In practical situations A and A' may be generated elsewhere, or A' may be directly obtained from A at the input to this circuit so that the essential hazard outlined here may have to be taken into account. In general, the occurrence of hazards is minimized if propagation delay differences between different paths through the circuit are minimized. This may be done, at the expense of speed, by deliberately adding delay in some feedback paths.

If the possibility of hazards is ignored, then experienced engineers will be aware that "Murphy" will undoubtedly strike in the worst possible way and at the worst possible time.

A further example of asynchronous sequential circuit analysis, introducing a "flow graph" concept

A circuit used for phase detection (due to Kikkert) is given as Figure 6.22 and will be seen to comprise four 2I/P *Nand* gates and four inverters, the latter being included to ensure hazard free operation. The signals to be compared in phase are input as A and B. Two output signals are taken, labelled F and G, which are used to indicate early or late arrival of B with respect to A.

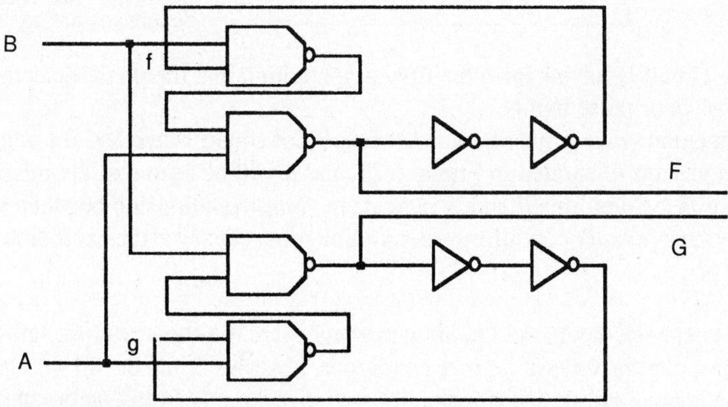

Figure 6.22 Phase detector circuit

1. Derive the logic (excitation) equations for the two output signals

$$F = ((B'+f').A)'$$

whence: $$F' = AB' + f'.A$$

similarly: $$G' = A'.B + g'.B$$

2. Draw up an excitation map for the output functions (Table 6.19).

Table 6.19 Excitation map for the phase detector

fg \ FG	AB 00	01	11	10
00	11	10	(00)	01
01	11	10	(01)	(01)
11	(11)	10	(11)	01
10	11	(10)	(10)	01

3. Determine the stable states. (Shown circled in Table 6.19).
4. Set out the circuit behavior as a "flow graph"*
   This is presented as Table 6.20.

Table 6.20 Flow graph for the example

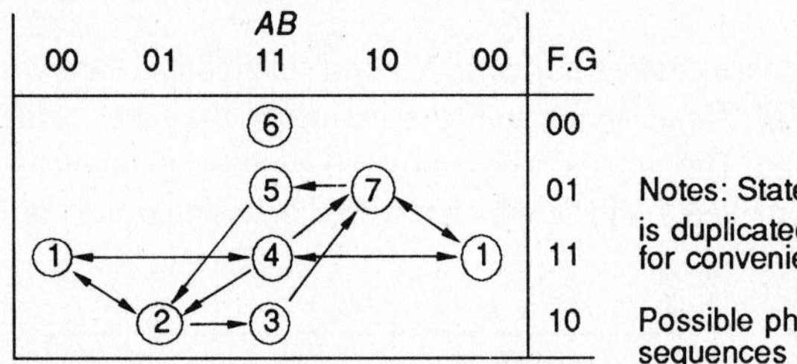

5. Deduce circuit behavior from the flow graph noting that the phase detector may have square wave or pulse inputs.

   Simulation waveforms of some sequences for square wave and for negative going pulse inputs are illustrated in Figure 6.23, and it will be seen that the relative average values of the waveforms *F* and *G* indicate the phase relationship between *A* and *B*.

   Checking for *race* conditions we may look more closely at the excitation map, Table 6.19.

First, let us check the $AB = 11$ column in which there is a choice of four stable states and obviously no potential for *critical race* conditions. If we move into the $AB = 11$ column from either of the immediately adjacent columns, we move directly from a stable state to another

* Footnote: The logic signal flow graph is a technique due to Kikkert (see Appendix 1.)

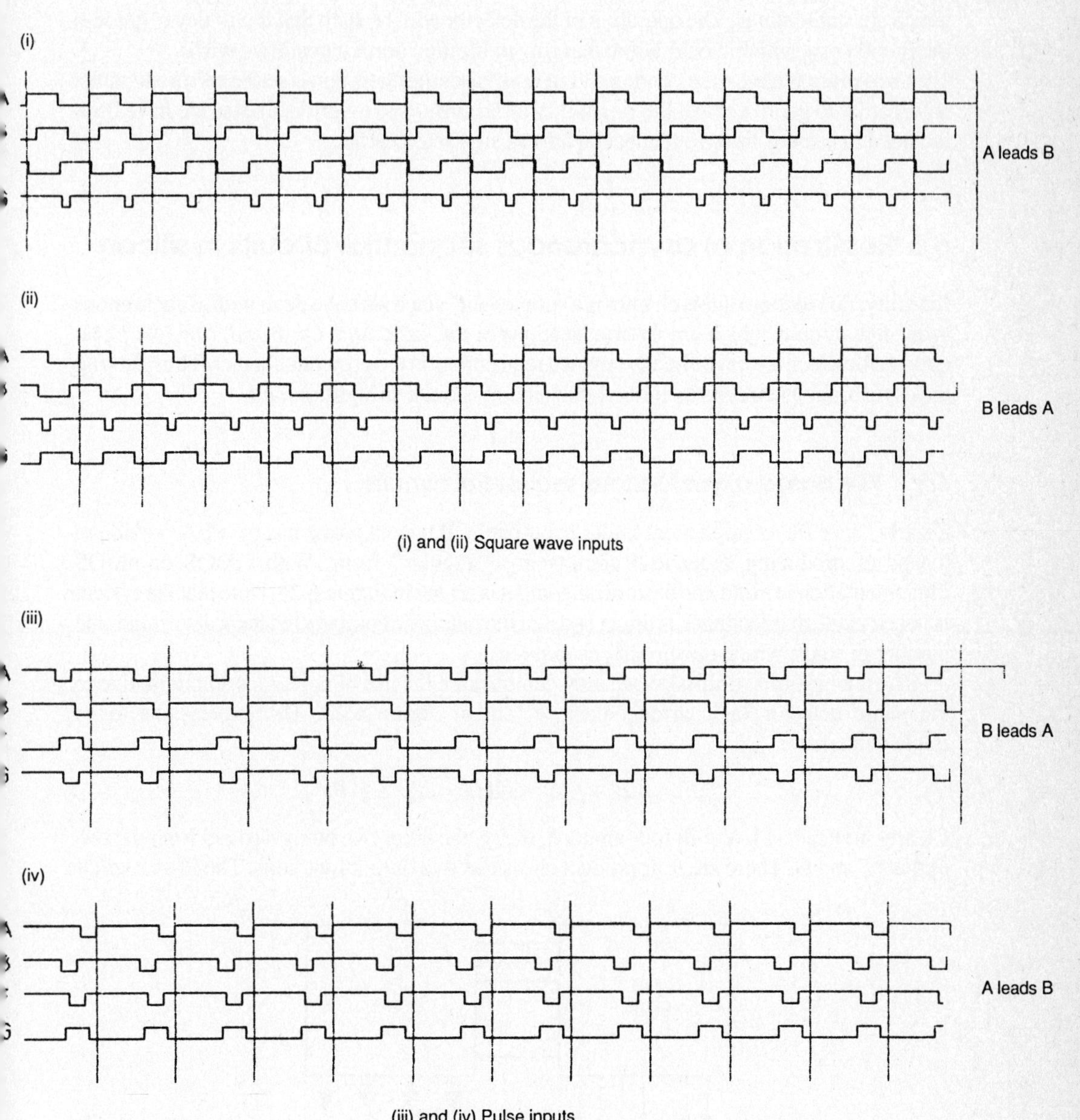

Figure 6.23 Simulated waveforms indicating phase detector behavior

stable state (2 to 3 or 7 to 5) having the same output conditions, that is, FG and fg remain the same so there is no ambiguity nor race in taking up the new state. Further, if we allow a direct transition from the stable state in the $AB = 00$ column for which $fg = 11$ to the $AB = 11$ column, we move to stable state 4 for which fg remains at 11. If the changes in A and B are simultaneous then we end up in stable state $FG = fg = 11$, but if the changes in A and B are not coincident and f or g changes first, then we take up either $FG = 10$ or $FG = 01$, both of which are stable states. The operation of the detector will be such that it will never get into the $fg = 00$ row which would allow that row to become don't cares if we wish.

Two columns have *race* conditions but in all cases the operation is stable and predictable.

In order to guard against the possibility of *hazards*, two inverters in cascade have been included in each of the two feedback paths as suggested earlier.

6.5 Realization of asynchronous sequential circuits in silicon

In reality, this section of this chapter is a "non-event" since we have dealt with asynchronous sequential circuits which are interconnections of the basic *And, Or, Nand, Nor* (etc.) gate circuits for which we have already covered realizations in the previous chapter. Thus, turning the asynchronous circuits examined into silicon is quite straightforward.

6.5.1 PLA based asynchronous sequential circuits

Clearly, since PLAs implement multiple output SOP type expressions, the PLA provides a means of producing sequential circuits in a "regular" form. With CMOS or nMOS implementation in mind, the basic arrangement is set out in Figure 6.24. Note that the system is not clocked, that feedback is direct and that the number of outputs fed back determines the number of states which the circuit can assume.

To conclude this discussion we may demonstrate the use of a PLA as an alternative to the phase detector logic arrangement set out in Figure 6.22. The expressions to be implemented are:

$$F' = A.B' + f'.A \quad \text{and } G' = A'B + g'.B.$$

Clearly we need a PLA with four inputs A, B, f, g, the latter two being derived from the two outputs F and G. There are four product terms and two output functions. The PLA *And/Or*

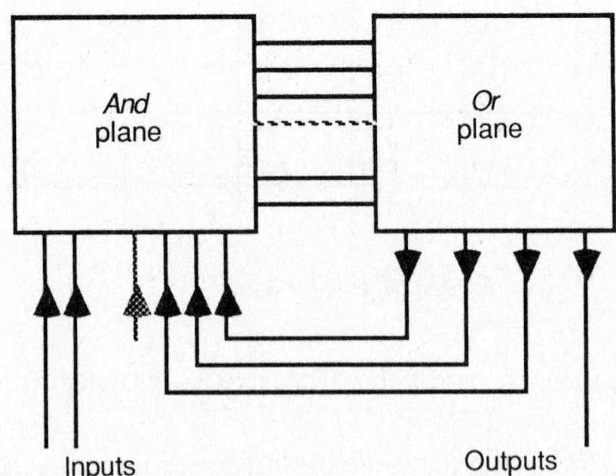

Figure 6.24 PLA based asynchronous sequential circuit model

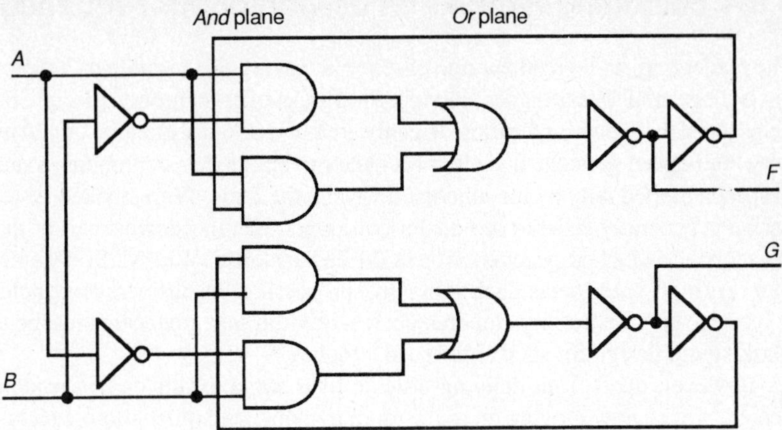

Figure 6.25 PLA structure for phase detector

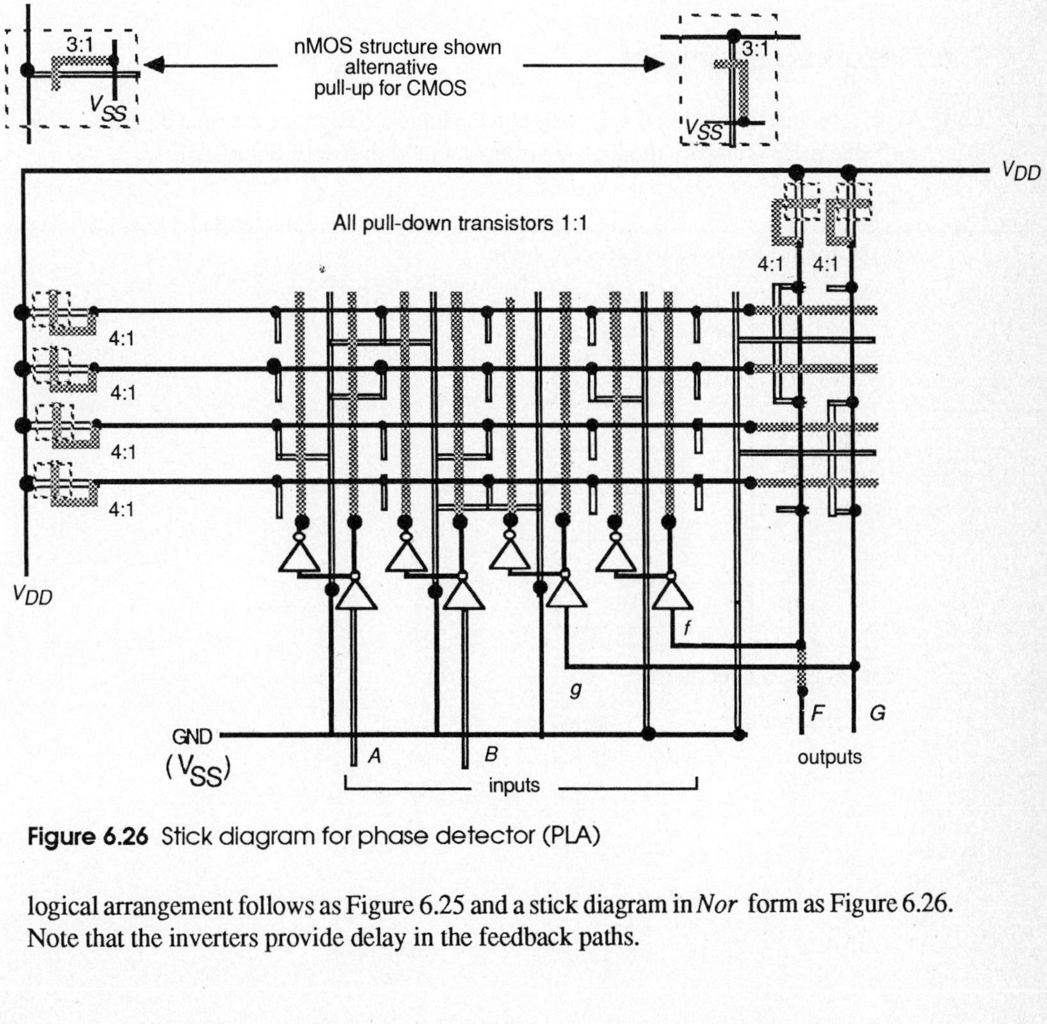

Figure 6.26 Stick diagram for phase detector (PLA)

logical arrangement follows as Figure 6.25 and a stick diagram in *Nor* form as Figure 6.26. Note that the inverters provide delay in the feedback paths.

6.6 Concluding remarks on asynchronous sequential circuits

The reader may well breathe a sigh of relief in leaving this hazardous and race prone topic, but it is essential for a designer to have a good grasp of asynchronous design. For one thing,it helps to set the scene for the more friendly area of clocked sequential circuit design and, for very high-speed systems, it is often necessary to employ asynchronous systems which run at speeds limited only by the inherent delays in the logic. With clocked systems, the clock period is generally fixed to be equal to, or longer than the slowest part of the circuitry. In consequence, clocked systems are hazard-free but slower. With VLSI systems now pushing into very high-speed areas, such as in signal processing, the business of asynchronous design is likely to be of increasing importance. It is obvious that great care must be taken with the mask layout design for such high speed circuitry.

However, apart from dallying a while over some examples of asynchronous circuit design, we are now moving on to the more tractable and much more widely used clocked sequential circuitry.

6.7 Worked examples

1. With reference to section 6.4.3, complete the logical design for the state diagram given as Table 6.15. Compare the logical arrangement with that of Figure 6.17.

   Solution:

   Set out the primitive state transition table.

Row	AB 00	01	11	10	O/P Z
(a)	①	5	-	2	0
(b)	1	-	3	②	0
(c)	-	4	③	2	1
(d)	1	④	3	-	1
(e)	1	⑤	6	-	0
(f)	-	5	⑥	2	0

Now make row mergers:

Row entries					*Merged row*
(a)	①	5	-	2	
(b)	1	-	3	②	① 5 3 ②
(a)	①	5	-	2	
(e)	1	⑤	6	-	
(f)	-	5	⑥	2	① ⑤ ⑥ 2

(continued)

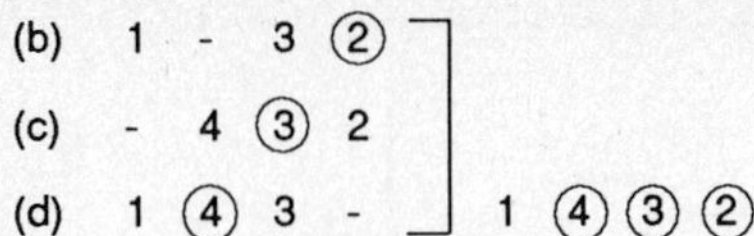

Clearly choose mergers (a), (e), (f) and (b), (c), (d) to give two rows only. Set out a merged state transition table:

Row	*AB* 00	01	11	10	
(h)	(1)	(5)	(6)	2	(a), (e), (f)
(j)	1	(4)	(3)	(2)	(b), (c), (d)

A single variable only, say X, is needed to identify the two rows. Whence the excitation map follows:

X	*AB* 00	01	11	10
x 0	0	0	0	1
1	0	1	1	1

From the map, $X = Bx + AB'$

Now mapping for output Z we have:

Z	*AB* 00	01	11	10
x 0				
1		1	1	

whence: $Z = B.x$

The complete circuit follows as Figure 6.27 and will be seen to be simpler than the design presented in Figure 6.17. Clearly this is due to the slight change to specification and the consequent reduction from three rows to two rows in the state transition table.

2. A counter circuit is required to count 0 to 1 (ΔW) transitions on a single input line W. An output Z is to become1 following every second such transition, and Z is then to remain 1 until W returns to 0. Complete and "test" the logical design.

Solution:

Set out a diagram showing state transitions and output condition:

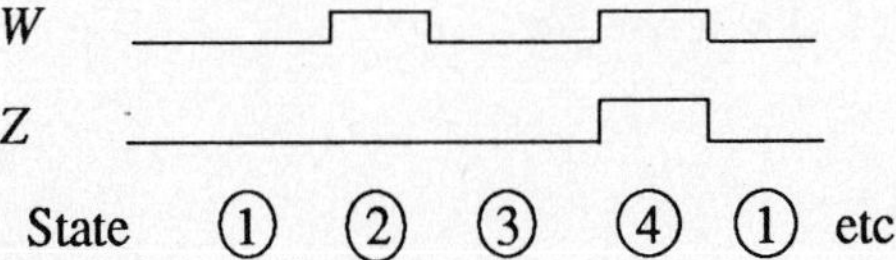

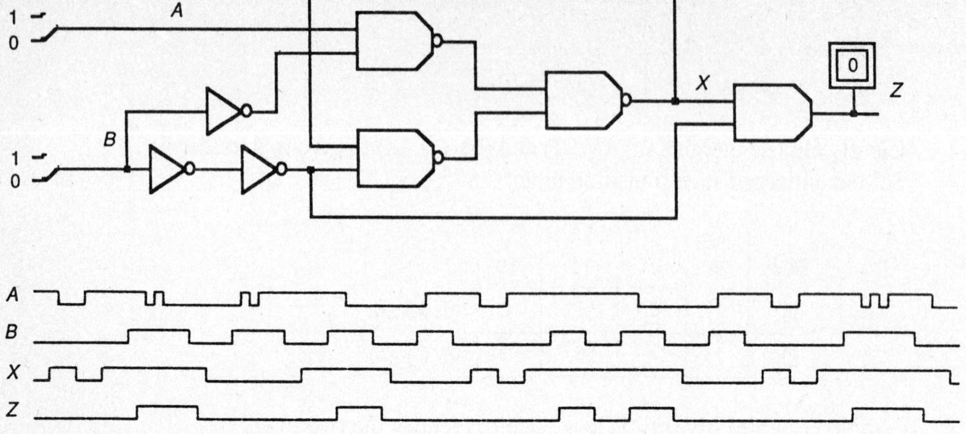

Figure 6.27 Question 1 solution

State ① ② ③ ④ ① etc.

allocate *XY* to states:

XY	00	01	11	10	00 etc.

Next a flow table:

xy \ *W*	0	1
00	①	2
01	3	②
11	③	4
10	1	④

Now excitation:

xy \ *W* (*XY*)	0	1
00	00	
01		01
11	11	
10		10

(a) stable states only

xy \ *W* (*XY*)	0	1
00	00	01
01	11	01
11	11	10
10	00	10

(b) full excitation map

From the excitation map we may write:

$$X = W'y + Wx\,;\quad Y = W'y + Wx'$$

and clearly, Z = state 4 = *WXY'*.
The circuit follows as Figure 6.28

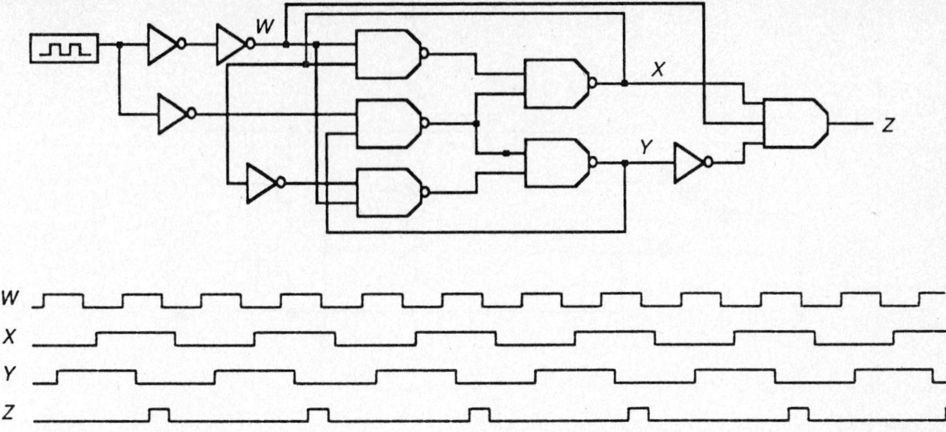

Figure 6.28 Circuit for question 2

3. A mechanical 2-position switch (Sw1 in Figure 6.29) is used with associated circuitry to switch control lines *A* and *B* so that either *A* =1 and *B* = 0, or *A* = 0 and *B* =1 the changeover being "bounce-free" as far as *A* and *B* are concerned. Generate logic to produce signal *Z* to switch nMOS transistors, as shown, to achieve this. Pull-up resistors may be assumed on *A* and *B* as in Figure 6.29.

Solution :

First identify possible states as shown in Figure 6.30.
Set out the primitive state transition table.

Row	*AB* 00	01	11	10	O/P Z
(a)	-	(1)	2	-	1
(b)	-	-	(2)	3	1
(c)	-	-	4	(3)	0
(d)	-	1	(4)	3	0

Merge (a) with (b) and (c) with (d), giving two rows as follows:

Row	*AB* 00	01	11	10	O/P Z	
(h)	-	(1)	(2)	3	1	(a), (b)
(j)	-	1	(4)	(3)	0	(c), (d)

A single variable only, say *X*, is needed to identify the two rows.
Whence the excitation map follows:

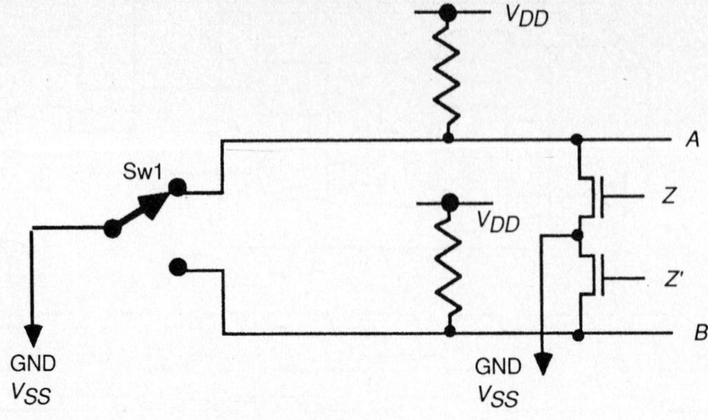

Figure 6.29 Switch and associated circuitry

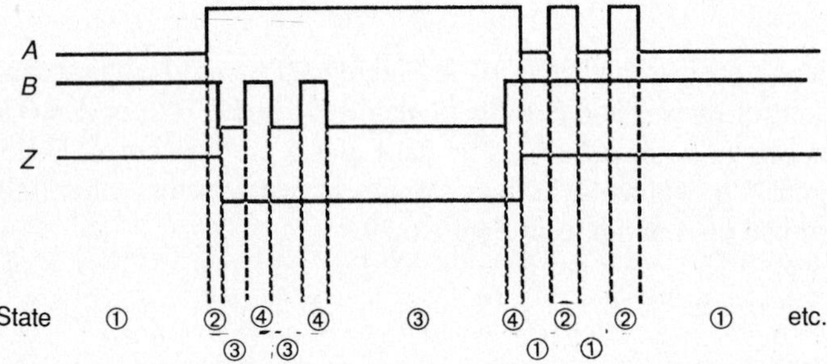

Figure 6.30 State identification

X	AB 00	01	11	10
x 0	-	0	0	1
1	-	0	1	1

From the map, $X = AB' + Ax$, $Z =$ states 1 or 2 so that $Z = Bx'$.
Circuit follows as Figure 6.31.

6.8 Tutorial 6

1. An asynchronous circuit is to be designed to respond to changes of logic level on a single input line *A* and give six distinct output states, 0, 1, 2, 3, 4, 5. In other words, the circuit is to count the edges present at input *A* in sequences of six and generate a separate output state for each state of the count.

   Design a suitable circuit and set out the logic circuit for the completed design.

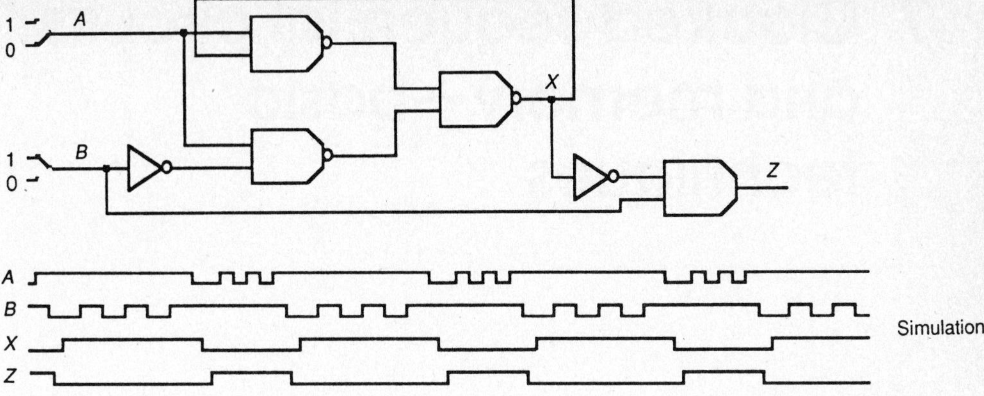

Figure 6.31 Final circuit for question 3

2. A circuit comprising two *Nand* gates is configured as in Figure 6.32. Analyze the behavior of the circuit and set out its characteristics in a way which would allow it to be used as a circuit element in the design of other circuits.

   What name would you give to this circuit?

3. Design an asynchronous circuit which will transmit one and one only clock pulse from a continuous stream of equally spaced clock pulses whenever a (manually operated) push button switch Sw1 is momentarily closed.

   The train of clock pulses is present on input line *W* and a pulse may be assumed to comprise a 0 to 1 transition, followed by a short period at logic 1, then a 1 to 0 transition. The input *W* will be at logic 0 between clock pulses which occur at regular intervals (e.g 100 μsec). There is no control over the length of closure of the push button switch.

4. A circuit has two inputs *P* and *R*, and is required to generate an output signal $Y = 1$ whenever the sequence PR = 00, then 01, then 11, is received. The output is to be coincident with the 11 condition of the required input sequence and is to be 0 under all other circumstances.

5. Design an asynchronous 3-bit counter to generate a 3-bit Gray code sequence at outputs *X*, *Y* and *Z*.

   A single input *W* is to cause advances in the state of the count on every logic level transition.

   Give the complete logic circuit for your design.

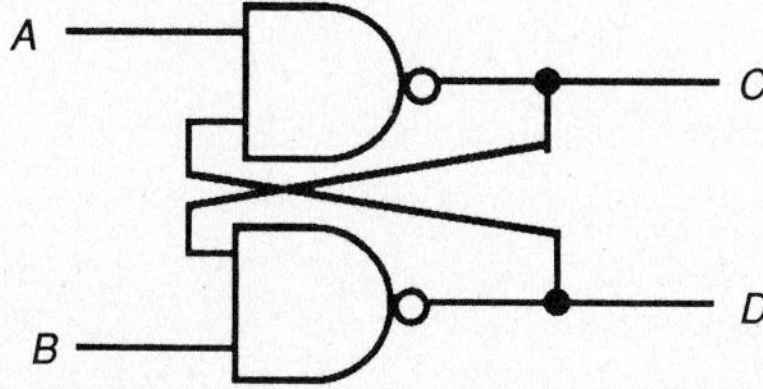

Figure 6.32 Circuit for analysis

7 Clocked sequential circuits and memory—basic techniques

Clocked sequential circuits and memories are widely used and readily designed. Quite often much of the design is centered around more or less standard circuit elements of which the most common type is the flip-flop.

To commence our considerations, we will examine some of the more common types of flip-flop and then base our design examples on such elements.

7.1 Some common types of flip-flop

7.1.1 The asynchronous RS flip-flop

We have already evolved a basic RS flip-flop (Chapter 6) and a *characteristic table* which provides a concise means of expressing the operation of the device. Table 7.1(a) is a characteristic table for the RS flip-flop (from section 6.3.2), and it should be remembered that the subscripts t and t+1 indicate the time at which an input to R or S occurs and the time at which the consequent output occurs (at Q and Q') respectively. Period "t+1" also covers the time interval until the next input condition occurs.

Table 7.1 RS flip-flop characteristics

R_t	S_t	Q_{t+1}	
0	0	Q_t	Output unchanged
1	0	0	
1	1	N.A.	Not allowed
0	1	1	

(a)

Q_t \ RS (Q_{t+1})	00	01	11	10
0	0	1	Ø	0
1	1	1	Ø	0

Ø is written for N.A. (or don't care)

(b)

Table 7.1(a) may be differently set out as in (b) and this then leads to an alternative way of expressing the device characteristics known as the *characteristic equation.* From the map, we have:

$$Q_{t+1} = (Q.R' + S)_t$$

The characteristic equation for RS flip-flop, where, *in this case*, t and t+1 have the meaning set out in this section.

7.1.2 The synchronous (clocked) JK flip-flop

The "not allowed " restriction placed on the RS device may well be most inconvenient, and the JK flip-flop was developed to overcome this problem. The basic idea is to use the current output conditions to gate the inputs to R and S so that they can never be 1 simultaneously, and for this purpose we may consider the addition of two *And* gates as in Figure 7.1. The two

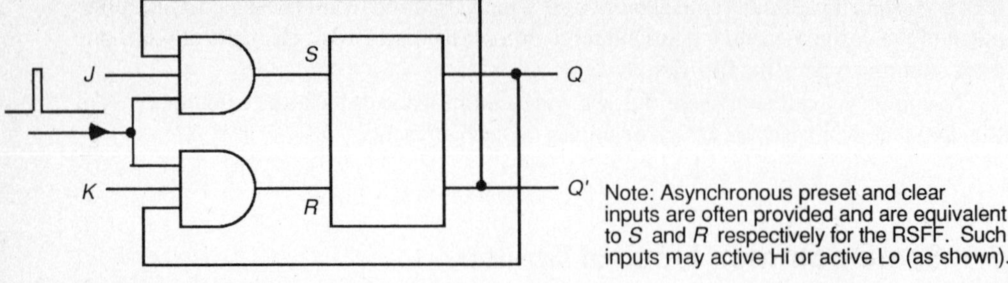

Figure 7.1 Development of JK flip-flop

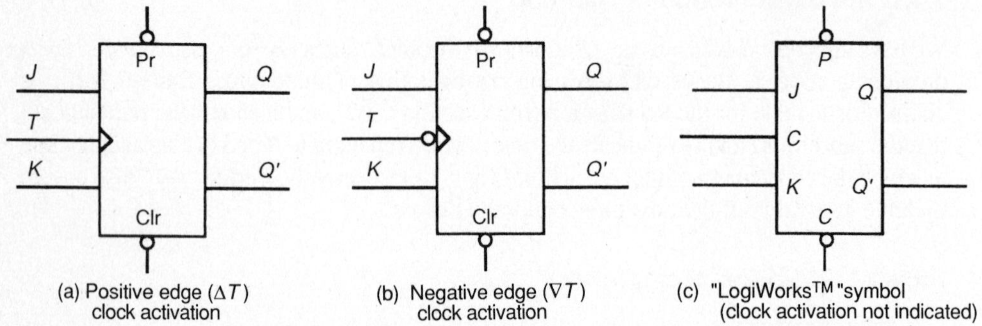

Figure 7.2 Commonly used symbols for JK flip-flop

input terminals to the flip-flop are now J and K and it will be seen that only one *And* gate can be active at any one time. This assumes that the outputs are always complementary which is the case. However, we are not out of the woods yet since a simultaneous input of 1's to both J and K would give rise to oscillations at Q and Q' as first one *And* gate then the other became active. The remedy is to use a "clock input" as the third input to both *And* gates as shown and if the clock signal consists of very short positive pulses, then for J and $K = 1$ simultaneously the outputs will change state once and once only for each clock pulse. In practice we do not usually have short positive pulses for clock signals but rather have square wave or related waveforms which have significant 1 and 0 periods. Flip-flop circuits are usually designed so that they respond to "edges" rather than logic levels, i.e. to ΔT or ∇T, if T is the clock signal. Commonly used symbols indicate the nature of the required clock signal as in Figure 7.2. The characteristics of the JK flip-flop are expressed in Table 7.2.

Table 7.2 JK flip-flop characteristics

J_t	K_t	Q_{t+1}	
0	0	Q_t	Output unchanged
1	0	1	Input to J
1	1	Q'_t	Output changes state
0	1	0	Input to K

(a)

Q_{t+1}	JK 00	01	11	10
Q_t 0	0	0	1	1
1	1	0	0	1

(b)

From the map (b), we have:

$$Q_{t+1} = (Q'J + Q.K')_t$$

The characteristic equation for JK flip-flop which defines the conditions under which the output Q will become 1. Such are known as *difference equations* due to the time difference between the effect (left-hand side) and the cause (right-hand side) of these equations.

Once again, we must properly define the significance of the time subscripts. For all clocked circuits we take t to denote the time at which *clock activation* occurs i.e., in coincidence with ΔT or ∇T, and "t+1" to indicate the time period *following activation* and prior to next activation. A *complementary characteristic equation for the JK F F* may be written as:

$$Q'_{t+1} = (Q'J' + Q.K)_t$$

which defines the conditions under which the output Q will become 0. *In clocked circuits, it is the time difference between cause and effect which allows elements to be cascaded in sequence by isolating input from output during clocking*, otherwise changes would propagate straight through cascaded stages. In VLSI designs, 2-phase clocks are often used to effect isolation.

JK flip-flop—designer's table

For design purposes a more convenient form of characteristic table may be set out as in Table 7.3. For any given present state of the output the flip-flop may be required to take up any particular next state and the table sets out the necessary inputs at J and K for each of the four possibilities. We will make use of this form in design work in this chapter because of the way in which entries transfer directly to flip-flop excitation maps as we shall see.

Table 7.3 JK flip-flop—designer's table

Present output Q_t	*Desired next O/P* Q_{t+1}	*Required inputs* J_t	K_t
0	0	0	Ø
0	1	1	Ø
1	0	Ø	1
1	1	Ø	0

Note "don't care" entries for J when present Q is 1 and for K when it is 0.

7.1.3. The D (data) flip-flop or latch (see Figure 7.3)

An important element is the D flip-flop or latch which is widely used to store (latch) a bit and may be used, for example, as flags and registers in digital system and computer architecture. In effect, the D FF uses the second and fourth rows only of the *JK* characteristic set out as Table 7.2(a). These are the two rows in which J and K are complementary and for

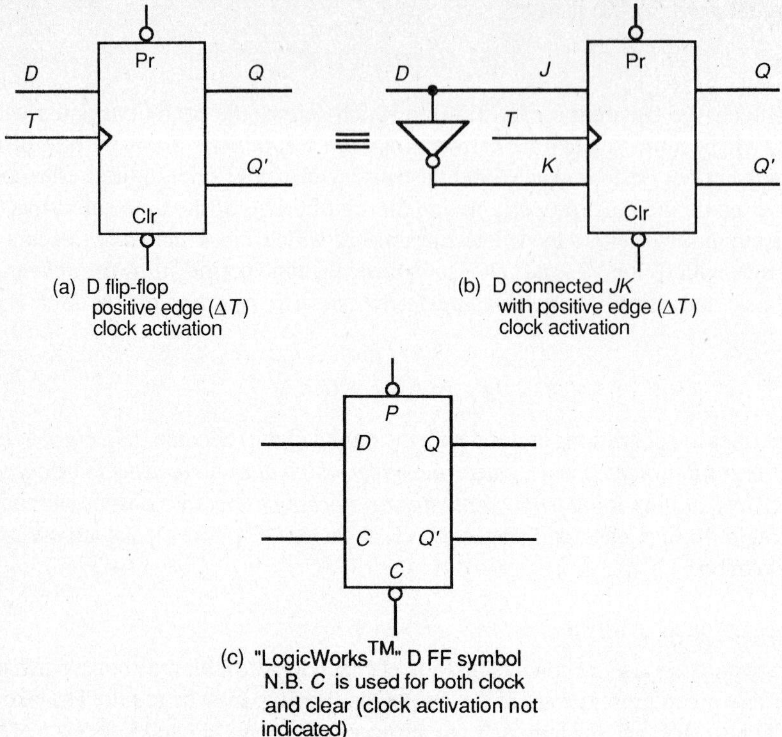

(a) D flip-flop positive edge (ΔT) clock activation

(b) D connected JK with positive edge (ΔT) clock activation

(c) "LogicWorks™" D FF symbol N.B. C is used for both clock and clear (clock activation not indicated)

Figure 7.3 The D (data register) flip-flop

this reason there is no mandatory need to implement clock edge activation and for isolated elements (e.g. latches) quite simple clocking arrangements can be used. The characteristic Table 7.4 and equation follows:

Table 7.4 D flip-flop characteristics

D_t	Q_{t+1}
1	1
0	0

The *characteristic equation for the D flip-flop is:*

$$Q_{t+1} = D_t$$

and the *complementary characteristic equation* is:

$$Q'_{t+1} = D'_t$$

D flip-flop—designer's table

Again, we may set out a convenient form for the designer as in Table 7.5.

Table 7.5 D flip-flop—designer's table

Present output Q_t	*Desired next O/P* Q_{t+1}	*Required inputs* D_t
0	0	0
0	1	1
1	0	0
1	1	1

The relationship between Q_{t+1} and D_t is plain to see and you will note that the present state (Q_t) of the output is immaterial.

7.1.4 The T (toggle) flip-flop (see Figure 7.4)

Another important element is the T flip-flop which is widely used in counters and related arrangements. In effect, the T FF uses the first and third rows only of the *JK* characteristic set out as Table 7.2(a). These are the two rows in which *J* and *K* have the same value. The characteristic table (Table 7.6) and equation are readily set out as follows:

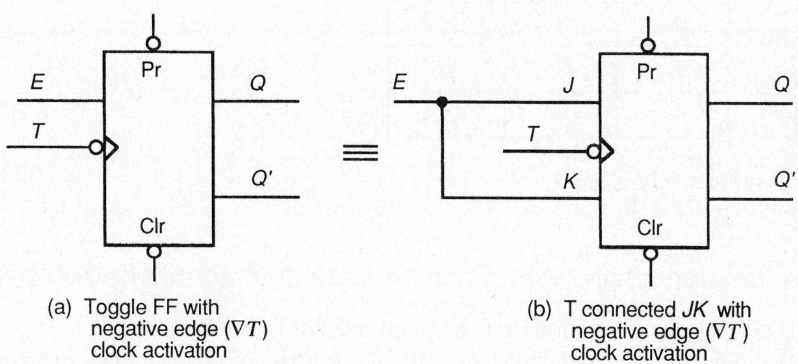

Figure 7.4 The T (toggle flip-flop)

Table 7.6 T flip-flop characteristics (with enable E*)

Q_t	Q_{t+1}	
0	1	Assuming E = 1
1	0	(a)

Q_t	Q_{t+1}	
1	1	Assuming E = 0
0	0	(b)

* The toggle enable input, (called E here), is most often called "T" but T has been used to denote the clock in this text.

Usually, the characteristics are given as in Table 7.6(a) since the T FF is often constructed from a JK FF with $J = K = 1$. On that basis the *characteristic equation* is readily written:

$$Q_{t+1} = Q'_t$$

and the *complementary characteristic equation* is:

$$Q'_{t+1} = Q_t$$

If an enable input is to be taken into account, we may write:

$$Q_{t+1} = E.Q'_t + E'.Q_t$$

which is the *characteristic equation for T flip-flop with toggle enable.*

T flip-flop—designer's table
Again, we may set out a convenient form for the designer as in Table 7.7.

Table 7.7 T flip-flop—designer's table

Present output Q_t	*Desired next O/P* Q_{t+1}	*Required inputs* E_t
0	0	0
0	1	1 *
1	0	1 *
1	1	0

* These two rows only if E = 1.

7.1.5 A summary of some common characteristic equations (Table 7.8)

Note that characteristic equations may be interpreted as follows:
Output Q *will become 1 at t+1* [left-hand side] *if* [=]*(conditions specified) apply at clock activation t* [right hand side].
The complementary equation gives the conditions for Q=0.

Table 7.8 Some common flip-flops

Flip-flop type	*Characteristic equation*	*Complementary equation*
RS (asynch.)	$Q_{t+1} = (Q.R' + S)_t$	$Q'_{t+1} = (Q'.S' + R)_t$
JK	$Q_{t+1} = (Q'.J + Q.K')_t$	$Q'_{t+1} = (Q'.J' + Q.K)_t$
D also JK with J=K'=D	$Q_{t+1} = D_t$	$Q'_{t+1} = D'_t$

Table 7.8 Some common flip-flops (continued)

T also JK with J=K=1	$Q_{t+1} = Q'_t$	$Q'_{t+1} = Q_t$
T(with Enable) also JK with J=K=E	$Q_{t+1} = E.Q'_t + E'.Q_t$	$Q'_{t+1} = E.Q_t + E'.Q'_t$

7.1.6 Designing an edge sensitive flip-flop

We are going to have a lot to say about *designing with clock edge sensitive circuits*, for example the JK flip-flop. Somewhere along the line such circuits have to be designed so we will digress slightly from our main purpose *to design a typical edge activated flip-flop* using our newly acquired knowledge of asynchronous sequential circuits.

The design process

Many of these circuits are required to respond to the falling edge (∇T) of the clock signal T. Knowing how we want the circuit to behave, we set out a state transition diagram (Figure 7.5), then a flow map and excitation map as in Table 7.9. We can then derive the logic circuit shown in Figure 7.6 (with the "LogicWorks™" simulation waveforms) for the completed design. The waveform shown as A clearly behaves as output Q (i.e. $Q = A$; $Q' = A'$).

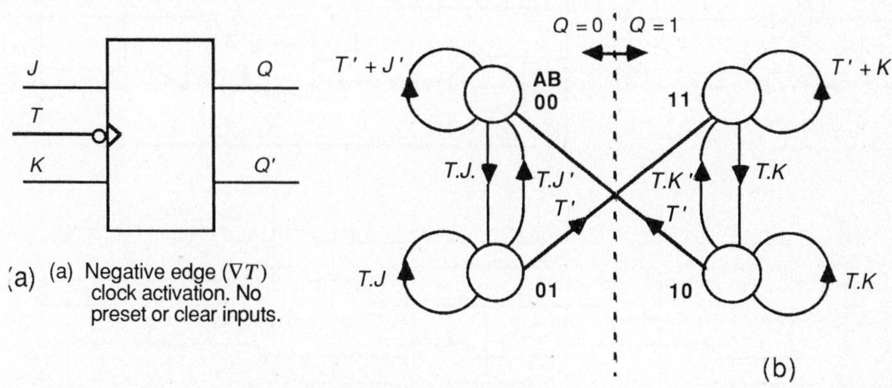

Figure 7.5 State transition diagram for JK flip-flop

Table 7.9 Design of ∇T activated JK flip-flop

	T'				T				
AB	JK 00	01	11	10	10	11	01	JK 00	
ab 00	①	①	①	①	2	2	①	①	00 ab
01	3	3	3	3	②	②	1	1	01
11	③	③	③	③	③	4	4	③	11
10	1	1	1	1	3	④	④	3	10

(a) Flow map for JK flip-flop design.

	T'				T				
	JK							*JK*	
AB	00	01	11	10	10	11	01	00	
ab 00	00	00	00	00	01	01	00	00	00 *ab*
01	11	11	11	11	01	01	00	00	01
11	11	11	11	11	11	10	10	11	11
10	00	00	00	00	11	10	10	11	10

(b) Excitation map for JK flip-flop design.

whence: $A = b.T' + a.T \qquad B = b.T' + a.K'.T + a'.J.T$

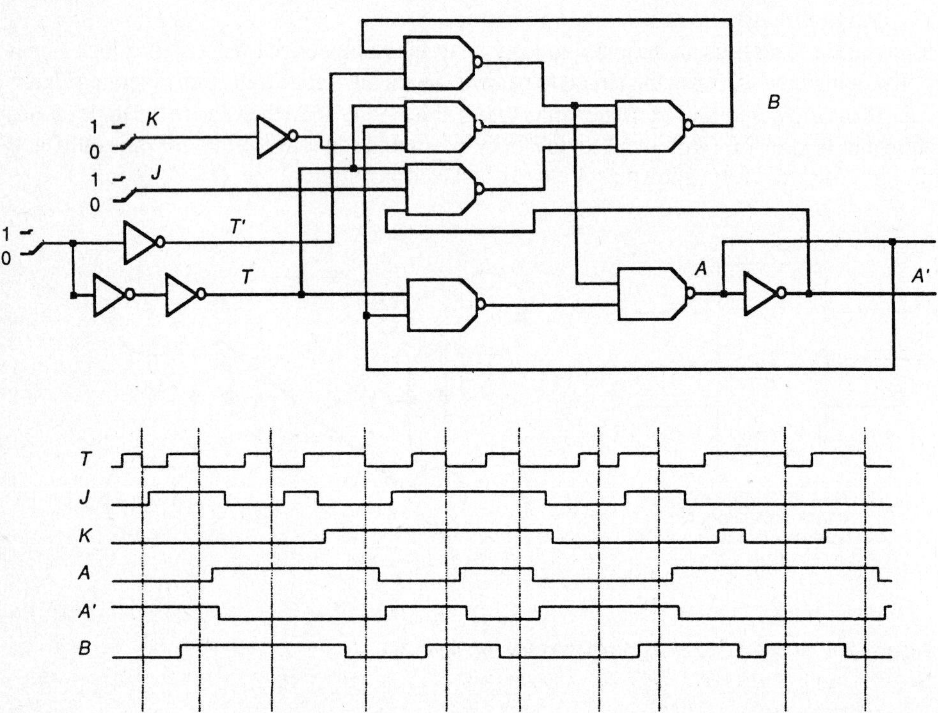

Figure 7.6 ∇*T* activated JK flip-flop circuit (with "LogicWorks" simulation waveforms)

A CMOS complementary logic implementation

The equations derived from Table 7.9 may be reorganized in a form which yields pull-up and pull-down configurations for secondary variables A and B as follows:

$$A = b.T' + a.T \quad \text{(pull-up)}\ ; \quad A' = b'.T' + a'.T \quad \text{(pull-down)}$$

$$B = b.T' + T.(a'.J + a.K') \quad \text{(pull-up)}\ ; \quad B' = b'.T' + T.(a'.J' + a.K) \quad \text{(pull-down)}$$

The circuit arrangement follows from these expressions and, not forgetting to add

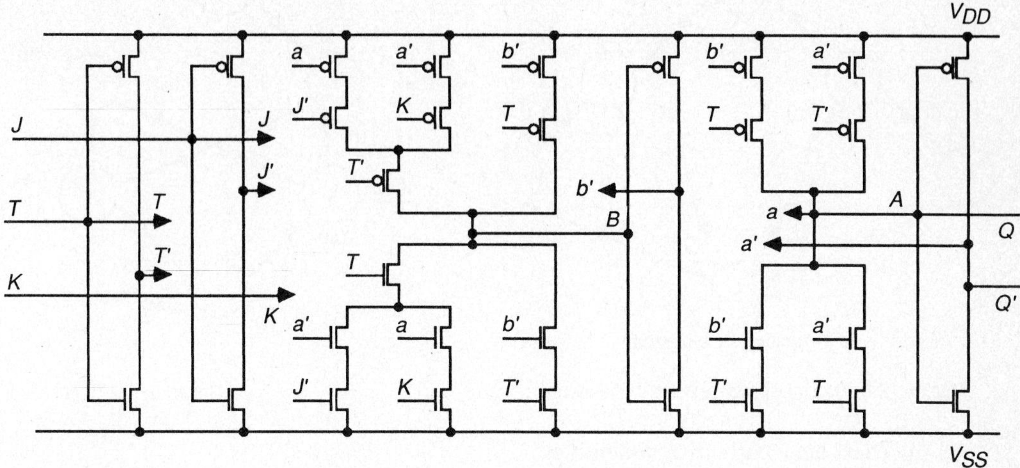

Figure 7.7 A CMOS complementary logic arrangement for the JK flip-flop

inverters as necessary, we may arrive at a CMOS complementary logic realization as in Figure 7.7. Other implementations also suited to VLSI are easily set out if required.

7.2 Clocked sequential circuit design

Once again, the easiest approach to explaining the design processes is to work through an example.

7.2.1 Example 1: Design of a clocked serial parity detector

1. Specification: The start of any design work should be to set out a specification and our example must meet the following requirements: Design a circuit to give an output Z = 1 whenever an odd number of 1's occurs in a *group of three* serially clocked bits on a single input line *W*. Groups do not overlap, so that bits 1,2,3 form the first group, bits 4,5,6 form the next group, bits 7,8,9 the next, and so on. You may assume that the serial clock is available to you for your circuit. *Z* is to remain 1 for one clock period and then revert to 0 for the next group evaluation.

2. Set out a primitive state transition diagram
This has been done as Figure 7.8(a) and we may explain the derivation of this diagram as follows:

(a) Assume the circuit to be in state 1 when the first bit appears on *W*.
Obviously, output Z = 0 since we are only looking at the first bit of three.

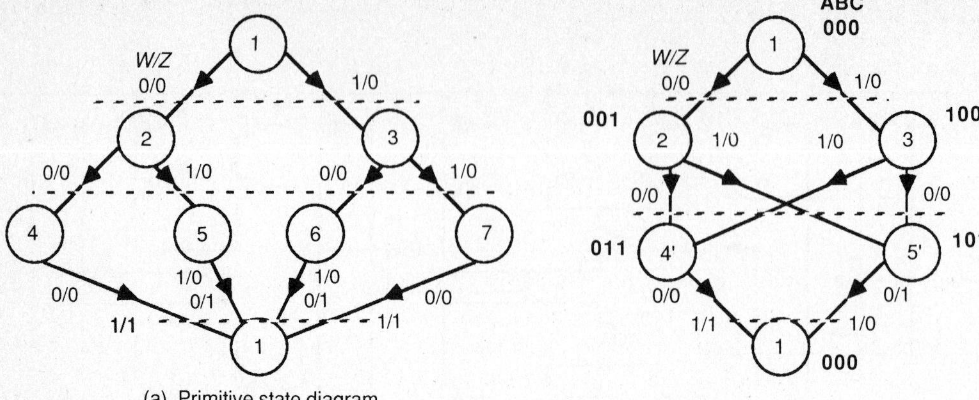

Figure 7.8 State diagrams for the example

(b) There are clearly two possibilities:
   (i) If $W = 0$, then take up state 2, or
   (ii) If $W = 1$, then take up state 3.
   The state of W and Z is given (W/Z) beside each link to the next state.

(c) The second bit is now on line W and we may be in state 2 or 3.
   Obviously, output $Z = 0$ since we are now looking at the second bit.
   If in state 2, then:
   (i) If $W = 0$, take up state 4, or
   (ii) If $W = 1$, take up state 5.
   If in state 3, then:
   (i) If $W = 0$, take up state 6, or
   (ii) If $W = 1$, take up state 7.

(d) The third, and last, bit of the group of three bits is now on line W and we may be in any one of states 4 to 7 inclusive.

   In any state we have two possibilities—either $W = 0$ or 1—and we are now in a position to specify a 0 or 1 for O/P Z, since we have now received a complete group of three bits and clearly the state we are now in has been determined by the sequence of 0s and 1's on W. For example, if we are in state 5 then the bit sequence on W has been 01. A further 0 will give an odd number of 1's, hence $Z=1$, but $W = 1$ will give $Z = 0$ as shown on the link to the next state.

(e) In all cases the next state will be state 1 to start the next group and Z must then be 0 again. In total, the primitive state diagram has seven states which would require a state vector three bits wide. We now attempt to reduce the number of states and hopefully the state vector width (in this case, if we can come down to four states or less).

3. Merge states if possible to reduce the overall complexity.
   Two or more states will merge if:
   (a) they each have the same next state(s), and
   (b) they have identical conditions on the links to the next state(s). In this case, 4 and 7 merge (to 4') and 5 and 6 merge (to 5').
   Draw the final (merged) state diagram - as in Figure 7.8(b).

Note that, for this example,the number of states has been reduced to five so that we have reduced complexity but not the state vector width.

4. Allocate secondary variables and state vector values to each state.

In this case we have chosen **ABC** as the secondary variables and have allocated discrete states of those variables (state vector values) to each state on the final state diagram. In this case we have allocated **000** to state 1 which is a reasonable thing to do so that the circuit may be readily "cleared" to state 1 following power up, etc.

The remaining allocations are arbitrary for clocked circuits since the timing of the clock signal is such that there can be no critical races or other timing related hazards. As a rough guide, a counting sequence based allocation sometimes leads to simple circuitry or, as is the case here, you can attempt to make allocations which only involve single variable changes between states (as is the case for asynchronous circuits). In any event there is no easy way to determine the best allocation and, if it is really essential to find the best, then this has to be on a "suck it and see" basis—by running through designs with alternative allocations. To cover all possibilities is very demanding, even for relatively small state vectors.

5. Set out the state transition table, which follows directly from the state diagram, with allocations. (The process should seem familiar since it parallels that for asynchronous circuits.)

For this example, the state transition table has been set out as Table 7.10 in a form convenient for subsequent programming of flip-flops to generate the required sequence.

At this point it is convenient to choose the circuit elements which will represent the secondary variables—*in this case choose JK flip-flops* since they are readily programmed—and set out in a diagram what we currently know about the circuit. (In this case, as Figure 7.9).

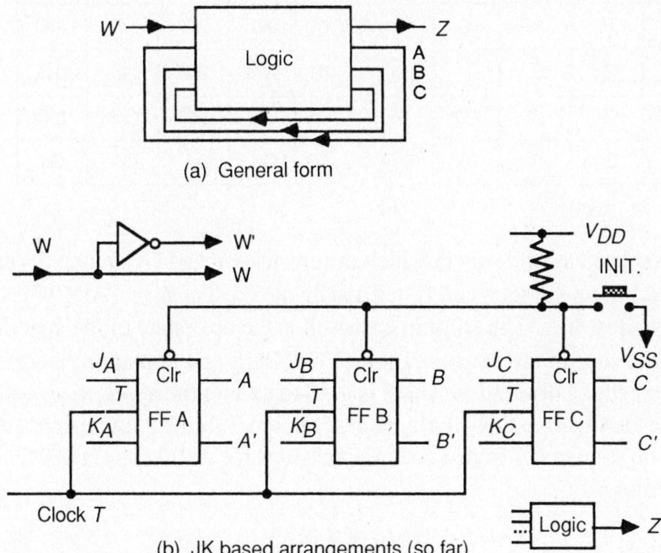

Figure 7.9 Form of circuit for example 1

Note: Q output of $FF\ A$ is named A, Q' output of $FF\ B$ is named B' etc.

Table 7.10 State transition table for example 1

Present state (time t) ABC	Next state (time t+1) W = 0 ABC	W = 1 ABC	Output Z
① 000	001	100	0
② 001	011	101	0
③ 100	101	011	0
④' 011	000	000	0 (W=0);1(W=1)
⑤' 101	000	000	1 (W=0);0(W=1)

Note: Three states of vector ABC are not used giving rise to don't cares (Ø).

6. Set out "programming" maps and hence logic for FF inputs.

We now take each of flip-flops A, B, C in turn and set out maps to show the required sequence of "next state outputs "and hence the required "JK input programming" as shown in Tables 7.11 to 7.13 inclusive.

Table 7.11 **FF A** (a) Next state sequence (b) JK input programming

(a)

A_{t+1}	$(CW)_t$ 00	01	11	10
$(AB)_t$ 00	0	1	1	0
01	Ø	Ø	0	0
11	Ø	Ø	Ø	Ø
10	1	0	0	0

(b)

J_AK_A	$(CW)_t$ 00	01	11	10
$(AB)_t$ 00	0Ø	1Ø	1Ø	0Ø
01	ØØ	ØØ	0Ø	0Ø
11	ØØ	ØØ	ØØ	ØØ
10	Ø0	Ø1	Ø1	Ø1

In order to appreciate the way in which the entries for J and K in map (b) are arrived at, consider map (a) in which each cell is defined by the present state, $(ABCW)_t$, of W and the secondary variables ABC. The entry in each cell is the next state of the specified variable, (e.g. $(A)_{t+1}$ in the case of the maps in Table 7.11). Each cell in map (a) therefore gives the present and next state information which is used to extract the appropriate conditions for J and K from the JK flip-flop designer's table set out in Table 7.3, in this case J_AK_A. These entries are made in map (b) from which expressions for J and K are directly obtained (as follows for J_A and K_A).

$$J_A = B'.W$$

$$K_A = W + C$$

Maps are then drawn for the other secondary variables (Tables 7.12 and13).

Table 7.12 FF B (a) Next state sequence (b) JK input programming

(a)

B_{t+1}	$(CW)_t$ 00	01	11	10
$(AB)_t$ 00	0	0	0	1
01	Ø	Ø	0	0
11	Ø	Ø	Ø	Ø
10	0	1	0	0

(b)

J_BK_B	$(CW)_t$ 00	01	11	10
$(AB)_t$ 00	0Ø	0Ø	0Ø	1Ø
01	ØØ	ØØ	Ø1	Ø1
11	ØØ	ØØ	ØØ	ØØ
10	0Ø	1Ø	0Ø	0Ø

From Table 7.12(b):

$$J_B = A'.C.W' + A.C'.W$$

$$K_B = 1$$

The logic linking the flip-flops is becoming apparent and we now need the conditions for flip-flop C.

Table 7.13 FF C (a) Next state sequence (b) JK input programming

(a)

C_{t+1}	$(CW)_t$ 00	01	11	10
$(AB)_t$ 00	1	0	1	1
01	Ø	Ø	0	0
11	Ø	Ø	Ø	Ø
10	1	1	0	0

(b)

J_CK_C	$(CW)_t$ 00	01	11	10
$(AB)_t$ 00	1Ø	0Ø	Ø0	Ø0
01	ØØ	ØØ	Ø1	Ø1
11	ØØ	ØØ	ØØ	ØØ
10	1Ø	1Ø	Ø1	Ø1

From Table 7.13 (b):

$$J_C = A + W'$$

$$K_C = A + B$$

The logic circuitry interconnecting the secondary variable flip-flops is now completely specified (as in Figure 7.10(a)) and there is a temptation to conclude that the design work is finished. However, in this case the secondary variables are just a means to an end and we have yet to provide the desired output Z from the circuit.

7. Derive the output functions to complete the design.
Turning back to the state transition Table 7.10 we see that Z is to be 1 in state 4' if $W = 1$ and, in state 5' if $W = 0$. The requirements may be mapped as in Table 7.14 (note that this is combinational logic and that Z is a function of secondary variables ABC, and input W).

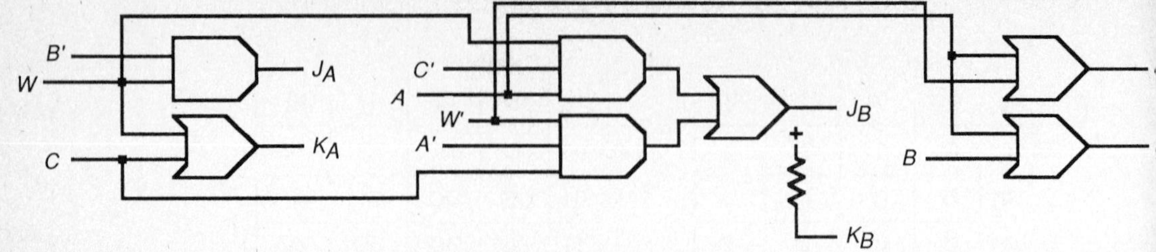

Figure 7.10(a) Logic circuitry for JK inputs

Table 7.14 Logic to produce output Z

Z	CW 00	01	11	10
AB 00	0	0	0	0
01	Ø	Ø	1	0
11	Ø	Ø	Ø	Ø
10	0	0	0	1

From this table

$$Z = B.W + A.C.W'$$

8. Check the completed design (logic lab., simulator etc.). The overall circuit is given as Figure 7.10(b) with simulations of its operation.

7.2.2 Summarized design procedure

Looking over the previous example, we can see that the following steps constitute a design process:

1. Set out a full, unambiguous specification.
2. Set out a primitive state transition diagram.
3. Merge states if possible to reduce the overall complexity.
   Draw the final(merged) state diagram.
4. Allocate secondary variables and state vector values to each state.
5. Set out the state transition table.

At this point it is convenient to choose the circuit elements which will represent the secondary variables.

6. Set out "programming" maps and hence derive logic equations for FF (or other chosen element) inputs.
7. Derive the output functions to complete the design.
8. Check the completed design (logic lab., simulator, etc.)

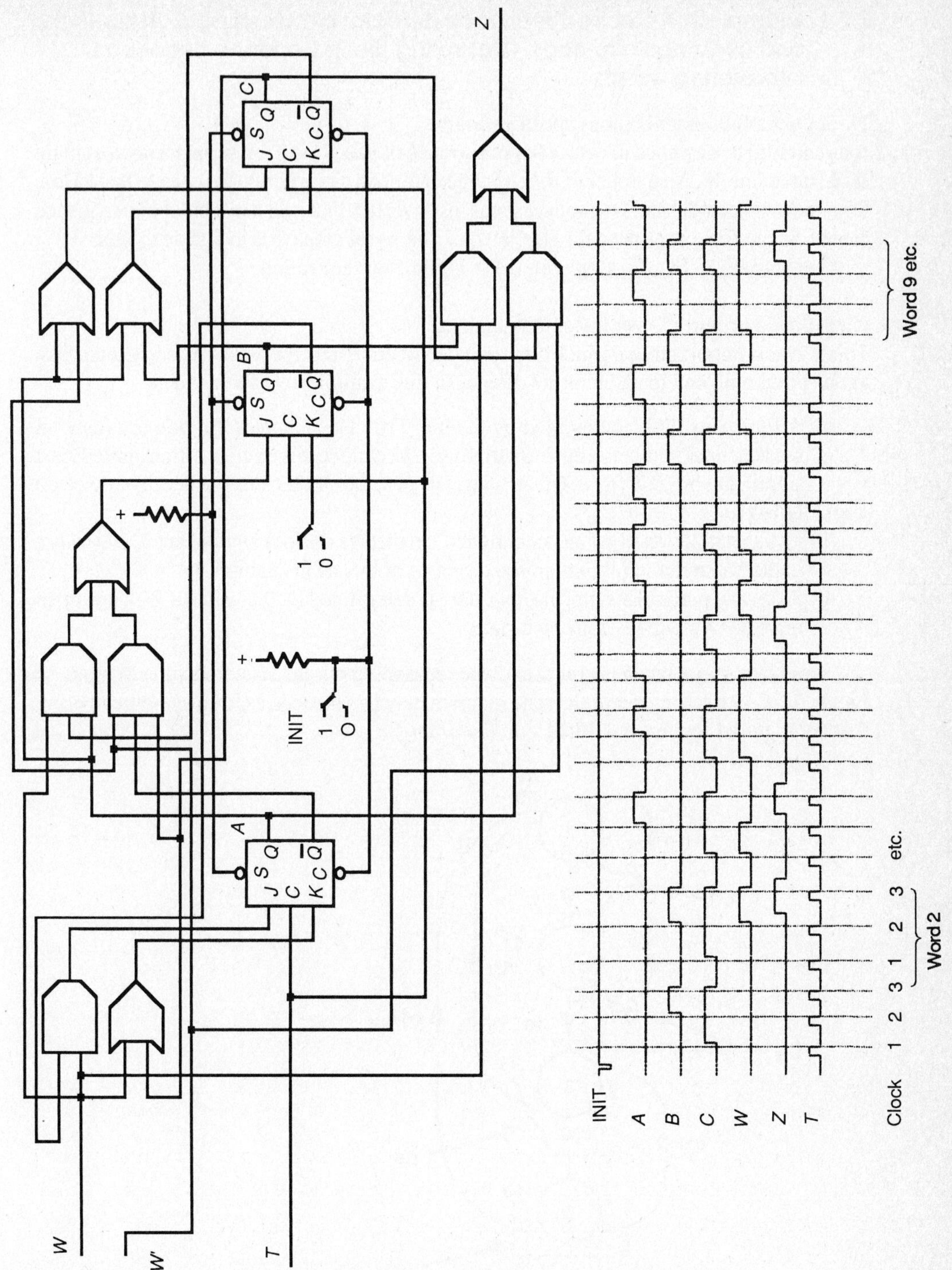

Figure 7.10(b) Completed circuit for example 1 (with stimulations)

7.2.3 Example 2: A clocked sequence detector circuit using (i) JK flip-flops, and (ii) Using D flip-flops. (Illustrating the importance of correctly interpreting words)

1. Set out a full, unambiguous specification.
A circuit is to be designed to detect any occurrence of Hex. D in 4-bit serial form on a single serial input line *W*. An output *Z* is to be produced for one clock period to indicate that a Hex. D has been detected. Hex. D sequences must not overlap, that is, the last bit of one sequence cannot be the first bit of the next Hex. D, etc. The serial clock *T* is available to you.

Compare (a) A JK flip-flop with (b) A D flip-flop realization.

2. Set out a primitive state transition diagram.
This is a most important step which turns the specification into a sequence. Clearly, a mistake at this point will lead to an incorrect design. In this example note three things:

(a) A 4-bit Hex. D is represented by the bits 1101 (see Chapter 2 if you are rusty on hexadecimal numbers), but in a serially clocked mode this number is transmitted least significant bit (*LSB*) *first* so that, with respect to the clock, the sequence to be detected is **1011.**
(b) The specification says "any occurrence" so that we are looking for Hex. D *anywhere* in the bit stream on *W*, *not* in fixed groups of bits as in example 1.
(c) From any particular state, the next state is determined by the value of *W*. Thus there must be two outlets from each state.

These factors are taken into account when reasoning out the state transition diagram of Figure 7.11. The setting out of the state diagram may be reasoned as follows (remembering that the required sequence is **1011** - *in that order*):

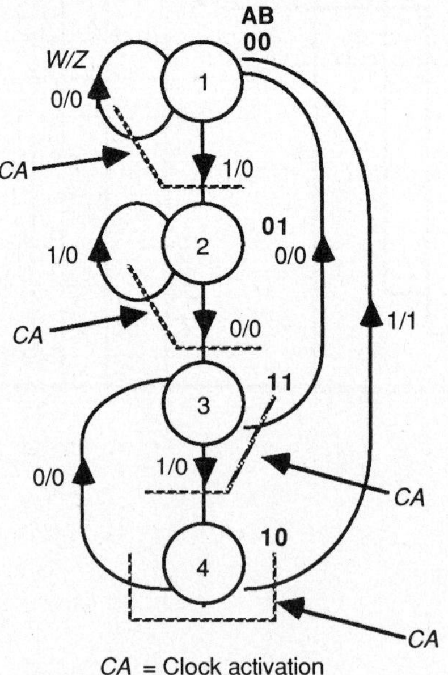

Figure 7.11 State transition diagram for example 2

(i) We may assume that the circuit is initialized into state 1 where it loops on itself until a **1** is received on W. This *could* be the start of a correct sequence, so, on the next clock activation we move to state 2. **Z** = **0** in both cases.

(ii) If in state 2, and the next value seen on *W* is **0** then the correct sequence is being detected and we move on to state 3 having so far detected **10**. On the other hand, if a **1** is detected on *W* then we loop on state 2 since this particular 1 may be the first bit of the sequence. **Z** = **0** in both cases.

(iii) If in state 3 and the next value seen on *W* is **1** then the correct sequence is being detected and we move on to state 4 having so far detected **101**. On the other hand, if a **0** is detected on *W* we have detected a **100** sequence which is incorrect and we have to return to state 1 to start over again. **Z** = **0** in both cases.

(iv) If in state 4 and the next value seen on *W* is **1** then the correct sequence **1011** has been detected and an output **Z** = **1** is produced and we return to state 1 ready to go through the detection sequence again. On the other hand, if a **0** is detected on *W* we have detected a **1010** sequence which is incorrect but the second **10** could be the first two bits of a correct sequence and so we loop back to state 3. **Z** = **0** in this case.

3. Merge states if possible to reduce the overall complexity.
Draw the final (merged) state diagram. No merging is possible here so that Figure 7.11 is it!

4. Allocate secondary variables and state vector values to each state. The diagram has four states so that a 2-bit state vector is needed.The variables **AB** have been nominated and the values allocated in this case appear in **bold** type beside each state in Figure 7.11 An arbitrary Gray code* allocation has been made here.

5. Set out the state transition table and choose the circuit elements. In this case the circuit elements are specified and are: (i) JK and (ii) D flip-flops.

Our next task is to set out the state transition table as in Table 7.15.

Table 7.15 State transition table for example 2

Present state	*Next state*		*Output*
	W = 0	*W* = 1	
$A\,B_t$	$A\,B_{t+1}$	$A\,B_{t+1}$	*Z*
① 00	00	01	0
② 01	11	01	0
③ 11	00	10	0
④ 10	11	00	0 (*W*=0);1(*W*=1)

6. Set out "programming" maps and hence derive logic equations for FF inputs.
   (a) JK based design: From the state transition table and using the designer's table for the JK flip-flop (Table 7.3), we now draw up maps for *J* and *K* inputs of flip-flops *A* and *B* as Tables 7.16 (a) and (b) respectively.

* Remember, Gray code is such that successive values are logically adjacent i.e. differing by only one literal.

Table 7.16 *J* and *K* "programming" maps for example 2

(a) FF A

J_AK_A	W = 0	W = 1
AB 00	0 Ø	0 Ø
01	1 Ø	0 Ø
11	Ø 1	Ø 0
10	Ø 0	Ø 1

(b) FF B

J_BK_B	W = 0	W = 1
AB 00	0 Ø	1 Ø
01	Ø 0	Ø 0
11	Ø 1	Ø 1
10	1 Ø	0 Ø

From the maps:

$$J_A = B.W' \qquad J_B = A.W' + A'.W$$

$$K_A = B.W' + B'.W \qquad K_B = A.$$

(b) D based design: From the state transition table (Table 7.15) and using the designer's table for the D flip-flop (Table 7.5), we now draw up maps for the D inputs of flip-flops *A* and *B* as Tables 7.17 (a) and (b) respectively.

Tables 7.17 D input "programming" maps for example 2

(a) FF A

D_A	W = 0	W = 1
AB 00	0	0
01	1	0
11	0	1
10	1	0

(b) FF B

D_B	W = 0	W = 1
AB 00	0	1
01	1	1
11	0	0
10	1	0

From the maps:

$$D_A = A'.B.W' + A.B'.W' + A.B.W$$

$$D_B = A.B'.W' + A'.B. + A'.W$$

We now have the logic circuit expressions for the two secondary variables.

7. Derive the output functions to complete the design.
In both cases the output Z is given in state 4 when $W = 1$, and since state 4 is defined by $AB = 10$ we may write:

$$Z = A.B'.W$$

for both (a) JK and (b) D based designs.

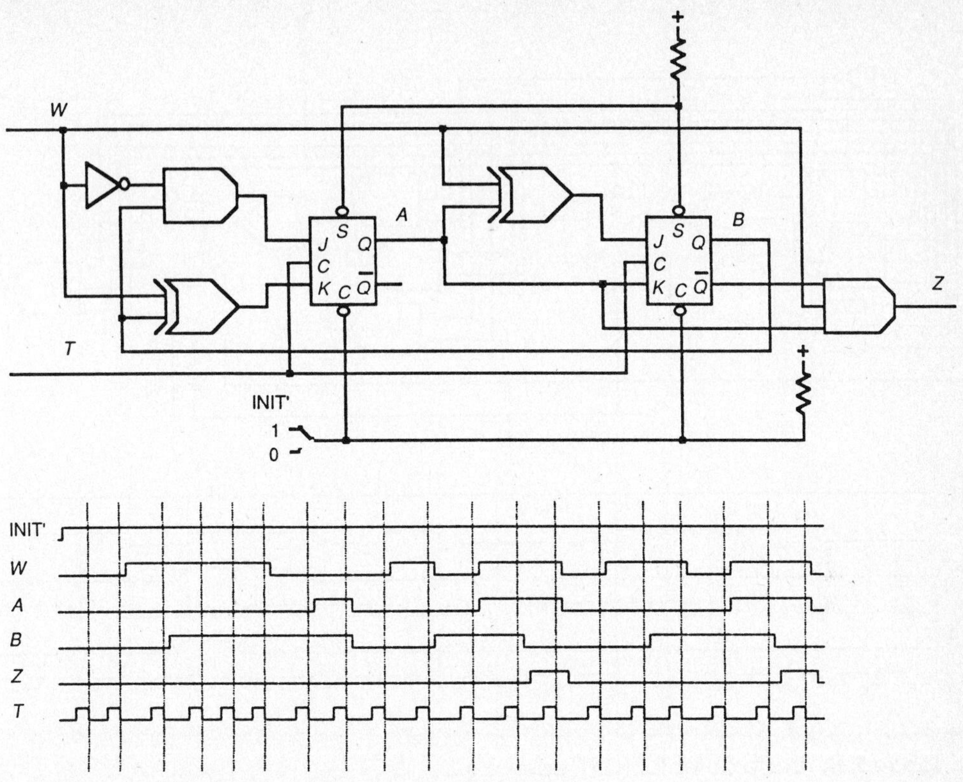

Figure 7.12 Code detector (JK FF based)

8. Check the completed design (logic lab., simulator etc.)

In this case the alternative designs (a) and (b) are given (with simulations) as Figures 7.12 and 7.13 respectively.

Note that the JK flip-flops used in Figure 7.12 are ∇T clocked, while the D flip-flops in Figure 7.13 are rising-edge (ΔT) clocked. This is typical of readily available packaged logic (e.g.TTL).

7.2.4 VLSI based realizations

A PLA based solution

The solution produced for example1and that for example 2 (a) employ JK flip-flops to represent the secondary variables. We have already seen that a JK flip-flop can be configured in silicon, e.g. Figures 7.6 and 7.7, but to say the least, the arrangement is not simple (28 transistors per JK FF) and we may well seek a less complex solution. One possibility is to use a PLA, and we will now rework example 2 with this in mind.

Steps (1) to (4) are as before, namely:

1. Set out a full, unambiguous specification.

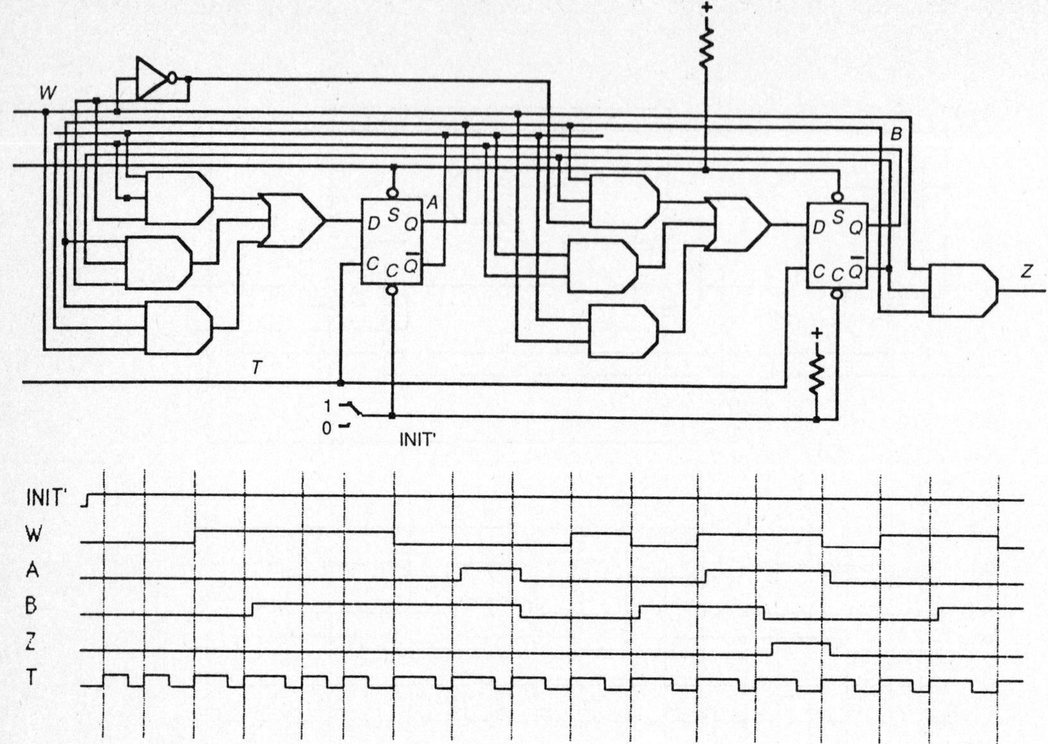

Figure 7.13 Code detector (D FF based)

2. Set out a primitive state transition diagram.
3. Merge states if possible to reduce the overall eomplexity. Draw the final (merged) state diagram.
4. Allocate secondary variables and state vector values to each state.
5. Set out the state transition table in a form which yields the *And* and *Or* plane equations for a PLA as in Table 7.18.

Table 7.18 State transition table for example 2 (PLA form)

	Present state AB and W $(A\ B)_p\ W$	Product term	Next state AB and output Z $(A\ B)_n\ Z$
①	00 0	-	00 0
	00 1	p_1	01 0
②	01 0	p_2	11 0
	01 1	p_3	01 0
③	11 0	-	00 0
	11 1	p_4	10 0
④	10 0	p_5	11 0
	10 1	p_6	00 1

Set out this way we may read the *And* terms (p_n) from the left side of the table—e.g. $p_1 = A'.B'.W$, $p_3 = A'.B.W$, $p_5 = A.B'.W'$ using $(A.B)_p$ etc.—and the *Or* terms from the "Next state and output" column on the right of the table. For example, looking down the A_n column we may pick out all the 1's to write:

$$A_n = p_2 + p_4 + p_5$$

Similarly:

$$B_n = p_1 + p_2 + p_3 + p_5$$

$$Z = p_6$$

where:

$p_1 = A'.B'.W$
$p_2 = A'.B.W'$
$p_3 = A'.B.W$
$p_4 = A.B.W$
$p_5 = A.B'.W'$
$p_6 = A.B'.W$

Note that two possible *And* terms are not formed (1st and 5th rows in the table) since there are no 1's for $(A.B)_n$ or Z in these rows and these *And* terms would not be used.

Thus the dimensions (v x p x z) of a PLA tailored to this problem are:

$$v = 3\ [(A\ B)_p \text{ and } W]\ \text{x}\ p = 6\ [p_1 \text{ to } p_6]\ \text{x}\ z = 3\ [(A\ B)_n \text{ and } Z].$$

6. Set out the PLA in stick diagram and/or mask layout form.
The general arrangement follows as Figure 7.14 and a stick diagram has been set out as Figure 7.15. Mask layout details can be determined from the standard cell developed earlier (Figure 5.22 (d)).

7. Check the correctness of the design using a design rule checker and check the design functionality using a logic/timing simulator.
The nature of these checks will be determined by available software.

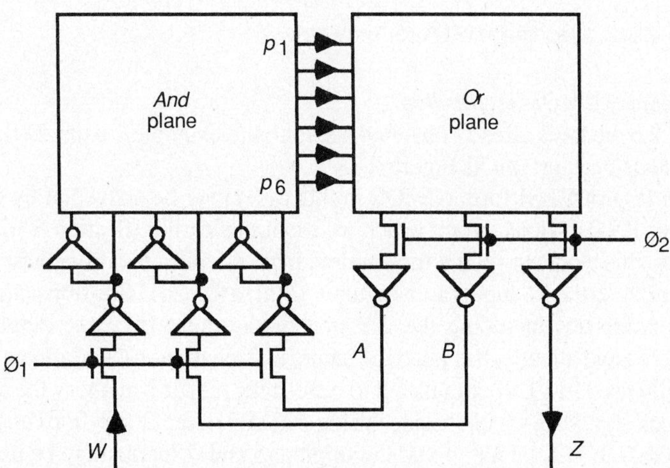

Figure 7.14 General form of PLA for example 2

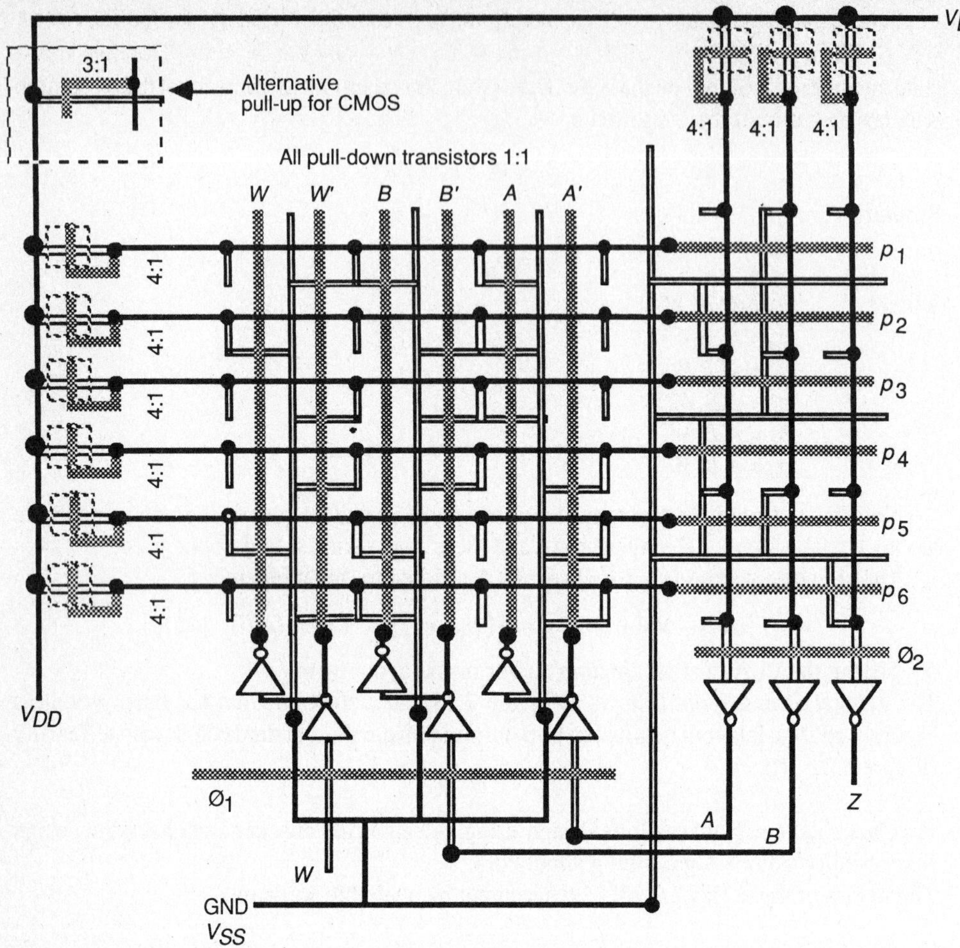

Figure 7.15 Stick diagram of PLA for example 2

A complementary CMOS D flip-flop

In section 7.2.3 we have already obtained a solution to example 2 using D flip-flops and a logic circuit has been set out as Figure 7.13.

A D flip-flop in VLSI form (CMOS in this case) may be arrived at by extending the asynchronous RS flip-flops circuit which we examined earlier (Section 6.3.2, Figure 6.7). In order to do this we must set out the required logic circuitry to add 2-phase clocking and additional gating to the RS inputs as in Figure 7.16(a) to form a D flip-flop which will exhibit the desired properties including the isolation of the input from the output after clock activation. We have already discussed the nature of 2-phase non-overlapping clocks which are commonly used in VLSI circuits but to refresh the reader's memory the characteristics of such a clock signal are set out beside the logic circuit of the D flip-flops under discussion (Figure 7.16(a)). It will be noted that the original S and R inputs may be used to provide asynchronous preset and clear facilities and that two output transmission gates are required if both Q and Q' outputs are needed (as shown).

A stick diagram, assuming complementary CMOS logic, is readily developed from the logic circuit as in Figure 7.16(b).

One possible mask layout has been drawn up and is set out as Figure 7.16(c). Although it is somewhat complex, it is much simpler than the JK flip-flop (Figures 7.6 and 7.7) comprising 16 transistors for each D flip-flop

A symbol and timing definitions for *t* and *t*+1 for this flip-flop appear in Figure 7.16(d) and the reader should note the timing details.

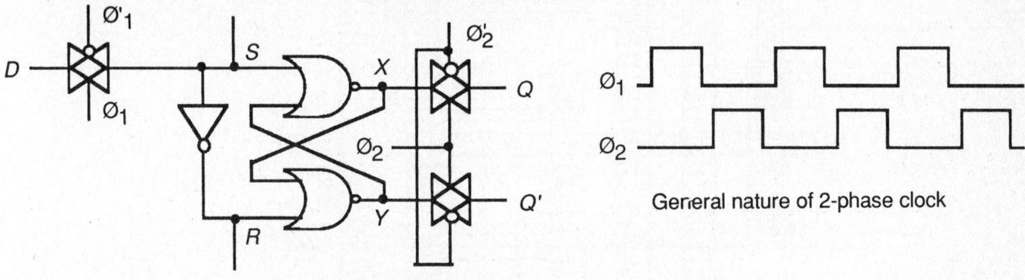

(a) Logic circuit for two-phase clocked D flip-flop

Figure 7.16 (a) Logic circuit for two-phase clocked D flip-flop

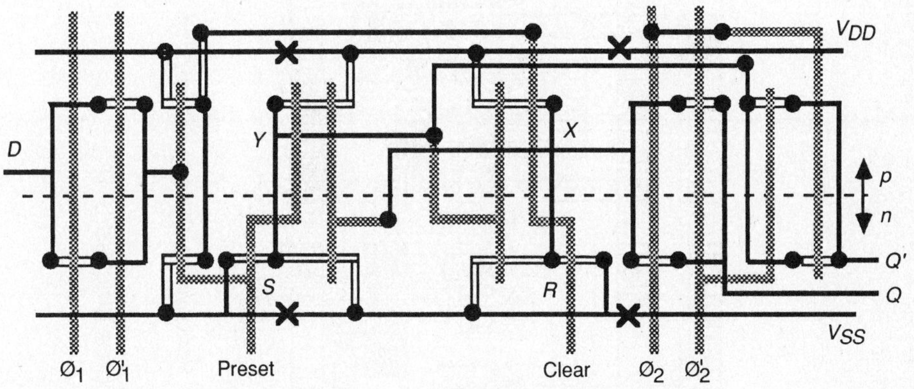

Figure 7.16 (b) Stick diagram for two-phase clocked D flip-flop (CMOS complementary logic)

A VLSI D flip-flop based solution to example 2

Clearly we could use this D flip-flop design to implement the circuit of Figure 7.13 by the addition of the inter-stage and output logic circuitry.

In order to simplify the design work, it could be advantageous to replace the *And/Or* interstage logic of Figure 7.13 by *Nand/Nand* logic as in Figure 7.17. The design of complementary CMOS *Nand* gates has been covered in section 5.2 and a mask layout appears in color plate 6.

An *And* gate might be designed for the output Z or again, a *Nand* gate followed by an *Inverter* could be used.

Thus we have seen that a PLA based and a D flip-flop based approach both provide a ready route to implementation in silicon.

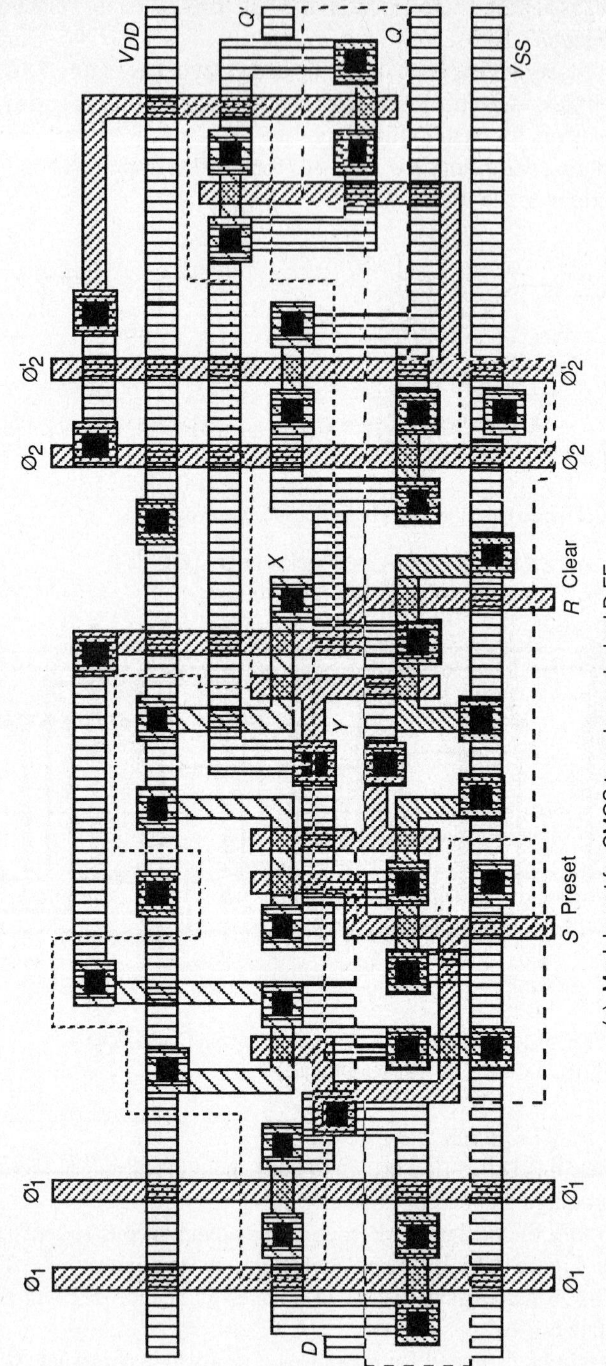

(c) Mask layout for CMOS two-phase clocked D FF

Figure 7.16(c) Mask layout for two-phase clocked D FF

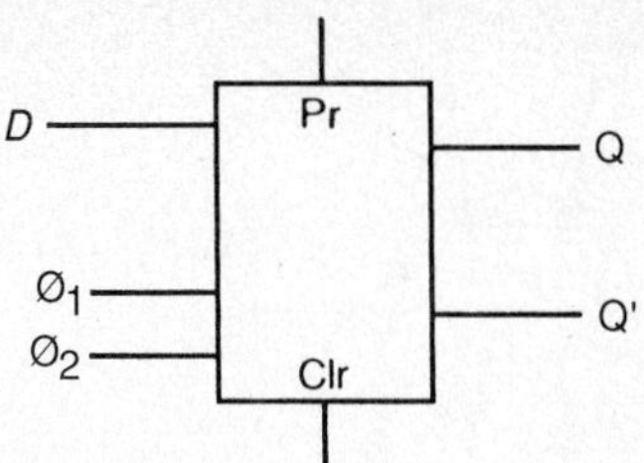

(d) Symbol for 2 phase clock D FF
Note: t is the time at which $\nabla \varnothing_1$ occurs.
$t + 1$ is the time at which the following $\Delta \varnothing_2$ occurs.

Figure 7.16(d) Symbol for two-phase clock D FF

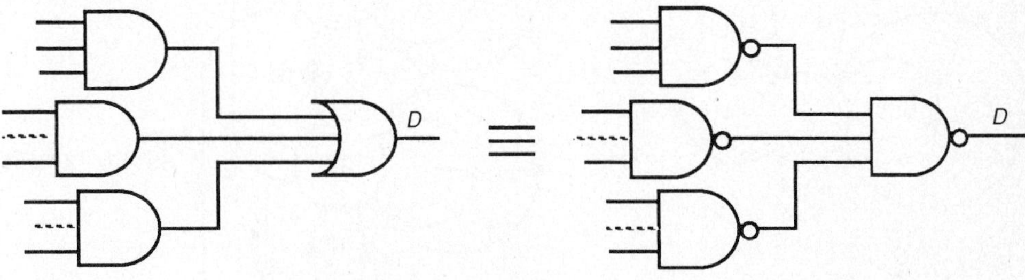

Figure 7.17 Interstage logic circuitry of Figure 7.13 with Nand equivalent

We have dealt with a complementary CMOS implementation of the D flip-flop and associated logic but clearly we could readily arrive at a pseudo-nMOS or at an nMOS based solution. You are invited to do so in Tutorial 7 which follows this chapter.

7.2.5 Mealy and Moore (finite state) machines

At this point it is convenient to discuss briefly the two commonly used types of sequential circuit—defined by Mealy and Moore* respectively.

In the most widely used and more general Mealy machine, the output at any time is a function of the present state (i.e. the state vector) of the circuit *and of the input variable(s)*.

It is this model that we have used in the examples so far presented in this chapter. In both cases output Z has been a function of the present state, defined by the state vector (secondary variables), and of input W.

In the Moore machine, the output at any time depends on the *present state alone*.

Counter circuits, which we are going to look at later, are good examples of natural Moore type circuits since the output at any time is clearly defined by the present state alone, i.e. the state of the count.

It may be shown that for every Mealy machine there is an equivalent Moore machine, and vice versa. To demonstrate this formally is outside the scope of this text but, in order to

*Mealy, G.H. "A method for synthesizing sequential circuits" *Bell System Tech J.*, 34 No.5, pp. 1045-80, Sept.1955.
Moore, E.F., "Gedanken experiments on sequential machines" in C.E. Shannon and J. McCarthy (eds), *Automation studies*, Princeton Uni. Press, 1956.

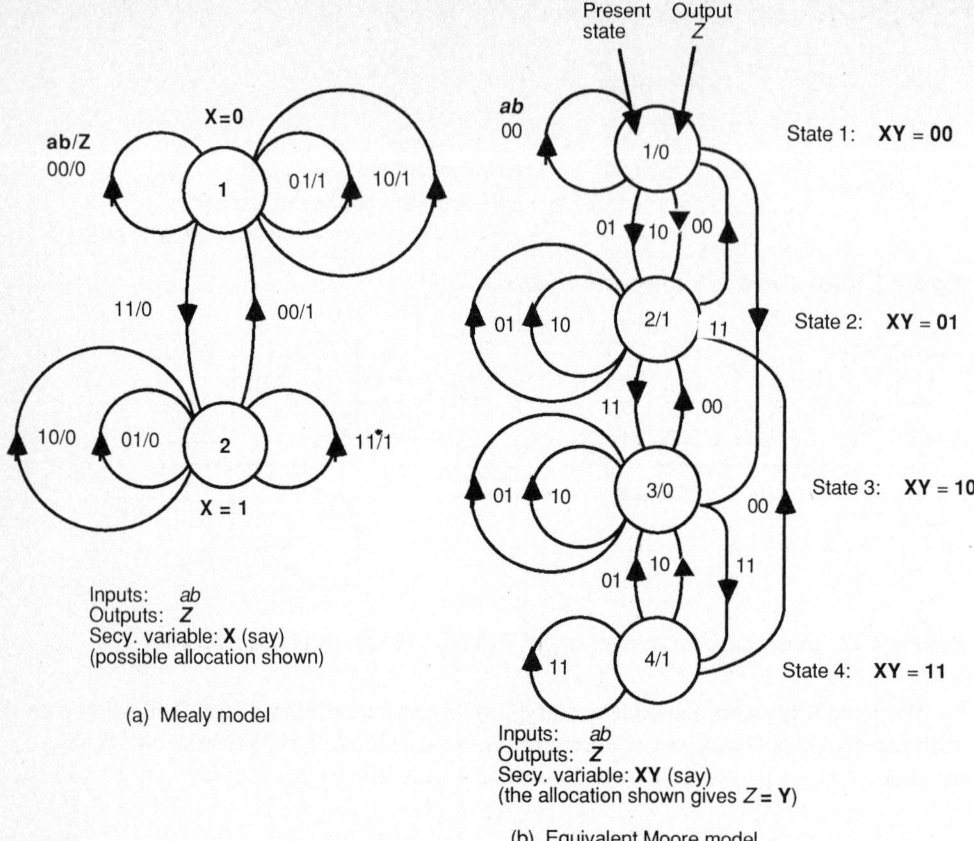

Figure 7.18 Mealy and Moore model state transition diagrams for the same sequential circuit

illustrate the translation, equivalent Mealy and Moore model state transition diagrams for one simple sequential circuit are given in Figure 7.18 (a) and (b) respectively. It will be noted that the Moore model requires more states than the Mealy model and this is usually, but not always, the case.

In simple terms, referring to Figure 7.18(a), it will be seen that there are two possible output conditions ($Z=0$ or 1), associated with state **1**. For the Moore model equivalent, each state is associated with one only output condition and so, state **1** is replaced by states 1 and 2. Similarly, Mealy state **2** is replaced by a further two states, 3 and 4, for the Moore circuits. The resulting state transition diagram is given as Figure 7.18(b). It will be noted that the output conditions can be written within the state symbol along with the state number. This cannot be done for the Mealy circuit.

Both diagrams share the common feature that there must be four outlets from each state since there are two input variables, **ab**, which means that four combinations must be covered.

To reinforce the idea of equivalence between the circuit types, the state transition tables are set out in Tables 7.19(a) Mealy, and (b) Moore. The PLA form has been used here by way of illustrating the VLSI implications but a form suitable for designs with flip-flop circuits is also readily arrived at.

Table 7.19(a) State transitions for Mealy type circuit

	Present state and inputs $(X)_p$ a.b	Product term	Next state $(X)_n$ and output Z $(X)_n$ Z
①	0 00	-	① 0 0
①	0 01	p_1	① 0 1
①	0 11	p_2	② 1 0
①	0 10	p_3	① 0 1
②	1 00	p_4	① 0 1
②	1 01	p_5	② 1 0
②	1 11	p_6	② 1 1
②	1 10	p_7	② 1 0

Table 7.19(b) State transitions for Moore type circuit

	Present state $(XY)_p$ a.b	Product term	Next state $(XY)_n$	Output Z $Z{=}Y_p$
①	00 00	-	① 00	0
①	00 01	p_1	② 01	0
①	00 11	p_2	③ 10	0
①	00 10	p_3	② 01	0
②	01 00	-	① 00	1
②	01 01	p_4	② 01	1
②	01 11	p_5	③ 10	1
②	01 10	p_6	② 01	1
③	10 00	p_7	② 01	0
③	10 01	p_8	③ 10	0
③	10 11	p_9	④ 11	0
③	10 10	p_{10}	③ 10	0
④	11 00	p_{11}	② 01	1
④	11 01	p_{12}	③ 10	1
④	11 11	p_{13}	④ 11	1
④	11 10	p_{14}	③ 10	1

7.3 Memory elements

A most important aspect of digital systems design is the ability to store or memorize individual bits or collections of bits, such as words (in the computer sense as well as text) or numbers. Data storage may be temporary (dynamic or volatile) or "permanent" (static or non-volatile).

7.3.1 General considerations

Large numbers of bits may be required to be stored in, for example, RAM and ROM arrays. In such cases, the area of silicon occupied by each stored bit is a critical factor determining the total number of bits, N, per chip or package. A further critical factor is the dissipation w per bit stored. Clearly, the total heat to be dissipated is largely determined by the product $N.w$. A further critical factor is the volatility or not of the stored data, whether this be a single bit or a large memory array.

Before dealing with some of the basic memory elements a few general remarks are in order on the three factors identified above.

1. *Area/bit stored.* For semiconductor memory, the storage mechanism is generally based on the transistor and a good guide to the relative area occupied is provided by the number of transistors.
2. *Dissipation/bit stored.* This depends on the technology being used, e.g. nMOS and pseudo-nMOS elements in some configurations can draw static rail to rail current and the dissipation, which is then also related to the number of transistors, is high in consequence. Other forms of both nMOS and CMOS memories draw current only on switching and dissipation is related to circuit capacitances and switching frequency.
3. *Volatility.* Static memories, such as flip-flops, are inherently non-volatile and will retain stored data until deliberately altered in circuit or until the power is switched off.

   Dynamic memories will retain data, for example, as charges on capacitors, for the short time it takes for charges to decay below some critical level. In MOS technology this decay time is usually several hundreds of µsecond, which for systems operating at MHz clock rates is a conveniently "long" period. Dynamic memory elements may have to be refreshed in many applications.

Another factor of concern is whether the data, once written, are to be retained permanently for read only memories, (ROM), or whether reprogrammable operation is required (EPROM) or whether there is to be full read/write operation, (RAM).

Every digital designer employs large amounts of memory and clearly expects to purchase it for next to no cost. An overriding factor for the commercial memory designer is thus the all important cost in cents/bit stored. The cost per bit must be very low for memories of any size, for example, a 64-kbit memory costing 1 cent / bit would imply an overall outlay of more than $650 which is generally unacceptable.

A further key factor is the speed of operation, often given as *access time*, which is the overall time taken to address the memory and present the stored data at the output. The various factors are usually in conflict and the design of commercially viable memories is a highly skilled business.

Clearly, the choice of memory element configuration and technology is important and we will now assess some commonly used elements in the light of the preceding discussion.

7.3.2 Some basic arrangements

Flip-flop circuits
We have already discussed various types of flip-flop. A flip-flop can be used to store one bit and clearly qualifies as a static read/write type device suitable for registers or flags. The most obvious type for data storage is the D flip-flop which, on clock activation, stores the logic level present on the D input. The JK can also be used to "remember" what its inputs were on clock activation and even the Toggle flip-flop can be regarded as a memory which records the fact that clock activation has taken place by a change of output state.

We may evaluate the potential of the flip-flop as a memory element in terms of the three parameters identified earlier.

1. *Area/bit stored*
Flip-flop designs, examined previously, are quite complex, the complementary CMOS JK design, for example, comprising 28 transistors and the CMOS D flip-flop 16 transistors. Since a T flip-flop is readily formed from a *D*, with the *D* input permanently tied to output *Q'*, it will be of the same complexity. All flip-flops are area hungry as can be seen from the *D* mask layout (Figure 7.16(c)) which occupies some 16500 μm^2 in 5 μm (feature size = 2 λ) technology. An estimate of area for another feature size *d* μm may be obtained by scaling the 5μm technology area by $(d/5)^2$.

Certainly nMOS and pseudo nMOS designs are less complex, 8 transistors instead of 16 for the D flip-flop, for example, but suffer from a much higher dissipation due to the static dissipation component.

2. *Dissipation/bit stored*
This depends on the technology, CMOS complementary designs having very small static power dissipation. However, nMOS and pseudo-nMOS designs will have appreciable static dissipation, for example, about 0.25mW per bit for 5 μm technology with 5 V supply rails. For fully scaled circuits, where supply voltage scales down directly as feature size, static dissipation is reduced by $(d/5)^2$. All flip-flop designs will also have switching dissipation which is proportional to f^2, where *f* is the switching frequency. Switching energy reduces as d^3 for fully scaled circuits.

3. *Volatility*
The flip-flop is a non-volatile storage element and will retain the bit stored as long as the power remains on. Clearly, the flip-flop designs which are based on cross coupled *Nor* or *Nand* gates, do not provide a "permanent" memory facility since data is lost on power down. The flip-flop is a good static memory element which is very suitable for flag bit storage and for registers, since it is basically fast in both read and write response times.

Storage by means of charges on capacitors
Some of the earliest electronic circuit memories were based on arrays of capacitors and the technique is widely used today: 1's and 0's can be stored as different polarity charges, or a 1 can be represented by a stored (often positive) charge and a 0 by zero charge or vice versa. The storage medium can be discrete capacitors or, commonly, the gate to channel capacitance C_g of nMOS or pMOS transistors. This particular approach is widely used in VLSI circuits

and in memory design generally, and it is useful to examine a few of the more popular methods for employing this concept.

First of all, consider the circuits set out in Figure 7.19(a). It can be seen that, using pass transistor or transmission gate (not shown) switches, it is possible to store either logic 0 or 1 voltage levels as charges on the input capacitance C_g of an n- or p-type transistor (or both simultaneously as at the input to a CMOS complementary inverter). In the diagram the charge is stored when a "store" signal is received, and once this signal becomes inactive again the charge is isolated and the gate of the transistor will remain at the stored level until the charge on C_g is modified sufficiently by leakage currents to destroy the stored level. Leakage currents are such that the level stored will typically remain "good" for an appreciable fraction of a millisecond at room temperature, that is, quite a long time in a high speed system context. The reader should, however, be aware that leakage current increases with increasing temperature (approx. double for each 10°C rise in temperature) and storage times are scaled down in proportion. A further hazard for the unwary lies in the loss of stored charges due to photo emission which occurs when light is incident upon the chip surface. This can happen when the proud designer views a masterpiece with the package lid raised.

This storage mechanism is nevertheless most useful and is widely used for dynamic memory. Some of the common arrangements follow.

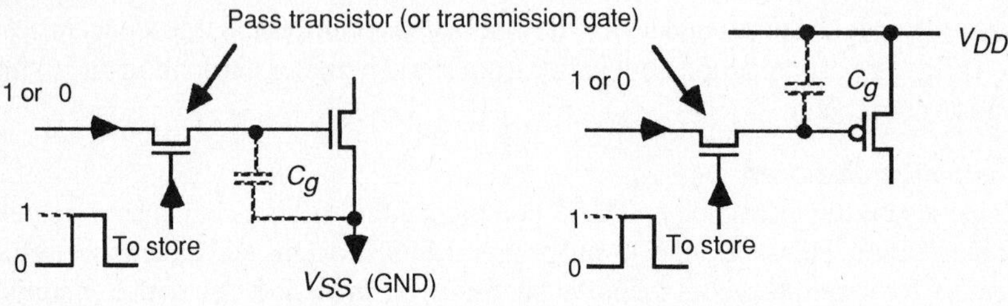

Figure 7.19(a) Bit storage on gate capacitance

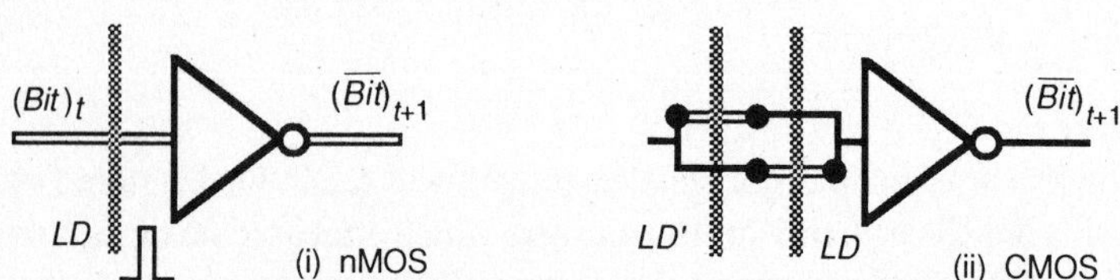

Figure 7.19(b) Basic inverting dynamic storage cells

Basic inverter based dynamic storage cells

Dynamic bit storage is readily formed from inverter(s) and pass transistor or transmission gate isolating switches as shown in Figures 7.19(b) and (c). It will be seen that either inverting or non-inverting storage may be configured, for example, in (i) nMOS or (ii) complementary CMOS technology. It will be seen that bit storage takes place when (in this

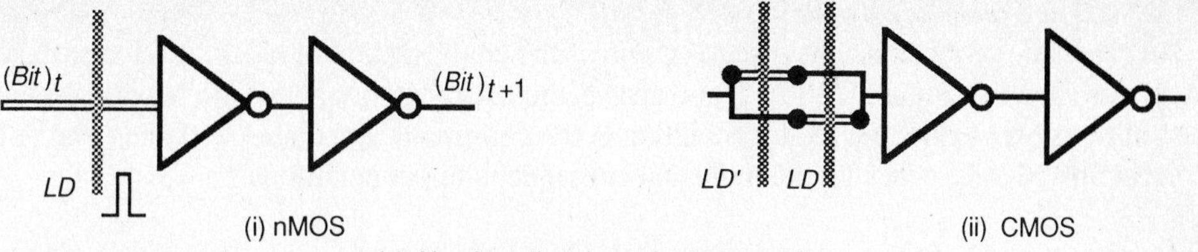

(c) Non-inverting dynamic storage cell

Figure 7.19(c) Non-inverting dynamic storage cells

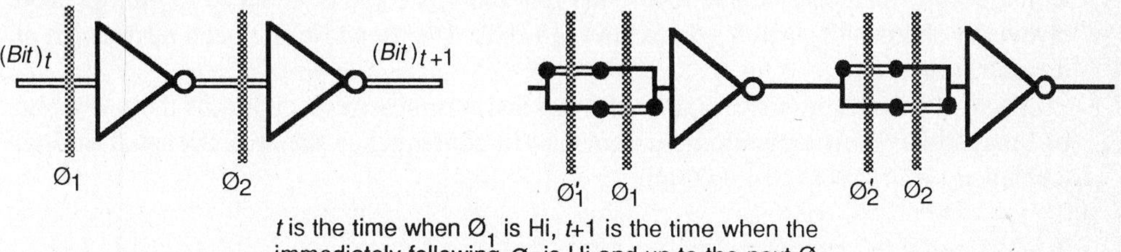

t is the time when $Ø_1$ is Hi, *t*+1 is the time when the immediately following $Ø_2$ is Hi and up to the next $Ø_1$.

(d) Clocked non-inverting dynamic storage cells

Figure 7.19(d) Clocked non-inverting dynamic storage cells

case) the LD (load) signal becomes active and the reader should particularly note the definitions for *t* and *t*+1 associated with this form of storage as given in the note beneath Figure 7.19(b). We may now assess the potential of this form of storage in terms of the three key parameters.

1. *Area/bit stored*

The nMOS inverting cell is the simplest, comprising only three transistors. In 5 μm technology such a cell would occupy an area of some 3000 μm^2 but this could be reduced in multiple cell layouts by sharing V_{DD} and V_{SS} lines between cells. The nMOS non-inverting cell is greater in area by about 40 percent.

The CMOS inverting cell comprises four transistors but space is occupied by the transmission gate and by the p-well and p+ areas and by separation requirements. In consequence the area occupied in 5 μm technology is in excess of 7000 μm^2. Adding a further Inverter to produce a non-inverting cell could add about 35 percent to this area.

2. *Dissipation/bit stored*

In nMOS 5 μm technology an 8:1 ratio inverter in the "on" state can have a typical static dissipation ranging between 0.28 and 0.75 mW. An Inverting cell will dissipate significant power in the "on" state (i.e. when storing a 1) having very little static dissipation when storing a 0. For the nMOS non-inverting cell, a further Inverter (4:1 ratio) is added thus adding dissipation in the 0 stored state also (0.5 to 1.0 mW typically).

For the CMOS cells there is negligible static dissipation. For all cells there will be switching dissipation related to the frequency of switching.

3. *Volatility*

Clearly, the cells all constitute volatile memory but, typically, can retain stored data for up to 0.5 msec.

Clocked non-inverting dynamic storage cells
An extension of the concepts discussed allows the configuration of clock isolated storage cells as shown in Figure7.19(d). These arrangements are important and are widely used in shift registers. Again, the reader should note the definitions associated with the t and t+1 subscripts, that is, denoting the input and consequent output conditions.

1. *Area/bit stored*
The nMOS cell is the simplest, comprising six transistors. In 5 µm technology such a cell would occupy an area of some 5000-6000 μm^2 but this could be reduced in multiple cell layouts by sharing V_{DD} and V_{SS} lines between cells. The pseudo-nMOS cell has an area of the same order.

The CMOS cell comprises eight transistors and extra space is occupied by the p-well and p+ areas and by their separationrequirements. In consequence the area occupied in 5µm technology is in excess of 13000 μm^2.

2. *Dissipation/bit stored*
In nMOS 5 µm technology an 8:1 ratio inverter in the "on" state can have a typical static dissipation ranging between 0.28 and 0.75 mW. A cell will dissipate significant power when storing a 1 or a 0 since either one or other inverter will be on and both have the same 8 :1 ratio. For the CMOS cell there is negligible static dissipation. For all cells there will be switching dissipation related to the frequency of switching.

3. *Volatility*
Clearly, all cells all constitute volatile memory but usually this is of no consequence since bits are clocked in or shifted at intervals which are negligible in comparison with the 0.5 msec. or so decay time.

A 3-transistor dynamic storage cell
The quest is on for bigger and better memories and clearly this implies a need for smaller, faster and lower dissipation cells.

A move away from the inverter-based dynamic storage so far considered in this section is beneficial in at least two of these respects, area and power dissipation, and is faster for stored 0's but slightly slower for stored 1's than the inverting dynamic cell discussed earlier. A comparison with this cell is appropriate since the 3-transistor cell is also inverting.

The nature of the cell may be deduced from the circuit of Figure 7.20 (a) and (c). It will be seen that the bit to be stored is "written" from the bus via Tr1 which is a series switch opened by the *WR* signal. The bit is stored on the gate capacitance C_g of Tr2 and can be "read" (complemented) back onto the bus when *RD* is activated, thus turning Tr3 on. The bus is precharged to logic 1 through the pull-up structure and, if a 1 was stored then both Tr2 and Tr3 will conduct during *RD* and the bus will be pulled down to the logic 0 condition. If a 0 was stored, then the bus will remain at 1 since Tr2 will not conduct. Note that ratio rules must be observed between the pull-up transistor and Tr2 and Tr3 in series for nMOS designs. This does not apply to the CMOS case where it is usual for all transistors to be minimum size 1:1 devices.

In order to assess area requirements, mask layouts for nMOS and CMOS designs have been developed in Figures 7.20 (b) and (d) respectively.

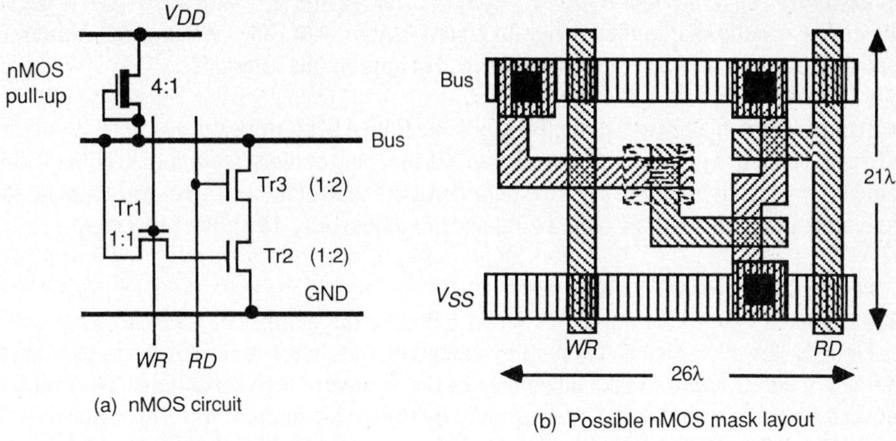

(a) nMOS circuit

(b) Possible nMOS mask layout

Note: RD and WR are mutually exclusive and coincident with $Ø_1$.

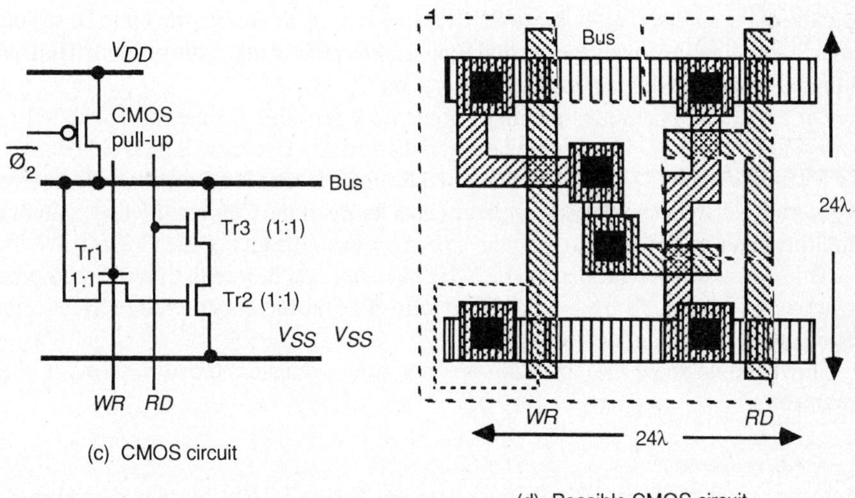

(c) CMOS circuit

(d) Possible CMOS circuit

Figure 7.20 Three transistor dynamic memory cell

1. *Area/bit stored*

The area required is considerably less than previous cells and, typically, a 5 µm design in CMOS technology will occupy about 3500µm². In nMOS designs the area is similar, but in both cases considerable space saving can result from sharing bus and V_{ss} rails in multiple cell arrays.

2. *Dissipation/bit stored*

There is no static power dissipation but there will be switching related dissipation as discussed earlier.

3. *Volatility*

The cell is volatile due to the change in stored charge with time. When used in memory applications the stored data must be refreshed at appropriate intervals.

In order to assess this cell as a potential candidate for RAM chip design, a simple calculation will reveal that, in 5 μm technology, a 5 mm X 5 mm chip could accommodate about 8 kbits of memory which is hardly likely to be regarded as a "state of the art" memory. Even in, say, 2 μm technology this size of chip would accommodate only 48 kbits of this cell.

A one transistor dynamic storage cell

This is indeed the "rock bottom" as far as reducing the number of transistors per cell is concerned. However, area is taken up by extending the source region of the transistor and adding a plate structure to take advantage of the relatively high capacitance per unit area between polysilicon and diffusion separated by thin oxide in order to form a capacitor C_M (see Figure 7.21(c)) of reasonably significant value in comparison with the capacitance of the Column *RD/WR* line. This runs through a large number of cells and thus has considerable length and appreciable capacitance in consequence.

The action of this cell may be examined with relation to Figure 7.21(a). On write (*WR*) operations the appropriate Column *RD/WR* line is used to present the bit to be stored at the drain D and simultaneously the desired *Row sel.* line is activated with a logic 1. The transistor is turned on and the bit is stored as a charge on C_M.

On read (*RD*) operations the appropriate *Row sel.* line is energized with a 1 and the particular Column *RD* line having been precharged to a 0 level will then either remain at 0 if a 0 was stored on C_M, or will be disturbed from the 0 state if a 1 was stored. Although C_M is generally small in comparison with the capacitance of the Column *RD* line, suitable sense amplifiers can be designed to read the stored bit from the end of the Column *RD* line.

This cell exhibits a property which we have not met before in that the read process is destructive of stored 1's. Thus, apart from refreshing the memory periodically, we must also regenerate a stored 1 after reading it.

Having considered the operation we may now evaluate the cell in terms of the three parameters.

1. *Area/bit stored*

From the circuit of Figure 7.21(a) a stick diagram, Figure 7.21(b), is readily developed which leads to the mask layout shown in Figure 7.21(d).The area per bit in 5 μm technology is around 1250 μm² which compares favorably with other cells. To relate this to practical reality, we can consider a 5 mm X 5 mm chip which would hold some 18 kbits of this cell. If we move to 2 μm technology, the chip would hold more than 100 kbits. As you can see, we have now moved into the big league.

2. *Dissipation/bit stored*

There is no static power dissipation but there will be switching related dissipation as C_M and the Column *RD/WR* line, etc., charge and discharge.

3. *Volatility*

The cell is volatile due to the change with time of the charge stored in C_M. When used in

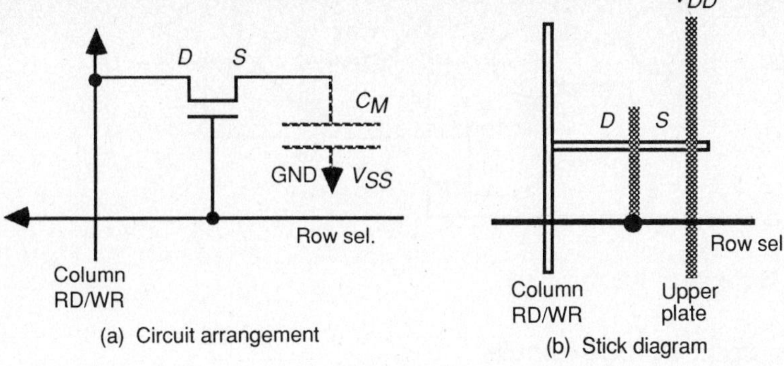

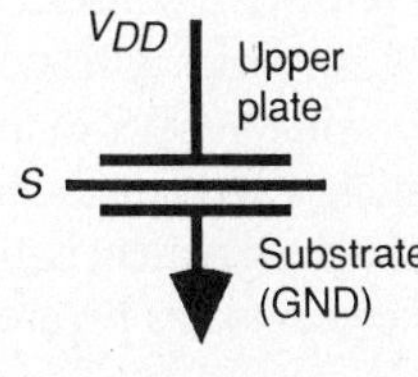

(c) Eqivalent circuit for C_M

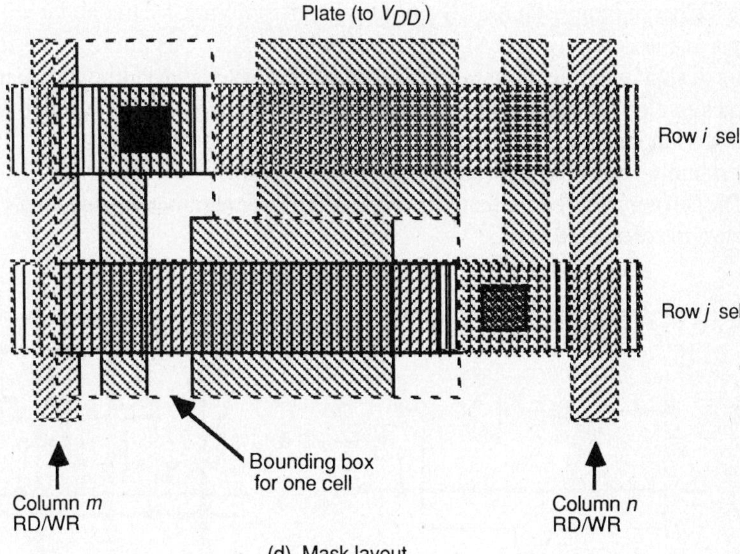

(d) Mask layout

Figure 7.21 One-transistor dynamic memory cell

memory applications the stored data must be refreshed at appropriate intervals. A stored 1 must also be regenerated after it is read.

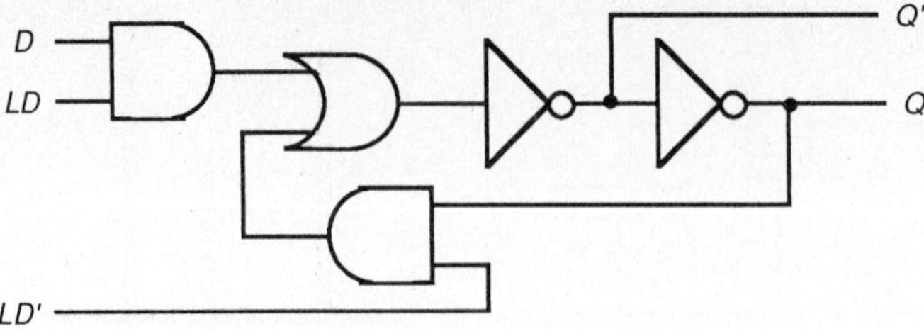

Figure 7.22 Gate logic latch circuit

7.3.3 Some static storage circuits

A gate logic based latching circuit

We have already seen, in Chapter 6, that positive feedback can be used to produce a circuit, comprising two inverters (see Figure 6.19(a)), which will latch and remain in either an output = 0 or output = 1 state. The addition of two simple gates to this circuit will produce a usable latch which can be loaded at will, and which will retain the bit thus stored until deliberately altered or until the power is interrupted. The arrangement is given as Figure 7.22 and it should be noted that another form of latch was presented as Figure 6.19(b).

An nMOS logic based latch

An arrangement similar in principle to the preceding one is readily configured in nMOS or in CMOS technology. An nMOS circuit is given in Figure 7.23(a) and it will be noted that this particular cell may be read onto either or both of two bus lines. This is to allow the use of this design in applications where it forms the registers providing the inputs to an adder. In that situation the cell may be read into either or both inputs of the adder simultaneously. As is commonly the case, the cell may be written to from only one of the buses. A corresponding mask layout follows as Figure 7.23(c).

The cell may now be evaluated in terms of the three parameters used for all VLSI memory cell designs presented here.

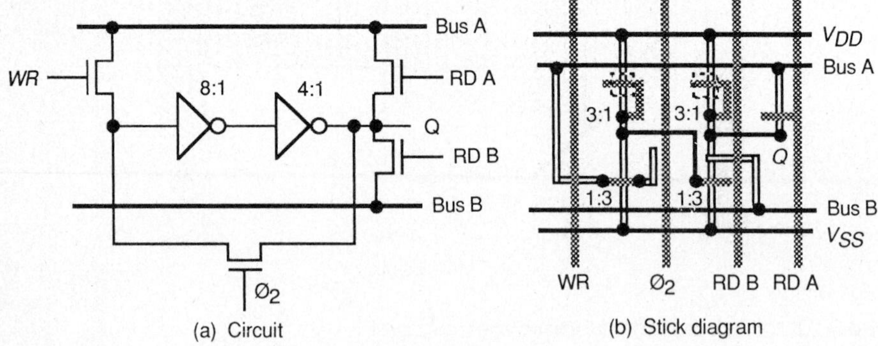

(a) Circuit

(b) Stick diagram

Note: RD and WR are mutually exclusive and coincident with $\varnothing_1$.

Figure 7.23 (a) and (b) nMOS pseudo-static latch

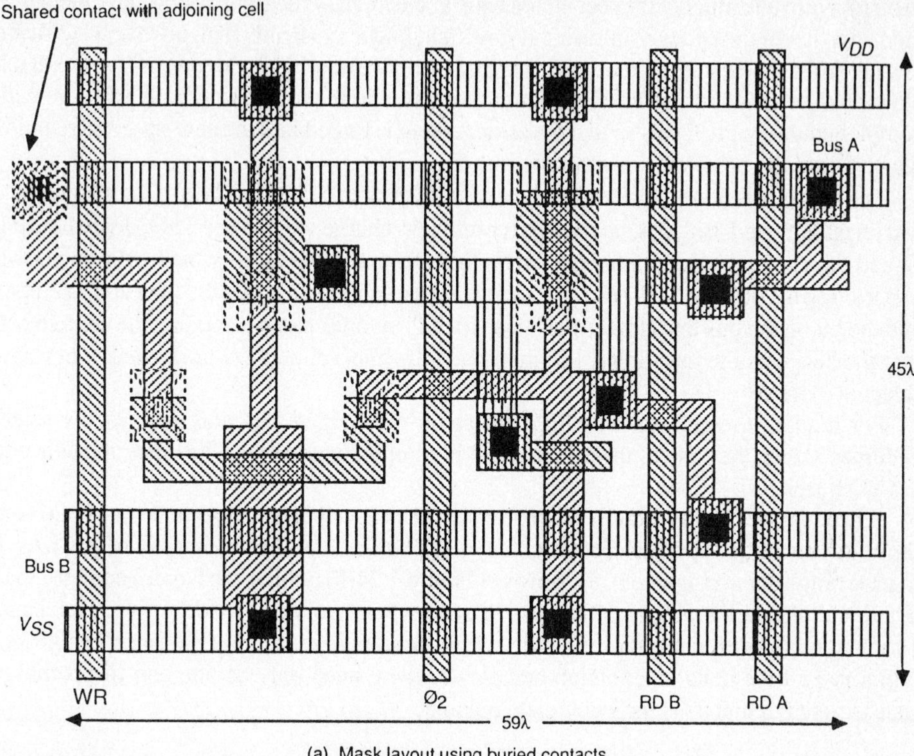

(a) Mask layout using buried contacts

Figure 7.23(c) nMOS pseudo-static latch—Mask layout using buried contacts

1. *Area/bit stored*

The cell is quite large, comprising eight transistors. Area assessment follows from the mask layout and, for the 2-bus layout, each bit will occupy some 16000 μm^2 in 5 μm technology. This particular 2-bus design is aimed at registers and not RAM but, to put the size in perspective, a 5 mm X 5 mm chip would hold just over 1.5 kbits. A CMOS equivalent of this cell would comprise twelve rather than eight transistors and is even larger so that neither is suitable for high density RAM design.

3. *Volatility*

This type of cell self-refreshes on every $Ø_2$ clock signal (the output is fed back to the input to recharge C_g) and we may therefore describe the cell as pseudo-static . In other words, stored data will remain valid as long as the clock keeps going, and the clock period is shorter than the decay time of the charge on C_g. Clearly, data will also be lost on power-down.

Although large in area, this cell is quite widely used for registers, etc., because it will retain data as long as required without any refreshing. It also has the property of non-destructive read out.

A static CMOS memory cell for complementary bus lines

The storage mechanism is one with which we are by now familiar, two inverters connected in series with the output of the second fed back to the input of the first. This arrangement will latch in either of the two possible output states and latches so firmly that, up to now, we have always broken the feedback loop before trying to input a new bit. In this design, the feedback loop remains intact but, to write, we simultaneously drive the inputs to both inverters with complementary logic levels so that the arrangement is forced into the new state and remains latched until the next write operation.

The way this is done may be explained with reference to Figure 7.24(a). Complementary bus lines, Bit_n and Bit'_n get the required column precharge via Tr3 and Tr4 to logic1 during $Ø_2$ and then have the desired bit written on them from the I/O buses which pulls one down to 0 leaving the other at 1. When row select is active (during $Ø_1$), switches Tr1 and Tr2 open and the bit on the bus lines forces the "nose to tail" inverter pair comprising the storage cell into the desired state in which it will then remain while Tr1 and Tr2 are closed (when row select is inactive).

For reading the cell, the bus lines are again precharged to 1 and then, if row select becomes active, the state of the selected cell is communicated to the bus lines and hence to the I/O buses.

Since we want each storage cell to be as small as possible, and also to be fast in driving the relatively large capacitances presented by the bit line and I/O bus lines jointly, we have a sense amplifier arrangement, as shown in Figure 7.24(b), which has larger geometry than a cell and assists by taking up the desired state when "sense" is made active. This sinks current much more quickly than is possible with the storage cell alone. One sense "amplifier" will serve all the cells in a column and clearly there need only be one pair of precharge transistors (Tr3 and Tr4) for a complete column.

1. *Area/Bit stored*

Ignoring the common elements, we have two inverters and "Tr1 and Tr2" in each cell, that is, a total of six transistors. A possible mask layout is presented as Figure 7.25(c) and it may be seen that a cell has an approximate area of $50\,\lambda \times 40\,\lambda = 2000\,\lambda^2$, which in 5 μm technology ($\lambda = 2.5$ μm) corresponds to an area per bit of 12500 μm^2. Thus, a 5 mm X 5 mm chip would hold less than 2 kbits of memory—allowing for common features, (row and column select, precharge, etc.)

2. *Dissipation/bit stored*

The storage element is two CMOS inverters which draw virtually zero static current in either state. The cell will exhibit switching dissipation which is proportional to the rate at which read and write operations take place.

3. *Volatility*

The cell is static and the read-out is non-destructive, data will be retained so long as power is maintained. Data can be maintained "permanently" if a back-up battery is used to maintain the stored information when the operating supplies are switched off (see Figure 4.5).

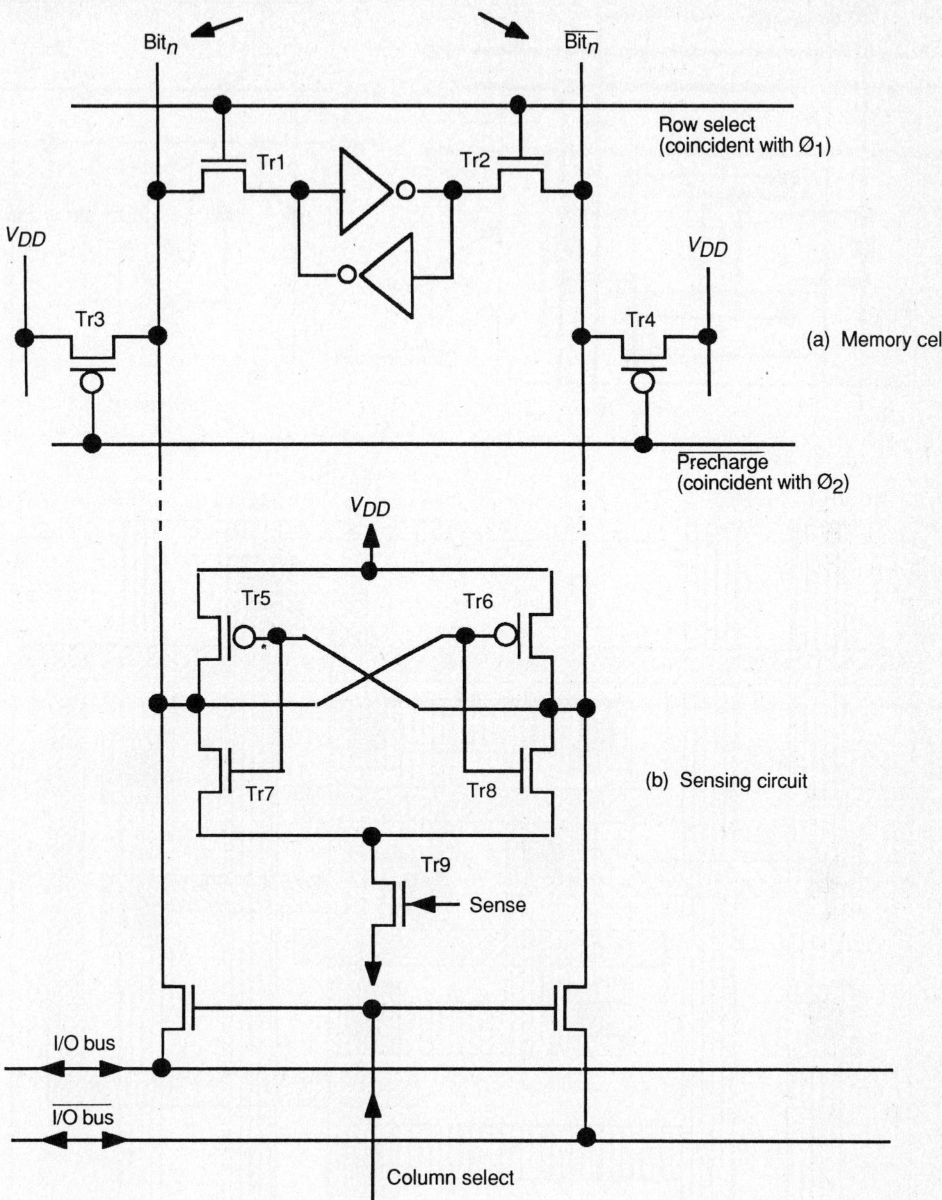

Figure 7.24 A CMOS static memory cell with complementary bus lines

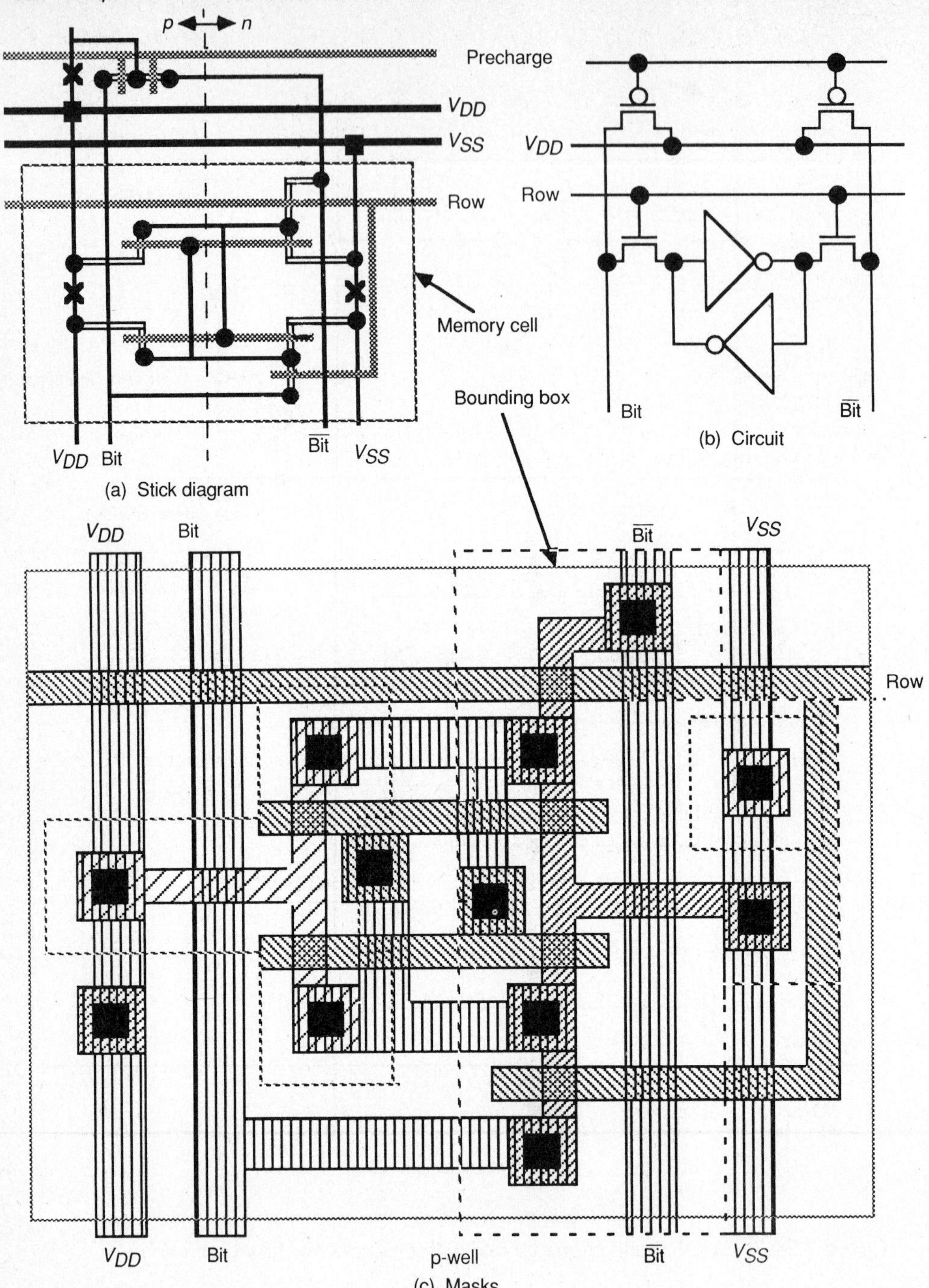

Figure 7.25 CMOS static memory cell layout

7.4 Remarks and observations

This chapter has covered the basic elements and some basic design processes for clocked sequential circuits (often referred to as finite state machines). It has been shown that memory and sequence dependence are closely related and we have introduced a range of memory elements. Although the treatment is not exhaustive the reader may well feel exhausted at this point and with this in mind the next chapter will apply what has been set out here to the design of some of the most common digital circuit subsystems. This will serve to further reinforce the examples in this chapter and provide useful case studies in design .

We have dwelt on the process of design (or synthesis) which is most often the task which falls upon the designer, but *analysis* sometimes has to be undertaken. Analysis is usually readily undertaken by putting design processes "in reverse" and this process is illustrated in in Chapter 9.

However, before relaxing a little, it is time once again for the reader to look over some further worked examples and to try out newly acquired skills by tackling tutorial work.

7.5 Worked examples

1. Develop a clocked D flip-flop from the RS flip-flop characterized in Table 7.1. Set out the logical arrangement and the D flip-flop symbol appropriate to your design.

   Solution:

   The RS flip-flop symbol and characteristics are reproduced here as Figure 7.26(a) and it may be seen that, if we always arrange for R and S to be complementary, the characteristics reduce to those of Figure 7.26(b). Renaming S as D we have the appropriate D flip-flop characteristic as shown.

   But this is not clocked so we may add a clock signal, 0-1-0, as shown in Figure 7.26(c) which is the final circuit.

   A little thought will reveal the clock activation as ΔT if the clock pulse is very short, or as ∇T if the clock pulse is long.

   A suitable symbol then depends on the length of clock pulse or indeed if it is a.c. rather than d.c. coupled as shown.

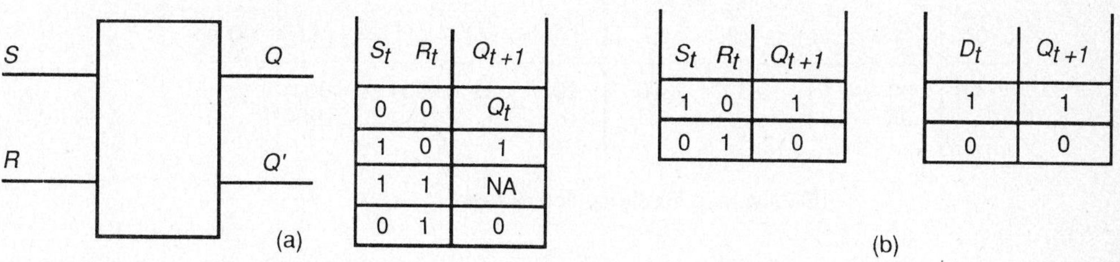

S_t	R_t	Q_{t+1}
0	0	Q_t
1	0	1
1	1	NA
0	1	0

S_t	R_t	Q_{t+1}
1	0	1
0	1	0

D_t	Q_{t+1}
1	1
0	0

Figure 7.26(a) and (b) Development of "D" from "RS" flip-flop

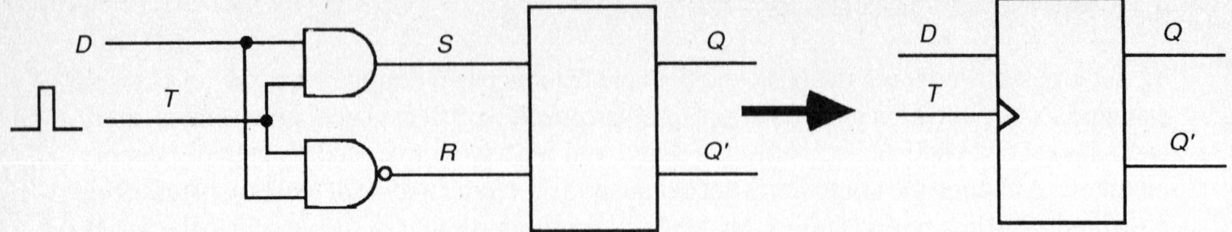

Figure 7.26(c) Logic for D FF, **(d)** Symbol (Short clock pulse)

2. (a) Design a positive clock edge (ΔT) activated JK flip-flop with J, K, T inputs and an asynchronous (Clr)' input.
   Set out a suitable logic circuit for your design.
   (b) Set out a circuit diagram for implementing your design in CMOS complementary logic.

   Solution:

   (a) The state transition diagram is set out as Figure 7.27. From the diagram we may set out maps as follows in Table 7.20.

Table 7.20 Maps for a JK flip-flop design

		T'				T				
	AB	JK 00	01	11	10	10	11	01	JK 00	
ab	00	①	①	①	①	①	①	①	①	00 ab
	01	1	1	1	1	1	1	1	1	01
Clr'										Clr'
	11	1	1	1	1	1	1	1	1	11
	10	1	1	1	1	1	1	1	1	10
	10	3	④	④	3	1	1	1	1	10
	11	③	4	4	③	③	③	③	③	11
Clr										Clr
	01	1	1	②	②	3	3	3	3	01
ab	00	①	①	2	2	①	①	①	①	00 ab
	AB	00 JK	01	11	10	10	11	01	00 JK	
		T'				T				

(a) Flow map for JK flip-flop design.

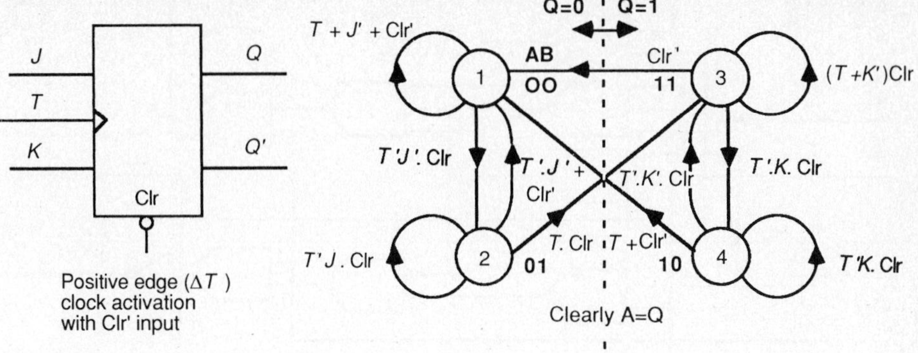

Figure 7.27 State transition diagram for JK flip-flop

Table 7.20 (continued)

		T'				*T*					
	AB	*JK* 00	01	11	10	10	11	01	*JK* 00		
ab	00	00	00	00	00	00	00	00	00	00	*ab*
	01	00	00	00	00	00	00	00	00	01	
Clr'											Clr'
	11	00	00	00	00	00	00	00	00	11	
	10	00	00	00	00	00	00	00	00	10	
	10	11	10	10	11	00	00	00	00	10	
	11	11	10	10	11	11	11	11	11	11	
Clr											Clr
	01	00	00	01	01	11	11	11	11	01	
ab	00	00	00	01	01	00	00	00	00	00	*ab*
	AB	00 *JK*	01	11	10	10	11	01	00 *JK*		
		T'				*T*					

(b) Excitation map for JK flip-flop design.

Whence, $A = \text{Clr}.(a.T' + b.T)$ $\quad B = \text{Clr}(b.T + a.K'.T' + a'.J.T')$

One possible logic diagram with simulations follows as Figure 7.28(a).

(b) A CMOS circuit diagram follows as Figure 7.28(b).

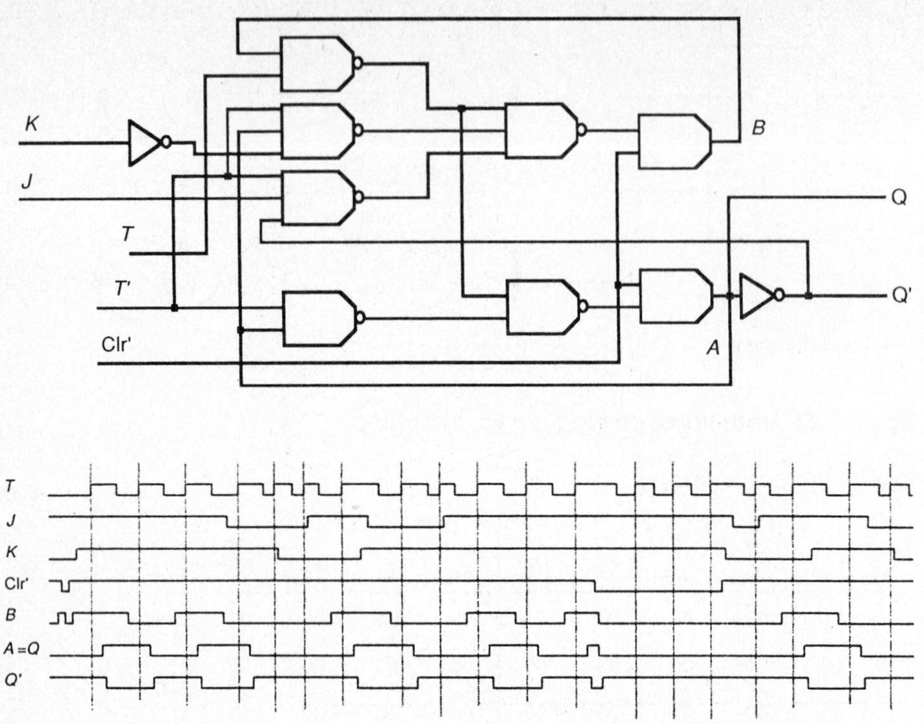

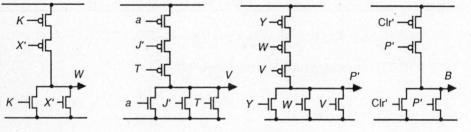

Figure 7.28(a) ΔT clocked JK with Clr' (LogicWorks simulations as shown)

First A

$Q = A = \text{Clr}.(aT' + bT) = \text{Clr}\,(X + Y)$ where $X = aT'$; $Y = bT$ $A = \text{Clr}.Z$ where $Z = X + Y$

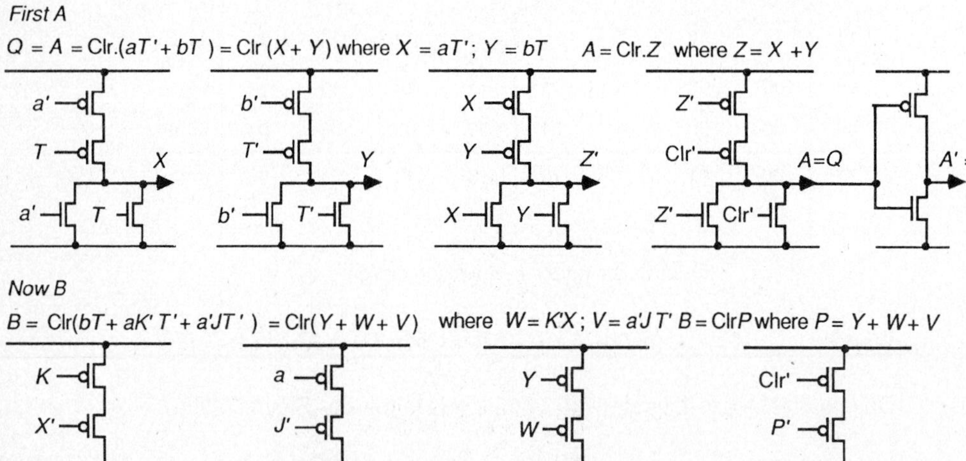

Now B

$B = \text{Clr}(bT + aK'T' + a'JT') = \text{Clr}(Y + W + V)$ where $W = K'X$; $V = a'JT'$ $B = \text{Clr}P$ where $P = Y + W + V$

K
X'
W
K
X'
a
J'
T
V
a
J'
T
Y
W
V
P'
Y
W
V
Clr'
P'
B
Clr'
P'

Figure 7.28(b) A possible CMOS complementary logic circuit for JK with CLr'

3. A clocked sequential circuit is required to detect the occurrence of either zero (000) or three (011) in 3-bit words appearing in serial form on an input line W. Note that word 1= bits 1, 2, 3, word 2 = bits 4, 5, 6, etc., in time sequence on W. Also note that words appear LSB first, MSB last.

   The clock is available to you.

   Design a circuit to meet these needs using JK flip-flops.

Solution:

Set out a state transition diagram and merge states where possible. This appears as Figure 7.29.

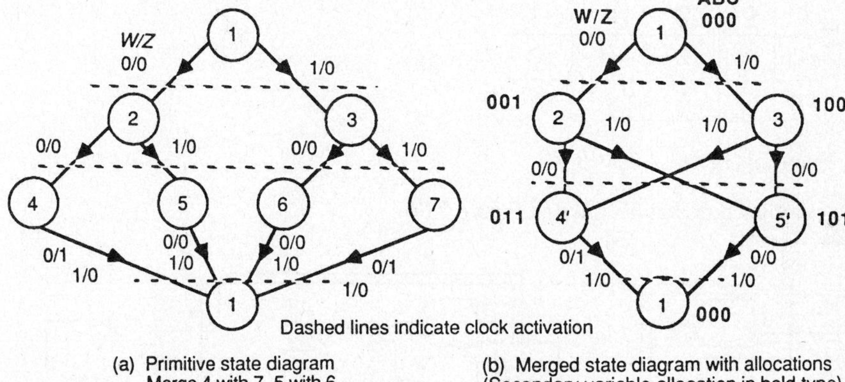

Figure 7.29 State diagrams for the solution to Question 3

Make state vector allocations as shown (**bold**).
Set out the state transition table, Table 7.21, as follows:

Table 7.21 State transition table for the solution to Question 3

Present state	*Next state*		*Output*
	W = 0	*W* = 1	
A BC	*A BC*	*A BC*	*Z*
(1) 000	001	100	0
(2) 001	011	101	0
(3) 100	101	011	0
(4') 011	000	000	1 (*W*=0); 0(*W*=1)
(5') 101	000	000	0

Note: Three states of vector ABC are not used giving rise to don't cares (Ø).

Setting out programming maps we get:

$$J_A = B'W \qquad K_A = W + C$$

$$J_B = A'CW' + AC'W \qquad K_B = 1$$

$$J_C = A + W' \qquad K_C = A + B$$

Now for the output Z, we set out a map as in Table 7.22:

Table 7.22 Map for output *Z*

Z	CW 00	01	11	10
AB 00	0	0	0	0
01	Ø	Ø	0	1
11	Ø	Ø	Ø	Ø
10	0	0	0	0

From this table $Z = B.W'$

The completed design follows as Figure 7.30.

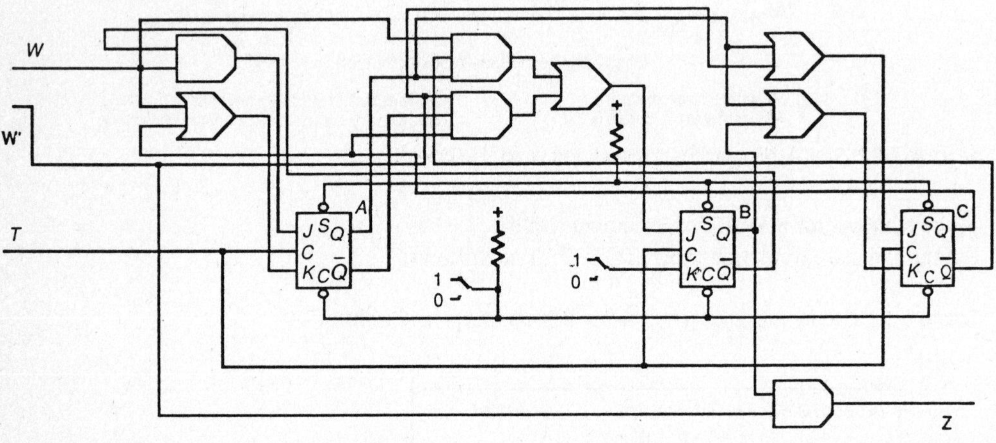

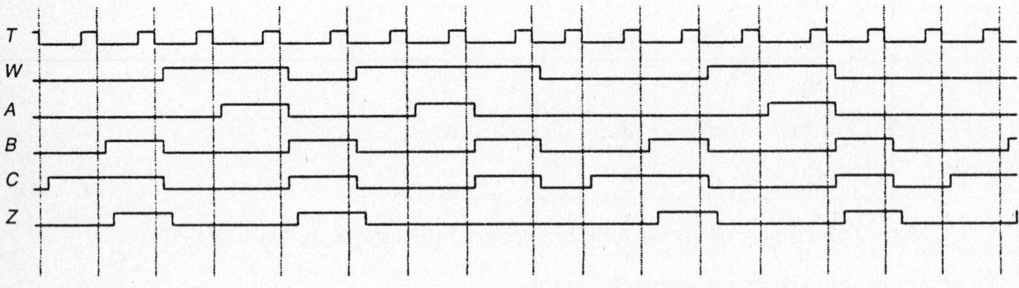

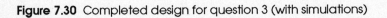

Figure 7.30 Completed design for question 3 (with simulations)

4. Derive a PLA based solution for the requirement set out in question 3. Take your design to the stick diagram stage but do not attempt a mask layout.

Solution:

Set out the state transition table, Table 7.23, as follows:

Table 7.23 State transition table in PLA form

	Present state ABC W	P_k	Next state ABC_n	Output Z	
①	000 0	p_1	001	0	
①	000 1	p_2	100	0	
②	001 0	p_3	011	0	
②	001 1	p_4	101	0	
③	100 0	p_5	101	0	
③	100 1	p_6	011	0	
④'	011 0	p_7	000	1	
④'	011 1	-	000	0	No need to form $p_{8,9,10}$
⑤'	101 0	-	000	0	all 0's for next ABC and Z
⑤'	101 1	-	000	0	

Derive the PLA equations from the transition table *not* making use of don't cares from unused states, to help avoid the possibility of problems from illegal states. The equations are:

And plane:

$$p_1 = A'B'C'W'\,;\; p_2 = A'B'C'W\,;\; p_3 = A'B'CW'\,; p_4 = A'B'CW\,;$$

$$p_5 = AB'C'W'\,;\; p_6 = AB'C'W\,;\; p_7 = A'BCW'\,.$$

Or plane:

$$A_n = p_2 + p_4 + p_5 \quad \text{(subscript } n \text{ denotes next.)}$$

$$B_n = p_3 + p_6$$

$$C_n = p_1 + p_3 + p_4 + p_5 + p_6$$

$$Z = p_7$$

A stick diagram follows as Figure 7.31.

5. (a) Convert the Mealy model state transition diagram given as Figure 7.32 into the equivalent Moore model form.
(b) Compare implementations of the two forms of the circuit, using JK flip-flops.

Solution:

(a) The equivalent Moore model state transition diagram is given as Figure 7.33.

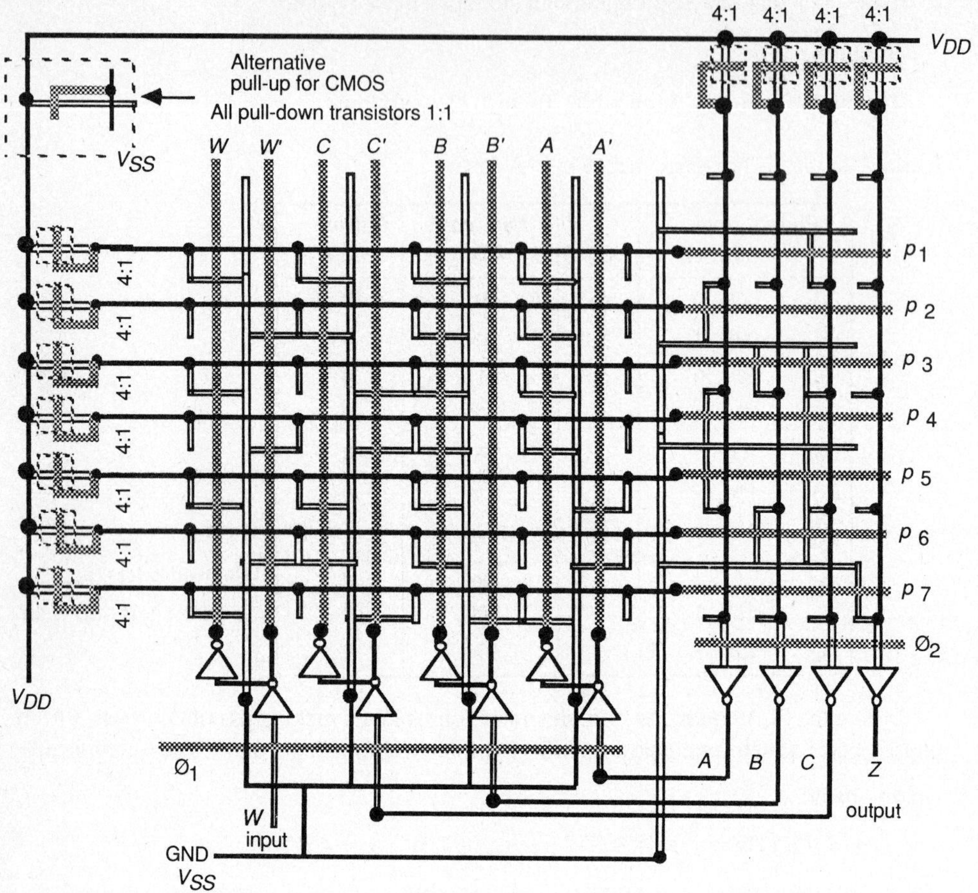

Figure 7.31 Stick diagram of PLA for Question 4

(b) This circuit has already been designed, in the Mealy model form, in section 7.2.3 of the text and the JK flip-flop realization was given as Figure 7.12.

For the Moore model circuit we have one extra state (5 in all) and therefore need a three bit state vector **ABC** allocated as suggested in Figure 7.33.

We may then set out a state transition table, Table 7.24 as follows:

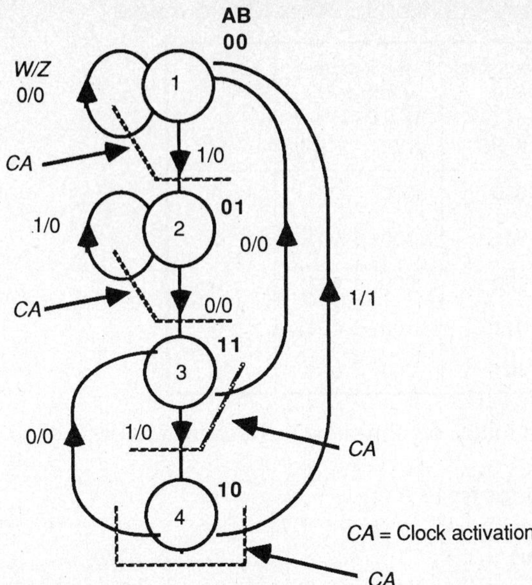

Figure 7.32 Mealy model state diagram for Question 5

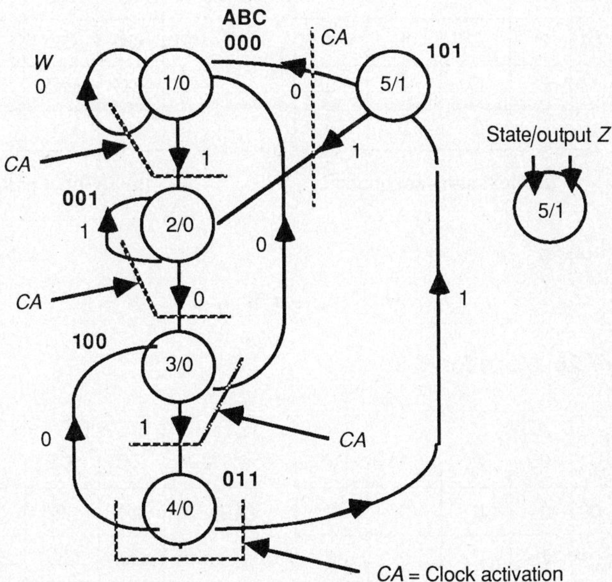

Figure 7.33 Equivalent Moore model state diagram for Question 5

Table 7.24 State transition table for Moore model

Present state (time t)	Next state (time t+1)		Output
	W = 0	W = 1	
A BC	A BC	A BC	Z
① 000	000	001	0
② 001	100	001	0
③ 100	000	011	0
④ 011	100	101	0
⑤ 101	000	001	1

Maps then follow for flip-flops A, B, and C (Tables 7.25 to 27 respectively)

Table 7.25 Maps for FF A

A_{t+1} / $(AB)_t$ \ $(CW)_t$	00	01	11	10
00	0	0	0	1
01	Ø	Ø	1	1
11	Ø	Ø	Ø	Ø
10	0	0	0	0

(a) Next state sequence

J_AK_A / $(AB)_t$ \ $(CW)_t$	00	01	11	10
00	0Ø	0Ø	0Ø	1Ø
01	ØØ	ØØ	1Ø	1Ø
11	ØØ	ØØ	ØØ	ØØ
10	Ø1	Ø1	Ø1	Ø1

(b) JK input programming

Whence:

$$J_A = CW' + B \qquad K_A = 1$$

Table 7.26 Maps for FF B

B_{t+1} / $(AB)_t$ \ $(CW)_t$	00	01	11	10
00	0	0	0	0
01	Ø	Ø	0	0
11	Ø	Ø	Ø	Ø
10	0	1	0	0

(a) Next state sequence.

J_BK_B / $(AB)_t$ \ $(CW)_t$	00	01	11	10
00	0Ø	0Ø	0Ø	0Ø
01	ØØ	ØØ	Ø1	Ø1
11	ØØ	ØØ	ØØ	ØØ
10	0Ø	1Ø	0Ø	0Ø

(b) JK input programming.

From map (b)

$$J_B = A.C'.W \qquad K_B = 1$$

Table 7.27 Maps for FF C

C_{t+1}	$(CW)_t$ 00	01	11	10
$(AB)_t$ 00	0	1	1	0
01	Ø	Ø	1	0
11	Ø	Ø	Ø	Ø
10	0	1	1	0

(a) Next state sequence.

$J_C K_C$	$(CW)_t$ 00	01	11	10
$(AB)_t$ 00	0Ø	1Ø	Ø0	Ø1
01	ØØ	ØØ	Ø0	Ø1
11	ØØ	ØØ	ØØ	ØØ
10	0Ø	1Ø	Ø0	Ø1

(b) JK input programming.

From map (b)

$$J_C = W \qquad K_C = W'$$

$$Z = \text{state } 5 = AB'C$$

The circuit then follows as shown in Figure 7.34.

7.6 Tutorial 7

1. With reference to example 2 (section 7.2.3), repeat the design with the following state allocations:

   State 1 $AB = 00$
   State 2 $AB = 01$
   State 3 $AB = 10$
   State 4 $AB = 11$

   Carry out the design for both JK and D flip-flops and compare the results with those obtained with the allocations used in the text.

2. From the logic circuit given as Figure 7.16(a), develop either (a) an nMOS or (b) a CMOS pseudo-nMOS version of the 2-phase clocked D flip-flop. Your work should include a stick diagram and a corresponding mask layout. Compare the results with the complementary CMOS version in the text.

3. Mealy and Moore model state transition diagrams for a clocked sequential circuit are given in Figure 7.18. Using JK flip-flops, design circuits to realize this circuit for (a) the Mealy model and (b) the Moore model. Compare the resulting arrangements and, if you can, simulate the operation of both versions.

4. A state transition diagram for example 1 in section 7.2.1 of the text has its state transition diagram given as Figure 7.8. Determine whether this is a Mealy or a Moore model and then develop the alternative state transition diagram. From the diagram, set out the state

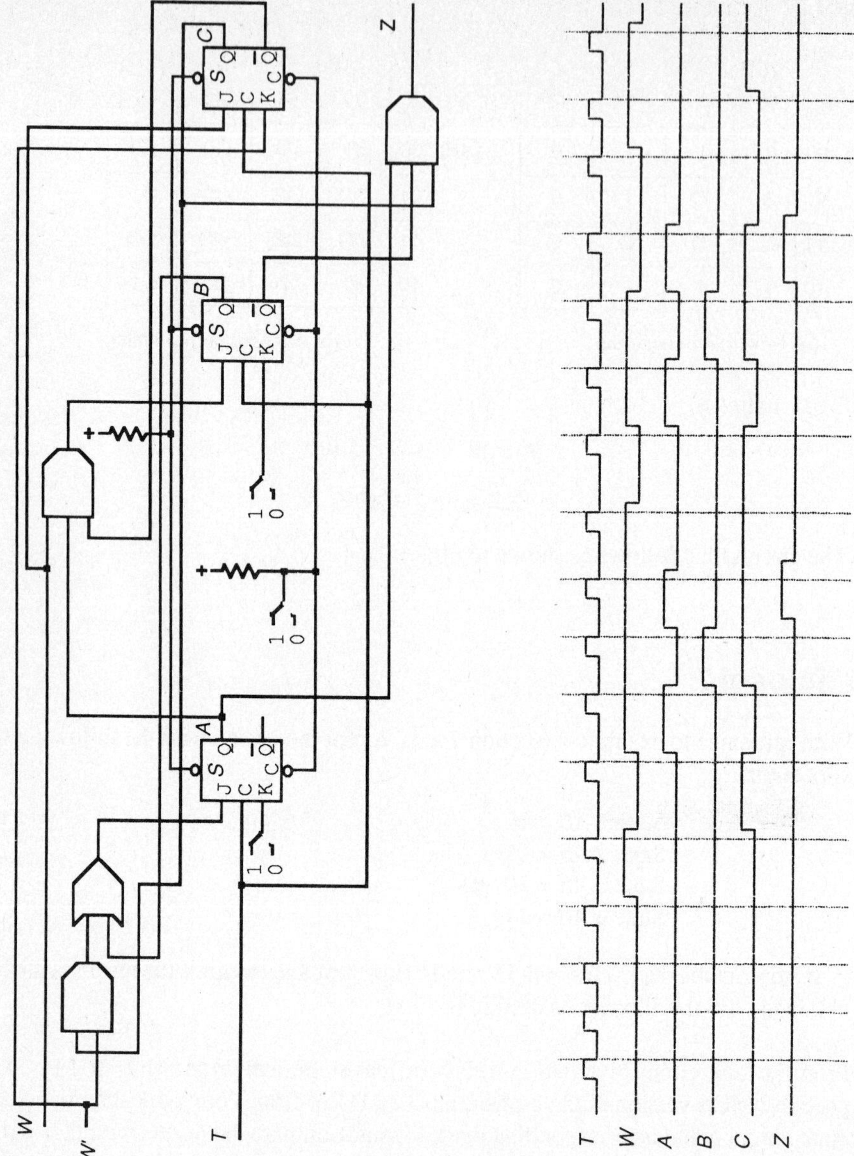

Figure 7.34 Circuit for Moore model (Question 5)

transition table for a flip-flop based implementation. Then set out this table in a form better suited to PLA based implementation.

5. Section 7.3.3 of the text describes an nMOS pseudo-static latch circuit (Figure 7.23). Develop a stick diagram and mask layout for a complementary CMOS version of this cell . Compare areas with the mask layout in the text.

6. A static CMOS memory cell is discussed in section 7.3.3 (Figures 7.24 and 7.25) of the text . A dynamic version of this cell can be configured by replacing the two inverters by two (n-type) cross-coupled transistors. Derive a suitable arrangement and set out a mask layout. Compare areas with Figure 7.25(c).

8 Some commonly applied clocked subsystems

8.1 Introductory remarks

In this chapter we are to apply the design techniques and processes dealt with earlier to design some commonly required subsystems such as counters, shift registers etc. Some types of circuit have already been covered in earlier chapters and we will design and analyze others, using alternative design techniques, in the next chapter. Overall, the aim is to cover representative designs of the more popular subsystems used in digital system design. It is hoped that this will be useful for applications and reinforce the work already covered up to this point. This is a chapter of worked examples so there will be no separate section of examples.

8.2 Counters

A very widely applied class of subsystem is the counter. Various forms of standard binary and decimal counters are used, and many and varied special purpose arrangements can be designed to order as required. Counters may be made up of flip-flop circuits or we may have specially configured VLSI circuits to perform the task. To start the ball rolling we will look at a very simple arrangement of T flip-flops or T-connected JK flip-flops to form a 4-bit counter.

8.2.1 A ripple-through 4-bit binary counter

A simple way of fully specifying the requirements is to set out a truth table, Table 8.1, in which the four bits of the counter are denoted ABCD (MSB to LSB in left to right order) and in which the clock is the input to be counted. This input defines the t and t+1 periods as clock activation advances the count from one state to the next. Falling edge (∇T) activation of the flip-flop clock inputs is assumed in this case. A state transition diagram (Moore model), Figure 8.1, may be readily derived from Table 8.1.

Table 8.1 4-bit binary count sequence

	A B C D		Base 10 count
	0 0 0 0		0
	0 0 0 1		1
	0 0 1 0		2
Time ↓	0 0 1 1	t	3
	0 1 0 0	t+1	4
	0 1 0 1	etc.	5
	0 1 1 0		6
	0 1 1 1		7
	1 0 0 0		8
	1 0 0 1		9
	1 0 1 0	t	10
	1 0 1 1	t+1	11
	1 1 0 0	etc.	12
	1 1 0 1		13
	1 1 1 0		14
	1 1 1 1		15
	0 0 0 0		0 ↓ Repeat sequence

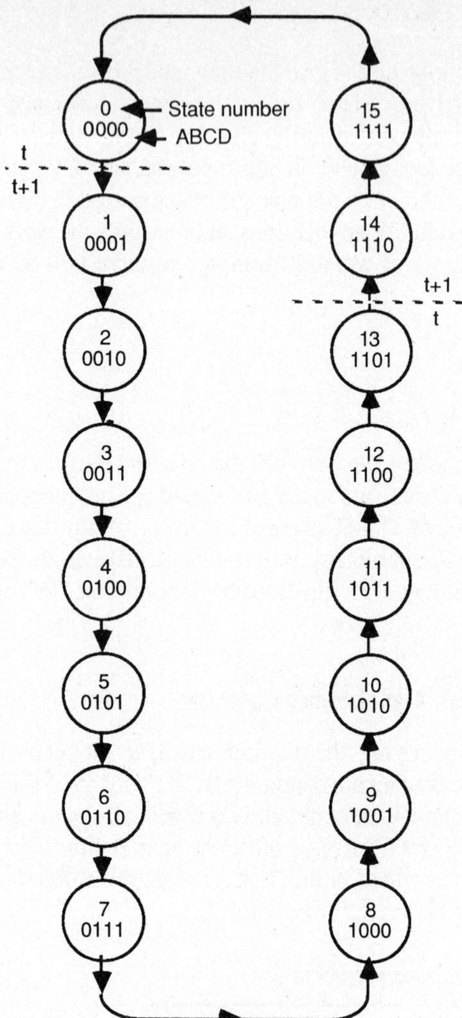

Figure 8.1 State diagram for 4-bit binary counter

The action of the counter is based on the count input signal providing the clock input to flip-flop D and the Q output of each flip-flop then providing the clock input to the next more significant flip-flop. For example, output *D* provides the clock input signal to flip-flop *C*, output *C* provides the clock input to flip-flop *B* and so on. The basis for this may be seen by inspection of Table 8.1. It will be seen that whenever *D* changes from 1 to 0, stage *C* is required to change, that is, clock activation of FF *C* is required on ∇D. Similarly, you will see that clock activation of FF *B* is required on ∇C, etc. The state of the count is thus set as clock activation "ripples through" the cascaded stages. The interconnection of four flip-flops (JKs in this case) as in Figure 8.2, constitutes a very simple 4-bit binary counter circuit. Clearly the complete count sequence constitutes a binary sequence of $2^4 = 16$ states, after which the sequence repeats.

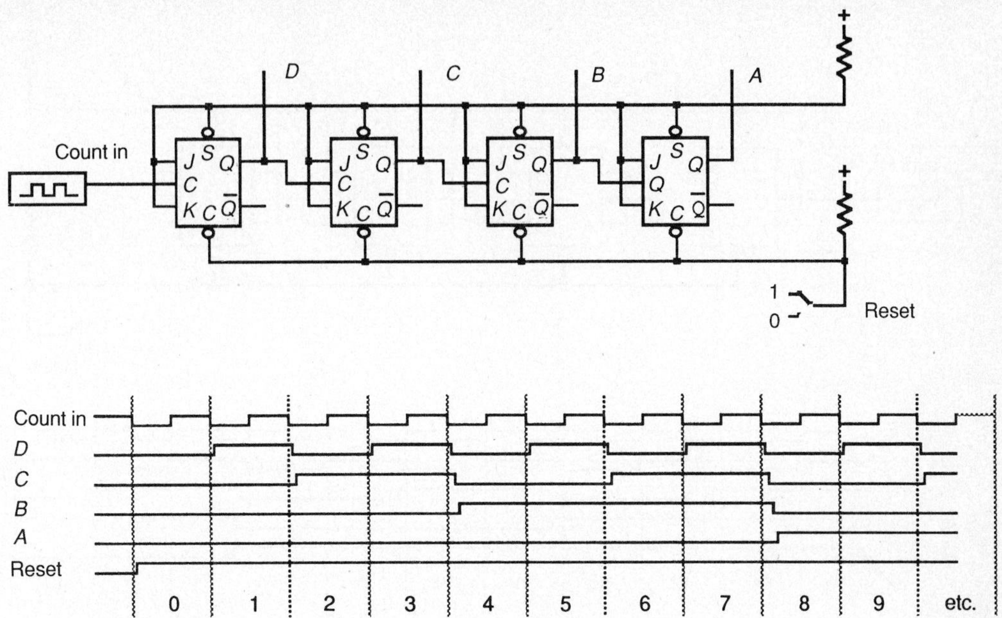

Figure 8.2 4-bit ripple-through counter with simulation waveforms

If we weight outputs as follows: $A = 2^3$, $B = 2^2$, $C = 2^1$, $D = 2^0$, then the addition of the weighted binary bits for each state of the counter will generate the sequence 0 to 15_{10} where the subscript$_{10}$ indicates base 10 (see Chapter 2).

The simulation which accompanies Figure 8.2 clearly demonstrates the ripple-through effect, as it may be seen that stages of the counter do not change in synchronism but are subject to cascaded delay effects.

Some experiments with the 4 bit ripple-through counter

The very simple arrangement of Figure 8.2 invites the attention of the enquiring mind.

For a start, what will happen if, instead of taking the clock inputs of flip-flops *C*, *B* and *A* from the *Q* output of the preceding stage, we take the clock inputs from *Q*'?

This has been done in Figure 8.3 and the simulation results show what happens—*a backward count* sequence 15,14,13, ... 2,1,0,15 ... etc.

We could of course have reasoned this from the truth table but, in any event, this opens up the possibility of reversing or up/down counters.

Next, we might ask ourselves how we might go about producing a *counter operating on the decimal scale* since so far we have produced only a binary count.

Once again, an intuitive approach can be taken and, noting that the flip-flops have active Lo clear (C') inputs, we can arrange for the natural 4-bit count sequence to be shortened to cycle through 0,1,2, ... 7,8,9,0,1, ... etc, that is, a ten state count. This is readily done by detecting, with a *Nand* gate, a count of 10 and using the output of the gate to clear the count to zero as soon as the count steps on to 10 as shown in Figure 8.4. The simulation results quite clearly show that the counter momentarily takes up state 10 but this may be unimportant in many applications.

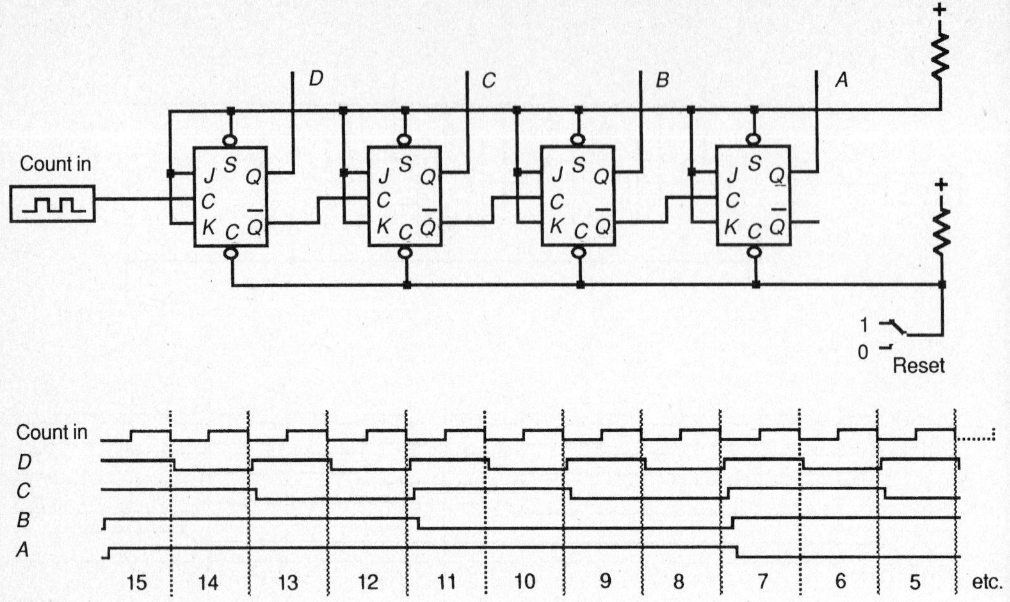

Figure 8.3 4-bit ripple-through counter with modified interconnections

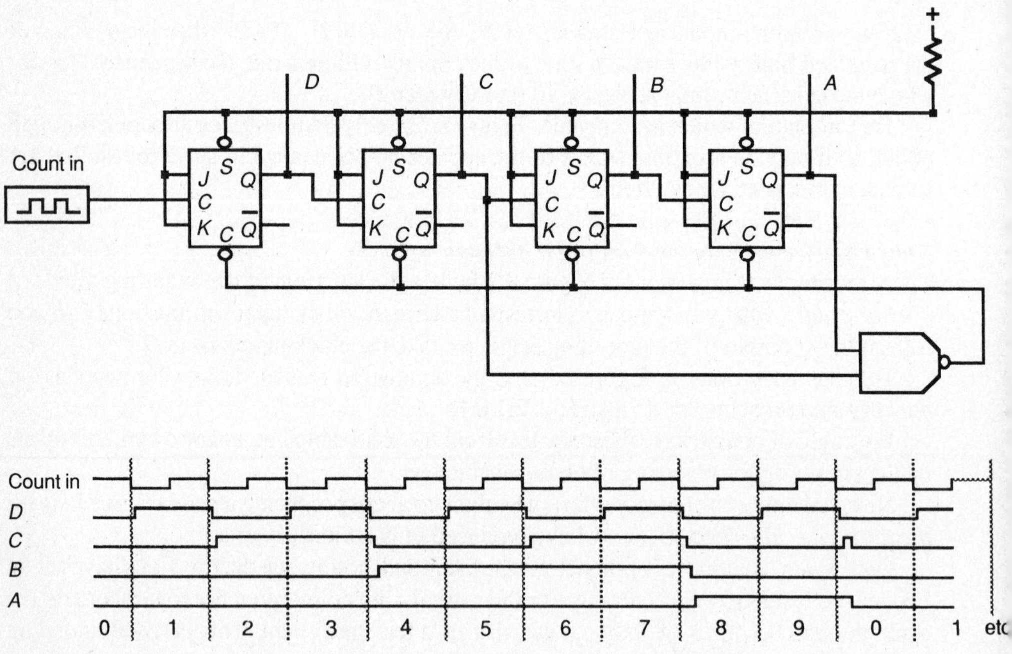

Figure 8.4 4-bit ripple-through counter modified for decimal count

Note that, as far as the flip-flops are concerned, *we have designed our circuits using the clock inputs and asynchronous inputs alone*, *J* and *K* inputs having been tied to logic 1 to make the JK behave as a T flip-flop.

It should be noted that, in all the above design work and experimentation we have not used our carefully developed formal design processes. This example has been deliberately chosen to illustrate the fact that *the design processes do not cater for designs making use of the clock inputs nor the asynchronous inputs to a flip-flop*. All characteristic equations and the characteristic and designer tables presented so far are developed around the clocked inputs only—not the clock input itself nor the preset and clear inputs.

Most designs can be tackled without this limitation causing too many difficulties, but techniques are available for expressing characteristics in a way which includes all inputs in characteristic expressions. One such approach is to use *transition equations* which have been described in Appendix 2 to this text.

Limitations of a ripple-through counter

Although the arrangements shown are delightfully simple, especially if T flip-flops are used, there are difficulties in application due to the cascaded delay effects apparent in the simulation and in practice.

Careful inspection of the waveforms will show that, in changing from one state of the count to another, the counter may take up other intermediate states as one stage then another changes as the required clock activations ripple through the cascaded stages. In order to illustrate this, Figure 8.5 shows a counter with (for example) a zero count detecting circuit attached. The simulation waveforms clearly show that the detector circuit indicates a zero count on the correct count but also, momentarily, during three counter state changes. Any further circuitry activated by the zero detector could be set off on four occasions during a complete count cycle. A similar problem applies in detecting other states for which more than one stage changes and this would mostly be completely unacceptable. Thus ripple-through counter arrangements are to be used with caution.

VLSI versions of the ripple-through counter

The attraction of the ripple-through counter in packaged logic is the relative simplicity of the arrangement. This simplicity is lost in a VLSI implementation although, clearly, the same configuration of JK flip-flops will do the job. However, we have seen that the VLSI version of a JK flip-flop is by no means simple and occupies an appreciable area.

As an alternative, we could perhaps use a simpler D flip-flop design connected as a T flip-flop and a suitable 4-bit, for example, logic circuit with positive edge clock activation is given in Figure 8.6.

A VLSI version of a ΔT clocked D flip flop

Let us start the design process by setting out a state transition diagram as in Figure 8.7 (refer to Figure 7.5 to compare with the JK flip-flop design in the previous chapter).

From this diagram the flow and excitation maps are readily derived as in Table 8.2 and the expressions to be implemented and the logic circuit (Figure 8.8(a)) follows.

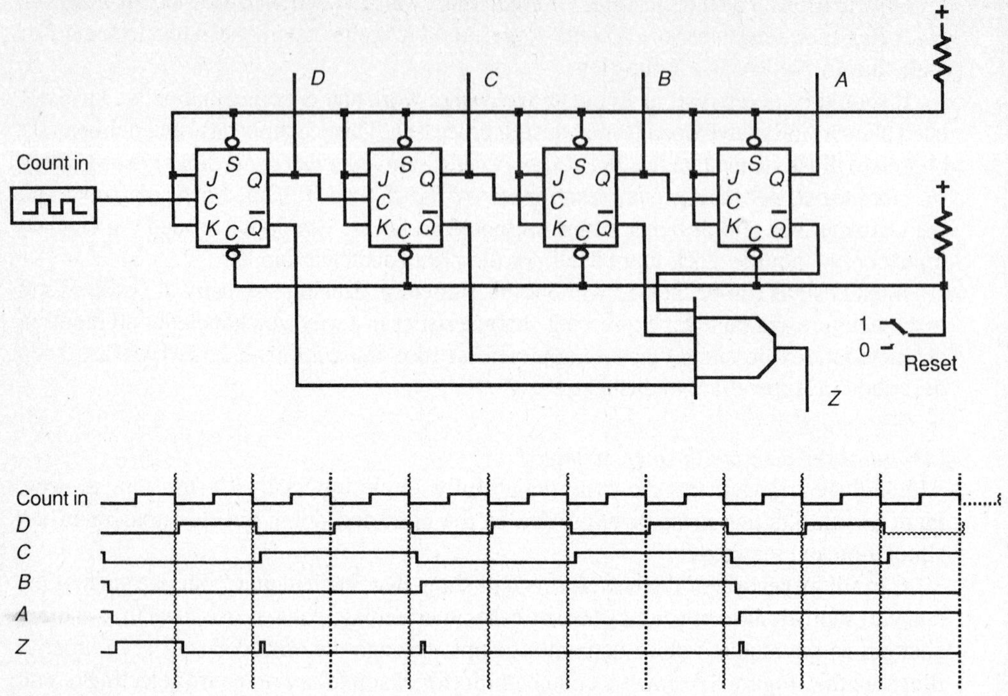

Figure 8.5 Ripple-through counter with zero detector

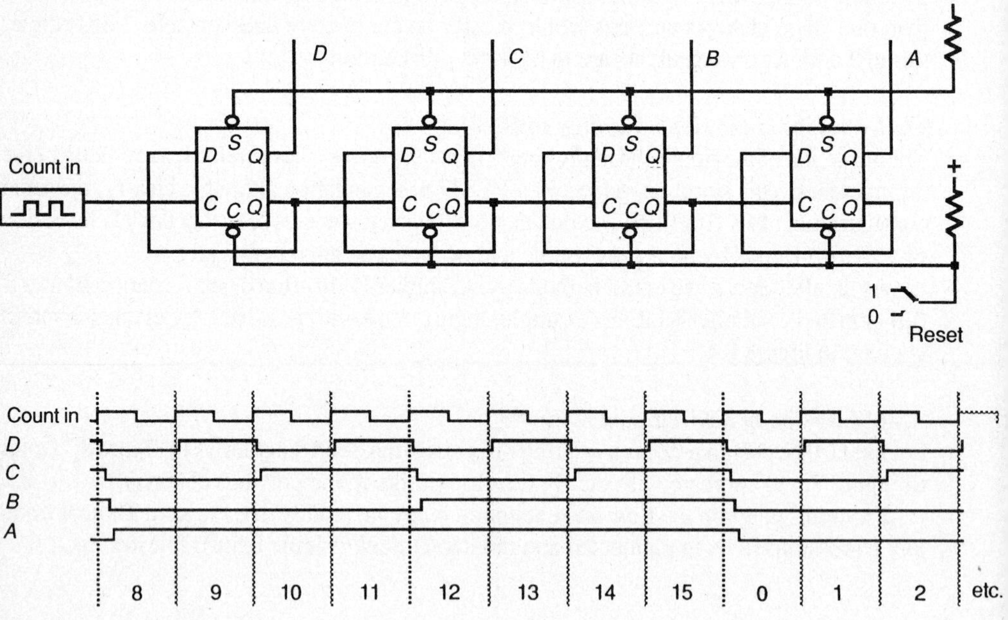

Figure 8.6 ΔT activated D flip-flop based ripple-through counter

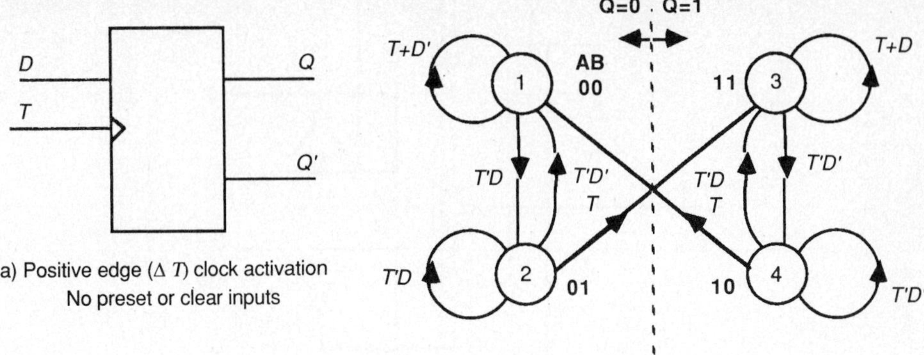

Figure 8.7 State transition diagram for Δ*T* clocked D flip-flop

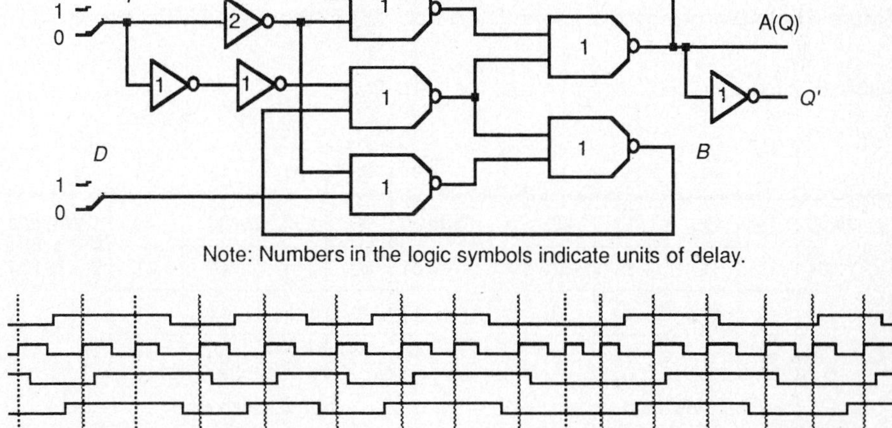

Figure 8.8 Implementations of the D flip-flop **(a)** Logic circuit with simulation

The logic is given, with simulations, in *Nand* gate form but the circuit can be rearranged as convenient for implementation in nMOS, pseudo-nMOS or in complementary logic CMOS forms. An arrangement suited to VLSI implementation is based on inverters and transmission gates and the configuration is set out as Figure 8.8(b).

8.2.2 An up/down synchronous 4-bit binary counter

To overcome the deficiencies of the ripple-through counter and to return to formal design processes, we will now examine the design of a *synchronous* counter, that is, one in which all state changes take place simultaneously.

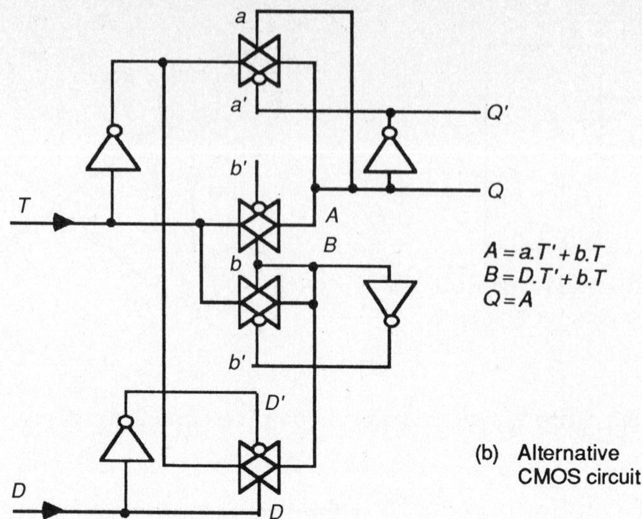

Figure 8.8 Implementations of the D flip-flop **(b)** Alternative CMOS circuit

Table 8. 2 Flow and Excitation maps for D flip-flop

(a) Flow map

ab \ *DT*	00	01	11	10
00	(1)	(1)	(1)	2
01	1	3	3	(2)
11	4	(3)	(3)	(3)
10	(4)	1	1	3

(b) Excitation map

ab \ *DT* (*AB*)	00	01	11	10
00	00	00	00	01
01	00	11	11	01
11	10	11	11	11
10	10	00	00	11

Whence:
A = a.T' + b.T
B = D.T' + b.T

We have earlier set out a design procedure (section 7.2.2) and this will now be followed.

1. Set out a full, unambiguous specification.
2. Set out a primitive state transition diagram.
3. Merge states if possible to reduce the overall complexity. Draw the final(merged) state diagram
4. Allocate secondary variables and state vector values to each state.

The starting point is a truth table and state transition diagram which will be the same as for the 4-bit ripple-through counter. In this case steps **1-4** are thus covered in the title of the subsystem to be designed together with Table 8.1 and Figure 8.1. We further specify control signal *U* such that *U* = 1 = up count, *U* = 0 = down count.

5. Set out the state transition table.
[At this point it is convenient to choose the circuit elements which will represent the

output variables. There are no separate secondary variables since the counter is a Moore type circuit.]

Let us choose JK flip-flops and set out the state transition Table 8.3 in a form suited to this particular realization.

Table 8.3 State transition table for 4-bit synchronous up/down counter

Present state $(A\ B\ C\ D)_p$	*Next state up* $(A\ B\ C\ D)_{nu}$	*Next state down* $(A\ B\ C\ D)_{nd}$
0 0000	1 0001	15 1111
1 0001	2 0010	0 0000
2 0010	3 0011	1 0001
3 0011	4 0100	2 0010
4 0100	5 0101	3 0011
5 0101	6 0110	4 0100
6 0110	7 0111	5 0101
7 0111	8 1000	6 0110
8 1000	9 1001	7 0111
9 1001	10 1010	8 1000
10 1010	11 1011	9 1001
11 1011	12 1100	10 1010
12 1100	13 1101	11 1011
13 1101	14 1110	12 1100
14 1110	15 1111	13 1101
15 1111	0 0000	14 1110

6. Set out "programming" maps and hence derive logic equations for the JK FF inputs. First, check for the obvious and, in this case, inspection of Tables 8.1 and 8.3 reveals that in either direction of the count $D_{t+1} = D'_t$, that is, FF D behaves as a toggle circuit and thus the connections must be $J_D = K_D = 1$, (where J_D indicates the J input of FF D etc.) So far, so good, but scanning the other columns, we are not so lucky and must now resort to exercising our mapping skills for each of the other flip-flops.

   To add to our worries, we see that we are dealing in each case with a 5-variable map, the variables being $(ABCD)_t$ and control signal U, where $U = 1$ signals an up count and $U = 0$, (i.e. U'), a down count.

   Tackling the stages in turn, we may start with a maps for C_{t+1} as Table 8.4 from which we derive the expressions for J_C and K_C. Note that we have also derived an expression for C_{t+1} which is not immediately relevant but which will be used in work to be done in the next chapter.

   Similar maps and expressions for the other two variables appear with Tables 8.5 and 8.6. Note that in all cases the particular variable being dealt with is highlighted in bold type in the maps, and that the entries in maps (b) are obtained by translating the required present and next state information of maps (a) using the JK FF designer's table, Table 7.3. Also, in filling in maps (b), note that when the selected variable is already 1, all entries for J will be Ø (don't cares). Similarly for K when the selected variable is 0.
7. Derive the output functions to complete the design.
   This step is not required since this is a Moore type circuit and vector $ABCD$ is the output count.

Table 8.4 (a) Next state map for flip-flop C

	U' (down)				U (up)				
	$(CD)_t$							$(CD)_t$	
C_{t+1}	00	01	11	10	10	11	01	00	C_{t+1}
$(AB)_t$ 00	1		1		1		1		00 $(AB)_t$
01	1		1		1		1		01
11	1		1		1		1		11
10	1		1		1		1		10

(Note: Selected variable is highlighted in bold type)

Whence: $C_{t+1} = [U'(C'.D' + C.D) + U(C.D' + C'.D)]_t$

Table 8.4 (b) Flip-flop C input "programming"

	U' (down)				U (up)				
	CD							CD	
J_cK_c	00	01	11	10	10	11	01	00	J_cK_c
AB 00	1Ø	0Ø	Ø0	Ø1	Ø0	Ø1	1Ø	0Ø	00 AB
01	1Ø	0Ø	Ø0	Ø1	Ø0	Ø1	1Ø	0Ø	01
11	1Ø	0Ø	Ø0	Ø1	Ø0	Ø1	1Ø	0Ø	11
10	1Ø	0Ø	Ø0	Ø1	Ø0	Ø1	1Ø	0Ø	10

Whence: $J_C = U'.D' + U.D = K_C$

Table 8.5 (a) Next state map for flip-flop B

	U' (down)				U (up)				
	$(CD)_t$							$(CD)_t$	
B_{t+1}	00	01	11	10	10	11	01	00	B_{t+1}
$(AB)_t$ 00	1					1			00 $(AB)_t$
01		1	1	1	1		1	1	01
11		1	1	1	1		1	1	11
10	1					1			10

Whence: $B_{t+1} = [U'(B(C + D) + B'C'D') + U(B(C' + .D') + B'CD)]_t$

Table 8.5 (b) Flip-flop B input "programming"

	U′ (down)				U (up)				
$J_B K_B$	CD 00	01	11	10	10	11	01	CD 00	$J_B K_B$
AB 00	1Ø	0Ø	0Ø	0Ø	0Ø	1Ø	0Ø	0Ø	00 AB
01	Ø1	Ø0	Ø0	Ø0	Ø0	Ø1	Ø0	Ø0	01
11	Ø1	Ø0	Ø0	Ø0	Ø0	Ø1	Ø0	Ø0	11
10	1Ø	0Ø	0Ø	0Ø	0Ø	1Ø	0Ø	0Ø	10

Whence: $J_B = U'.C'.D' + U.C.D = K_B$

Table 8.6 (a) Next state map for flip-flop A

	U′ (down)				U (up)				
A_{t+1}	$(CD)_t$ 00	01	11	10	10	11	01	$(CD)_t$ 00	A_{t+1}
$(AB)_t$ 00	1								00 $(AB)_t$
01						1			01
11	1	1	1	1	1		1	1	11
10		1	1	1	1	1	1	1	10

Whence: $A_{t+1} = [U'(A(B + C + D) + A'B'C'D') + U(A(B' + C' + D') + A'B.C.D)]_t$

Table 8.6 (b) Flip-flop A input "programming"

	U′ (down)				U (up)				
$J_A K_A$	CD 00	01	11	10	10	11	01	CD 00	$J_A K_A$
AB 00	1Ø	0Ø	0Ø	0Ø	0Ø	0Ø	0Ø	0Ø	00 AB
01	0Ø	0Ø	0Ø	0Ø	0Ø	1Ø	0Ø	0Ø	01
11	Ø0	Ø0	Ø0	Ø0	Ø0	Ø1	Ø0	Ø0	11
10	Ø1	Ø0	Ø0	Ø0	Ø0	Ø0	Ø0	Ø0	10

Whence: $J_A = U'B'C'D' + U.B.C.D. = K_A$

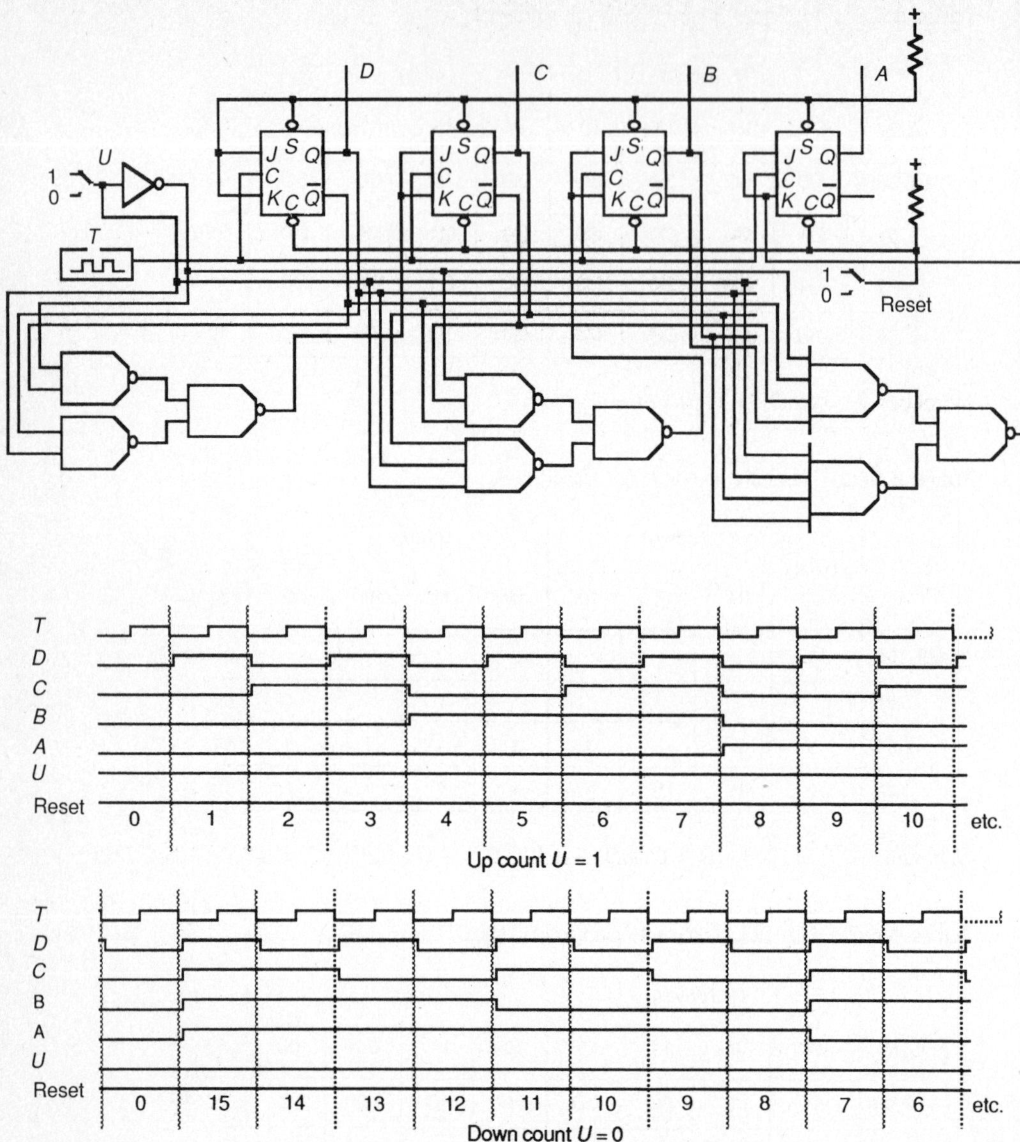

Figure 8.9 4-bit synchronous up/down counter with simulations

8. Check the completed design (logic lab., simulator, etc.)
   The complete circuit has been "wired up" using the simulator and the circuit and simulations for an up and a down count are given as Figure 8.9.
   Note that the changes of state are now synchronous with the clock T for all stages (allowing for the propagation delay through each flip-flop).

Finally, it should be recognized that this counter could be further supplemented by the addition of "parallel load" or "preset count" facilities. Typically, this type of input is implemented through the asynchronous preset and clear inputs and you will be asked to tackle the design of suitable logic circuitry in the tutorial work following this chapter.

8.2.3 An up/down counter (incrementer/decrementer) for VLSI

Clearly, we could produce a VLSI realization of the design of section 8.2.2. This would use four JK flip-flops, five 2 I/P, two 3 I/P and two 4 I/P *Nand* gates as in Figure 8.9. However, the transistor count would be quite high—4 X 28 + 5 X 4 + 2 X 6 + 2 X 8 = 160—for a complementary CMOS realization. Thus we may seek other ways of implementing an up/down count, say, once again, a 4-bit binary count. One possibility is to use a 4-bit adder circuit and a 4-bit register, as in Figure 8.10. On "up count" or "increment", 1 is added to the "count" held in the register on every input pulse (or appropriate edge) of the signal being counted. It is readily shown that the arrangement will "down count" or "decrement" if a simple adjustment is made to the inputs as shown in the figure.

It will be seen that the system comprises two main subsystems, an adder and a register, and we will now look at each in turn.

A 4-bit parallel adder

We have already examined the design of an adder in section 5.4 and a CMOS circuit design for an adder element (1-bit slice) was arrived at in Figure 5.39. The design comprises 20 transistors, which would mean 80 transistors for a 4-bit adder. Before accepting this we should examine alternative realizations.

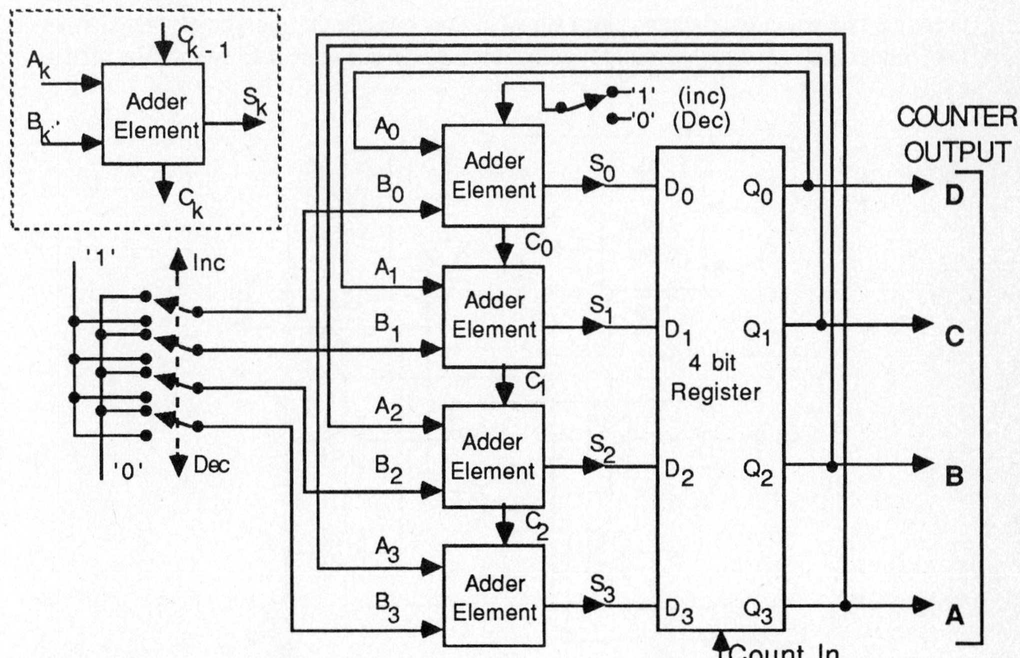

Figure 8.10 A 4-bit incrementer/decrementer (up/down counter)

The truth table for an adder element is given as Table 5.2 and it is possible to derive expressions for the adder element in a form differing from that used previously. We may write:

for the sum S_k: If $A_k = B_k$ then $S_k = C_{k-1}$

else, $S_k = (C_{k-1})'$

and, for new carryC_k: If $A_k = B_k$ then $C_k = A_k$

else, $C_k = (C_{k-1})'$

These expressions are readily implemented in, say, CMOS transmission gate logic as shown in Figure 8.11. It will be seen that each adder element will now comprise 20 transistors which is the same as for the design of Figure 5.39, but mask layout designs would have to be carried out to determine if there is any area advantage to be had.

Thus, from the area point of view, either of the two approaches discussed could be chosen. It now remains to design the 4-bit clocked register in which the count will be held.

A 4-bit parallel clocked register and 2-phase clock generator circuit

Let us make an assumption that the period between any two successive count inputs never exceeds a few hundred microseconds, and if this is so then we can use a dynamic register cell to hold the count between inputs. A very simple cell, comprising two inverters and two transmission gates, as in Figure 8.12, can be used. A mask layout for this cell follows in the section of 8.3.2 dealing with a CMOS VLSI version of the shift register. However, this cell needs a 2-phase non-overlapping clock signal to store bits and this must be generated from the "count input". A simple circuit for doing this is given as Figure 8.13 and will be seen to

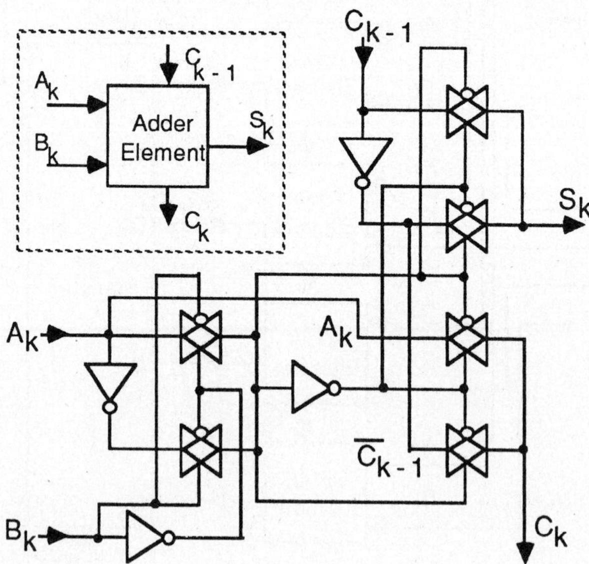

Figure 8.11 Possible CMOS adder element

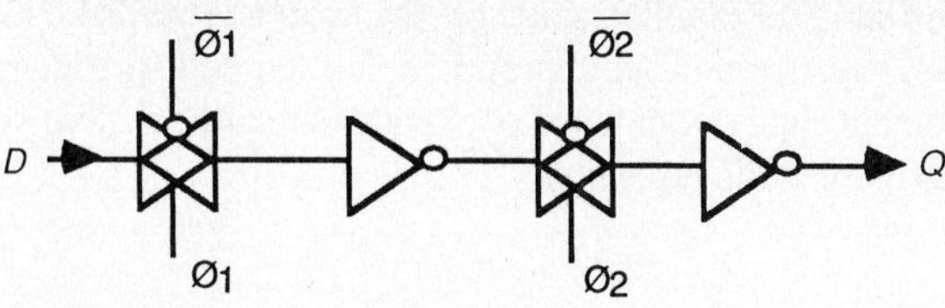

Figure 8.12 Dynamic register cell

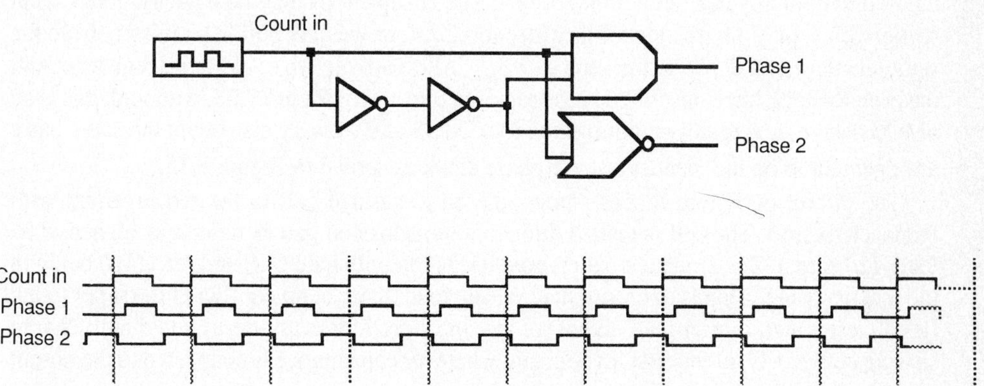

Figure 8.13 Two-phase non-overlapping "clock" signal generator

comprise two inverters (to generate an underlap period) and an *And* and a *Nor* gate, each with two inputs as shown.

The simulation waveforms clearly show the two phases generated and in complementary logic the circuit will comprise 12 transistors.

The 4-bit register with the input circuitry will thus comprise 12 + 4 X 8 = 44 transistors. Overall the complete incrementer/decrementer, as in Figure 8.10, will comprise 44 + 4 X 20 = 124 transistors. This is a distinct improvement on the JK flip-flop based counter, discussed earlier, which had some 160 transistors. Other possibilities present themselves and the reader is encouraged to explore some for comparison.

Many counter arrangements are possible, some others being included in the examples in the next chapter of this text.

8.3 Registers

Registers are storage arrays in which working data, vectors, operands, etc., are *temporarily stored*. A register generally holds a number of bits such as a nibble (4-bits), or a byte (8-bits), or several bytes which often represent words (in the computer sense), or instructions, or numbers or characters. We have seen registers employed with adders to hold operands and to accumulate results and, in the last section, to hold the current state of a count in an incrementer/decrementer.

Registers may be serial or parallel, static or dynamic, and we have already set out a *4-bit parallel dynamic register* and an appropriate storage cell in Figures 8.10 and 8.12. Clearly, the arrangement can be contracted or expanded to any sensible number of bits, and we may readily configure static registers.

8.3.1 A static 4-bit parallel register, D FF and VLSI based

This may be based on the D flip-flop as shown in Figure 8.14 and the arrangement can be tailored to hold any desired number of bits. The D flip- flops may be edge triggered, as in section 8.2.1, or 2-phase clocked as in section 7.2.4, or we may build up registers from the static memory cell described in section 7.3.3. Alternatively we could build registers from the pseudo-static latch described in the same section and in Figure 7.23. Although this is an nMOS design, it is readily reconfigured for CMOS and we may also adapt the same basic arrangement to be independent of a 2-phase clock as shown in Figure 8.15.

The circuit of Figure 8.15(a) shows the basic storage cell to be two inverters with feedback overall. The cell is loaded through a transmission gate activated by an active Hi load (*Ld*) signal. The feedback path is conditional on both load (*Ld*) and read (*Rd*) being in the inactive state. This is to ensure that, during load, there is no conflict at the input to the first inverter and, during read, to ensure that the stored bit is not destroyed due to charge sharing between C_g of the first inverter and whatever capacitance is connected to the output (*Out*) of the cell.

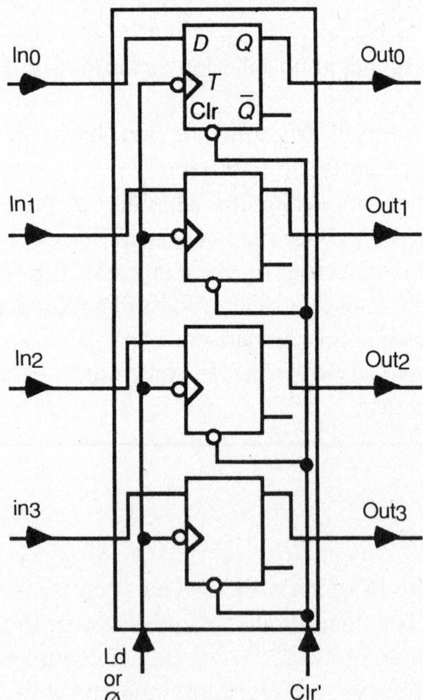

Figure 8.14 4-bit parallel register using D flip-flops

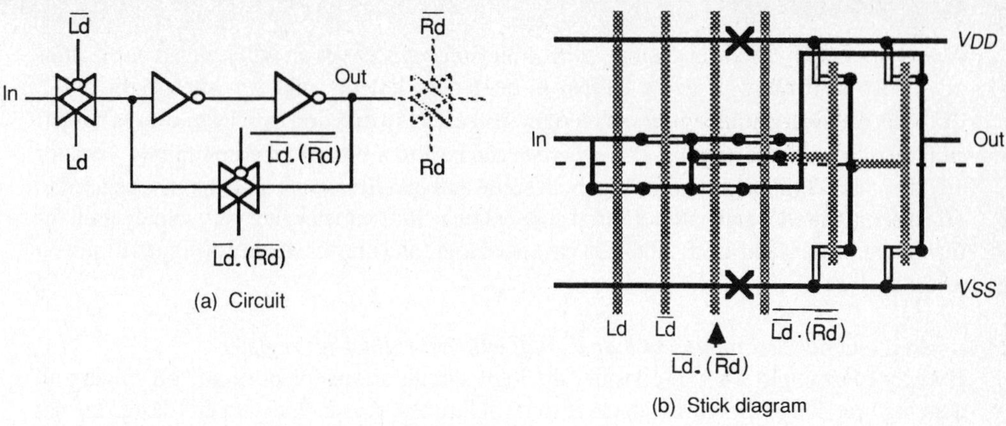

(a) Circuit

(b) Stick diagram

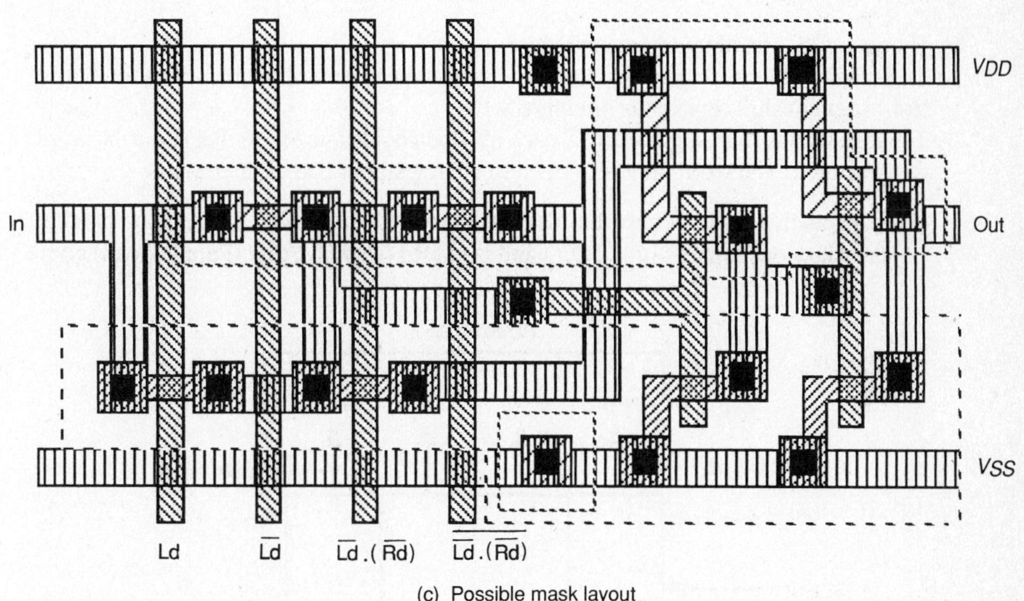

(c) Possible mask layout

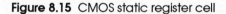

Figure 8.15 CMOS static register cell

Note that a further transmission gate may well be used, as suggested in the figure, to read the output as required.

A suitable stick diagram has been developed (Figure 8.15(b)) leading to the mask layout of Figure 8.15(c). Cell dimensions are 74λ X 41λ but this could be reduced in a multi-cell environment by combining V_{DD} and/or V_{SS} lines with those of neighboring cells.

8.3.2 Shift registers

A very commonly used subsystem is the n-bit shift register which will hold an n-bit word or number and allow it to be shifted m-bit places left or right (or often right only). Considering binary numbers, an m-bit right shift effects a division by 2^m while an m-bit left shift effects multiplication by 2^m, (where m can be 1 to n bits and n is commonly 4 or 8 or 16, etc., bits). Shift registers having both serial and parallel inputs and outputs are used to effect serial/parallel and parallel/serial conversions. Shift registers are very widely used for manipulating data and are available in packaged logic and may be readily designed in silicon as we shall see.

A serial and parallel input and output, shift left/shift right 4-bit register

By way of example we will consider the logic circuit arrangement of an "all singing all dancing" version of the shift register. If some of the resources included in this design are not required, the design is readily simplified to suit the needs and such exercises are left to the reader. A 1-bit shift per serial clock activation is assumed here and this is most often the case although it is possible to design for shifts other than one bit. (You might risk an educated guess on the possibility of being asked to do this in tutorial work!)

1. Set out a full, unambiguous specification
   To start the design, let us consider the general requirements (specification) for a 4-bit (for example) shift register as in Figure 8.16.
   The register has four stages, *ABCD*, each of which holds one bit, so that a four-bit word can be stored. The word may be written to the register in one of three ways:

   (a) In parallel, in which case the data is connected in parallel to the four parallel asynchronous inputs, AP_{in} ... DP_{in} and is loaded when *PARLD* (Parallel load) goes

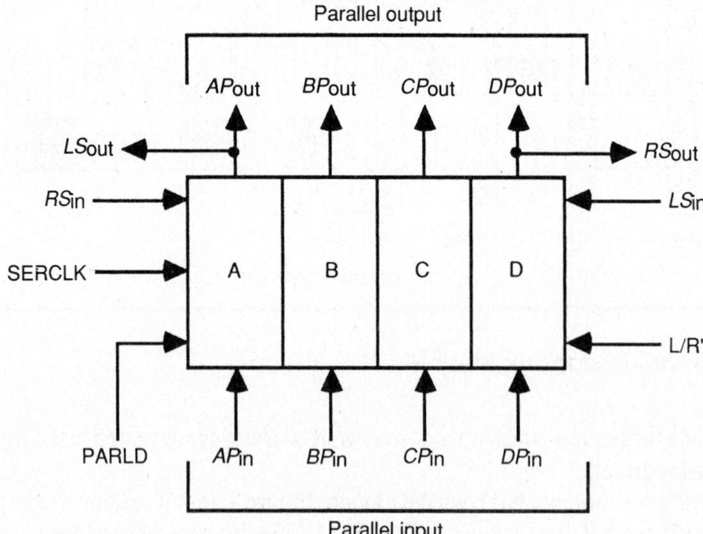

Figure 8.16 Basic shift register (4-bit) arrangement

Hi. The word thus stored is then available at the parallel outputs, AP_{out} ... DP_{out}, or may be clocked out left or right by the serial clock (SERCLK) from the left or right serial outputs $LS_{out}(AP_{out})$ or $RS_{out}(DP_{out})$ respectively.

(b) In series from the RS_{in} (right shift in) input, with left/right control line L/R' set Lo, a bit being entered into stage A when serial clock activation takes place. On the next clock activation, the next bit from RS_{in} is clocked into stage A while the bit originally held in A is clocked right into stage B, etc., until four bits are clocked in, at which time the first bit will reside in stage D.

(c) In series from the LS_{in} (left shift in) input, with left/right control line L/R' set Hi, a bit being entered into stage D when serial clock activation takes place. On the next clock activation, the next bit from LS_{in} is clocked into stage D while the bit originally held in D is clocked left into stage C, etc., until four bits are clocked in, at which time the first bit will reside in stage A.

Words or numbers held in the register may be read out in parallel form from AP_{out} ... DP_{out}, or may be shifted left or right by activating the serial clock (SERCLK). Shifting has the effect of either multiplying (left shift) or dividing (right shift) a number held in the register by 2 for each activation of SERCLK, provided that no bits "fall off" either end of the register, via LS_{out} or RS_{out}, during shifting.

2. Set out a primitive state transition diagram.
3. Merge states if possible to reduce the overall complexity.
   Draw the final(merged) state diagram.
4. Allocate secondary variables and state vector values to each state.
5. Set out the state transition table.

Steps 2-5 are not needed as the straightforward requirements for serial shifting are readily expressed algebraically as we shall see. The parallel inputs are usually via asynchronous inputs and can be added after the serial design is complete.

[At this point it is convenient to *choose the circuit elements which will represent the output variables*. There are no separate secondary variables since the register is a Moore type circuit.]

A most convenient circuit element for each stage is a clocked D flip-flop which also has asynchronous preset and clear inputs.

6. (a) Derive logic equations for the D FF inputs

This is easily done by, first, setting out the interconnections which will give the required shift of bits between stages as follows:

Right shift: $A_{t+1} = (RS_{in})_t ; B_{t+1} = A_t ; C_{t+1} = B_t ; D_{t+1} = C_t$

Left shift: $A_{t+1} = B_t ; B_{t+1} = C_t ; C_{t+1} = D_t ; D_{t+1} = (LS_{in})_t$

where t is the time at which serial clock (*SERCLK*) activation occurs.

Since each stage transfers the bit at its D input to output Q of that stage on clock activation, we may write:

$$D_A = L'.(RS_{in})_t + L.(B)_t.$$

$$D_B = L'.(A)_t \quad + \ L.(C)_t.$$

$$D_C = L'.(B)_t \quad + \ L.(D)_t.$$

$$D_D = L'.(C)_t \quad + \ L.(LS_{in})_t.$$

where L indicates left and L' indicates right shift.

(b) Derive connections for the preset and clear inputs (assuming active "Lo" inputs). For stage N:

when $NP_{in} = 1$ and $PARLD = 1$ then:

$(PresetN)' = 0 : \ (ClearN)' = 1;$

and, when $NP_{in} = 0$ and $PARLD = 1$ then:

$(PresetN)' = 1 : (ClearN)' = 0;$

Finally, when $PARLD = 0$, then:

$(PresetN)' = 1 : (ClearN)' = 1.$

These considerations lead to the following *Nand* operations:

$$(PresetN)' = NP_{in}.PARLD \ ; \quad (ClearN)' = (NP_{in})'.PARLD;$$

To clarify the interconnections, two stages of a register to this design have been drawn as Figure 8.17.

7. Derive the output functions to complete the design.
   This step is not required since this is a Moore type circuit and vector *ABCD is the parallel output of the register.*
8. Check the completed design (logic lab., simulator, etc.)
   In this case the complete 4-bit register has been "connected up" and its operation simulated as shown in Figure 8.18.

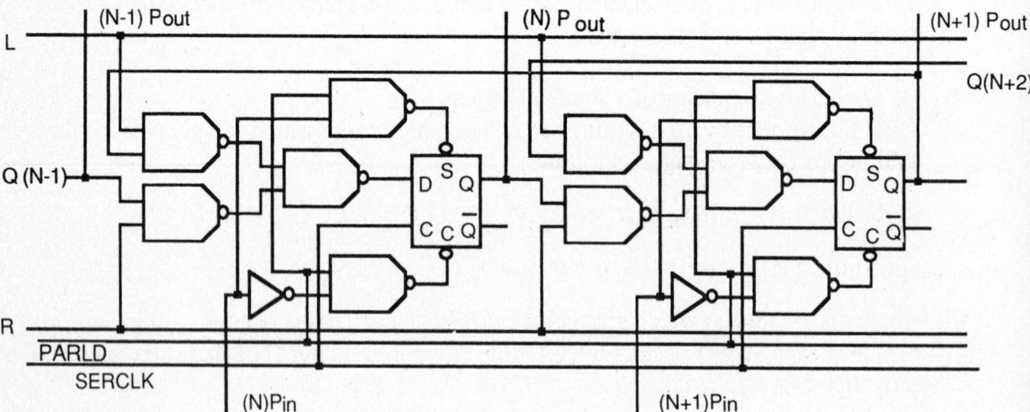

Figure 8.17 Two stages of the register – showing interconnections

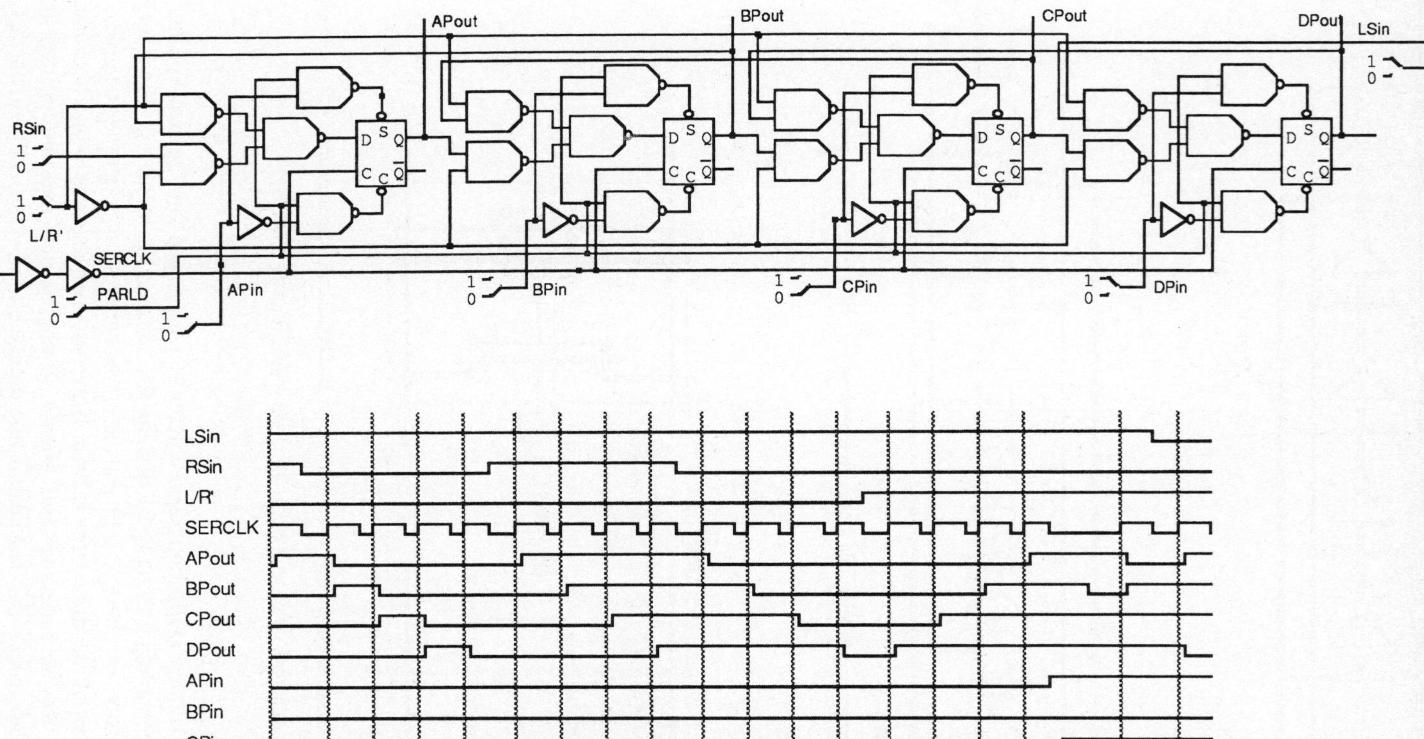

Figure 8.18 Complete 4-bit shift register with simulation waveforms

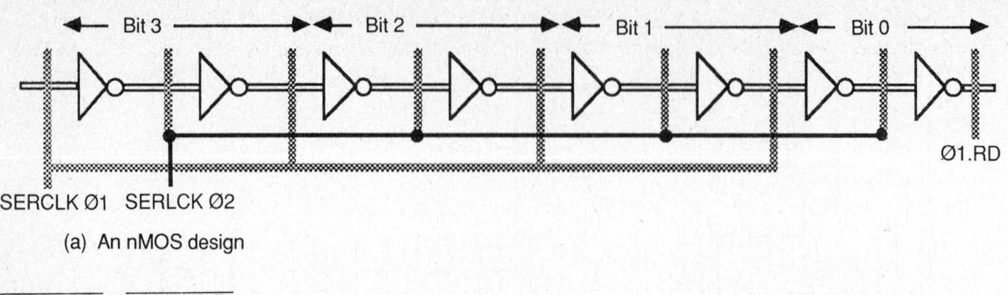

(a) An nMOS design

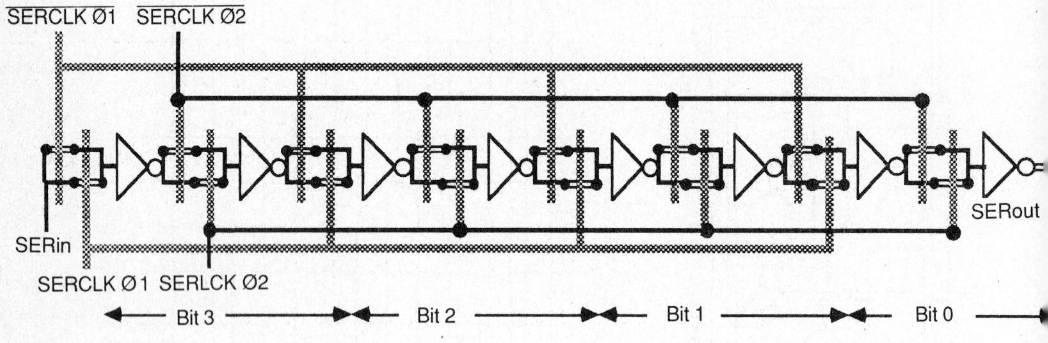

(b) A CMOS design

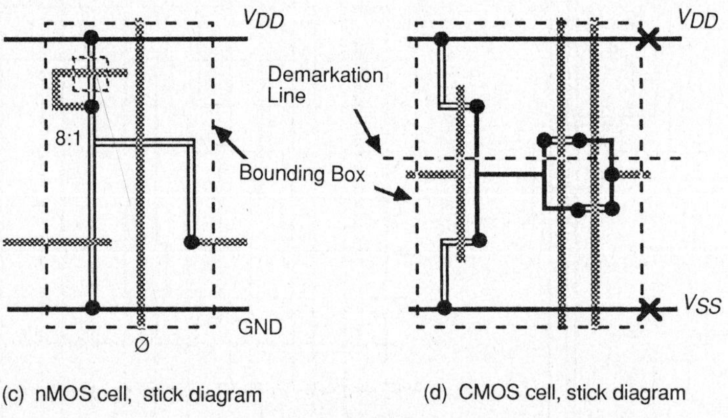

(c) nMOS cell, stick diagram

(d) CMOS cell, stick diagram

Figure 8.19 Shift registers and shift register cells (nMOS and CMOS)

VLSI designs for shift registers

Obviously we could realize D flip-flop based shift register stages and shift registers in either nMOS or CMOS technologies using D FF designs discussed earlier in this text. However, the number of transistors required per D FF and therefore per bit, is quite high and a complete 8-bit register, say, would be quite large in area. However, such registers would be *static* which may well be necessary in some cases. Mostly we use registers, and shift registers in particular, to hold data *temporarily* so that dynamic storage is usually perfectly satisfactory with a consequent saving in area.

With this in mind, we may readily configure arrangements of dynamic storage elements as left shift or right shift registers, or indeed with left/right shift facilities as discussed earlier in this chapter. Taking a simple 8-bit right shift register by way of example, we come up with the arrangements set out in Figure 8.19, (a) nMOS and (b) CMOS, which in turn lead to the cells of (c) and (d) respectively. Clearly, complete registers of any size may be built up by replicating the appropriate cell and thus the amount of detailed design work is limited.

Mask layouts are readily produced from the stick diagrams and example cells are given in Figure 8.20 (a) for nMOS using butting contacts, (b) for nMOS with buried contacts and (c) for a complementary CMOS register.

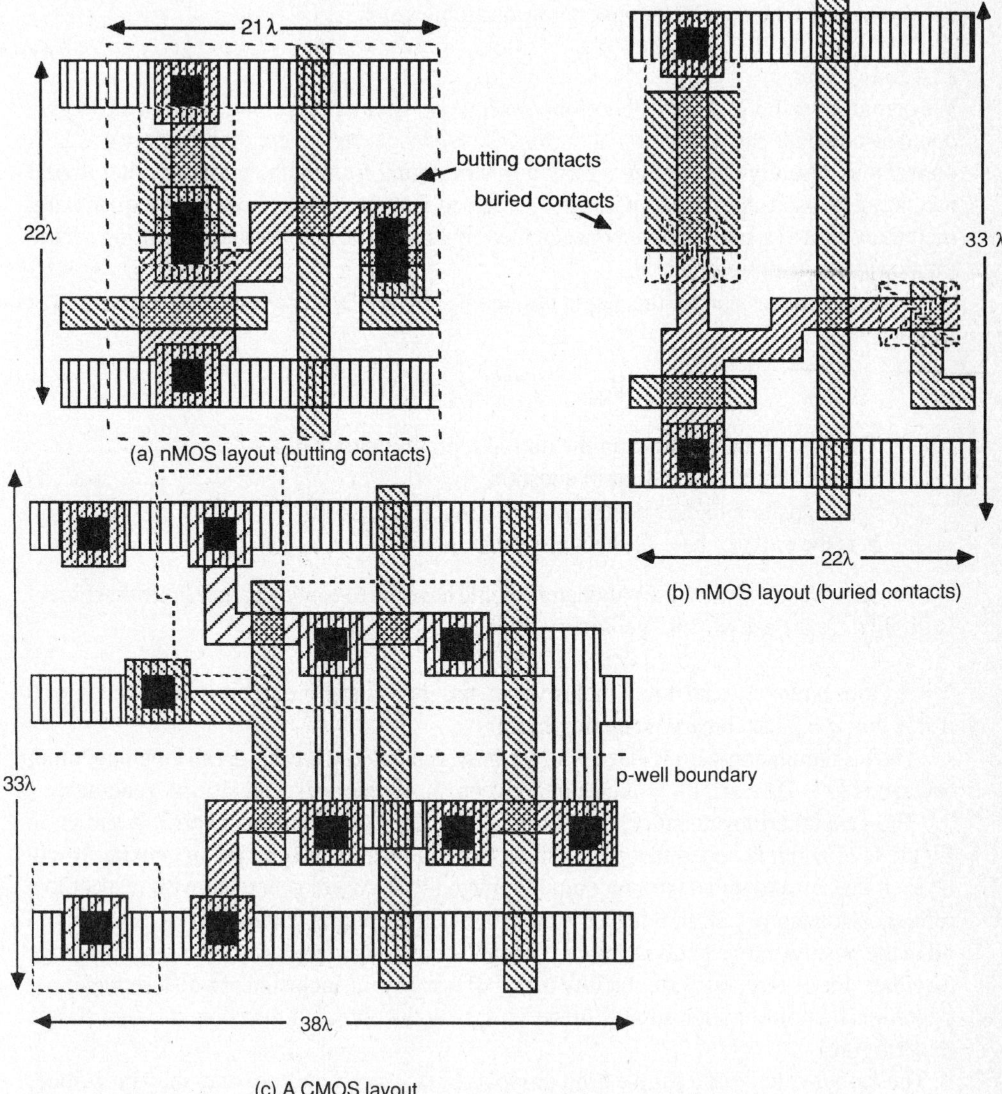

Figure 8.20 Mask layouts for nMOS and CMOS shift register cells

8.3.3 A successive approximation register (SAR) for an A/D (analog to digital) converter

There are many types of special purpose registers in digital systems architectures but it is beyond the purposes of this text to review the range of possibilities. However, the interface between the analog world and the digital system is a very important area and D/A (digital to analog) and A/D conversions are frequently needed. It is therefore appropriate to end this section on registers with the design of a successive approximation register for an A/D converter.

To start a discussion, we look first at the process of D/A conversion since an A/D converter may include a D/A converter in its architecture.

D/A conversion

We commonly refer to digitally controlled voltage sources as D/A converters but clearly we could also include the generation of current or resistance, etc., under digital control as D/A conversion. Usually conversion is from a binary digital form but decimal or other digital forms can be used. Associated with each weighted D/A (voltage) converter is a quantizing increment Δv and a settling time between feeding in a digital number and outputting a valid (analog) voltage representation.

A general expression for the output voltage V_{out} from a D/A converter may be written as follows:

$$V_{out} = \Sigma_{n=0}^{n=N-1} [r_n . q^n . \Delta v]$$

where, N is the number of digits in the digital representation word,
n is the weight of the digit in question,
r_n is a positive integer coefficient such that $0 \leq r_n \leq q - 1$,
q is the radix or base of the weighting.

For an 8-bit converter these parameters would have the following values (note that binary is implied by the use of "bit" i.e. **bi**nary digi**t**):

$N = 8$; thus n ranges from 0 to 7 inclusive ; r_n has the value 1 if that bit is switched in or 0 if it is not; $q = 2$ (the binary weighting radix).

The minimum non-zero voltage which can be generated is clearly Δv and the maximum voltage is $(2^N - 1)\Delta v$ and all voltages in between, at Δv intervals, can also be generated.

The general arrangement of, for example, an 8-bit D/A voltage converter is shown in Figure 8.21 and it is hoped that this is sufficient to illustrate the process in general. It will be seen that Δv and the maximum output voltage $(2^N - 1)\Delta v$ are generated with respect to a reference voltage V_{ref}. It should be further noted that the output voltage can be unipolar and all in the positive range or all negative, or bipolar covering a positive and a negative range on either side of zero, this latter facility being effected by the inclusion of a differential input operational amplifier with suitable offset voltage at the output of the converter (not shown in the figure).

The output voltage of a simple 8-bit unipolar binary weighted converter may be written:

$$V_{out} = K\,(0.ABCDEFGH)_2 \times V_{ref}$$

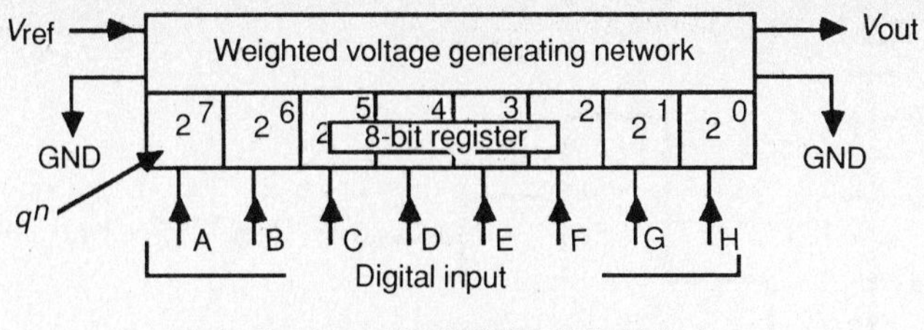

Figure 8.21 Typical D/A (voltage) converter

where K is a constant scaling factor such that $K < 1$ (usually), and$ABCDEFGH$ are the eight bits of the counter, (A = MSB).

The successive approximation A/D conversion process

In introducing the operation of a successive approximation A/D converter, we may consider the arrangement of Figure 8.22. A digitally controlled voltage, V_{out} is generated by a binary weighted network as is the case in D/A conversion. The output is set by an 8-bit register, in this case a successive approximation register (SAR), and the voltage generated at any time is compared with the incoming analog voltage V_a by a comparator which generates error signal, e.

This error signal is in turn fed back to control stage settings of the SAR and V_{out} in consequence. The control sequence may readily be explained with the aid of an example conversion, in this case a 5-bit converter with $\Delta v = 0.1$ V and an input $V_a = 1.88$ V. With reference to Figure 8.23, on receiving a "start" signal the SAR is reset to zero and V_{out} is also

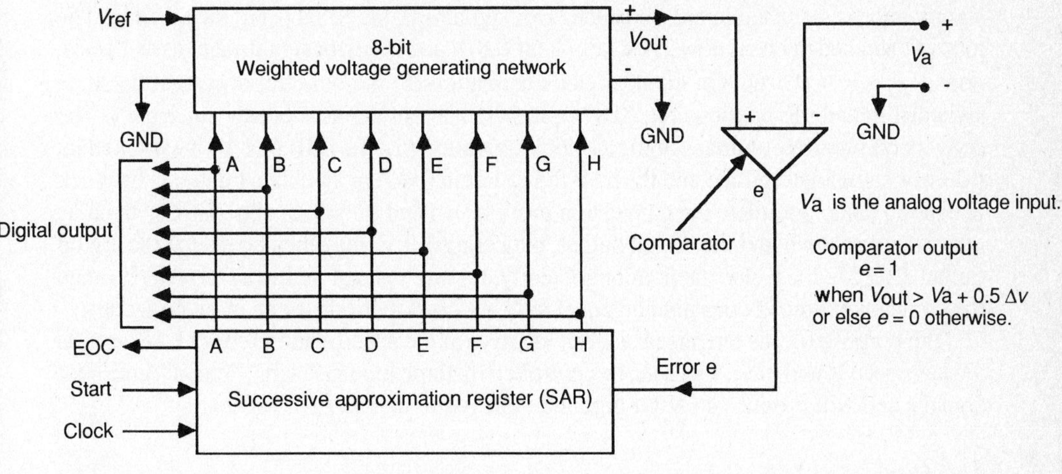

Figure 8.22 Arrangement of 8-bit successive approximation A/D converter

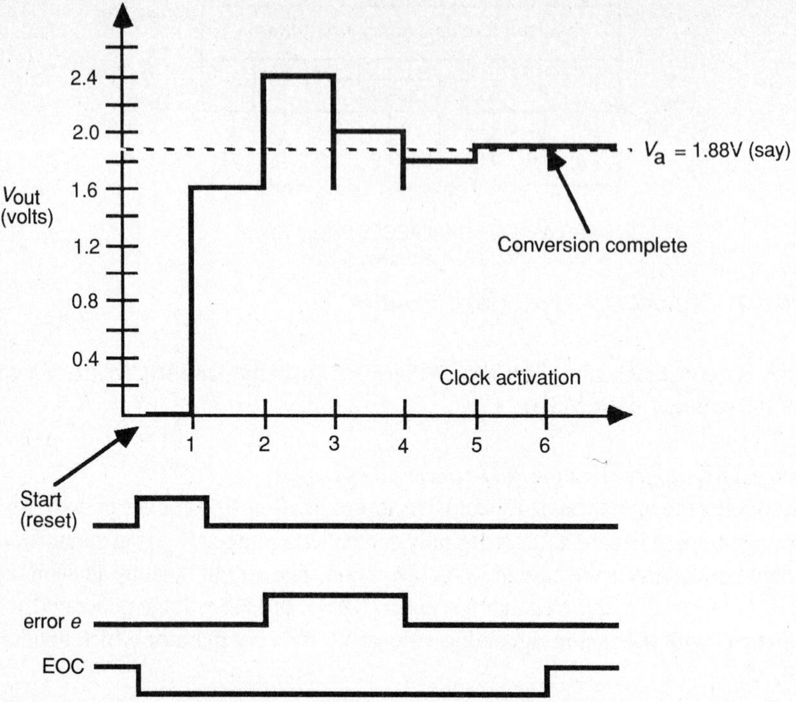

Figure 8.23 Example of successive approximation conversion

reset to zero. Then clock activation (1) takes place and the MSB of the register is set to 1 giving an output voltage V_{out} = 1.6 V, i.e., (Δv X 2^4) which is compared with V_a giving an error signal e=0 since $V_{out} < V_a$. On the next clock activation (2), the next bit in order of significance is switched in and since the error signal is 0, the MSB is left switched in. The total of switched in bits is now 2.4 V i.e., (1.6 + 0.8)V and the error signal changes to 1 since, now, $V_{out} > V_a + 0.5\Delta v$. On the next clock activation (3) the next bit, of weight 0.4 V, is switched in and the previous bit, (0.8 V), is switched out simultaneously since e =1. The conversion process continues until, on clock activation (5), the LSB (0.1 V) is switched in, the error signal determined and the LSB either left in (e=0) or switched out (e=1) by clock activation (6) to complete the conversion process. An end of conversion, (*EOC*), signal is generated as shown and the digital output, indicating 1.9 V, may then be read at the digital output after 5 + 1 = 6 clock activations. Clearly, for any voltage within a converter's range the conversion time is constant and equal to N + 1 clock periods for an N-bit converter.

The purpose of this exercise has been to arrive at the specification for the SAR and, as we have seen it will have N+1 bits, be controlled by three inputs, "start", "clock", and error signal e and will produce an N-bit digital output vector and an *EOC* signal.

The design of an SAR

In order to extract the logical requirements of an SAR let us tackle the design of a 3-bit register. This will be enough to determine the interconnection pattern without becoming too

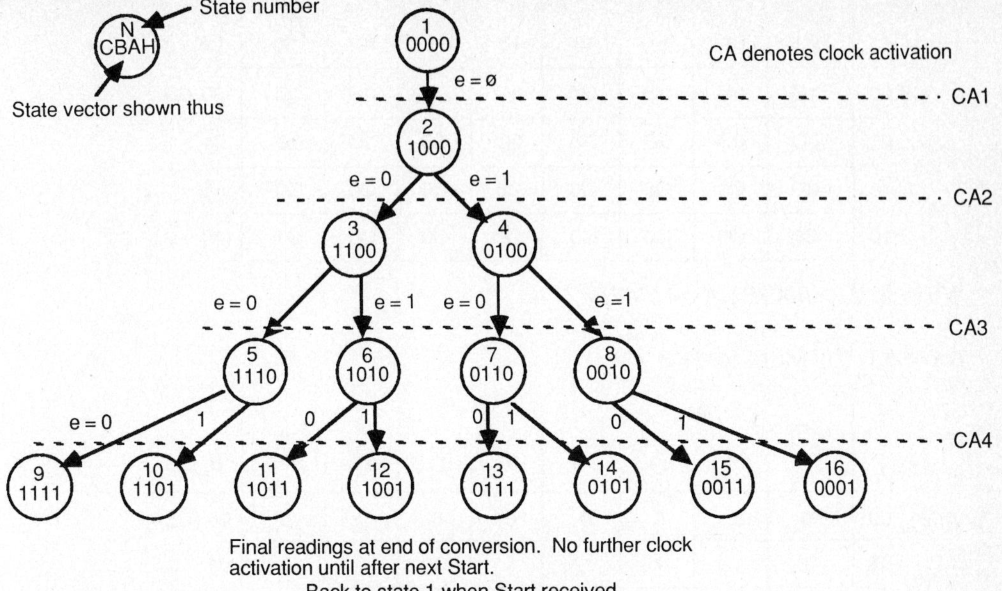

Figure 8.24 State transition diagram for 3-bit SAR

unwieldy. The requirements are best set out in the form of a state transition table and this has been done in Figure 8.24. An asynchronous start signal has been assumed but that all other operations are to be synchronous with the conversion clock. It is convenient to denote the three converter stages, *C B A*, where *C* is the MSB, and to allow a further stage *H*, denoting hold, at the end of conversion. The comparator error output signal is denoted *e*.

The maps for the transitions of each of the four flip-flops can be derived directly from the state transition diagram and it is a straightforward next step to arrive at the connections to J and K inputs through applying the designer's table for the JK FF. These steps are set out in Table 8.7 from which the expressions to be implemented in connecting the flip-flops are obtained.

Table 8.7 (a) Maps for FF C

	e'				*e*				
C_{t+1}	$(AH)_t$ 00	01	11	10	10	11	01	$(AH)_t$ 00	C_{t+1}
$(CB)_t$ 00	1	0	0	0	0	0	0	1	00 $(CB)_t$
01	0	0	0	0	0	0	0	0	01
11	1	1	1	1	1	1	1	1	11
10	1	1	1	1	1	1	1	0	10

	e'				e				
$J_C K_C$	AH 00	01	11	10	10	11	01	AH 00	$J_C K_C$
CB 00	1Ø	0Ø	0Ø	0Ø	0Ø	0Ø	0Ø	1Ø	00 CB
01	0Ø	0Ø	0Ø	0Ø	0Ø	0Ø	0Ø	0Ø	01
11	Ø0	Ø0	Ø0	Ø0	Ø0	Ø0	Ø0	Ø0	11
10	Ø0	Ø0	Ø0	Ø0	Ø0	Ø0	Ø0	Ø1	10

Whence: $J_C = B'A'H'$; $K_C = B'A'H'e$

Table 8.7 (b) Maps for FF B

	e'				e				
B_{t+1}	$(AH)_t$ 00	01	11	10	10	11	01	$(AH)_t$ 00	B_{t+1}
$(CB)_t$ 00	0	0	0	0	0	0	0	0	00 $(CB)_t$
01	1	1	1	1	1	1	1	0	01
11	1	1	1	1	1	1	1	0	11
10	1	0	0	0	0	0	0	1	10

	e'				e				
$J_B K_B$	AH 00	01	11	10	10	11	01	AH 00	$J_B K_B$
CB 00	0Ø	0Ø	0Ø	0Ø	0Ø	0Ø	0Ø	0Ø	00 CB
01	Ø0	Ø0	Ø0	Ø0	0Ø	Ø0	Ø0	Ø1	01
11	Ø0	Ø0	Ø0	Ø0	Ø0	Ø0	Ø0	Ø1	11
10	1Ø	0Ø	0Ø	0Ø	0Ø	0Ø	0Ø	1Ø	10

Whence: $J_B = C.A'.H'$; $K_B = A'.H'.e$

By similar mapping for FF A: $J_A = B.H'$; $K_A = H'.e$

Table 8.7 (c) Maps for FF H

	e'				e				
H_{t+1}	$(AH)_t$ 00	01	11	10	10	11	01	$(AH)_t$ 00	H_{t+1}
$(CB)_t$ 00	0	1	1	1	1	1	1	0	00 $(CB)_t$
01	0	1	1	1	1	1	1	0	01
11	0	1	1	1	1	1	1	0	11
10	0	1	1	1	1	1	1	0	10

	e'				e				
J_HK_H	AH 00	01	11	10	10	11	01	AH 00	J_HK_H
CB 00	0Ø	Ø0	Ø0	1Ø	1Ø	Ø0	Ø0	0Ø	00 CB
01	0Ø	Ø0	Ø0	1Ø	1Ø	Ø0	Ø0	0Ø	01
11	0Ø	Ø0	Ø0	1Ø	1Ø	Ø0	Ø0	0Ø	11
10	0Ø	Ø0	Ø0	1Ø	1Ø	Ø0	Ø0	0Ø	10

Whence: $J_H = A \; K_H = 0$

The logic diagram and the waveforms for sample "conversions" are set out in Figure 8.25.

The connections for larger converters are easily deduced from a 3-bit design. For example, if a fourth stage D is added then the connections to J_C are extended to:

$$J_C = D.B'.A'.H'$$

whilst stage D would be connected as follows:

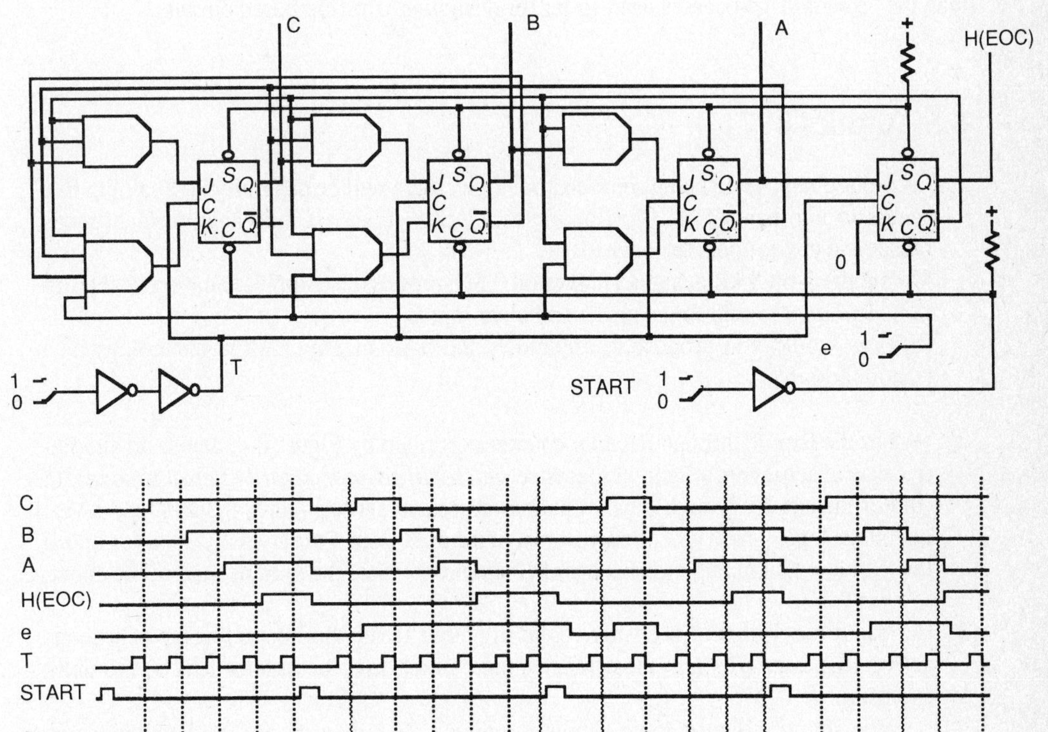

Figure 8.25 3-bit SAR logic with simulated conversion waveforms

$$J_D = C'.B'.A'.H'$$

$$K_D = C'.B'.A'.H'.e$$

and so on for further stages.

It should be noted that a more compact design for the same 3-bit SAR is worked through using alternative design processes in Appendix 2.

8.4 Observations

The purpose of this chapter has been to reinforce the reader's grasp of seqnential circuit design processes by working through example designs. It is hoped that the material is also useful in providing details of some commonly used clocked subsystems which are typical designs to which variations, that is, more bits, etc., can be readily made. The work is further reinforced by the tutorial questions which follow .

The concept of using "real world" needs to illustrate the design process is carried through to the next chapter, where further useful designs are presented but using an alternative approach to design and analysis which the author has found to be more direct and often easier than the "standard" processes used so far for designing flip-flop based circuits.

8.5 Tutorial 8

1. (a) Sometimes, when using semi-custom logic, there will be difficulty in obtaining free access to numbers of JK flip-flop elements, particularly if individual asynchronous preset and clear inputs are needed.
   Taking the 3-bit SAR designed in section 8.3.3, redesign for implementation via D flip-flops *without* asynchronous preset and clear inputs.
   (b) How would you propose to implement the SAR in, say, CMOS technology as a custom design?

2. (a) For the Ripple through *decimal* counter described by Figure 8.4, devise an alternative way of achieving a decimal sequence *which will not momentarily step onto count 10.*
   (b) For the ripple-through binary counter arrangements of Figures 8.2 and 8.3, devise interstage logic which will allow reversal of the direction of the count by a single control line U. Note that changing the state of U should not alter the current state of the count.

3. Carry out the design of a positive edge triggered D flip-flop including asynchronous active Hi preset (*Pr*) and clear (*Clr*) inputs which may be assumed to be mutually exclusive.

   You should give the logic circuit and then a stick diagram for implementation in CMOS complementary logic.

4. For the synchronous up/down counter of section 8.2.2, design a "parallel load" (or "preset count") facility.

5. (a) Design a 4-bit shift left/shift right register in which the shift can be one bit or two bits under the control of a signal "Sh2".
(b) For the arrangement (a), design a stick diagram for an element from which such a register of n-bits could be formed.

9 Characteristic and application equations - a difference equation approach to clocked sequential circuit design and analysis

9.1 Introduction

In the design of a clocked sequential digital system one can recognize three interrelated factors:

1. *The specification and representation of system requirements* or desired behavior of interconnected elements (e.g. specification in words, state diagram, state transition tables, etc.)

2. *The representing of the characteristics of the elements with which design of a system is to take place* (e.g. specification in words, logic symbols, characteristic tables, characteristic equations).

3. *The design techniques to be used* to realize (1) with (2). Some techniques have already been considered.

If the representations used in (1) and (2) are similar and compatible, it is possible to streamline the design processes (3) to a point where they become simple and straightforward.

Such compatible representations may be based upon the *difference equation* which, as we have already seen, sets out the conditions in one time interval $t+1$ against those determining them in the preceding time interval t. Characteristic equations for devices, such as flip-flops can be written in this form.

We will also show that digital subsystem and system performance may be readily characterisd in difference equation form. *Applications equations* may be written which are then highly compatible with device characteristics and straightforward design and analysis processes follow.

In this chapter, we will depart from the use of the "prime" to denote the *complement* or *not* operation. Instead, we will use the "bar" for this purpose so that the reader may also become familiar with this commonly used operator.

9.2 A summary of some common characteristic equations (as in Table 9.1)

Note that characteristic equations may be interpreted as follows:

Output Q will become 1 at $t+1$ if (conditions specified) apply at clock activation t.
The complementary equation gives the conditions for $Q=0$.

Table 9.1 Some common flip flops

Flip-flop type	*Characteristic equation*	*Complementary equation*
RS (Asynch.)	$Q_{t+1} = (Q.\bar{R} + S)_t$	$\bar{Q}_{t+1} = (\bar{Q}.\bar{S} + R)_t$
JK	$Q_{t+1} = (\bar{Q}.J + Q.\bar{K})_t$	$\bar{Q}_{t+1} = (\bar{Q}.\bar{J} + Q.K)_t$
D also JK with J=K'=D	$Q_{t+1} = D_t$	$\bar{Q}_{t+1} = \bar{D}_t$
T also JK with J=K=1	$Q_{t+1} = \bar{Q}_t$	$\bar{Q}_{t+1} = Q_t$
T(with Enable)* also JK with J=K=E	$Q_{t+1} = E.\bar{Q}_t + \bar{E}.Q_t$	$\bar{Q}_{t+1} = E.Q_t + \bar{E}.\bar{Q}_t$

* Note that this Enable input is commonly denoted by "T" but T is used for the clock in this text.

Other variations of the clocked flip-flop can be readily accommodated by this method of characterization.

9.3 Application equations

Consider now the requirement for interconnected sequential circuit elements to form a digital system or subsystem. One commonly required subsystem is the binary counter. Let us, for purposes of illustration, consider the requirements for a 4-bit synchronous up/down binary counter which we have already examined in section 8.2.2 and in Table 8.2 in particular.

An alternative way of expressing the requirements is by way of a state transition diagram which is presented here as Figure 9.1.

We may equally well express this information as a set of *application equations*, one for each stage (bit) of the counter. This has a general form as follows, using stage A as an example:

$$A_{t+1} = (\text{function of } ABCD)_t$$

The reader will recognize this as having the form of a difference equation discussed earlier and in Chapter 7.

We can express the requirements for each stage in this way and, in fact, we have already derived these equations for this counter from the first map in each of Tables 8.4, 8.5 and 8.6. The *application equations* in question for each stage are as follows (reproduced from the aforementioned tables and the text of section 8.2.2)

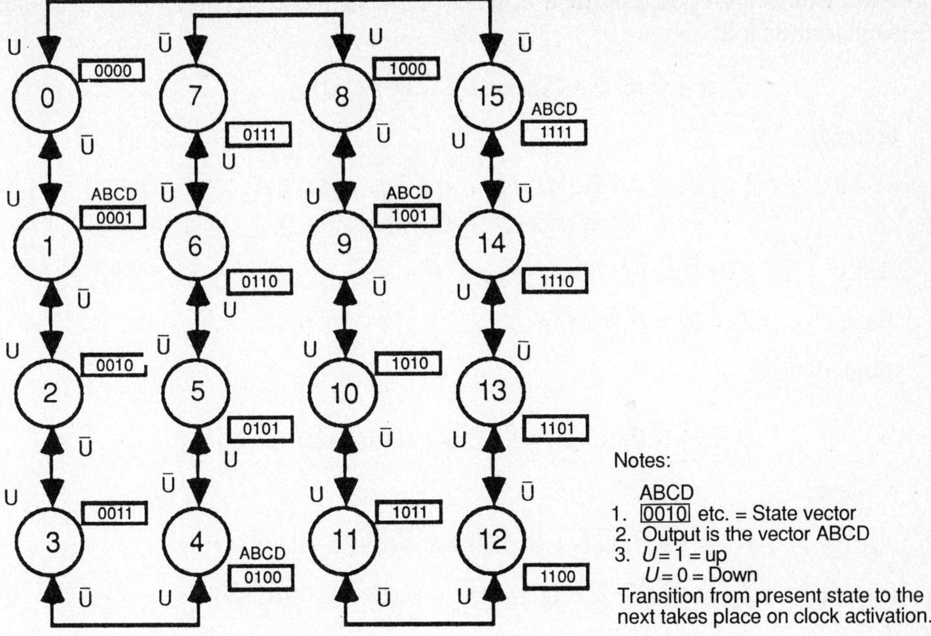

Figure 9.1 State diagram for 4-bit up down synchronous binary counter

$$D_{t+1} = \bar{D}_t.$$

$$C_{t+1} = [\,\bar{U}.(\bar{C}.\bar{D} + C.D.) + U.(\bar{C}.D + C.\bar{D})\,]_t.$$

$$B_{t+1} = [\,\bar{U}.(B.(C+D) + \bar{B}.\bar{C}.\bar{D}) + U.(B(\bar{C}+\bar{D}) + \bar{B}.C.D)\,]_t.$$

$$A_{t+1} = [\,\bar{U}.(A.(B+C+D) + \bar{A}.\bar{B}.\bar{C}.\bar{D}) + U.(A(\bar{B}+\bar{C}+\bar{D}) + \bar{A}.B.C.D)\,]_t.$$

Clearly, stage D is a toggle connected flip-flop so that:

$$J_D = K_D = 1$$

Now note the form of the JK characteristic expression as set out in Table 9.1:

$$Q_{t+1} = (\bar{Q}.J + Q.\bar{K})_t$$

If we can manipulate the application equations to be in the same form as the FF characteristic equation (the JK FF in this case), namely:

$$(\text{variable})_{t+1} = [(\overline{\text{variable}}).(\text{expression 1}) + (\text{variable}).(\text{expression 2})]_t$$

then we can directly derive the expressions for J = (expression 1) and $\bar{\text{K}}$ = (expression 2) from the application equations.

Having already disposed of stage D, let us rearrange the expression for C_{t+1} into the required form as follows:

$$C_{t+1} = [\, \bar{U}.(\bar{C}.\bar{D} + C.D.) + U.(\, \bar{C}.D + C.\bar{D})\,]_t$$

becomes:

$$\begin{aligned} C_{t+1} &= [\, \bar{C}.(\, \bar{U}.\bar{D} + U.D\,) \;+\; C.(\, \bar{U}.D + U.\bar{D}\,)\,]_t \\ &= \;\bar{C}.(\text{expression 1}) \;+\; C.(\text{expression 2}) \end{aligned}$$

Hence: $J_C = \bar{U}.\bar{D} + U.D$ and $\bar{K}_C = \bar{U}.D + U.\bar{D}$

Thus: $J_C = K_C = \bar{U}.\bar{D} + U.D$

Similarly from:

$$B_{t+1} = [\, \bar{U}.(B.(C{+}D) + \bar{B}.\bar{C}.\bar{D}) + U.(B(\bar{C}{+}\bar{D}) + \bar{B}.C.D)\,]_t.$$

we have,

$$B_{t+1} = [\, \bar{B}.(\bar{U}.\bar{C}.\bar{D} + U.C.D) + B(\bar{U}(C{+}D) +.U.(\bar{C}{+}\bar{D})\,]_t.$$

whence: $J_B = \bar{U}.\bar{C}.\bar{D} + U.C.D$ and $\bar{K}_B = \bar{U}.(C{+}D) + U(\bar{C}{+}\bar{D})$

Thus: $J_B = K_B = \bar{U}.\bar{C}.\bar{D} + U.C.D$

Finally from:

$$A_{t+1} = [\, \bar{U}.(A.(B{+}C{+}D) + \bar{A}.\bar{B}.\bar{C}.\bar{D}) + U.(A(\bar{B}{+}\bar{C}{+}\bar{D}) + \bar{A}B.C.D)\,]_t$$

we have:

$$A_{t+1} = [\, \bar{A}.(\bar{U}.\bar{B}.\bar{C}.\bar{D} + U.B.C.D) + A.(\bar{U}.(B{+}C{+}D) + U(\bar{B}{+}\bar{C}{+}\bar{D})\,]_t$$

whence, $J_A = K_A = \bar{U}.\bar{B}.\bar{C}.\bar{D} + U.B.C.D$

Comparing these expressions for the J and K inputs with those obtained from the second maps in each of Tables 8.4-8.6 we can see that we have come to the same results.

The process seems to be rather lengthy but this is due to the form in which we took the application equations from the maps of Tables 8.4-8.6. The process is very effective if we take application equations in the required form from the mapping (or directly from the state transition table in some cases).

For other types of flip-flop we can readily take application equations in a form which matches the characteristic equation.

9.4 Application equation based design procedure for clocked elements

Looking over the previous example, and allowing for Mealy as well as Moore circuits, we can see that the following steps form a design process:

1. Set out a full, unambiguous specification.
2. Set out a primitive state transition diagram.
3. Merge states if possible to reduce the overall complexity.
   Draw the final (merged) state diagram.
4. Allocate secondary variable or output state vector values to each state.
5. Set out the state transition table and choose the circuit elements which will represent the secondary or output variables. This will determine the form required for application equations
6. Set out suitable maps or otherwise derive application equations in the desired form for the FF (or other chosen element) and hence obtain the required input expressions.
7. Derive the output functions (if necessary).
8. Check the completed design (logic lab., simulator etc.)

Note that the procedure closely parallels that we have used before but omits mapping to obtain the FF input expressions.

9.5 Examples of the design of clocked flip-flop based circuits using the application equation approach

9.5.1. Example 1: A binary coded decimal (8421) up/down synchronous counter

This example serves the triple purpose of demonstrating design methods, setting out the design of a commonly required counter arrangement, and introducing considerations in avoiding "lock out" in illegal states.

1. *Set out a full, unambiguous specification.*
   In this case the specification is largely defined above but we may add that JK flip-flops are to be used, the counter outputs are to be $DCBA$* weighted 8421 respectively, and that the up/down control is exercised via a control line U (U=1=up). We may now move directly to step 5.

5. *Set out the state transition table and choose the circuit elements which will represent the output variables.* This will determine the form required for application equations.
   No intermediate steps are required here and the counter specification translates directly into the state transition table, Table 9.2. We have already specified that JK flip-flops are to be used.

6. *Set out suitable maps or otherwise derive application equations in the desired form for the FF (or other chosen element) and hence obtain the required input expressions.*
   This follows after setting out Table 9.2.

* The order of variables is not set by any hard and fast conventions. Other examples may use D as the least significant bit, or indeed, different letters entirely may be used.

Table 9.2 State transitions for BCD counter

Present state vector $(D\,C\,B\,A)_t$	Clock ↓	Next state vectors U=1 $(D\,C\,B\,A)_{t+1}$	U=0 $(D\,C\,B\,A)_{t+1}$	
0 0 0 0		0 0 0 1	1 0 0 1	Legal (designed) states
0 0 0 1		0 0 1 0	0 0 0 0	
0 0 1 0		0 0 1 1	0 0 0 1	
0 0 1 1		0 1 0 0	0 0 1 0	
0 1 0 0		0 1 0 1	0 0 1 1	
0 1 0 1		0 1 1 0	0 1 0 0	
0 1 1 0		0 1 1 1	0 1 0 1	
0 1 1 1		1 0 0 0	0 1 1 0	
1 0 0 0		1 0 0 1	0 1 1 1	
1 0 0 1		0 0 0 0	1 0 0 0	
1 0 1 0		0 Ø Ø Ø	0 Ø Ø Ø	Not used (illegal) states
1 0 1 1		0 Ø Ø Ø	0 Ø Ø Ø	
1 1 0 0		0 Ø Ø Ø	0 Ø Ø Ø	
1 1 0 1		0 Ø Ø Ø	0 Ø Ø Ø	
1 1 1 0		0 Ø Ø Ø	0 Ø Ø Ø	
1 1 1 1		0 Ø Ø Ø	0 Ø Ø Ø	

Ø = Don't care

Note that the present state conditions shown as "not used" or "illegal" in Table 9.2 should never occur, thus, the next state conditions may be written as don't cares. However, should a glitch or power up condition result in the circuit assuming one of these states, then means must exist to return the circuit to a legal state. hence we make 0 entries in the next state D columns in the "Not used" area since all vectors with D=0 are legal.

We have chosen a JK flip-flop based realization. Thus from Table 9.2 we may now derive the application equations for each variable in the form to match up with the JK characteristic equation.

By inspection: $A_{t+1} = \overline{A}_t$ (i.e. a "toggle FF")

Whence: $J_A = K_A = 1$

Now, for FF B we may set out a map as in Table 9.3

Table 9.3 Map for B_{t+1}

B_{t+1} \ $(BA)_t$	00	01	11	10
$(DC)_t$ 00		U	$\overline{U}$	U
01	$\overline{U}$	U	$\overline{U}$	U
11	Ø	Ø	Ø	Ø
10	$\overline{U}$		Ø	Ø

and from the map:

$$B_{t+1} = \{\bar{B}(U\bar{D}A + \bar{U}\bar{A}\,(C{+}D)) + B\,(U\bar{A} + \bar{U}A)\}_t$$

which is in the form:

$$[B_{t+1} = \{\bar{B}\ (\text{expression 1}) + B\ (\text{expression 2})\}_t]$$

Thus:

$$J_B = U\bar{D}A + \bar{U}\bar{A}\,(C{+}D)$$

$$\bar{K}_B = U\bar{A} + \bar{U}A$$

$$\therefore \qquad K_B = UA + \bar{U}\bar{A}$$

Logic circuitry for the inputs is set out as Figure 9.2.

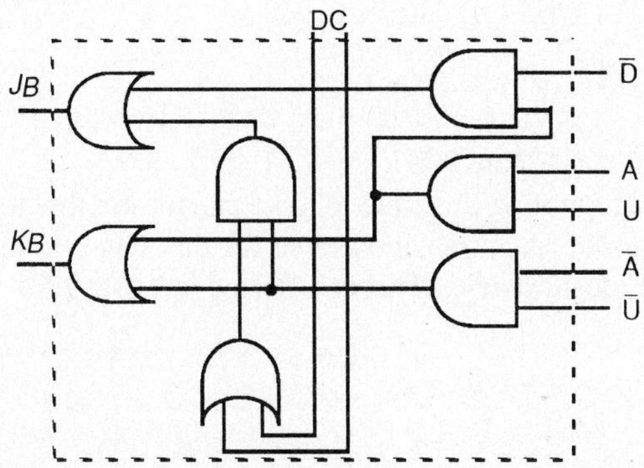

Figure 9.2 Logic circuit for FF B

Similarly from Table 9.2 we may draw a map (Table 9.4) for C_{t+1}.

Table 9.4 Map for C_{t+1}

C_{t+1}	$(BA)_t$ 00	01	11	10
$(DC)_t$ 00			U	
01	U	$U\ \bar{U}$	$\bar{U}$	$U\ \bar{U}$
11	Ø	Ø	Ø	Ø
10	$\bar{U}$		Ø	Ø

From Table 9.4:

$$C_{t+1} = \{\bar{C}(UBA + \bar{U}\bar{D}\bar{A}) + C(U(\bar{B} + \bar{A}) + \bar{U}(B + A))\}_t$$

thus: $$J_C = UBA + \bar{U}\bar{D}\bar{A}$$

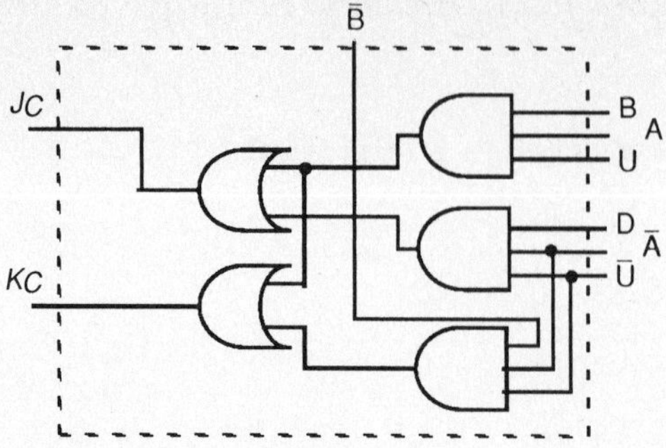

Figure 9.3 Logic circuitry for FFC

$$\overline{K}_C = U(\overline{B} + \overline{A}) + \overline{U}(B + A)$$

$$\therefore \quad K_C = UBA + \overline{U}\overline{B}\overline{A}$$

The required logic circuitry is set out as Figure 9.3.

Finally, stage D may be designed, noting this time that we cannot use the don't care conditions and have written 0s into the state transition table for this stage (to avoid the possibility of becoming trapped in illegal states). This is reflected in Table 9.5.

Table 9.5 Map for D_{t+1}

D_{t+1}	$(BA)_t$ 00	01	11	10
$(DC)_t$ 00	$\overline{U}$			
01			U	
11				
10	U	$\overline{U}$		

From the table:

$$D_{t+1} = \{\overline{D}(UCBA + \overline{U}\overline{C}\overline{B}\overline{A}) + D(U\overline{C}\overline{B}\overline{A} + \overline{U}\overline{C}\overline{B}A)\}_t$$

Thus: $$J_D = UCBA + \overline{U}\overline{C}\overline{B}\overline{A}$$

$$\overline{K}_D = U\overline{C}\overline{B}\overline{A} + \overline{U}\overline{C}\overline{B}A$$

$$\therefore \quad K_D = (\overline{U} + C + B + A) \ . \ (U + C + B + \overline{A})$$

$$K_D = C + B + UA + \overline{U}\overline{A}$$

The required logic circuitry is set out as Figure 9.4

All stages have thus been designed and a complete logic circuit is given as Figure 9.5. Step 7 is not required since *DCBA* is the output of the counter.

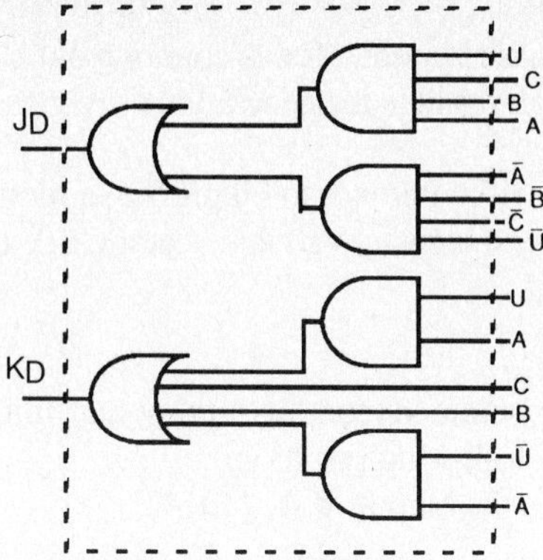

Figure 9.4 Logic circuit for FFD

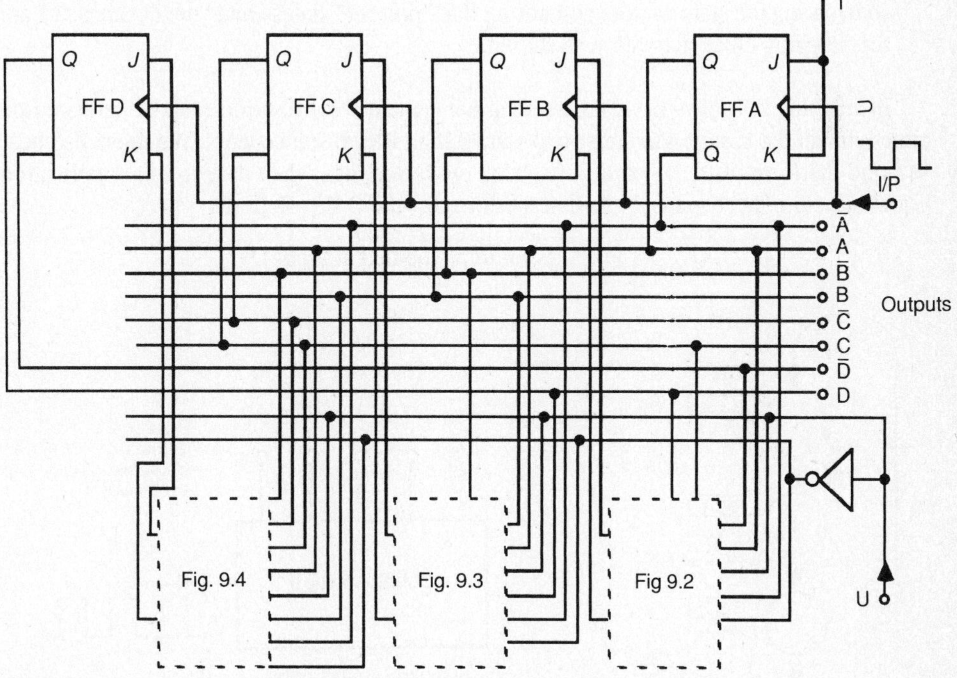

Figure 9.5 Complete arrangement of BCD up/down counter

Step 8 should now be carried out, checking the count sequence and for operation in recovering from illegal states.

9.5.2 Example 2: The design of a base 5 counter for up counts only

The objectives here are to introduce counting to a base other than 10 or 2 and to examine the effect of the choice of alternative flip-flops and of an alternative secondary variable allocation strategy.

Base 5 counter specification and design

Using (a) JK and then (b) D flip-flops, design a base 5 counter for up counting only. Further, let us specify that entering any "illegal state" shall result in the generation of zero as the next count.

The design requirements may be envisaged with reference to Figure 9.6, noting that the five states of the circuit are identified by a binary secondary variable vector CBA as shown and that base 5 outputs 0, 1, 2, 3, 4 are required.

We will design the circuit in two stages.

1. The sequential circuit generating the binary state vectors (assuming that initial reset facilities are available to force state 0 on or following power up).
2. A decoder to convert the binary vectors into the outputs **0**, **1**, **2**, **3**, **4**.

1. *Sequential circuit design*

   From Figure 9.6 (state diagram) we may directly map the next states of each variable by considering the state vectors and noting that "present" states t and "next" states t+1 are separated by clock activation as shown.

In mapping in Table 9.6, we must remember to make 0 entries to correspond with unused states so that the circuit will step on to state 0 if an illegal state occurs. We must also bear in mind the form of the JK and D flip-flop characteristics when deriving the application equations from these maps. Note that the form for the D flip-flop is:

$$Q_{t+1} = [\text{ (expression implemented at input D) }]_t$$

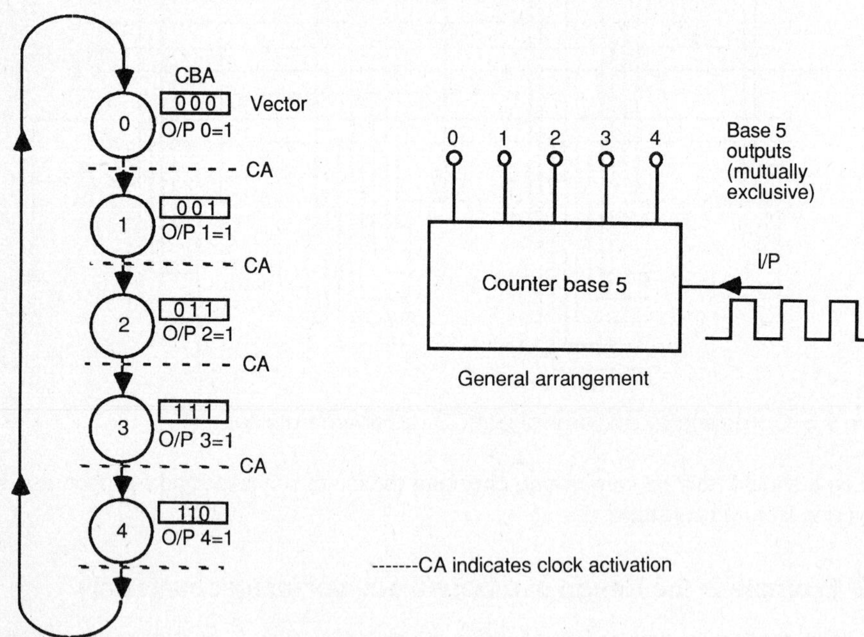

Figure 9.6 General arrangement and state diagram for a base 5 counter

Table 9.6 Maps for each stage

(a)

A_{t+1}	$(BA)_t$ 00	01	11	10
C_t 0	1	1	1	0
1	0	0		

(b)

B_{t+1}	$(BA)_t$ 00	01	11	10
C_t 0		1	1	0
1	0	0	1	

(c)

C_{t+1}	$(BA)_t$ 00	01	11	10
C_t 0			1	0
1	0	0	1	

Input expressions (Table 9.7) may be derived as follows:

Table 9.7 Input expressions for (a) JK and (b) D flip-flops

(a) For the JK flip-flop	*(b) For the D flip-flop*
From Table 9.6(a):	
$A_{t+1} = (\bar{A}(\bar{B}\bar{C}) + A.\bar{C})_t$	$A_{t+1} = \bar{C}(\bar{B} + A)_t$
whence:	whence:
$J_A = \bar{B}\bar{C};\ K_A = C$	$D_A = \bar{C}(\bar{B} + A)$
and from Table 9.6 (b):	
$B_{t+1} = (\bar{B}(\bar{C}A) + BA)_t$	$B_{t+1} = A(\bar{C} + B)_t$
whence:	whence:
$J_B = \bar{C}A;\ K_B = \bar{A}$	$D_B = A(\bar{C} + B)$
and from Table 9.6(c):	
$C_{t+1} = (\bar{C}(BA) + C(BA))_t$	$C_{t+1} = (BA)_t$
whence:	whence:
$J_C = BA;\ K_C = (\bar{B} + \bar{A})$	$D_C = BA$

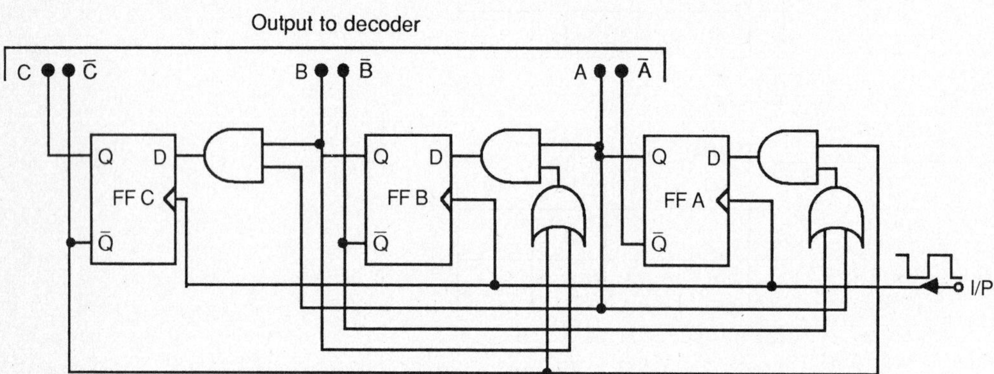

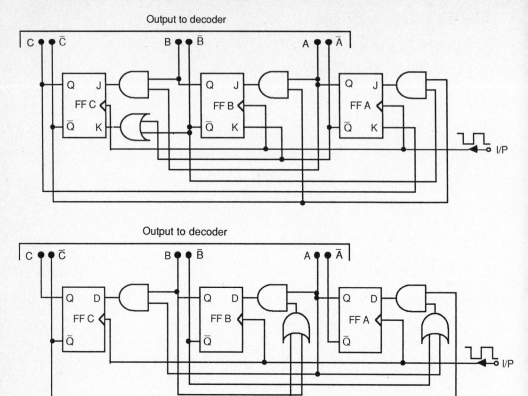

Figure 9.7 (a) JK flip-flop and (b) flip-flop realizations of the sequential logic

The resulting *sequential* circuit arrangements (a) with JK flip-flops and (b) with D flip-flop are given in Figure 9.7.

2. It now remains to *design the decoder* which converts the state vectors *CBA* into individual base 5 outputs **0, 1, 2, 3, 4.** For convenience, all outputs can be mapped on a single map as in Table 9.8.
   Note that the map utilises the "don't care" or illegal states since the sequential circuit design has catered for the return from illegal states.

Table 9.8 Outputs versus vector CBA

Outputs	*BA* 00	01	11	10
C 0	**0**	**1**	**2**	Ø
1	Ø	Ø	**3**	**4**

From Table 9.8:

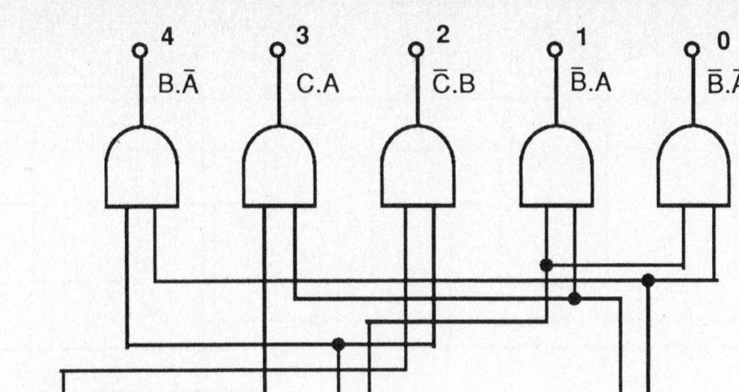

Figure 9.8 Decoder for base 5 counter

$0 = \bar{B}.\bar{A}$ (or alternatively $= \bar{C}.\bar{A}$); $1 = \bar{B}.A$;
$2 = \bar{C}.B$; $3 = C.A$; $4 = B.\bar{A}$ (or alternatively $= C.\bar{A}$).

The decoder circuit to complete the design process is given as Figure 9.8.

Base 5 counter—an alternative allocation (the "one hot" method of design)
The design process used previously results in a two part realization. One part as in Figure 9.7(a) or (b), and the other part the decoder of Figure 9.8.

For a small number of states, as we have here, the following may be an attractive way to approach the design. Referring to the state diagram of Figure 9.6, rather than using three flip-flops to provide a state vector which we then decode, we could allocate one flip-flop per state. The design process would then be such that only the one flip-flop corresponding to the current base 5 count is set to 1 (i.e. only one FF is "hot"). Thus, no decoding would be required, the outputs being directly taken from the Q outputs of the five flip-flops. Both are Moore model circuits.

D flip-flops are well suited to this design and denoting each flip-flop by its associated state number (e.g. FF 0, FF 1 etc. having inputs and outputs D_0, Q_0, D_1, Q_1, etc.) we may obtain the application equations by reasoning as follows:

$$(Q_1)_{t+1} = (Q_0)_t \quad \text{thus,} \quad D_1 = Q_0$$
$$(Q_2)_{t+1} = (Q_1)_t \quad \text{thus,} \quad D_2 = Q_1$$
$$(Q_3)_{t+1} = (Q_2)_t \quad \text{thus,} \quad D_3 = Q_2$$
$$(Q_4)_{t+1} = (Q_3)_t \quad \text{thus,} \quad D_4 = Q_3$$
$$(Q_0)_{t+1} = (Q_4)_t \quad \text{thus,} \quad D_0 = Q_4$$

The very simple interconnections which result are shown in Figure 9.9 and some readers may recognize the "ring counter" configuration in which a single 1 is propagated from stage to stage in an end around fashion. Simple reset facilities are as shown in the figure which set the counter into the state 0 condition initially.

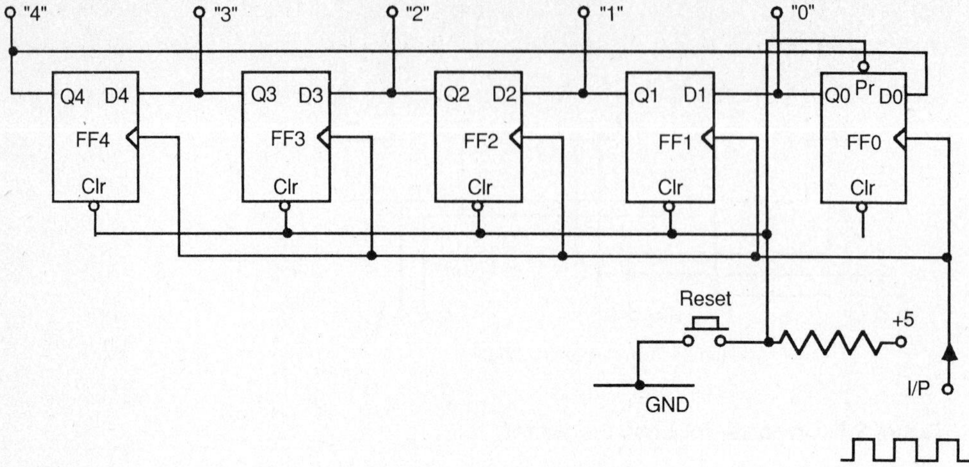

Figure 9.9 Alternative design for example 2 (base 5 counter)

The relative complexities and costs of each approach may be assessed by comparing Figure 9.7(a) or 9.7(b) plus Figure 9.8 with Figure 9.9. For small designs with relatively few states, the "one hot" approach may well be an economical one.

9.5.3 Example 3: A 6-bit serial code detector

Up to now we have concentrated on counter circuits to simply illustrate approaches to design. In order to make the illustration more general, a further sequential circuit design will now be carried through starting from a specification of requirement.

1. *Specification*
A serial code detector is required which will detect any occurrence of the 6-bit sequence 011010 on a single imput line *W*.
→ Time

Bits are clocked on *W* by a clock signal which is available for the detector circuit also. Correct sequences do not overlap, that is, the last 0 of a correct sequence cannot be the first 0 of the next sequence. The circuit is to produce the two outputs: $Y = 1$ (Hi) when the first three or more bits of a correct sequence have been detected; $Z = 1$ (Hi) whenever a complete correct sequence is detected.

Steps 2 - 5
From the specification, a state diagram, Figure 9.10 may be drawn and the corresponding state transition table, Table 9.9, set out accordingly. Secondary variables *ABC* have been allocated to generate the state vector for the six states of the circuit. The allocations made here are arbitrary except for state 1 which has been allocated $ABC = 000$ to allow for easy reset or initialization.

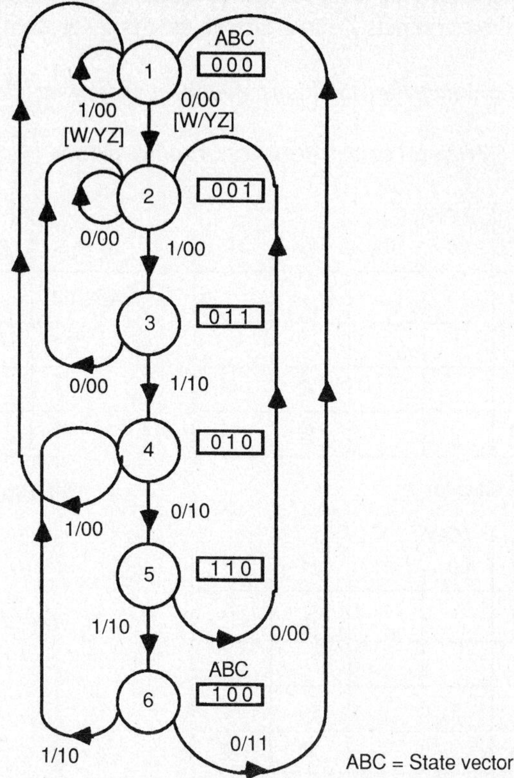

Figure 9.10 State diagram for code detector (example 3)

Table 9.9 State transition table for code detector

State	*Present ABC*	*Input W*	*Next state*	*Next ABC*	*O/P YZ*
①	000 000	0 1	② ①	001 000	00 00
②	001 001	0 1	② ③	001 011	00 00
③	011 011	0 1	② ④	001 010	00 10
④	010 010	0 1	⑤ ①	110 000	10 00
⑤	110 110	0 1	② ⑥	001 100	00 10
⑥	100 100	0 1	① ④	000 010	11 10

In this case it is convenient to deal with the secondary variable (state vector) generation first and then produce outputs Y and Z later in the design process.

6. *We may now map the transitions of each secondary variable separately as in Table 9.10.*

Table 9.10 Maps for secondary variable transitions

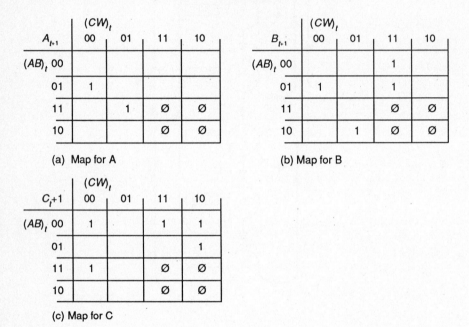

A_{t+1} \ $(CW)_t$	00	01	11	10
$(AB)_t$ 00				
01	1			
11		1	Ø	Ø
10			Ø	Ø

(a) Map for A

B_{t+1} \ $(CW)_t$	00	01	11	10
$(AB)_t$ 00			1	
01	1		1	
11			Ø	Ø
10		1	Ø	Ø

(b) Map for B

C_t+1 \ $(CW)_t$	00	01	11	10
$(AB)_t$ 00	1		1	1
01				1
11	1		Ø	Ø
10			Ø	Ø

(c) Map for C

For the moment we will make free use of the don't care states (due to the two unused combinations of ABC) but we will check later on the results of getting into an unused state and modify the design if necessary.

Assuming JK flip-flop based realization of the circuit, we may write application equations and the consequent J and K input connections to each of flip-flops A, B and C as follows:

$$A_{t+1} = (\bar{A}(B\bar{C}\bar{W}) + A(BW)_t \qquad B_{t+1} = ((\bar{B}(CW + AW) + B(\overline{A + \bar{C}W + C\bar{W}})_t$$

$$J_A = B\bar{C}\bar{W} \qquad K_A = \bar{B} + \bar{W} \qquad J_B = W(A + C) \qquad K_B = A + \bar{C}W + C\bar{W}$$

$$C_{t+1} = (\bar{C}(\bar{A}\bar{B}\bar{W} + AB\bar{W}) + C(\bar{W} + \bar{B}))_t$$

$$J_C = \bar{W}(\bar{A}\bar{B} + AB) \quad K_C = WB.$$

Now, from the application equations check the performance of the circuit if an "illegal" state is assumed. Illegal conditions of ABCW are:

$$ABCW \qquad ABC\bar{W} \qquad A\bar{B}CW \qquad A\bar{B}C\bar{W}$$

From each application equation by substitution of these "illegal" values of $ABCW$ we may check the next state of the secondary variables following the assumption by the circuit of any of the "illegal" states. The prognosis is as follows:

$ABCW$ will be followed by $A\bar{B}\bar{C}$ i.e. state (6)		
$ABC\bar{W}$ " " " " $\bar{A}\bar{B}C$ i.e. state (2)	OK - a legal state is	
$A\bar{B}CW$ " " " " $\bar{A}BC$ i.e. state (3)	assumed in each case	
$A\bar{B}C\bar{W}$ " " " " $\bar{A}\bar{B}C$ i.e. state (2)		

For a start, we can now see that "glitching" or powering up into an illegal state will *not* result in a lock-out situation (in which the circuit remains in illegal states alone). It will be seen that, following the taking up of an illegal state, the circuit will next assume a legal state.If states (2), (3) or (6) are entered and then followed by the rest of a correct sequence, this will generate outputs Y and Z and thus may give a false indication of a correct sequence but correct operation will eventually follow.

If this is not acceptable, the remedy is to go back to Tables 9.10 (a), (b) and (c) and replace the don't care map entries for $ABCW$, $ABC\bar{W}$, $A\bar{B}CW$ and $A\bar{B}C\bar{W}$ with 0s.

In this case the change will affect the application equation for flip-flop A, flip-flop B and flip-flop C and the expressions will be slightly more complex.

7. *Now to generate the outputs Y and Z.*
From Table 9.9 we may map Y and Z in terms of the present state of ABC and W as shown in Table 9.11.

Table 9.11 Map for *Y* and *Z*

Outputs $(AB)_t$ \ $(CW)_t$	00	01	11	10
00				
01	Y		Y	
11		Y	Ø	Ø
10	Y Z	Y	Ø	Ø

$Y = A\bar{B} + AW + BCW + \bar{A}B\bar{C}\bar{W}$

$Z = A\bar{B}\bar{W}$

The complete circuit may then be set out as in Figure 9.11

8. Circuit operation should then be checked by "bread boarding" or simulation.

9.6 Application equation based design for clocked VLSI circuits

In the design of VLSI systems, much emphasis is put on "regularity" and the consequent reduction in detailed design effort.

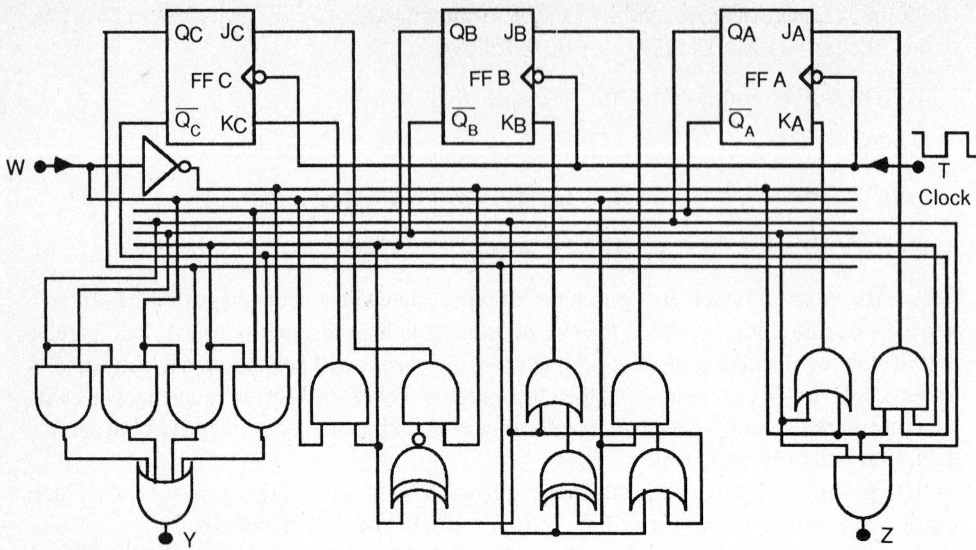

Figure 9.11 Circuit arrangement

In consequence, while it is possible to design sequential circuits around the VLSI realizations of JK or D (etc.) flip-flops using circuit designs such as those we have already developed, it is not always a good approach from the regularity point of view.

A favoured approach is to base our designs on the PLA structure and an applications equation approach is very convenient. Examples will serve to illustrate this point.

9.6.1 A PLA based version of a binary coded decimal (BCD) (8421) synchronous up/down counter (as in section 9.5.1)

We already know the requirements and have earlier derived applications equations in JK FF compatible form. Referring back to the maps of Tables 9.3-9.5, we may derive the application equations in the *simplest SOP form* of expression from the maps as follows:

$$A_{t+1} = \overline{A}_t.$$

$$B_{t+1} = [U\overline{D}\overline{B}A + UB\overline{A} + \overline{U}C\overline{B}\overline{A} + \overline{U}BA + \overline{U}D\overline{B}\overline{A}]_t.$$

$$C_{t+1} = [UC\overline{B} + \overline{U}D\overline{B}\overline{A} + U\overline{C}BA + CB\overline{A} + \overline{U}CA]_t.$$

$$D_{t+1} = [UD\overline{C}\overline{B}\overline{A} + U\overline{D}CBA + \overline{U}\overline{D}\overline{C}\overline{B}\overline{A} + \overline{U}D\overline{C}\overline{B}A]_t.$$

These expressions are directly realizable in a PLA and are of the form:

(next state of output) = (product terms of present state *Or*—ed together)

which maps directly onto the *And* (product) and *Or* planes. For nMOS or CMOS the required PLA will then have the form shown in Figure 9.12.

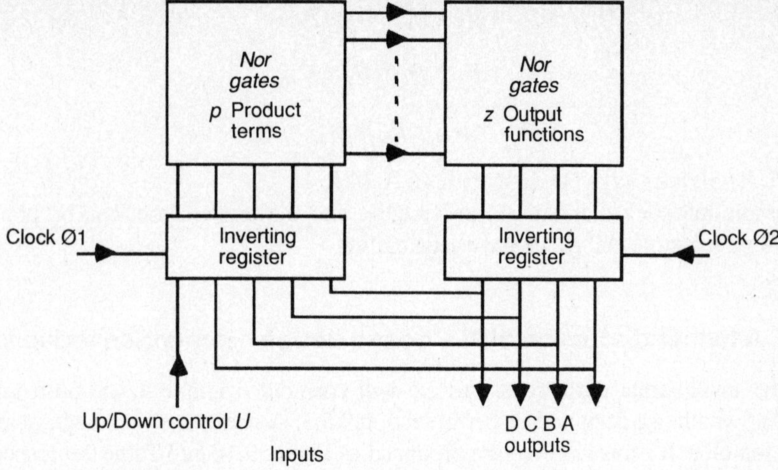

Figure 9.12 PLA based BCD counter

From the applictions equations the product (p_n) terms required are:

$p_1 = \bar{A}$ from expression for A_{t+1}

$p_2 = U\bar{D}\bar{B}A$ from expression for B_{t+1}

$p_3 = UB\bar{A}$ " " " "

$p_4 = \bar{U}C\bar{B}\bar{A}$ " " " "

$p_5 = \bar{U}BA$ " " " "

$p_6 = \bar{U}D\bar{B}\bar{A}$ from expressions for B_{t+1} and C_{t+1}

$p_7 = UC\bar{B}$ " " " C_{t+1}

$p_8 = CB\bar{A}$ " " " "

$p_9 = U\bar{C}BA$ " " " "

$p_{10} = \bar{U}CA$ " " " "

$p_{11} = UD\bar{C}\bar{B}\bar{A}$ from expression for D_{t+1}

$p_{12} = U\bar{D}CBA$ " " " "

$p_{13} = \bar{U}\bar{D}\bar{C}\bar{B}\bar{A}$ " " " "

$p_{14} = \bar{U}D\bar{C}\bar{B}A$ " " " "

Then, for the "*Or*" plane:

$$A_{t+1} = p_1$$

$$B_{t+1} = p_2 + p_3 + p_4 + p_5 + p_6$$

$$C_{t+1} = p_6 + p_7 + p_8 + p_9 + p_{10}$$

$$D_{t+1} = p_{11} + p_{12} + p_{13} + p_{14}$$

The PLA will thus have the dimensions 5 X 14 X 4 .

Circuit and stick diagrams follow readily and the mask layout can be produced by replicating a simple cell as we have seen earlier.

9.6.2 A further discussion of PLA based design—dimension reduction

In order to illustrate our discussions we will consider example 3, the 6-bit serial code detector, which we designed earlier in section 9.5.3. A state transition diagram and a state transition table for this circuit were produced as Figure 9.10 and Table 9.9 respectively.

A PLA can be used to implement this circuit and it is clear that it would have the form indicated in Figure 9.13. We could derive the product terms and output *Or* functions directly from Table 8.9, but first consider if there is a possibility of reducing the PLA dimensions v X p X z which determine the area in silicon.

The dimensions v and z are determined by the number of inputs specified, the number of outputs required and the width (in bits) of the state vector. Provided there are no redundant inputs or outputs, then v and z are minimized if we carry out our state merging process correctly. Thus our main target for reduction must be the dimension p, that is, the number of product terms.

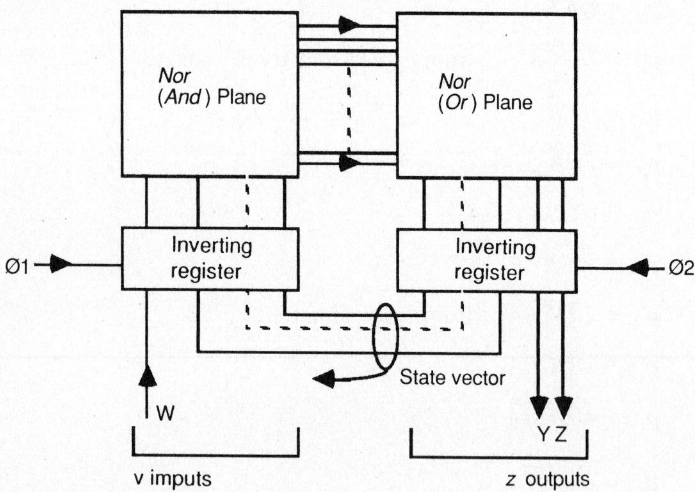

Figure 9.13 State diagram for 6-bit serial code detector

Looking again at Table 9.9 and recognizing that each row in the table constitutes a potential product term to be formed we may further see that there is no need to form a product term in a row where next *ABC* and *YZ* are all 0 since this product term will not appear in any output function. Thus, our task is to maximize the number of rows having next *ABC* and *YZ* all = 0.

To set the scene we can count the number of such rows in Table 9.9 and we will see that there are two. The complete table has 12 rows so that we would have to form 12 - 2 = 10 product terms using the conditions set out in the table.

A useful way of maximizing the number of "all 0" next state/output rows is to delay making state allocations of the state vector until we have examined a skeleton state transition table as in Table 9.12. Note that the state transition diagram, with new secondary variable allocations, is also reproduced as Figure 9.14. Now examine Table 9.12 for all rows in which *YZ* = 00 and determine the most frequently occuring "next state" associated with those rows. In this case state 2 occurs 4 times with *YZ* = 00. If we now allocate the state vector *ABC* = 000 to state 2, we will eliminate the need to form four product terms leaving us 12 - 4 = 8 product terms in this example. This is clearly an improvement on the previous conditions.

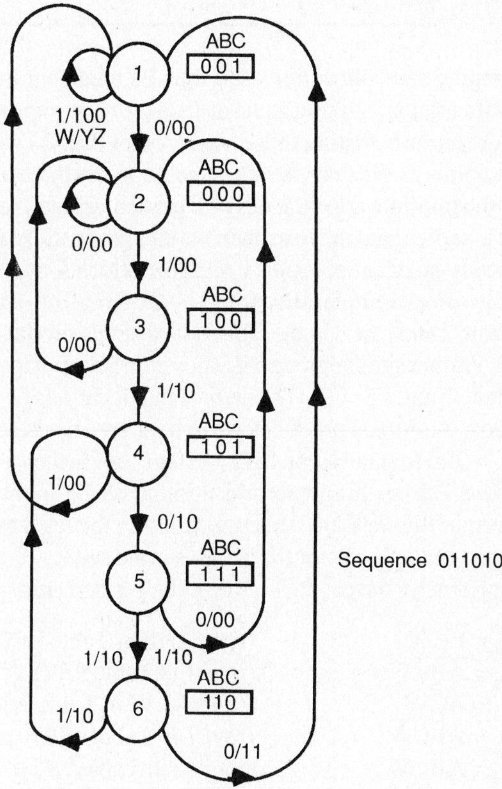

Figure 9.14 State diagram with improved allocations

Table 9.12 Part state transition table

State	Present ABC	Input W	Next state	Next ABC	O/P YZ
①		0 1	② ①		00 00
②		0 1	② ③		00 00
③		0 1	② ④		00 10
④		0 1	⑤ ①		10 00
⑤		0 1	② ⑥		00 10
⑥		0 1	① ④		11 10

The remaining state allocations can then be made but in so doing we should look for further simplifications. Looking again at Table 9.12 we may seek possible further grouping among product terms by listing those states (other than 2) which appear more than once in the next state column. We may see that state 1 appears three times and state 4 appears twice. If for any of the multiple appearances of a particular state, input *W* and output *YZ* have the same values in more than one row, then we can group those rows into one product term by making logically adjacent secondary variable allocations to the present states associated with those rows. For example, allocating say *ABC* = 001 to state 1 and 101 to state 4, we can satisfy both row 2 and row 8 in the table with a single product term $\overline{B}CW$ ($\overline{A}\overline{B}CW$ grouped with $A\overline{B}CW$). Further similar grouping is possible between rows 6 and 12, both of which have *W*=1, next state 4, and *YZ* = 10. Thus adjacent allocations of vector *ABC* to present states 3 and 6 will allow a common product term to be shared by these two rows. For example, if we allocate *ABC* = 100 to state 3 and *ABC* = 110 to state 6, then a single product term $A\overline{C}W$ will serve both rows. The resulting state allocations with the one remaining state being arbitrarily fixed are given in Figure 9.14. This will result in the need to form 8 - 2 = 6 product terms which is a good reduction from the 10 we started with.

For completeness, the product terms to be formed are:

$p_1 = \overline{B}CW$ (rows 2&8 in Table 9.12)
$p_2 = \overline{A}\overline{B}\overline{C}W$ (row 4 in Table 9.12)
$p_3 = A\overline{C}W$ (rows 6&12 in Table 9.12)
$p_4 = A\overline{B}C\overline{W}$ (row 7 in Table 9.12)
$p_5 = ABCW$ (row 10 in Table 9.12)
$p_6 = ABC\overline{W}$ (row 11 in Table 9.12)

You are invited to check these terms by filling out a complete version of Table 9.12 using the revised state allocations. The application equations for each secondary variable *A*, *B* and *C* and for outputs *Y* and *Z* may be written in SOP form from the table entries.

The procedure we have followed may be summarized as follows:

1. Draw up the state diagram, check for and make all possible mergers.
2. Draw up the framework of the state transition table—as in Table 9.12.
3. Identify (e.g. with an asterisk) all rows in which all outputs are = 0. For example, seven rows in table 9.12.
4. Find the most common next state associated with the rows identified in 3. Allocate an all zeroes state vector to this next state. Thus state 2 will be allocated the vector ABC = 000 in the example.
5. Make other appropriate state vector (secondary variable) allocations to the remaining states taking advantage of grouping where input(s), (W), the next state, (vector ABC), and the output(s), (YZ) are common to two or more rows as discussed in the text.

9.7 Analysis of clocked sequential circuits using an application equation based approach

9.7.1 Analysis of JK flip-flop based designs

The procedure for analysis is quite straighforward and is similar to the design procedure but worked through in reverse.

1. From the logic diagram, extract the logic expressions implemented on each J and K input.

2. Remembering that, for each flip-flop generating the sequence (state vector or secondary variable), the application equation is of the form:

   $$Q_{t+1} = (\bar{Q} \text{ (expression for } J) + Q \text{ (expression for } \bar{K}))_t$$

   The application equation for each stage may be derived.

3. Noting the configuration of any "initialization" or "reset" facility, if provided, we may determine one state of the circuit. If not, a valid state may be deduced or determined by trial and error.

4. Using this initial state as the first "present state", generate the next state, then the next state and so on to generate the entire state transition table. The next state of each variable is determined by entering the present state values in the right-hand side of each application equation to evaluate the next condition of that variable.

5. If required, generate the state diagram from the state transition table and/or write the specification in words.

   An example serves to illustrate the process and analysis must start from the logic diagram of the circuit to be analyzed (see Figure 9.15).

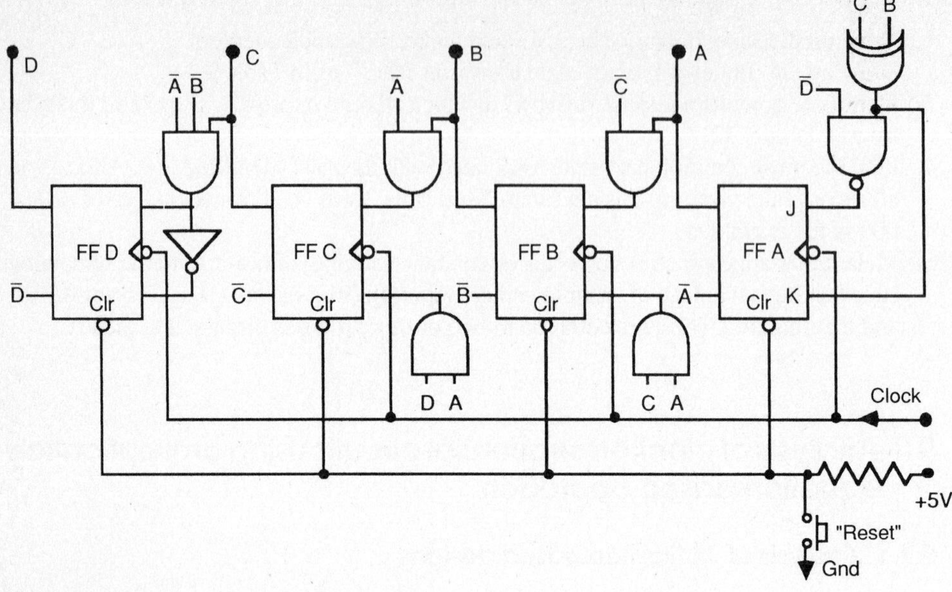

Figure 9.15 Circuit 1 to be analyzed

1. Extract expressions for $\bar{J}$ and $\bar{K}$ which are:

$J_A = D + CB + \bar{C}\bar{B}$ $\quad K_A = C\bar{B} + \bar{C}B$ $\quad \bar{K}_A = CB + \bar{C}\bar{B}$

$J_B = \bar{C}A$ $\quad K_B = CA$ $\quad \bar{K}_B = (\bar{C} + \bar{A})$

$J_C = B\bar{A}$ $\quad K_C = DA$ $\quad \bar{K}_C = (\bar{D} + \bar{A})$

$J_D = C\bar{B}\bar{A}$ $\quad K_D = \overline{C\bar{B}\bar{A}}$ $\quad \bar{K}_D = C\bar{B}\bar{A}$

2. Derive the application equations in the form:

$Q_{t+1} = \{\bar{Q}\ (J \text{ conditions}) + Q\ (\bar{K} \text{ conditions})\}_t$

$A_{t+1} = \{\bar{A}\ (D + CB + \bar{C}\bar{B}) + A\ (CB + \bar{C}\bar{B})\}_t$ $\qquad$ whence $A_{t+1} = (CB + \bar{C}\bar{B} + D\bar{A})_t$

$B_{t+1} = \{\bar{B}\ (\bar{C}A) + B\ (\bar{C} + \bar{A})\}_t$ $\qquad$ whence $B_{t+1} = (B\bar{A} + \bar{C}A)_t$

$C_{t+1} = \{\bar{C}\ (B\bar{A}) + C\ (\bar{D} + \bar{A})\}_t$ $\qquad$ whence $C_{t+1} = (B\bar{A} + C\bar{A} + \bar{D}C)_t$

$D_{t+1} = \{\bar{D}\ (C\bar{B}\bar{A}) + D\ (C\bar{B}\bar{A})\}_t$ $\qquad$ whence $D_{t+1} = (C\bar{B}\bar{A})_t$

3. From the reset facility we may see that a valid state of the circuit is $DCBA = 0000$.

4. State transition table generation. Using the reset condition and the application equations the complete sequence may be deduced as in Table 9.13.

Table 9.13 State transition tables for the example

	D	C	B	A	
Reset condition →	0	0	0	0	Present (t)
	0	0	0	1	Next (t+1)
	0	0	1	1	t
	0	0	1	0	t+1
	0	1	1	0	
	0	1	1	1	t
	0	1	0	1	t+1 etc.
	0	1	0	0	
	1	1	0	0	
	1	1	0	1	
	0	0	0	0	
		↓ Repeat			

Equations used to derive the state transition sequence are:

$$A_{t+1} = (CB + \bar{C}\bar{B} + D\bar{A})_t \qquad (1)$$

$$B_{t+1} = (B\bar{A} + \bar{C}A)_t \qquad (2)$$

$$C_{t+1} = (B\bar{A} + C\bar{A} + \bar{D}C)_t \qquad (3)$$

$$D_{t+1} = (C\bar{B}\bar{A})_t \qquad (4)$$

For example, when $DCBA = 0000$ only **(1)** can be satisfied so that the next condition is $DCBA = 0001$. When $DCBA = 0001$, equations **(1)** and **(2)** are satisfied, so that next $DCBA = 0011$, etc. Successive entries are thus made in the table (Table 9.13) as shown.

5. The specification can readily be determined from Table 9.13 and it will be seen that the circuit is a Gray coded ten position counter circuit with outputs *DCBA*.

   Checks on the illegal (unused) states will reveal that any illegal state will return a value of 0 for D_{t+1} and thus generate a legal next state for this particular design.

9.7.2 Analysis of VLSI (PLA based) designs

A straightforward procedure is again possible and it will be assumed that a mask layout and/or that a stick diagram (for nMOS or CMOS designs) or symbolic diagram is available.

1. Either from the layout and/or stick or symbolic diagram derive a circuit diagram of the finite state machine (unless you are very proficient at reading directly from mask layouts, etc.)

2. From (1) derive the *Or* plane and *And* plane equations in the form:

   Or plane: $Z_n = p_1 + p_k + p_n + —$

   And plane: $p_1 = A\bar{B}C$ —

   $p_k = \bar{A}BC$ —

   etc.

   Note: It may be necessary to allocate variables *ABC*, *XYZ*, etc., to identify state vector inputs and outputs.

3. Write application equations as the overall PLA equations in SOP form:
   e.g. $Z_n = A\bar{B}C \ldots + \bar{A}B\bar{C} + \ldots + \ldots$

4. Find or deduce an intitial condition for the circuit, e.g. the "reset" or "initialized" state.

5. Starting from this initial state and by substitution in the application equations, derive the next state conditions—and hence, the following state, and so on—to generate the state transition table.

6. Hence, if necessary, draw up the state diagram.

7. Deduce the circuit specification from (5) and/or (6).

An example will serve to illustrate the process of analysis. Let us start with an nMOS PLA based circuit for which the stick diagram is set out in Figure 9.16.

1. Derive a circuit diagram from the stick diagram—(This step is optional.)

2. Derive the *Or* plane equations by identifying the product terms as p_1 — p_k etc., as shown and allocating variables to the output and inputs of the PLA as follows: inputs *A*; secondary variables (state vector) *XYZ*; output *B*. The *And* plane product terms may then be read from the stick or circuit diagram. In this case the equations are:

   Or plane:

   $$B_n = p_9$$

   $$X_n = p_1 + p_6 + p_7 + p_8 + p_9 + p_{10} + p_{11}$$

   $$Y_n = p_2 + p_3 + p_5 + p_7 + p_8 + p_9 + p_{10} + p_{11}$$

   $$Z_n = p_1 + p_2 + p_4 + p_9 + p_{10} + p_{11}$$

   And plane:

   $$\bar{p}_1 = A + \bar{X} + \bar{Y} + \bar{Z}$$

   $$\therefore \quad p_1 = \bar{A}XYZ,$$

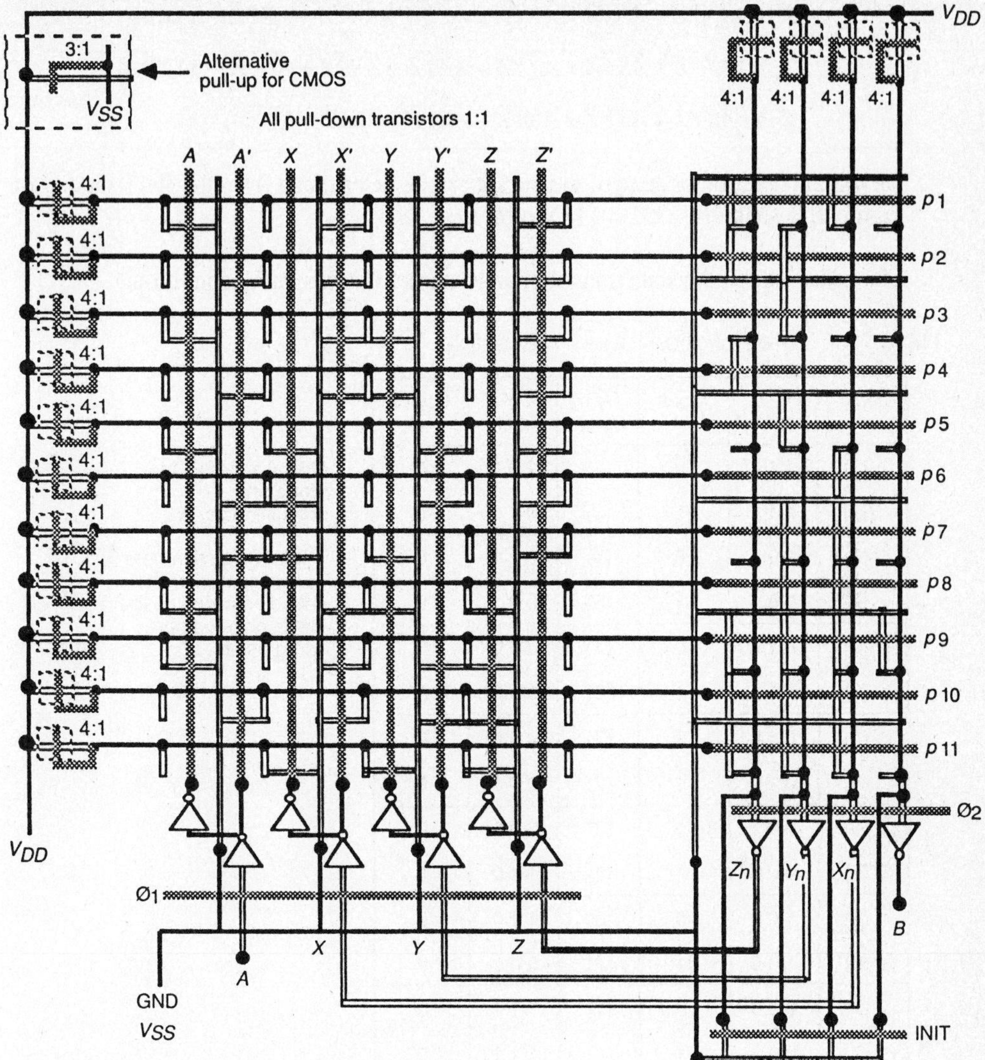

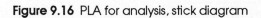

Figure 9.16 PLA for analysis, stick diagram

$$\overline{p}_2 = \overline{A} + \overline{X} + \overline{Y} + \overline{Z}$$

$\therefore \quad p_2 = AXYZ$

similarly $p_3 = \overline{A}X\overline{Y}Z;\; p_4 = AX\overline{Y}Z;\; p_5 = \overline{A}\overline{X}YZ;\; p_6 = A\overline{X}YZ$

$p_7 = A\overline{X}\overline{Y}Z;\; p_8 = \overline{A}X\overline{Y}\overline{Z};\; p_9 = \overline{A}XY\overline{Z};\; p_{10} = AXY\overline{Z};\; p_{11} = \overline{X}\overline{Y}\overline{Z}$

3. The application equations follow by substitution:

$B_n = \overline{A}XY\overline{Z}$ (output function)

$$X_n = (\bar{A}XYZ + A\bar{X}YZ + A\bar{X}\bar{Y}Z + \bar{A}X\bar{Y}\bar{Z} + \bar{A}XY\bar{Z} + AXY\bar{Z} + \bar{X}\bar{Y}\bar{Z})$$

$$Y_n = (AXYZ + \bar{A}X\bar{Y}Z + \bar{A}\bar{X}YZ + A\bar{X}\bar{Y}Z + \bar{A}X\bar{Y}\bar{Z} + \bar{A}XY\bar{Z} + AXY\bar{Z} + \bar{X}\bar{Y}\bar{Z})$$

$$Z_n = (\bar{A}XYZ + AXYZ + AX\bar{Y}Z + \bar{A}XY\bar{Z} + AXY\bar{Z} + \bar{X}\bar{Y}\bar{Z})$$

4. An initial condition or state of the circuit can be determined from the INIT facility of Figure 9.16 such that $XYZ = 111$ initially.

5. The generation of the state transition table starts from this initial condition.

Table 9.14 State transition table for example

State	*X Y Z present*	*Input A*	*Next state*	*X Y Z next*	*Ouput B*
(1)	1 1 1	0	(2)	1 0 1	0
(1)	1 1 1	1	(3)	0 1 1	0
(2)	1 0 1	0	(4)	0 1 0	0
(2)	1 0 1	1	(5)	0 0 1	0
(3)	0 1 1	0	(4)	0 1 0	0
(3)	0 1 1	1	(6)	1 0 0	0
(4)	0 1 0	0	(7)	0 0 0	0
(4)	0 1 0	1	(7)	0 0 0	0
(5)	0 0 1	0	(7)	0 0 0	0
(5)	0 0 1	1	(8)	1 1 0	0
(6)	1 0 0	0	(8)	1 1 0	0
(6)	1 0 0	1	(7)	0 0 0	0
(7)	0 0 0	0	(1)	1 1 1	0
(7)	0 0 0	1	(1)	1 1 1	0
(8)	1 1 0	0	(1)	1 1 1	1
(8)	1 1 0	1	(1)	1 1 1	0

Initial state = 111 for both conditions of A.

Note: State numbers are allocated as the need emerges as the table development progresses.

6. Generate the state diagram from the table as shown in Figure 9.17.
State vector shown thus [101] etc. Conditions on the links from state to state indicate the state of input A during $\emptyset_1$ and the corresponding output B during and following the next $\emptyset_2$ clock signal.

7. Deduce the circuit specification.
From Figure 9.17 it will be seen that the circuit is designed to detect the 4-bit sequences $\underset{\rightarrow \text{Time}}{0110}$ or $\underset{\rightarrow \text{Time}}{1100}$ appearing on a single clocked input line A. If either sequence occurs,

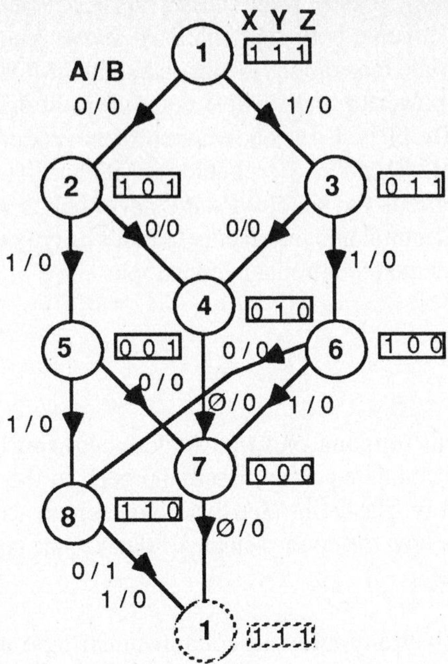

Figure 9.17 State diagram for the example

output *B* will go High for one clock period. If the bit stream on input *A* is interpreted as a series of 4-bit *BCD* numbers (LSB first) then, the circuit will respond to *BCD*6 or *BCD*3 only.

There can be no "illegal states" since all eight combinations of the 3-bit state vector are used.

9.8 Conclusions

It has been shown that the *application equation* approach to design and analysis is logical and straightforward. The processes are well suited to design with a range of clocked flip-flops and to design with PLAs, since an application equation in SOP form directly translates to the *And* and *Or* plane connections of the PLA. The processes of analysis are equally direct. It may be argued that the *application equation* approach is well suited to both systems using packaged logic and to custom design as in VLSI.

There are other clocked sequential system design techniques such as the ASM (algorithmic state machine)* approach and the use of transition equations (TEs) which have not been treated in this text although the second method is outlined in Appendix 2. However, the objectives have been to provide a set of working tools and techniques to serve the designer of both packaged logic and custom VLSI systems. It has also been the intention to teach the design techniques through examples and the examples used have been chosen both to

* Christopher Clare, "Designing logic systems using state machines", McGraw-Hill, New York, 1973.

provide details of useful, widely used subsystems and to compare approaches where appropriate. We have covered both combinational and sequential logic and have related these topics to appropriate realizations in silicon in both nMOS and CMOS technologies.

Perhaps the most powerful and flexible component in digital systems design is the microprocessor (μP). The μP is also a good example of the products of digital systems design and of the application of LSI and VLSI technology. Although it needs a textbook to deal with any one of the many currently available μPs, it is nevertheless worthwhile coming to grips with some of the architectural and interfacing features of typical μPs. The final chapter of this text is therefore devoted to introducing that topic.

9.9 Tutorial 9

1. A 4-bit binary counter (up count only) is to be designed using D flip-flops and *Nand* gates *only*. Use an applications equation based approach to the design and implement an overall "reset" facility. The D flip -flop to be used is characterized by Figure 9.18. Draw waveforms to show how the count sequence relates to the count input clock. Comment on the arrangement.

2. Using JK flip-flops with any necessary combinational logic, use an applications equation based approach to design an up/down synchronous 2421 *BCD* counter. Compare this with the design in section 9.5.1. You are reminded to allow for illegal states and to *note the 2421 stage weighting* (rather than 8421 as in the text).

3. Use an applications equation based approach to design a 4-bit Gray code generator using a CMOS (pseudo-nMOS) PLA. Carry your design through to the stick diagram level but do not attempt a full mask layout. You should, however, suggest the nature of a cell from which the bulk of the PLA could be realized by replication.

4. (a) Set out the characteristic equations for (i) JK and (ii) D flip-flops. Explain the significance of any subscripts you have used.
(b) Using (i) JK and then (ii) D flip-flops, design a *base 6* counter as defined in Figure 9.19. Set out logic diagrams for both and then compare the two.

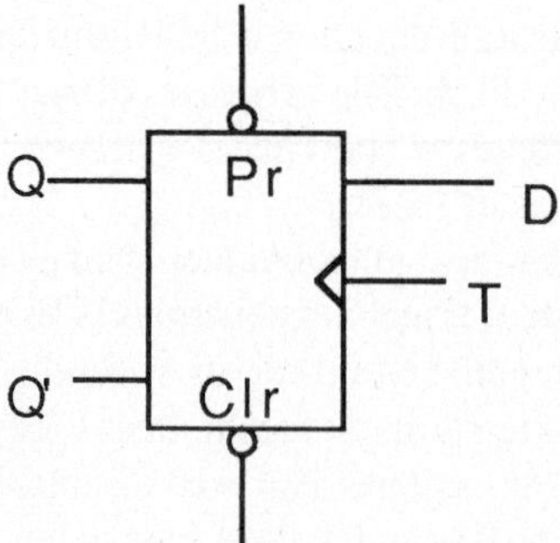

Figure 9.18 D flip-flop for tutorial 9, question 1

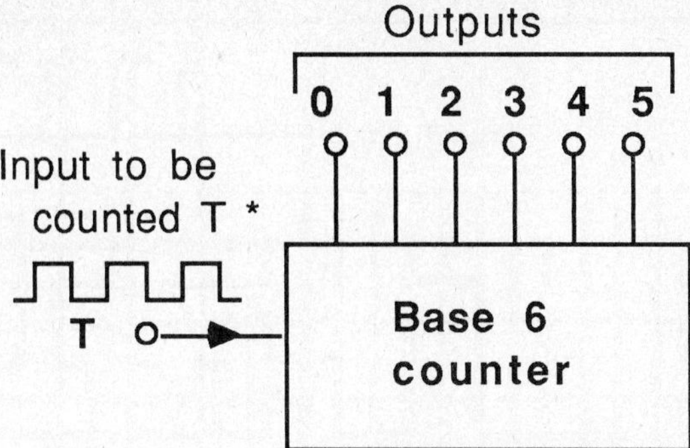

Figure 9.19 General arrangement and state diagram for a base-6 counter

5. (a) Analyze the behavior of the circuit given as Figure 9.20. Set out the state transition diagram and explain the operation of the circuit in words.
   (b) Are there any "illegal states"? If so, what happens if the circuit "glitches" or powers up into any of these states?

6. Analyse the behavior of the PLA based finite state machine set out in stick diagram form as Figure 9.21. Having derived the state transition table, examine the secondary variable allocations and determine whether you can improve on the design by making a better allocation.

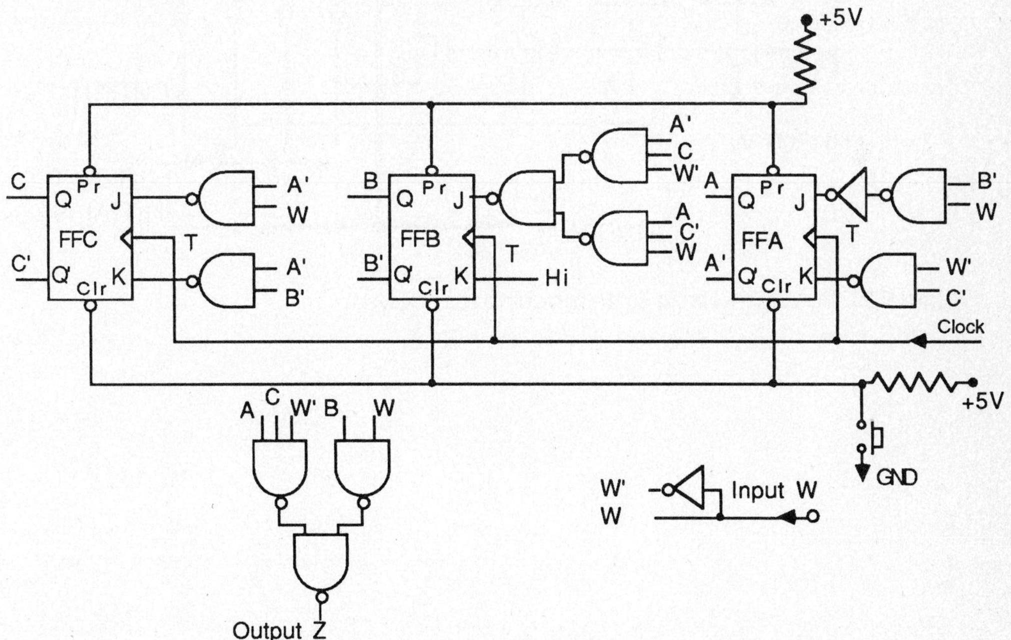

Figure 9.20 Flip-flop based sequential circuit arrangement for analysis

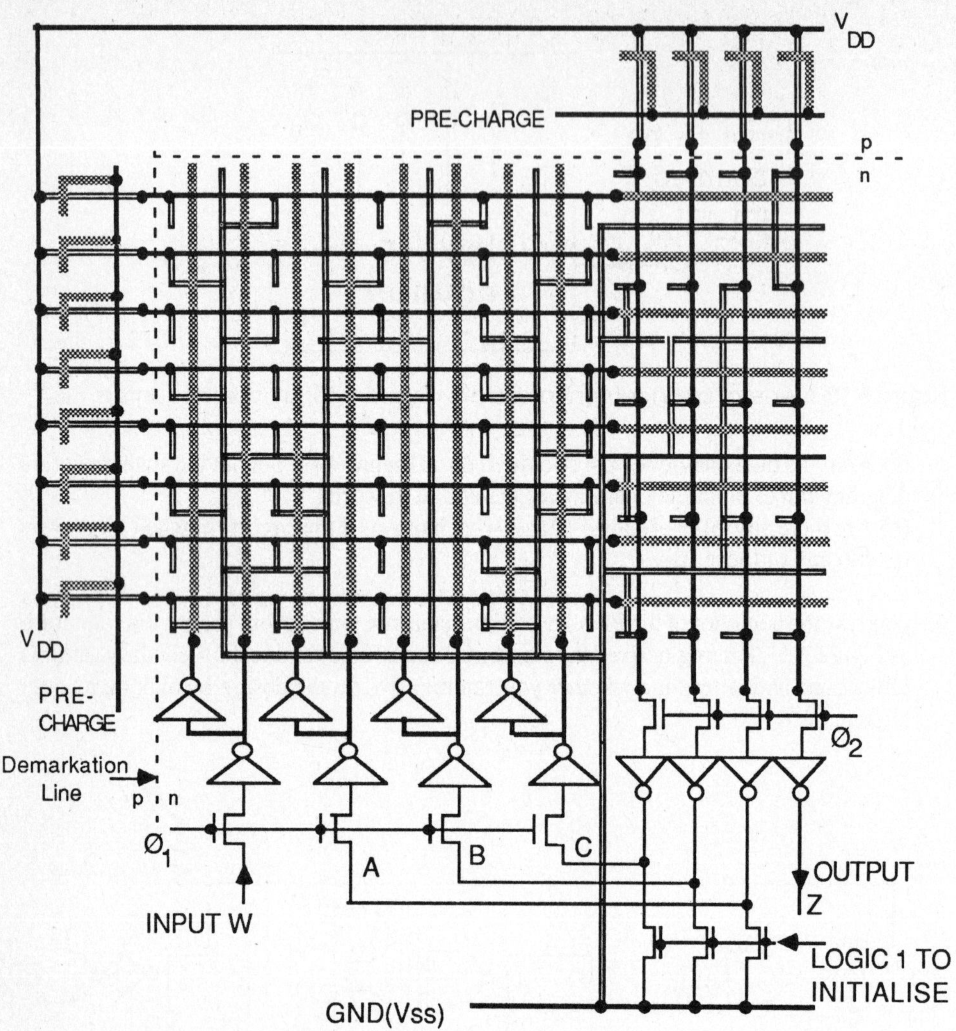

Figure 9.21 PLA based finite state machine for analysis

10 Basic microprocessor architecture and organization with interfacing techniques

10.1 Introduction

Microprocessors are products of both digital system and VLSI (or LSI) design and are also most useful devices to apply in designing digital systems. It is thus appropriate to conclude this text with a brief overview of key aspects of microprocessors. Most people will have been consciously aware of the microprocessor (μP) as part of a microcomputer (μC) or personal computer (PC). Indeed computing as such is a wide area of microcomputer application, but another equally wide area lies in the use of microprocessors, and also microcomputers as subsystems which may be utilized to great advantage in digital system design.

It is useful to distinguish between a microprocessor and a microcomputer and this is readily done with reference to Figure 10.1. The figure sets out the main architectural blocks and general organization of a microcomputer. The processing part of the architecture (lower half of the diagram) is the microprocessor or central processing unit (CPU) and is commonly available as a suitably packaged silicon chip. To form a microcomputer we need to add memory and input/output (I/O) facilities as indicated. This often involves adding other packaged silicon chips although single chip microcomputers are also readily available.

Our main interest in computers in this text has been through a consideration of some of the various digital subsystems, such as registers and adders, which are used in computer architecture. We are now recognizing the microprocessor and the microcomputer in their role as system components or subsystems for the digital designer. Considering the processing element, the microprocessor, we may see that it constitutes a very powerful and flexible component .

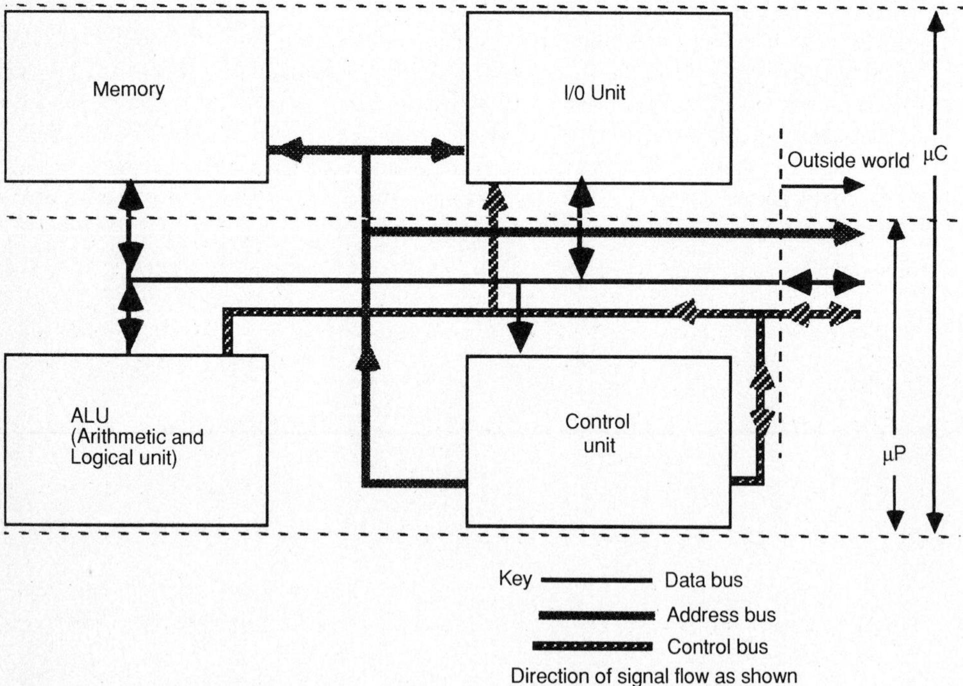

Figure 10.1 General arrangement of a microcomputer

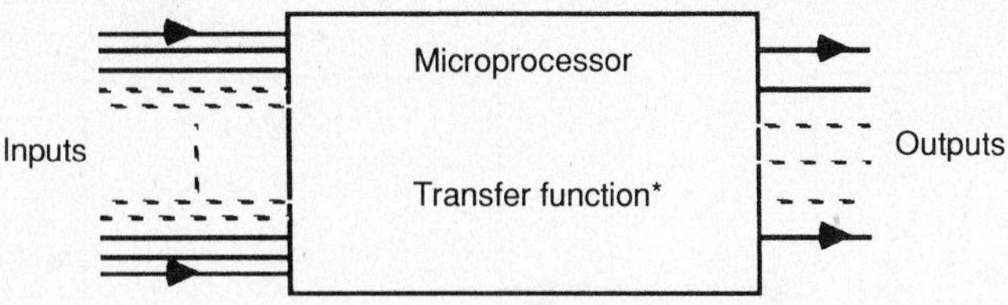

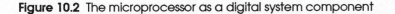

* Determined by program execution sequence

Figure 10.2 The microprocessor as a digital system component

The μP may be regarded as a device which will accept a large number of logic signals as input, process those signals in a way determined by the program which is being executed by the μP, and produce a large number of processed responses to the inputs. It may therefore be regarded as a component with a programmable and dynamically alterable transfer function as suggested in Figure 10.2.

Clearly, it is almost the ideal digital system component—almost, because of the relatively long response time between receiving inputs and producing the corresponding output(s).This is due to the fact that programs are executed in a sequential manner, that is, as a set of instructions executed one after the other. Delays may also be experienced in the input and output regions, particularly if large numbers of signals are being handled by a μP with a small word length—for example, eight bits.

In order to deal with real world situations, and where it is necessary to delve into detail, we will do so by referring to one or other or both of two commonly used microprocessors. For 8-bit architecture we will refer to the Zilog Z80™ μP and for larger 16-bit architecture we will use the Motorola 68000™ as the specific examples. Both are in widespread use and typify the general characteristics of a wide range of devices.

Many readers may have had considerable experience in the use of computers, large and small, but this may well have been through the use of high level languages such as Fortran, Pascal, C, etc. Such languages hide the characteristics of the processor from the user and processor details are therefore of little direct interest. However, in making effective use of the microprocessor as a subsystem, it is essential to have a working knowledge of its architecture in order to employ it effectively. We must also come to terms with its machine language instruction set so that we understand and can determine exactly the steps which the μP will take in performing a given task.

Two characteristics common to the vast majority of digital processors in general and microcomputers in particular is the overall architecture and the use of binary arithmetic. We have already considered the relevant aspects of arithmetic in Chapter 2 so we will now go on to look at architecture and organization in more detail.

10.2 General microcomputer architecture and organization

A convenient way of discussing these topics is to expand the simple diagram given in Figure 10.1 by including more detail, as in Figure 10.3, and briefly discuss each of the main features.

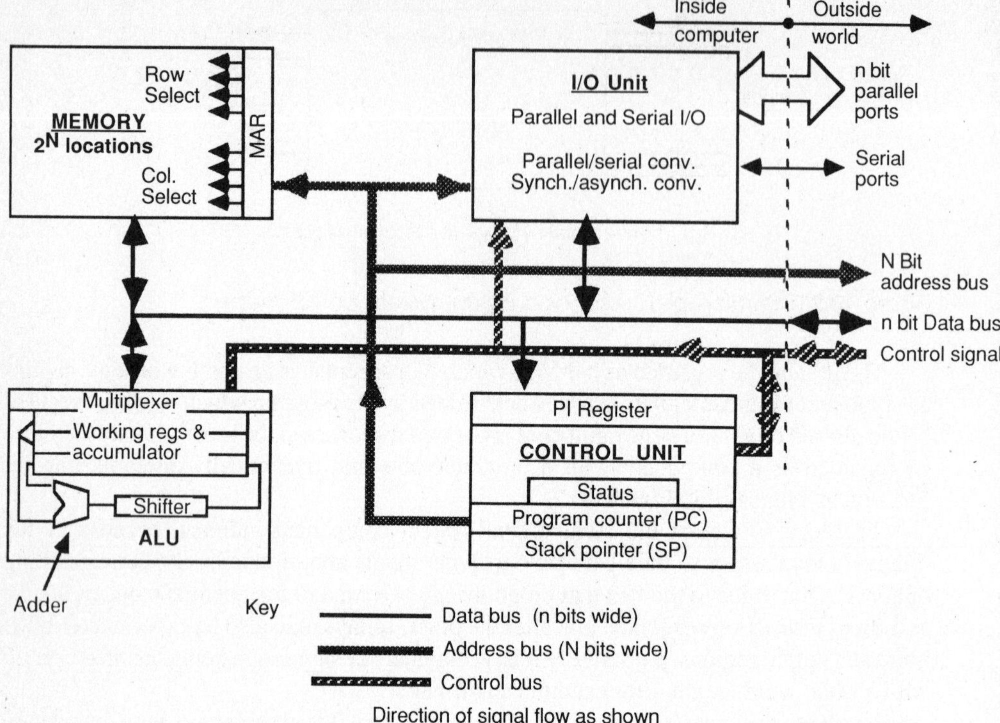

Figure 10.3 A more detailed microprocessor architecture

10.2.1 Main Memory

This is invariably a fast random access memory (RAM), some possible designs for which were discussed in Chapter 7. The number of addressable locations is determined by (or determines) the *number N of address lines*. Commonly, small μP systems, such as the Z80, use $N = 16$ giving $2^{16} = 64K$ addressable locations, (where K in this context implies $2^{10} = 1024_{10}$). Larger μPs use larger values of N and the Motorola 68000 for example has $N = 24$ giving an addressing range of $2^{24} = 16M$, (where $M = K^2 = 2^{20} = 1048576_{10}$).

The next key factor is the *number n of bits* stored in each location and commonly this may be one word as for the Z80, or 1 byte as for the Motorola 68000. In both of these particular cases n happens to be eight bits since the Z80 word length = 1 byte = 8 bits and although the word length is 16 bits, the addressing system of the 68000 is byte oriented. Eight bits is a convenient size since many RAM chips are organized in byte fashion but addressing may still be done as words of 16 or even 32 bits by addressing more than 1 byte at a time.

The address information supplied to memory is decoded in the memory address register (MAR) and used to generate row and column select signals to address individual RAM locations as we have seen earlier.

10.2.2 The I/O facilities

No computer of any sort spends all its time on internal deliberations, so that at least a limited form of interface to the outside world is necessary. When we are using the μP or μC as a component in a system, the I/O facilities are critical in determining the ease with which we can interconnect the system.

Typically, we require *parallel ports* to which we can connect input signals or from which we may take output signals. For complex interconnections, we may need a large number of lines and thus need a significant number of ports, each individually addressable by the processor.

Other transfers may be in *serial* form, for example, those to and from a VDU, for which bits are read in and out on single lines at, typically, anything from 1200 bits/sec to 19600 bits/sec. [The term "Baud rate" is often used in place of bits/sec but really there can be a difference since Baud rate is defined as the number of symbols being transmitted or received per second.] Several serial ports may be required in a typical configuration and the consequent *serial / parallel conversions* are effected and parallel data communication takes place over the data bus.

Clearly, the rate at which various devices external to the μP will operate may well not be synchronous with the μP clock so a further feature of the I/O region is to effect *synchronous / asynchronous translations.*

Yet other transfers may be made directly over the data bus and we will look at ways of establishing interfaces for this purpose.

10.2.3 The arithmetic and logical unit (ALU)

This is the "*number crunching* " part of the architecture and it is here that all arithmetic and all logical operations on operands and data take place. Typical operations include: add, subtract, *And, Or, Excl.Or,* complement, negate, move (or load or copy), compare, increment, decrement, shift, etc.

The heart of the ALU is an *adder* through which the arithmetic and logical operations are performed as we have seen in Chapter 5.

There must be at least one register (*the accumulator*) and usually a number of other *working registers* in which operands, etc., are temporarily held while being worked on. The Z80 has 14 working registers and the 68000 has 17 as we shall see. Although this doesn't seem to be a big difference in architectural resources, the Z80 registers are each eight bits while those of the 68000 are each 32 bits long and only half the Z80 registers can be used at one time.

There will also be a *complementer* and a *shifter* (shift register) associated with the ALU architecture.

10.2.4 The control unit

This is the unit which interprets (decodes) instructions and generally controls other parts of the overall architecture.

Important features here are the *program counter (PC)* and the *present instruction (PI)* register (sometimes called the instruction register (IR)).

The former is of *N*-bit capacity and holds the address of the next instruction to be executed and increments this address after each instruction is fetched—hence the name program *counter*.

A further *N*-bit register which may be present is the *stack pointer (SP)* register which holds the address of a location in memory, identifying the top entry in stack.

The *PI register* is generally one word wide and receives instructions from memory and holds each instruction while it is being decoded to determine the particular operation to be performed. Control signals are then generated and sent to all parts of the computer to effect the desired operation.

10.2.5 The buses

The most general form of organization is to interconnect the four main subsytems, described above, using a 3-bus stucture as shown in Figure 10.3. The three buses are:

1. The *data bus* which is *n* bits (usually one word) wide and is bidirectional over most of its extent as shown. The data bus is the main highway over which all internal inter-unit transfers take place and it also provides for data flow between the processor or computer and peripherals in the "outside world". The bus width *n* and the nature of the bus are critical factors in interfacing with other digital systems.
2. The *address bus* is unidirectional and *N* bits wide. It carries memory and I/O addresses between units of a μC and also supplies I/O address information to the outside world as shown. The I/O addresses may be a subset of the full address bus, for example, the Z80 address bus to memory is 16 bits wide but only half (the least significant 8 bits) are used for I/O addressing. Other processors, such as the 68000, do not separate I/O from memory addressing. The choice of *N* determines the addressing range of the processor.
3. The *control bus* carries control information between the control unit and other units and the outside world. It is often not as obviously bus structured as the other two and may have unidirectional control lines into (e.g. interrupt inputs) as well as out of the control unit. In some processors, some control lines are bidirectional (e.g. reset and halt for the 68000 processor).

A knowledge of the buses is vital to the designer if full use is to be made of the particular μP to be interfaced with.

10.2.6 Key registers

There are certain registers with which the user must directly interact and some others of which the designer needs a knowledge to better understand the processor operation.

1. *Working registers (and accumulator(s))*.

These registers hold operands and data being processed at the time. The user must have a knowledge of their extent and disposition since they are directly addressed through the instruction set. For example, in the Z80 instruction set we could write ADD A, B which will add the contents of 8-bit register B to the 8-bit contents of register A (the accumulator) and return the result to register A. Using the 68000 instruction set, we could write ADD.L D0, D6 which will add all 32 bits of the contents of register D0 to the 32-bit contents of register D6 and return the result to register D6. All 32 bits are involved in this case because we have written ADD.*L* indicating a long word (32 bit) operation.

Other registers may be provided to allow ready manipulation of address information such as accessing tables of data in memory. The Z80 provides two registers, IX and IY, which are used as index registers and the 68000 has eight address registers, A0 to A7, which help to provide the very powerful addressing capabilities of the 68000.

As a user we need to know the register architecture, the way in which sources and destinations for operations are determined and the operand sizes which can be operated on etc.

2. *Status and flag register*

This register holds information on conditions arising within the processor as a result of operations in the ALU and sometimes also indicates the current state of the processor itself. The Z80 has an 8-bit flag register which records, as a result of arithmetic or logical operations, whether a *carry* has been generated, whether *overflow* has occurred, whether the result is *negative,* whether the result is *zero* and the *parity* of the result. Each condition is recorded as a "flag" or setting of a particular bit of the register and flags are set to 1 if the particular condition arises and reset to 0 if not.

The 8-bit flags register of the Z80 can be regarded as being part of the ALU since that is where flag setting operations originate. Operations which merely move or copy operands, such as LD D,L (copy contents of L to D), do not affect the flags which will therefore retain their previous setting.

The 68000 has a larger,16-bit, register known as the status register. Half of this register is devoted to *condition codes* which are the same as the Z80 flags except that parity indication is not given, but an extra flag, *carry extend* (a copy of the carry bit used in extended arithmetic operations), is provided. The other half of the status register records the *mode* (trace or not) and the processing *state* (user or supervisor) and the current interrupt priority level. In the 68000 some condition codes are affected by data movements as well as by arithmetic or logical operations. Since the processor status information as well as condition codes (or flags) appear in the status register, it is conveniently regarded as being part of the control unit as shown.

The user should have knowledge of flag or status register settings.

3. *Program counter (PC) and stack pointer(s) (SP)*

These are registers which hold memory addresses and are therefore 16 bits wide for the Z80 and 24 bits wide (actually, 32-bit registers are used) for the 68000. The *program counter* points to the next instruction word in memory and indexes or counts on to the next instruction word address whenever an instruction word is fetched from memory.

The *stack pointer register(s)* hold an address indicating the current top of stack which is incremented or decremented whenever the stack is used.

For the Z80 there is one 16-bit SP which points to the *lowest* memory address occupied by the stack, this being the "top" entry since the stack grows downwards in memory as entries are made.

The 68000 has two 32-bit address registers devoted to keeping the two stack pointers (USP and SSP) which are maintained for the user and the supervisor stacks respectively. These stacks also grow downwards in memory and the pointers point to the top entry, that is, the lowest address currently occupied by a stack. Other user-generated stacks may be created using other 68000 address registers and may be made to grow upwards or downwards in memory.

In any event, the user must have knowlege of and control of the stack pointer settings so that the stacks are conveniently and safely located.

4. *Other registers*

In general, there are two or three other registers of which the user should be aware. They are the *memory address register (MAR)* which holds the address currently being accessed in memory, the *present instruction (PI) register* which holds the instruction word currently being decoded and, perhaps, a register holding some form of interrupt vector, as in the Z80.

We shall set out the actual register configuration, for each of the two processors being considered, in a later section of this chapter.

10.3 Interfacing with a microprocessor

A very simple interface model of the μP, when used as a component, is set out in Figure 10.4. It will be seen that the μP has two interfaces:

1. A *software interface* which allows programming of the device behavior. This usually results in a set of instructions in memory which the processor will execute in order to behave in the desired way.

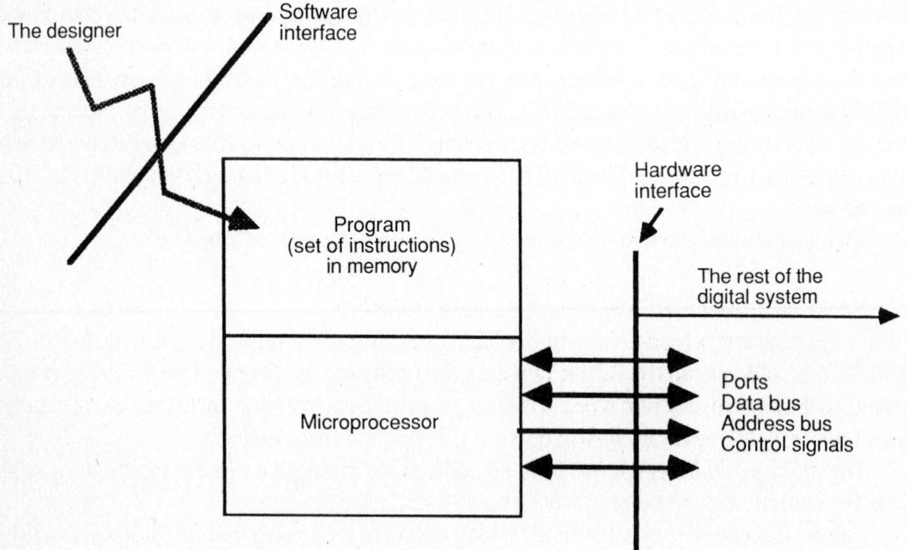

Figure 10.4 Interfacing with a microprocessor

2. A *hardware interface* which provides all the necessary interconnections between the µP and the other parts of the overall digital system.

Both areas are critical for successful integration and operation.

10.3.1 The software interface

The µP may be interfaced with RAM alone or, more commonly, with a combination of both RAM and ROM. If RAM alone is provided, hardware facilities must be designed to allow the desired instructions for the µP to be loaded into the RAM at or following power-up.

ROM alone is not usually practical since the needs of temporary storage of operands and stack operations require a minimum amount of read/write memory (RAM).

In either case, the program must be written, tested and debugged, and then stored in a non-volatile form somewhere accessible by the µP. If the microprocessor is interfaced to sufficient memory, it is possible to temporarily run an assembler program to allow writing and testing of machine language programs "on line". More commonly, the program is written elsewhere, either using the facilities provided by a microcomputer outfit using the same µP, or using the resources of a microcomputer development system (MDS), or using another computer system running suitable cross software (i.e. software on a host machine which produces code for the desired target µP). In all but the first case, high level language software, for example, Pascal, C, etc., may be used to write the non-time-critical sections of the program. However for fast real-time applications machine language software must be written. This could also be the case if a very limited size of memory is to be used in the final system.

In any event, the program must be transferred to the target µP memory after debugging. This is normally done by relocating the desired program in a ROM (EPROM, etc.) and connecting the ROM into the system containing the µP. Alternatives include the "down line loading" of the program after the target system and processor are powered up.

In order to assist in effecting these operations, the µP based system may well have its own elementary "operating system" often referred to as a *monitor program*.

For very small system requirements, the program could be "hard wired" rather than using a ROM, but this is becoming increasingly unusual due to the comparatively low cost of memory.

In any event, the user should have a knowledge of "the programmer's model" of the architecture and of the instruction set and addressing modes for the particular µP to be employed as part of the digital system.

The next subsections illustrate such details for the Z80 and 68000 µPs.

10.3.2 The Z80 programmer's model—registers, instruction set and addressing modes

The registers of interest to the user are set out in Figure 10.5. The six 8-bit working registers (B,C,D,E,H,L) are duplicated as is the 8-bit accumulator (register A) and the 8-bit flag register (F). The changeover from one working register set to the other is achieved by executing an EXX instruction which exchanges the working register set in use for the other set. The register set which is thus dropped out of use will retain its current contents and its

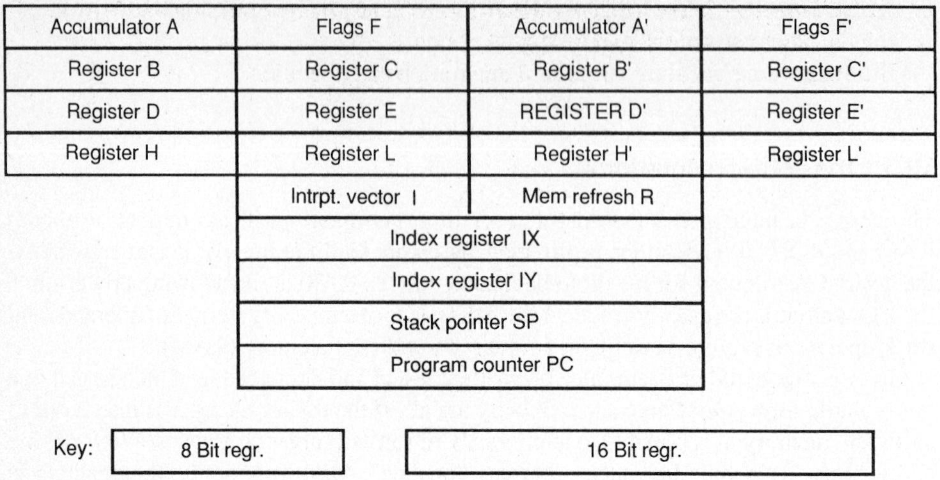

Accumulator A	Flags F	Accumulator A'	Flags F'
Register B	Register C	Register B'	Register C'
Register D	Register E	REGISTER D'	Register E'
Register H	Register L	Register H'	Register L'
	Intrpt. vector I	Mem refresh R	
	Index register IX		
	Index register IY		
	Stack pointer SP		
	Program counter PC		

Key: 8 Bit regr. | 16 Bit regr.

Figure 10.5 Z80 registers – programmer's model

use may be resumed by executing another EXX instruction. The accumulator and flag register as a pair (AF) may be similarly exchanged for an alternative set, this time by executing a EX AF, AF' instruction.

It should be further noted that the working registers may be used in pairs for 16-bit operations. The pairs are: B with C, D with E, and H with L in each register set. The instructions which evoke 16-bit operations clearly include the double register reference, for example, LD BC,09A5H would load the 4-digit hexadecimal (16-bit) number 09A5 into register pair BC. Note the general Z80 format where, if a source and destination are specified, they follow the instruction in the order *destination, source.*

Other "visible" registers are the 16-bit *program counter (PC)* and *stack pointer (SP)* which, as discussed earlier, hold memory addresses representing the location of the next instruction and the top entry of stack respectively.

We also have the use of *two index registers IX and IY* which provide a useful resource in addressing tables or blocks in memory, and may have initial address entered in them using, for example, the LD IX,1234 or LD IY,1234 form of instruction where 1234 represents a typical address.

Finally, the *interrupt vector register (I)* is also of concern but this will be dealt with in a later section on interrupts.

Other aspects of interest to the programmer may be established with reference to the Z80 instruction set which is summarized in Table 10.1 published here by courtesy of Zilog Inc. Instructions may occupy one to four memory locations and fall into types as suggested in the table. An alternative grouping is often used as follows:

Z80 Instruction types

1. Data transfers.
2. Arithmetic.
3. Logical and shift(rotate).
4. Branching (jumps, calls, restarts, returns)

5. Stack.I/O, machine control.
6. Exchange, block transfer, search.
7. Bit manipulation.

Table 10.1 is a summary of the Z80, Z80A instruction set showing the assembly language mnemonic and the symbolic operation performed by the instruction. A more detailed listing appears in the Z80-CPU technical manual, and assembly language programming manual. In the table the instructions are divided into the following categories:

8-bit loads	Miscellaneous Group
16-bit loads	Rotates and shifts
Exchanges	Bit Set, Reset and Test
Memory Block Moves	Input and Output
Memory Block Searches	Jumps
8-bit arithmetic and logic	Calls
16-bit arithmetic	Restarts
General purpose Accumulator & Flag Operations	Returns

In the table the following terminology is used.

b ≡ a bit number in any 8-bit register or memory location
cc ≡ flag condition code
NZ ≡ non zero
Z ≡ zero
NC ≡ non carry
C ≡ carry
PO ≡ Parity odd or no overflow
PE ≡ Parity even or overflow
P ≡ Positive
M ≡ Negative (minus)
d ≡ any 8-bit destination register or memory location
dd ≡ any 16-bit destination register or memory location
e ≡ 8-bit signed 2's complement displacement used in relative jumps and indexed addressing
L ≡ 8 special call locations in page zero. In decimal notation these are 0, 8, 16, 24, 32, 40, 48 and 56
n ≡ any 8-bit binary number
nn ≡ any 16-bit binary number
r ≡ any 8-bit general purpose register (A, B, C, D, E, H, or L)
s ≡ any 8-bit source register or memory location
s_b ≡ a bit in a specific 8-bit register or memory location
ss ≡ any 16-bit source register or memory location
subscritpt "L" ≡ the low order 8 bits of a 16-bit register
subscript "H" ≡ the high order 8 bits of a 16-bit register
( ) ≡ the contents within the ( ) are to be used as a pointer to a memory location or I/O port number

8-bit registers are A, B, C, D, E, H, L, I and R
16-bit register pairs are AF, BC DE and HL
16-bit registers are SP, PC, IX and IY

Addressing Modes implemented include combinations of the following:

Immediate	Indexed
Immediate extended	Register
Modified Page Zero	Implied
Relative	Register Indirect
Extended	Bit

Table 10.1 Z80 CPU, Z80A CPU, Instruction set

	Mnemonic	Symbolic Operation	Comments
8-bit Loads	LD r, s	r ← s	s ≡ r, n, (HL), (IX+e), (IY+e)
	LD d, r	d ← r	d ≡ (HL), r (IX+e), (IY+e)
	LD d, n	d ← n	d ≡ (HL), (IX+e), (IY+e)
	LD A,s	A ← s	s ≡ (BC), (DE), (nn), I, R
	LD d, A	d ← A	d ≡ (BC), (DE), (nn), I, R
16-bit Loads	LD dd, nn	dd ← nn	dd ≡ BC, DE, HL, SP, IX, IY
	LD dd, (nn)	dd ← (nn)	dd ≡ BC, DE, HL, SP, IX, IY
	LD (nn), ss	(nn) ← ss	ss ≡ BC, DE, HL, SP, IX, IY
	LD SP, ss	SP ← ss	ss = HL, IX, IY
	PUSH ss	(SP-1) ← ss_H; (SP-2) ← ss_L	ss = BC, DE HL, AF, IX, IY
	POP dd	dd_L ← (SP); dd_H ← (SP+1)	dd = BC, DE, HL, AF, IX IY
Exchanges	EX DE, HL	DE ↔ HL	
	EX AF, AF'	AF ↔ AF'	
	EXX	(BC, DE, HL) ↔ (BC', DE', HL')	
	EX (SP), ss	(SP) ↔ ss_L, ((SP+1) ↔ ss_H	ss ≡ HL, IX, IY
Memory Block Moves	LDI	(DE) ← (HL), DE ← DE+1 HL ← HL+1, BC ← BC-1	
	LDIR	(DE) ← (HL), DE ← DE+1 HL ← HL+1, BC ← BC-1 Repeat until BC = 0	
	LDD	(DE) ← (HL), DE ← DE-1 HL ← HL-1, BC ← BC-1	
	LDDR	(DE) ← (HL), DE ← DE -1 HL ← HL - 1, BC ← BC-1 Repeat until BC = 0	

Group	Mnemonic	Symbolic Operation	Comments
Memory Block Searches	CPI	A-(HL), HL ← HL+1 BC ← BC-1	
	CPIR	A-(HL), HL ← HL+1 BC ← BC-1, Repeat until BC = 0 or A = (HL)	A-(HL) sets the flags only. A is not affected
	CPD	A-(HL), HL ← HL-1 BC ← BC-1	
	CPDR	A-(HL), HL ← HL-1 BC ← BC-1, Repeat until BC = 0 or A = (HL)	
8-bit ALU	ADD s	A ← A + s	
	ADC s	A ← A + s + CY	CY is the carry flag
	SUB s	A ← A - s	
	SBC s	A ← A - s - CY	s ≡ r, n, (LH)
	AND s	A ← A . s	(IX+e), (IY+e)
	OR s	A ← A+s	
	XOR s	A ← A ⊕ s	
	CP s	A - s	s = r, n (HL) (IX+e), (IY+e)
	INC d	d ← d + 1 `	
	DEC d	d ← d - 1	d = r, (HL) (IX+e), (IY+e)
16-bit Arithmetic	ADD HL, ss	HL ← HL + ss	
	ADC HL, ss	HL ← HL + ss + CY	} ss ≡ BC, DE HL, SP
	SBC HL, ss	HL ← HL - ss - CY	
	ADD IX, ss	IX ← IX + ss	ss ≡ BC, DE, IX, SP
	ADD IY, ss	IY ← IY + ss	ss ≡ BC, DE, IY, SP
	INC dd	dd ← dd + 1	dd ≡ BC, DE, HL, SP, IX, IY
	DEC dd	dd ← dd - 1	dd ≡ BC, DE, HL, SP, IX, IY
GP,ACC : FLAG	DAA	Converts A contents into packed BCD following add or subtract	Operands must be in packed BCD format
	CPL	A ← $\overline{A}$	
	NEg	A ← 00 - A	
	CCF	CY ← $\overline{CY}$	
	SCF	CY ← 1	
Miscellaneous	NOP	No operation	
	HALT	Halt CPU	
	DI	Disable Interrupts	
	EI	Enable Interrupts	
	IM 0	Set interrupt mode 0	8080A mode
	IM 1	Set interrupt mode 1	Call to 0038_H
	IM 2	Set interrupt mode 2	Indirect Call

	Mnemonic	Symbolic Operation	Comments
Rotators and Shifts	RLC s	CY ← 7 ← 0 (s), bit 7 → bit 0	s r, (HL) (IX+e), (IY+e)
	RL s	CY ← 7 ← 0 (s), CY → bit 0	
	RRC s	7 → 0 (s) → CY, bit 0 → bit 7	
	RR s	7 → 0 (s) → CY, CY → bit 7	
	SLA s	CY ← 7 ← 0 (s) ← 0	
	SRA s	7 → 0 (s) → CY, bit 7 retained	
	SRL s	0 → 7 → 0 (s) → CY	
	RLD	A: 7 4|3 0; (HL): 7 4|3 0	
	RRD	A: 7 4|3 0; (HL): 7 4|3 0	
Bit	Bit b, s SET b,s RES b, s	$Z \leftarrow \overline{s_b}$ $s_b \leftarrow 1$ $s_b \leftarrow 0$	Z is zero flag s r, (HL) (IX+e), (IY+e)
Input and Output	IN A, (n)	A ← (n)	
	IN r, (C)	r ← (C)	Set flags
	INI	(HL) ← (C), HL ← HL + 1 B ← B - 1	
	INIR	(HL) ← (C), HL ← HL + 1 B ← B - 1 Repeat until B = 0	
	IND	(HL) ← (C), HL ← HL - 1 B ← B - 1	
	INDR	(HL) ← (C), HL ← HL - 1 B ← B - 1 Repeat until B = 0	
	OUT(n), A	(n) ← A	
	OUT(C), r	(C) ← r	
	OUTI	(C) ← (HL), HL ← HL + 1 B ← B -1	
	OUTIR	(C) ← (HL), HL ← HL + 1 B ← B -1 Repeat until B = 0	
	OUTD	(C) ← (HL), HL ← HL - 1 B ← B -1	
	OTDR	(C) ← (HL), HL ← HL - 1 B ← B -1 Repeat until B = 0	

	Mnemonic	Symbolic Operation	Comments
Jumps	JP nn JP cc, nn JR e	PC ← nn If condition cc is true PC ← nn, else continue PC ← PC + e	cc { NZ PO, Z PE, NC P, C M
	JR kk, e	If condition kk is true PC ← PC + e, else continue	kk { NZ NC, Z C
	JP (ss) DJNZ e	PC ← ss B ← B - 1, if B = 0 continue, else PC ← PC + e	ss = HL, IX, IY
Calls	CALL nn Call cc, nn	(SP -1) ← PC_H (SP-2) ← PC_L, PC ← nn If condition cc is false continue, else same as CALL nn	cc { NZ PO, Z PE, NC P, C M
Restarts	RST L	(SP -1) ← PC_H (SP-2) ← PC_L, PC_H ← 0 PC_L ← L	
Returns	RET RET cc	PC_L ← (SP), PC_H ← (SP+1) If condition cc is false continue, else same as RET	cc { NZ PO, Z PE, NC P, C M
	RETI	Return from interrupt, same as RET	
	RETN	Return from non- maskable interrupt	

Reproduced with permission of Zilog Inc. This material shall not be reproduced without the written consent of Zilog Inc.

Input and output operations are normally effected through special IN and OUT instructions. I/O operations may take place through whichever accumulator is currently in use by writing IN A, (*n*) or OUT (*n*), A where *n* is an 8-bit I/O address. In the first case, an 8-bit input is taken from the peripheral device at address *n* and placed in the accumulator A. For the corresponding OUT instruction, the 8-bit contents of A are sent to the peripheral device at address *n*.

Input and output operations may also be carried out using any working register or the accumulator. If, say, we wish to use register *L*, we may do so by writing IN L, (C) or OUT (C), L where the desired I/O address is assumed to be the contents of register C.

Repetitive (or block) IN or OUT operations are also provided by single instructions INIR or INDR, OTIR or OTDR, where register pair HL have been previously loaded with a base address in memory for the data to be input or output, and register B loaded with an 8-bit number which will be counted down to zero to determine the number of INs or OUTs to be executed. Data is taken to or from memory locations starting at (HL), indexing up (INIR, OTIR) or down (INDR, OTDR) as each I/O operation is completed.

An important factor in the software interface is the range of addressing modes available through the instruction set. The *addressing modes* are the various ways in which the location of source and/or destination of operands, or the next instruction address, may be carried by instructions. The modes available in the Z80 are listed in Table 10.1 and a brief illustration of each may help convey the concepts involved. *In this discussion all numbers are assumed to be hexadecimal unless otherwise stated* although many Assemblers would require an H suffix for a hex number.

1. *Immediate addressing*: In this mode the 8-bit operand follows in the location immediately after the instruction in memory, for example: LD E, 2C. In this case the source operand is 2C (hex) and this will follow the instruction code in memory. As a result of executing this instruction, the 8-bit number 2C will be loaded into register E.
2. *Extended immediate addressing*: In this mode a 16-bit operand follows in the two (8-bit) locations immediately following the instruction in memory. The 16-bit number is held in two memory addresses, least significant byte first in the lower address, for example: LD DE, 123C. In this case the source operand is 123C and this will follow the instruction code in memory as two bytes, 3C then 12. As a result of executing this instruction, the 16-bit number 123C will be loaded into register pair DE.
3. *Modified page zero*: In this mode, which is confined in the Z80 to restart instructions, a page zero (lowest 256 locations in memory) address is carried by the numerical code associated with each of the eight possible restart instructions. The numerical code is supplied as bits 3,4,5 of the instruction word. The result is used as a page zero address, for example the execution of an RST 28 instruction will cause the next instruction to be taken from location 00101000 $=40_{10}$(= 28H). At the same time, since RST instructions act like calls to subroutines, a return address (the current contents of the program counter PC) will be placed on the stack.
4. *Relative addressing:* This is an important feature which allows for relocatable code to be written by making loop addresses independent of absolute addresses in memory. Rather than specifying an actual address, an 8-bit twos complement number is given as the second word of an instruction, this acting as a +/- displacement from the current PC setting. For example, consider this short program:

Address	*Instruction mnemonic*	*Machine code*	*Remarks*
1000	LD B,24	06	Load 24H into B.
1001		24	
1002	LD A,B	78	Copy (B) to A
1003	OUT (01),A	D3	Output (A) to
1004		01	I/O addr. 01.
1005	DEC B	05	(B) -1 to B.
1006	JR NZ,-6	20	Jump -6 relative to
1007		FA	PC after this instr.
			if (B) $\neq$ 0.
1008	PC setting after executing JR instruction.		

Note: (i) (B) indicates the contents of B, etc., [All addresses, codes and numbers are Hex.]

This program executes a loop 24H times and on each pass through the loop it will output (B), to the I/O device at I/O address 01 then decrement (B) and test for a non-zero result. The JR instruction could be replaced by JP NZ,1002 which, similarly, tells the processor to jump to location 1002 (absolute address) unless the zero flag is set, but if this was done the program would then only run in the actual memory addresses given above. Using the PC relative form of addressing, the program could be relocated in, say, locations 2000 - 2007 and would still run correctly.

5. *Direct absolute (sometimes called absolute extended) addressing*: In this mode the actual(16-bit) address in memory is carried as part of the instruction. For example: LD H, (1234) will copy the contents of memory address 1234H to the register H.
6. *Indexed addressing:* This mode makes use of the contents of index register IX or IY as a base address and adds a signed displacement to this base address in order to arrive at the address to be used. For example: LD (IX - 24), A will copy the contents of register A to the memory address given by (IX)-24H.
7. *Register addressing*: In this mode the register to be used as source and/or destination is specified in the instruction. For example: ADD HL, BC will add the 16-bit contents of register pair BC to the contents of register pair HL returning the sum to register pair HL.
8. *Implied addressing*: In this mode no address information is specified, this being implied in the instruction itself. For example: CPL this is the complement instruction which will complement (A). Register A is not actually specified in the instruction.
9. *Register indirect*: In this mode the address given in an instruction is *the address of the address* to be used. For example: LD (HL), C will use the 16-bit number contained in register pair HL as the address in memory to which it will copy (C), that is, the 8-bit contents of register C.
10. *Bit addressing*: This mode allows testing, setting or resetting of any nominated bit (0 to7) of an 8-bit word in a register or memory location. For example: SET 2, (HL) will set bit D2 to "1" in the memory location addressed by the 16-bit word contained in register pair HL.

Clearly, the foregoing discussion is only an overview but it is hoped that the essential features and concepts have been at least touched on. The user should at all times consult the manufacturer's data when using or applying a microprocessor.

10.3.3 The 68000 programmer's model—registers, instruction set and addressing modes

The registers of interest to the user are set out in Figure 10.6. There are *8 data registers,* D_0 *to* D_7, which may be used as the source or destination for data operations on bytes (8 bits), words (16 bits) or long words (32 bits). Data is taken from or stored in register bits 0 to 7, or bits 0 to 15, or all thirty-two bits 0 to 31 respectively, as shown. Data registers may also be used as index registers.

Certain instructions will also operate on individual bits of bytes or words in data registers or memory locations.

The register array also includes *9 Address registers,* A_0 *to* A_7 *and* A_7'. A_7 and A_7' are normally used as user and supervisor mode stack pointers respectively, and the remaining seven address registers may be used as base address or further stack pointers or as index registers. Their use will become clearer when we examine addressing modes.

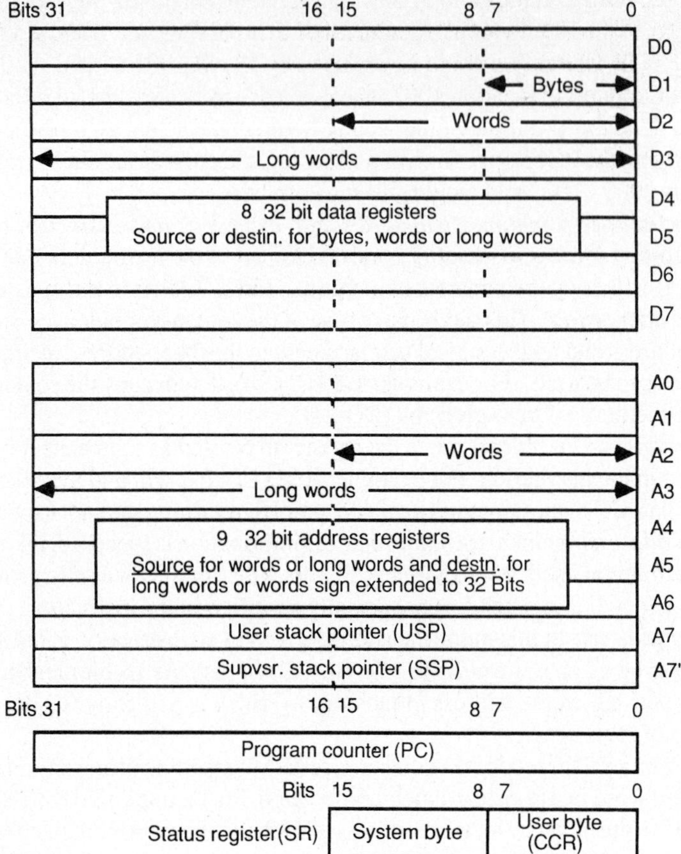

Figure 10.6 68000 registers - programmer's model

There is also a *program counter (PC)* of 32 bits but only twenty-four bits are in fact used. In relating this to the address bus remember that bit A_0, as such, is not used since the strobes UDS' and LDS' are used to address even and odd bytes within each 16-bit word. Word addresses are, therefore, always even so that PC_0 is always taken as 0. The PC indexes by 2 each time an instruction word is fetched. The PC contents may be set to access any given address by executing a JMP instruction.

Finally there is a *16-bit status register* which is subdivided into two bytes as shown.

The upper (MS) byte is the *system byte* which has two bits, S and T indicating the processor mode of operation—(trace or not) and state (user or supervisor). Three further bits, I_0, I_1, I_2, indicate the current interrupt priority level.

The lower byte is the user byte or CCR (condition code register) and contains condition code bits (or flags) X,N,Z,V,C as shown. All unused bits of the status register may be taken as 0 .

A further point of interest is the *data structure in memory*. This may be determined with reference to bytes, words and long words as shown in Figure 10.7.

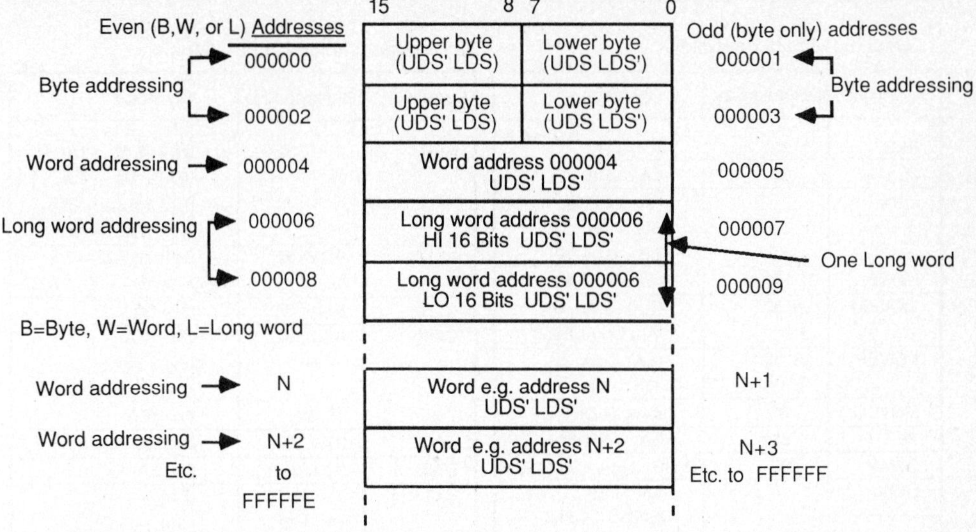

Figure 10.7 68000 data structure in memory

Note that *words and long words must always be addressed at even addresses.*

The instruction set of the 68000 is complex and powerful. Instructions may occupy one to five words in memory. An overview may be obtained from the summary, by courtesy of Motorola, given as Table 10.2.

It will be seen that the instructions may be classified as follows:

1. Data movement.
2. Integer arithmetic.
3. Logical.
4. Shift and rotate.
5. Bit manipulation.
6. Binary control.
7. Program control.
8. System control.

A notable difference from the Z80 is that I/O is dealt with in the same way as any other data movement (it is "memory mapped"). To properly examine the instruction set the reader is referred to the reference manual.* One reason for the power and success of the 68000 is the range of *addressing modes* available. They may be broadly classified as follows:

1. *Register direct* (three types): data register, address register and status register.
2. *Immediate* (two types): next word and quick.
3. *Register indirect* (five types): address register indirect (ARI), ARI with post-increment, ARI with pre-decrement, ARI with displacement (d), ARI with index and displacement (d).
4. *Absolute* (two types): short address and long address.
5. *Program counter* (two types): with displacement (d), and, with index and displacement (d).
6. *Relative mode* (for branching) (two types): 8-bit offset, and, 16-bit offset.

*Motorola, *M68000 16/32 bit microprocessor-Programmer's reference manual*, Prentice Hall.

Table 10.2 68000 Instruction set summary

1. Data movement operations

Instruction	Operand Size	Operation
EXG	32	Rx ↔ Ry
LEA	32	EA → An
LINK	-	An → SP @ - SP → An SP + d → SP
MOVE	8, 16, 32	(EA)s → EAd
MOVEM	16, 32	(EA) → An, Dn An, Dn → EA
MOVEP	16, 32	(EA) → Dn Dn → EA
MOVEQ	8	#xxx → Dn
PEA	32	EA → SP @ -
SWAP	32	Dn (31:16) ↔ Dn (15:0)
UNLK	-	AN → SP SP@ + → AN

Notes:

s = source

d = destination

[ ] = bit numbers

2. Integer arithmetic operations

Instruction	Operand Size	Operation
ADD	8, 16, 32 16, 32	Dn + (EA) → Dn (EA) + Dn → EA (EA) + #xxx → EA An + (EA) → An
ADDX	8, 16, 32 16, 32	Dx + Dy+ X → Dx Ax@ - Ay@ - + X → Ax@
CLR	8, 16, 32	0 → EA
CMP	8, 16, 32 16, 32	Dn - (EA) (EA) - #xxx Ax@ + - Ay@ + An - (EA)
DIVS	32 + 16	Dn/(EA) → Dn
DIVU	32 + 16	Dn/(EA) → Dn
EXT	8 → 16 16 → 32	$(Dn)_8 \rightarrow Dn_{16}$ $(Dn)_{16} \rightarrow Dn_{32}$
MULS	16*16 →32	Dn*(EA) → 32 Dn
MULU	16*16 →32	Dn*(EA) → Dn
NEG	8, 16, 32	0 - (EA) → EA
NEGX	8, 16, 32	0 - (EA) - X → EA
SUB	8, 16, 32 16, 32	Dn - (EA) → Dn (EA) - Dn → EA (EA) - #xxx → EA An - (EA) → An
SUBX	8, 16, 32	Dx - Dy - X → Dx Ax@ - - Ay@ - - X → Ax@
TAS	8	(EA) - 0, 1 → EA (7)
TST	8, 16, 32	(EA) - 0

Note:

[ ] = bit number

3. Logical operations

Instruction	Operand Size	Operation
AND	8, 16, 32	Dn ∧ (EA) → Dn (EA) ∧ Dn → EA (EA) ∧ #xxx → EA
OR	8, 16, 32	Dn ∨ (EA) → Dn (EA) ∨ Dn → EA (EA) ∨ #xxx → EA
EOR	8, 16, 32	(EA) ⊕ Dy → EA (EA) ⊕ #xxx → EA
NOT	8, 16, 32	~(EA) → EA

Note:

~ = invert; ∧ = and; ∨ = or.

4. Shift and rotate operations

Instruction	Operand Size	Operation
ASL	8, 16, 32	X/C ← [←] ← 0
ASR	8, 16, 32	[→] → X/C
LSL	8, 16, 32	X/C ← [←] ← 0
LSR	8, 16, 32	0 → [→] → X/C
ROL	8, 16, 32	C ← [←]
ROR	8, 16, 32	[→] → C
ROXL	8, 16, 32	C ← [←] ← X
ROXR	8, 16, 32	X → [→] → C

5. Bit manipulation operations

Instruction	Operand Size	Operation
BTST	8, 32	~bit of (EA) → Z
BSET	8, 32	~bit of (EA) → Z 1 → bit of EA
BCLR	8, 32	~bit of (EA) → Z 0 → bit of EA
BCHG	8, 32	~bit of (EA) → Z ~bit of (EA) → bit of EA

6. Binary coded decimal operations

Instruction	Operand Size	Operation
ABCD	8	$Dx_{10} + Dy_{10} + X \rightarrow Dx$ $Ax@ -_{10} + Ay@ -_{10} + X \rightarrow Ax@$
SBCD	8	$Dy_{10} - Dy_{10} - X \rightarrow Dx$ $Ax@ -_{10} + Ay@ -_{10} - X \rightarrow Ax@$
NBCD	8	$0 - (EA)_{10} - X \rightarrow EA$

7. Program control operations

Instruction	Operation
Conditional	
Bcc	Branch conditionally (14 conditions), 8- and 16-bit displacement
DBcc	Test condition, decrement and branch, 16-bit displacement
Scc	Set byte conditionally (16 conditions)
Unconditional	
BRA	Branch always, 8- and 16-bit displacement
BSR	Branch to subroutine, 8 and 16-bit displacement
JMP	Jump
JSR	Jump to subroutine
Returns	
RTR	Return to restore conditon codes
RTS	Return from subroutine

8. System control operations

Instruction	Operation
Privileged	
ANDI to SR	Logical AND to status register
EORI to SR	Logical EOR to status register
MOVE EA toSR	Load new status register
MOVE USP	Move user stack pointer
ORI to SR	Logical OR to status register
RESET	Reset external devices
RTE	Return from exception
STOP	Stop program execution
Trap Generating	
CHK	Check register against bounds
TRAP	Trap
TRAPV	Trap on overflow
Status Register	
ANDI to CCR	Logical AND to condition codes
EORI to CCR	Logical EOR to condition codes
MOVE EA to CCR	Load new condition codes
MOVE SR to EA	Store status register
ORI to CCR	Logical OR to condition codes

The conditional instructions provide setting and branching for:

CC - Carry Clear	LS - Low or Same
CS - Carry Set	LT - Less Than
EQ - Equal	MI - Minus
F - Never True	NE - Not Equal
GE - Greater or Equal	PL - Plus
GT - Greater Than	T - Always True
HI - High	VC - No Overflow
LE - Less or Equal	VS - Overflow

Let us now utilize each of the main addressing modes. This serves the double purpose of illustrating concepts and conveying aspects of the instruction set. A convenient class of instruction for our examples is the MOVE family of instructions. The general format is as follows:

MOVE.X <ea>,<ea>

where : MOVE is the operation;
.X indicates the data length - .B for byte, .W for word, .L for long word - default is to word length operation.
<ea> is the effective address calculated in the appropriate mode.

Where two addresses are given they are in the order *source, destination*. Data registers may be used as either the source or destination for bytes, words and long words the remaining bits being unaffected in byte and word operations. Address registers may be used as a source for words or long words and as a destination for long words *and for words which will be sign extended to 32 bits.*

1. *Register direct addressing*
   Examples: (Note: $ implies Hex; # indicates immediate data).

MOVE.B D0, D2	COPY BITS 0 TO 7 OF REGISTER D0 TO BITS 0 TO 7 OF REGISTER D2.
MOVE D1, $201234	COPY BITS 0 TO 15 OF REGISTER D0 TO MEMORY WORD ADDRESS 201234
MOVEA.L D0, A5	COPY ALL 32 BITS OF DATA REGISTER 0 TO ADDRESS REGISTER 5. [NOTE USE OF MOVEA WHEN MOVE IS *TO* AN ADDRESS REGISTER].
MOVE.L A0, USP	COPY ALL 32 BITS OF REGISTER A0 TO USER STACK POINTER.
MOVE SR, D4	COPY 16 BIT STATUS REGISTER TO BITS 0 TO 15 OF REGISTER D4 *
MOVE.B $1234, D2	COPY UPPER *BYTE* OF WORD ADDRESS 001234 TO BITS 0 TO 7 OF D2.
MOVE.W A0, D2	COPY BITS 0 TO 15 OF REGISTER A0 TO BITS 0 TO 15 OF REGISTER D2.
MOVEA.W D0, A2	COPY BITS 0 TO 15 OF REGISTER D0 *SIGN EXTENDED TO 32 BITS* TO A2.
MOVE #$4321, SR	WRITE IMMEDIATE DATA HEX 4321 TO STATUS REGISTER *
	* PRIVILEGED INSTRUCTION - SUPERVISER STATE ONLY.

2. *Immediate data addressing* (two types-immediate and immediate quick).
   Examples: (Note: "immediate" can only be source)

   (a) *Immediate addressing*

MOVEA #$ 8567,A0	MOVE 8567 SIGN EXTENDED TO LONG WORD FFFF8567 INTO A0.
MOVE #$ 8567,D0	MOVE WORD 8567 INTO BITS 0 TO 15 OF D0.

   (b) *Immediate quick addressing:*
   IMMEDIATE QUICK IS LIMITED TO 8-BIT DATA CARRIED IN THE INSTRUCTION WORD ITSELF. THIS 8-BIT DATA IS SIGN EXTENDED TO 32 BITS.

MOVEQ #$07,D4	MOVE BYTE 07 SIGN EXTENDED TO LONG WORD [00000007] INTO D4.

3. *Register indirect addressing* (Note: (An) signifies contents of address register *n*).
   There are five types as follows:

   (a) *Address register indirect* for which <*ea*> = (An)

   Examples:

MOVE.L (A3),D0	LONG WORD IN MEMORY ADDRESSES (A3) AND (A3)+2 TO DO.
MOVE.L (A3),$1234	LONG WORD IN MEMORY ADDRESSES (A3) AND (A3)+2 TO MEMORY ADDRESSES 001234 AND 001236.

MOVE D0,(A3) — BITS 0 TO 15 OF D0 TO MEMORY WORD ADDRESS (A3).

(b) *Address register indirect with post-increment* for which <*ea*> = (An) and (An) is then incremented by 1 for .B, 2 for .W, and 4 for .L operations.

Examples:

MOVE (A3)+,D0 — WORD IN MEMORY ADDRESSES (A3) TO D0. (A3)+2 TO A3

MOVE.B D0,(A4)+ — BYTE IN D0 BITS 0 TO 7 TO MEMORY *BYTE* ADDRESS (A4).
(A4)+1 TO A4.

(c) *Address register indirect with pre-decrement* for which <*ea*> = (An) pre-decremented by 1 for .B, 2 for .W, and 4 for .L operations.

Examples:

MOVE.L -(A3),D0 — (A3) - 4 to A3.
LONG WORD IN MEMORY ADDRESSES (A3)-4 AND (A3)-2 TO D0.

MOVE.B D0,-(A4) — (A4) -1 to A4.
BYTE IN D0 BITS 0 TO 7 TO MEMORY *BYTE* ADDRESS (A4) -1.

(d) *Address register indirect with displacement* for which <*ea*> = (An) added to a 16-bit *twos compl. displacement d.*

Examples:

MOVE $FF00(A3),D0 — WORD IN MEMORY ADDRESSES [(A3)-256] TO BITS 0-15 OF D0.

MOVE.L D0,$100(A4) — LONG WORD IN D0 TO MEMORY ADDRESSES [(A4)+256] AND [(A4)+258].

(e) *Address register indirect with index (and displacement)* for which <*ea*> = (An) added to (Rn) and also to 16-bit *twos compl. displacement d.* [where Rn is any data or address register which may be specified as word Rn.W or as long word Rn.L. Both Rn.W and d will be sign extended to 32 bits].

Examples: (All numbers are assumed to be hex.)

MOVE $FF00(A3,D1.L),D0 — WORD IN MEMORY ADDRESS<ea> (as below) TO BITS 0-15 OF D0.
If (A3)=00001234 and (D1)=00001000 then <ea>=00001234+00001000+FFFFFF00=00002134.
i.e. the source address will be $002134.

MOVE D0,$100(A3,A2.W) — WORD IN D0 TO MEMORY ADDRESS [<ea>] where, for (A3) as above and (A2, bits 0-15) = 8642, <ea>=00001234+FFFF8642+00000100=FFFF9976
i.e. the destination address will be $FF9976.

4. *Absolute addressing.*
There are two types as follows:

(a) *Absolute short addressing* in which a 16-bit address with the instruction is used to specify the effective address by *sign extension to 24 bits.*

Examples:

MOVE.L $2000,D0	LONG WORD FROM MEMORY LOCATIONS 002000 AND 002002 T0 D0.
MOVE D0,$8000	WORD FROM D0(0-15) TO LOCATION *FF8000* (FF DUE TO SIGN EXTENSION)

Note effect of sign extension—normally do not use short addresses >$7FFF.

(b) *Absolute long addressing* in which a 24-bit address is supplied with the instruction.

Examples:

MOVE.L $002000,D0	LONG WORD FROM MEMORY LOCATIONS 002000 AND 002002 T0 D0.
MOVE D0,$008000	WORD FROM D0(0-15) TO LOCATION *008000* .

5. *Program counter (PC) addressing.*
There are two types as follows:

(a) *PC with displacement (source only for MOVE).*

Example:

MOVE <LABEL1>,D0	THIS IS A TWO WORD INSTRUCTION, THE FIRST WORD CARRIES THE OPERATION CODE DETAILS WHILE THE SECOND WORD CARRIES A 16-BIT DISPLACEMENT WHICH IS THE DIFFERENCE BETWEEN THE CURRENT (PC), I.E. THE ADDRESS OF THE SECOND WORD, AND THE MEMORY LOCATION IDENTIFIED BY "LABEL1". THIS DISPLACEMENT IS "ADDED" TO THE CURRENT (PC) TO DETERMINE *<ea>* (source here).

(b) *PC with index (source only for MOVE).*

Example:

MOVE <LABEL2> (A3), D0	THIS IS A TWO WORD INSTRUCTION, THE FIRST WORD CARRIES OPERATION CODE DETAILS WHILST THE SECOND WORD IDENTIFIES THE INDEX REGISTER AND CARRIES AN 8-BIT DISPLACEMENT WHICH IS THE DIFFERENCE BETWEEN CURRENT (PC), I.E. THE ADDRESS OF THE SECOND WORD, AND THE MEMORY LOCATION IDENTIFIED BY "LABEL2". THIS DISPLACEMENT AND INDEX REGISTER CONTENTS IS "ADDED" TO THE CURRENT (PC) TO DETERMINE <ea> (source here).

6. *Branching relative to (PC):*
   There are two types- 8-bit offset and 16-bit offset.
   (a) *8-bit offset.* One word only, the offset being carried in the 8 LSB of the Op. word

   Example:
   BEQ <LABEL3> IF "Z" FLAG IS SET THEN PROCESSOR WILL BRANCH TO LABEL3. THE 8-BIT DISPLACEMENT IS THE DIFFERENCE BETWEEN CURRENT (PC) AND LABEL 3. ALLOWABLE DISPLACEMENT RANGE IS FROM +127 TO -128.

   (b) *16 bit offset:*
   Two words, the offset being carried in the next word after the Op. word.

   Example:
   BCC <LABEL4> IF "C" FLAG IS CLEAR THEN PROCESSOR WILL BRANCH TO LABEL4. THE 16-BIT DISPLACEMENT IS THE DIFFERENCE BETWEEN CURRENT (PC) AND LABEL4. ALLOWABLE DISPLACEMENT RANGE IS FROM $+2^{15}-1$ TO -2^{15} i.e.+/-32K approx.

Overall the 68000 has a powerful instruction set and a wide range of addressing modes. It is an example of a complex instruction set computer (CISC) and many would argue that its strength lies in these attributes. However, there is a growing trend towards reduced instruction set computers (RISCs) and a few words at this point will establish some of the concepts of basic RISC architecture.

10.3.4 Reduced instruction set computer (RISC) concepts

The overall organization and architecture has the same general form as that indicated in Figure 10.1, but the concepts used in designing the architectural subsystems are differently accentuated from the µPs we have so far considered.

Both the Z80 and the 68000 processors have powerful instruction sets and allow many different modes for addressing operands and memory. This implies a number of different ways of interpreting instruction words (a range of different instruction formats) and consequent complexity in decoding instructions. Such machines fall into the category of complex instruction set computers (CISCs), and, in consequence of the factors discussed, some individual instructions may take a number of machine cycles to decode and execute. A wide range of instructions are allowed to access memory which is also a comparatively slow process. Against this slowing down of instruction execution time, it may be argued that individual instructions can be powerful and perform the task of several instructions of a simpler set. Also, due to instruction variety and complexity, it is usual for the instructions of such machines to be microcoded, that is, to be built up of a number of individual (simple) micro-operations. This again is slower than having instructions hardwired, or directly decoded in logic.

An emerging school of thought is directed to the merits of computers with small and simple instruction sets. In consequence of simple instructions, each instruction executes rapidly. Although this inevitably means longer programs, it is argued that this is less important than the benefits of simple decoding and streamlined execution of each instruction.

RISC goals* may be seen as:

1. The execution of all instructions within one machine cycle.
2. All instructions should be the same size, (not varying in size from one to four words as in the Z80 or from one to five words for the 68000).
3. Accesses to memory to be strictly limited, for example, to simple LOAD and STORE type instructions.
4. Chosen instruction sets should readily support high level languages.

A consequence of these concepts is that arithmetic and logical and all other operations except LOAD and STORE are carried out on a register to register basis. Thus, a trend in RISC machines is to provide large register banks which generally allows all current operands to be kept out of memory.

A further consequence is that very few types of instruction are available, for example the RISC1 machine developed at the University of California at Berkeley implements just four types:

- Arithmetic and logical.
- Memory access.
- Branching.
- Miscellaneous.

This particular machine has a 32-bit word length and only 31 instructions so that the instruction format (and decoding) is straightforward as follows:

OP CODE(7)	SCC(1)	DEST(5)	SRCE1(5)	IMM(1)	SRCE2(13)

The key to the fields indicated for this instruction format is as follows:

Op Code—(self-explanatory)—7 bits allowed giving 128 possible Op codes;
SCC—set condition codes;
Dest and Srcel—two 5-bit addresses allowing for register selection as source and destination;
Imm—allows alternative interpretations of the Immediate 13-bit data following as Srce2.

The 13-bit Srce2 vector could be used, for example, to act as a signed displacement to a 32-bit address obtained from the contents of any one of the registers. Alternatively, a 5-bit subset of Srce2 could be used to address a register, allowing the instruction word to address three registers, representing, say, operand 1, operand 2 and the destination for the result of an arithmetic or logical operation. Such an instruction word format is known as a *three address format* for obvious reasons.

Such machines have proved fast in comparison with other contemporary microcomputers and minicomputers in certain situations, but have tended to perform rather poorly when carrying out floating point arithmetic operations.

However, RISC architecture may well perform effectively when interfaced with certain

* David A. Patterson, Carlo H. Sequin, "A VLSI RISC", *Computer*, Sept. 1982, pp. 8-18.

processes or in some instrumentation applications and it is certainly worth while establishing and maintaining a reasonable level of awareness in this area. Some references follow to allow further investigation.

Dennis A. Fairclough, "A unique microprocessor instruction set," *IEEE Micro.*, vol. 2, No.2, May 1982, pp. 8-19.

David A. Patterson, Carlo H. Sequin, "A VLSI RISC," *Computer*, Sept.1982, pp. 8-18.

David A. Patterson, Richard S. Piepho, "Assessing RISCs in high level language support" *IEEE Micro*, vol 2, No.4, Nov. 1982, pp. 9-19.

David A. Patterson, "Reduced instruction set computers", *Communications of the ACM*, vol. 28, No.1, Jan.1985, pp. 8-21.

Robert P. Colwell, et al., "Computers, Complexity and Controversy", *Computer*, Sep. 1985, pp. 8-20.

Kevin J. McNeley, Veljko M. Milutinovic, "Emulating a Complex Instruction Set Computer with a Reduced Instruction Set Computer", *IEEE Micro*, vol. 7, No.1, Feb. 1987, pp. 60-72.

Paul Chow, Mark Horowitz, "Architectural tradeoffs in the design of MIPS-X", *IEEE Computer Architecture News*, vol. 15, No.2, June 1987, pp. 300-8.

Charles E. Gimarc, Veljko M. Milutinovic, "A survey of RISC processors and computers of the mid-1980s", *Computer*, vol. 20, No. 9, Sept. 1987, pp. 59-84.

Barbara A. Naused, Barry K. Gilbert, "A 32-Bit, 200-MHz GaAs RISC", *IEEE Micro*, vol. 7, No. 6, Dec. 1987, pp. 8-20.

Borioje Furht, "A RISC architecture with two-size, overlapping register windows", *IEEE Micro*, vol. 8, No. 2, Apr. 1988, pp. 67-80.

10.3.5 The Hardware Interface

General hardware considerations

Referring back to the simple interfacing model given as Figure 10.4, we may see that the second interface region which must be considered is the "hardware interface". This links the microprocessor to the other digital hardware forming a system and, thus, designs can be less than effective if requirements are not properly understood or considered. It is here that we wish to derive signals from or send them into the μP. This requires coupling directly or indirectly to the *data bus* of the microprocessor in a way which does not interfere with the operation of the μP and also takes account of the synchronous behavior of the μP. A straightforward way of doing this is to treat the digital system as an addressable peripheral I/O device to the μP. This will allow us to send signals into and derive signals from the data bus whenever the μP program addresses the particular I/O address(es) allocated to the digital system. Use of the *interrupt facilities* will further allow the digital system to call the μP's attention to particular needs. Thus, the *data bus, address bus* and a minimum set of *control bus signals* will be required in the interface arrangements. Typically, a microprocessor will provide the general facilities indicated in Figure 10.8. It will be seen that an n-bit bidirectional data bus and the full address bus (or a subset for I/O addressing) are required, respectively, to allow data transmission to/from and selection of the digital system by the μP. In some cases,where I/O and memory are treated differently, we will need a control signal (e.g MEM/I/O') to indicate whether I/O or memory is being addressed. Two further mutually

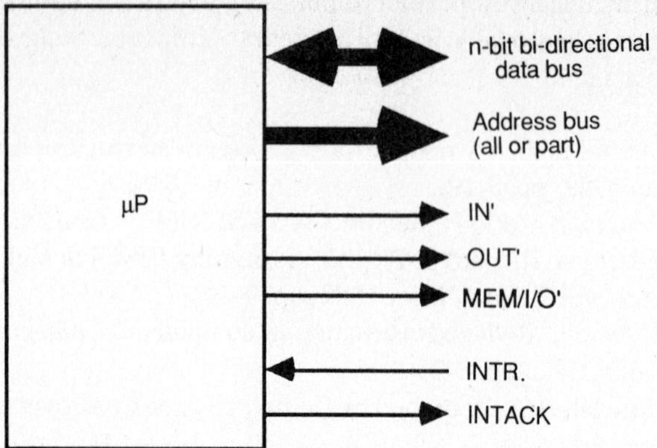

Figure 10.8 Typical interface signals

exclusive signals (which may be pulses and/or may be active Low as indicated) may be provided to indicate an IN (or READ) or OUT(or WRITE) operation and the timing of that operation.

Finally there will also be an interrupt input (INTR) to the μP and an interrupt acknowledge (INTACK) from the μP.

Leaving interrupts aside, there are three basic I/O interface requirements which must be met as follows:

1. *I/O Device selection.* Any given I/O device may be called on by the μP which puts the appropriate address momentarily on the address bus (having first set the MEM/I/O' signal to Low (0)). Input devices are distinguished from output devices by detecting a pulse on one or other of the IN' (READ') or OUT' (WRITE') control lines. Hardware must be configured to carry out the necessary *decoding* of all these signals in selecting the required I/O device.
2. *Buffering of input signals to the data bus.* Since the data bus is the only way of communicating with the μP(and its registers and memory), there may be a large number of different I/O devices connected in parallel to the data bus. Clearly, this will be a disastrous situation unless I/O devices are connected in such a way that they present a high impedance to the data bus unless actually selected. One only can be connected at one time and this is achieved by *gated buffering* activated by the selection process described in (1). The selection process ensures that only the required input device is connected to the data bus and then only for a brief period while the IN' signal is active. At that time the μP will read the data on the data bus.
3. *Latching of data output from the data bus.* When the μP wishes to send data to an output device, it will place the required address on the address bus. Then the data to be output is placed on the data bus momentarily while OUT' is active. Because the data appears only momentarily, there must be facilities in the output peripheral device to *latch* onto and *hold* the data presented momentarily on the bus. Flip-flop or register type circuits are generally used to perform this task when the selection process indicates the particular address and OUT'.

Because these requirements are so commonplace, the semiconductor manufacturers provide special circuit packages to perform parallel and serial I/O interfacing. We will consider the basic facilities provided by these packages before proceeding to examine hardware details of the Z80 and 68000 in particular.

10.3.6 Typical parallel and serial I/O packages

The requirements for parallel and for serial I/O are generally well catered for and a variety of LSI logic packages are available for implementing the basic interface hardware requirements set out in the preceding section.

Basic parallel interfacing

Various devices are in common use and are often classified as PPIs (programmable peripheral interfaces) or as PIOs (parallel I/O interface controller) or as PIAs (peripheral interface adapters). All have facilities to meet the three basic interface requirements and provide the user with two or more *ports* (usually 8-bits wide) to which input or output devices may be simply and directly connected. The parallel interface chips are programmed through the instruction set and I/O addressing facilities of the μP to set up the ports for the desired direction of data flow. In some cases, port lines may be individually set for input or output.

Consider, for example, the very popular Intel™ 8255A PPI device. The general architectural configuration is set out as Figure 10.9 and it will be seen that there are *three 8-bit parallel ports A, B and C* each of which may be individually programmed through a common *control register*. Further,in this case, port *C* is split into two individually programmable 4-bit nibbles. The way in which the CS' (chip select) input is activated determines the general I/O address range occupied by the PPI in the μP I/O space, while the pins A_1 and A_0 determine which port, or the control register is selected as shown in the table accompanying Figure 10.9. Pins A_1 and A_0 are usually connected to (I/O) address bus lines A_1 and A_0 respectively. RD' and WR' (read and write) signals serve the obvious purpose in effecting either input or output operation through the PPI when it is selected. Assuming that the I/O devices are addressed by the lowest eight address lines alone, then the decoding arrangements shown in Figure 10.9 would select port A on I/O address 08H, port B on I/O address 09H, port C on I/O address 0AH, and the control register on I/O address 0BH.

The *control word format* is also indicatedwith the figure and a simple illustration will serve to establish the basic way the PPI is used.

Let us say that we wish to take inputs from eight switches through port A and take outputs to an 8-bit D/A (digital to analog) converter and, separately, to 8 LED indicators through ports B and C respectively as suggested in the figure. Early on in the program, which would be run to utilize these resources, we must set up the PPI ports for the desired direction of data in each case: A = input, B and C = output. The required control word (from the format set out in Figure 10.9) is as follows:

Control word	1	0	0	1	0	0	0	0	= 90H

The required control instructions, in Z80 code, including the way in which each port could be addressed, are as follows (Hex addresses and data assumed):

```
LD A,90          control word into A
```

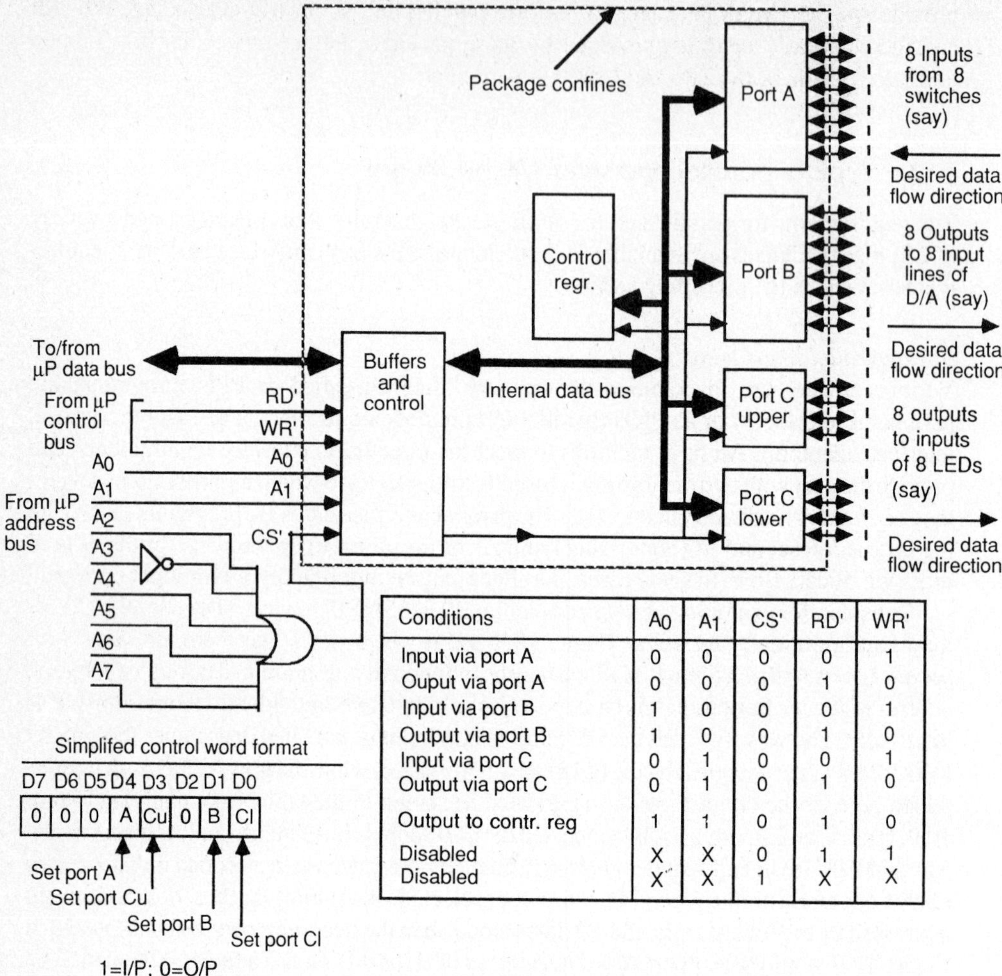

Conditions	A0	A1	CS'	RD'	WR'
Input via port A	0	0	0	0	1
Ouput via port A	0	0	0	1	0
Input via port B	1	0	0	0	1
Output via port B	1	0	0	1	0
Input via port C	0	1	0	0	1
Output via port C	0	1	0	1	0
Output to contr. reg	1	1	0	1	0
Disabled	X	X	0	1	1
Disabled	X	X	1	X	X

Figure 10.9 Simplified view of Intel™ 8255 parallel interface

```
OUT (0B),A      control word to control regr.
- - - - -       (PPI is now set up as desired)
IN A,(08)       to read switch settings into A.
- - - - -
OUT (09),A      (A) to D/A converter.
- - - - -
OUT (0A),A      (A) to LED display.
```

Note the very simple access to each port once set up.

Other packages have an 8-bit control register or data direction register associated with each port allowing individual port lines to be set for direction but their use is similarly straightforward.

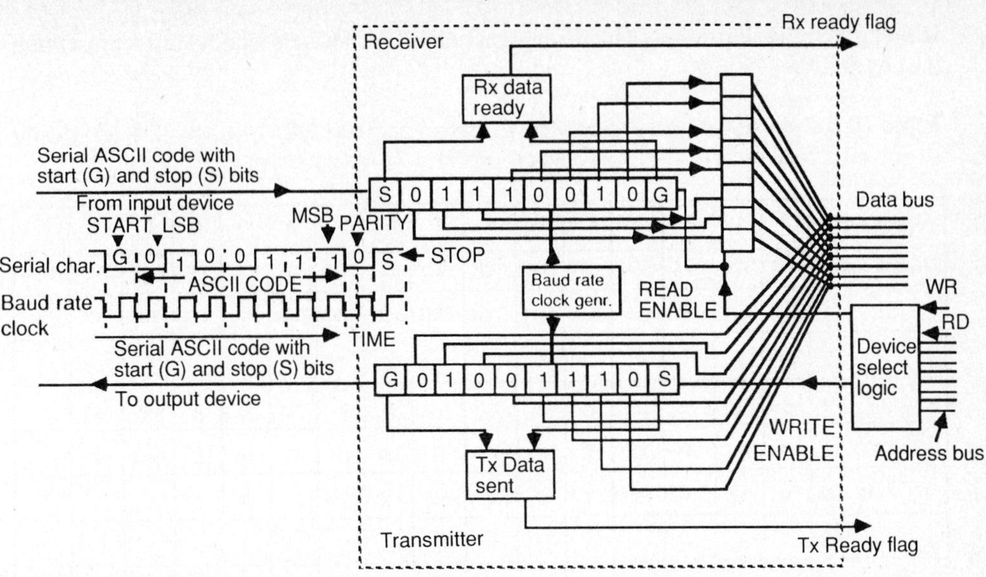

Figure 10.10 Basic serial interface operation (showing ASCII character *r*)

Basic interfacing with serial data

Once again, various devices are in common use and are generally classified as UARTs (universal asynchronous receiver/transmitters) or USARTs (universal synchronous/ asynchronous receiver/transmitter). In either case their purpose is most often concerned with the flow of data communications, usually in ASCII code (American Standard Code for Information Interchange). ASCII code is a 7-bit code which is generally shipped between devices in a serial fashion, that is, bit by bit against a clock signal. The rate at which information bits are clocked per second is known as the Baud rate and may typically vary from 110 Baud to 19.6 kiloBaud. ASCII code is used to represent alphabetic characters, numerals, punctuation marks, etc., and is summarized in Table 10.3. An eighth bit, the *parity bit*, is frequently added during transmission to always make the total number of logical 1's per character an even (even parity) or an odd (odd parity) number to allow a simple check on correct transmission. Particular manufacturers have their own names for these serial interface packages, such as PCIs (programmable communications interfaces) or as SIO (serial input/output) packages, or again, as ACIAs (asynchronous communications interface adapters). All have facilities to meet the three basic interface requirements and provide the user with one or more *serial port* to which input or output devices may be simply connected, usually through buffers.

The serial interface chips also perform the additional tasks of serial/parallel data conversions and Baud rate setting. The serial interface chips are programmed through the instruction set and I/O addressing facilities of the µP to set them up to the desired parameters.

A general simplified architectural configuration is set out as Figure 10.10 which shows how the basic serial to parallel conversions take place on input and output operations.

The interface basically consists of two 10-bit shift registers to effect the serial to parallel conversion for the receiver and the parallel to serial conversion for the transmitter. Ten bits* are required, seven bits, LSB first for the ASCII character (as in Table 10.3)+ 1 Parity bit

* At 110 Baud, 11 bits per character may be transmitted—as above but with 2 stop bits.

with start (G) and stop (S) bits which frame each character and help identify bits representing the character.

Table 10.3 7-bit ASCII code (parity bit to be added as MSB) (e.g. r = 72 even parity = F2 odd parity; s = F3 even parity = 73 odd parity)

LS Hex digit	0	1	2	3	4	5	6	7	8	9	A	B	C	D	E	F
MS hex digit 0	NUL	SOH	STX	ETX	EOT	ENQ	ACK	BEL	BS	HT	LF	VT	FF	CR	SO	SI
1	DLE	DC1	DC2	DC3	DC4	NAK	SYN	ETB	CAN	EM	SUB	ESC	FS	GS	RS	US
2	SP	!	"	#	$	%	&	'	(	)	*	+	,	-	.	/
3	0	1	2	3	4	5	6	7	8	9	:	;	<	=	>	?
4	@	A	B	C	D	E	F	G	H	I	J	K	L	M	N	O
5	P	Q	R	S	T	U	V	W	X	Y	Z	[	\	]	^	_
6	`	a	b	c	d	e	f	g	h	i	j	k	l	m	n	o
7	p	q	r	s	t	u	v	w	x	y	z	{	|	}	~	DEL

The figure shows the receiver having just received (in serial form) a 10-bit input representing character r with even parity. Subsequently this would be read out in parallel onto the data bus through buffers as shown.

The transmitter is shown with its register loaded in parallel from the data bus with character r ready to send. This will be clocked out in serial form at the chosen Baud rate and sent to an output device.

Flag logic, for example a receiver ready and a transmitter ready flag, is included to indicate the status of each subassembly. Suitable address line connection and decoding will be used to place this serial interface device in the I/O map of the microcomputer.

10.3.7 Memory and I/O space maps

We have just mentioned the I/O map and it is appropriate to consider briefly the topic of space maps. We have seen that all microprocessors have an addressable range of memory determined by the width of the address bus and the consequent width of the program counter. If this width is N bits, then 2^N locations are addressable. For the Z80, $N = 16$ so that the number of addressable memory locations, or addressable space, is 64K. For the 68000, N becomes 24 and the number of addressable byte locations is 16M. Both represent large amounts of memory and appreciable cost if all possible memory spaces are occupied.

In consequence, most microcomputers use only part of the available memory space. Further, areas of memory may be read/write (RAM) or read only (ROM or EPROM, etc.). The user needs to know where the various areas of memory are located and a convenient way of doing this is by *memory map*. One such map for a Z80 based microcomputer is set out as Figure 10.11 and it will be seen that areas of ROM and RAM and blank areas are readily identified in terms of location and extent. A similar memory map for a 68000 based system follows as Figure 10.12. In all cases, the computer system designer decides on the amount

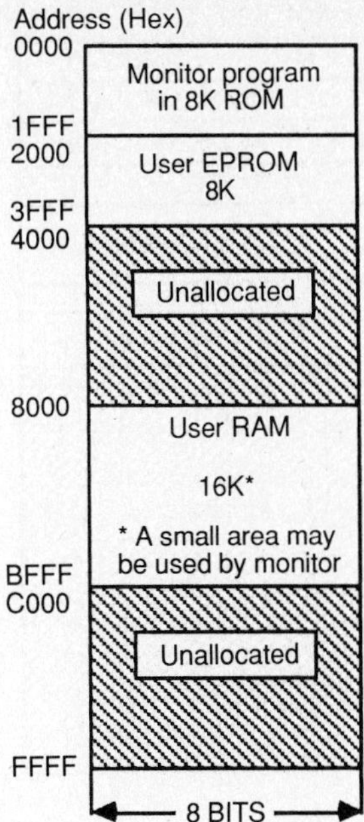

Figure 10.11 Possible Z80 memory map

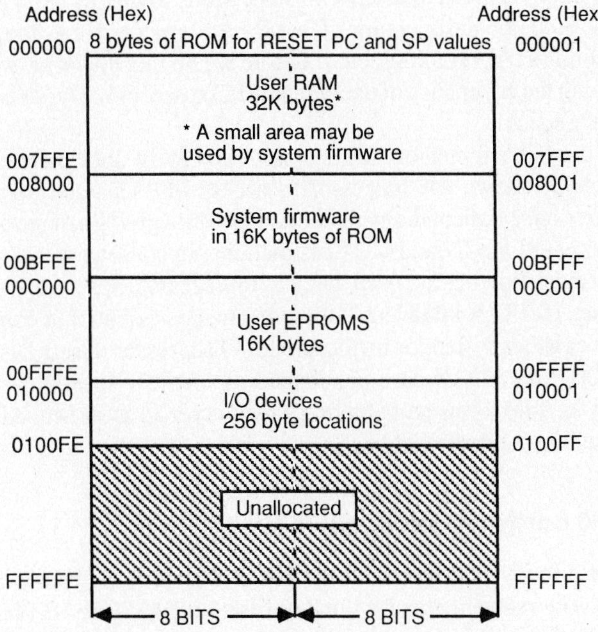

Figure 10.12 Possible 68000 memory map (NTS)

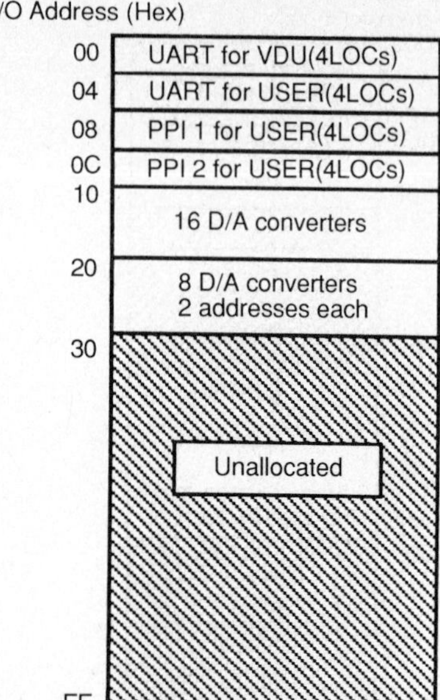

Figure 10.13 Possible I/O map for Z80 system

and location of usable memory and we will show by example how this is done. Frequently, the bottom of memory is given over to ROM since this is where the basic programs and data necessary for the fundamental operation of the μP are often stored (e.g., the monitor program, operating system, interrupt routines, etc.). Before we do this however, we will extend this concept to defining the occupancy of the available I/O space, but *only if this space is separate* from the memory space.

For the Z80, the I/O and memory are accessed separately since there is a control signal, MEM/(I/O)', which allows the processor to access I/O separately from memory. I/O addressing utilizes only eight of the available address lines giving a range of 256 addressable I/O locations accessed via IN and OUT instructions. In consequence a separate map, the I/O map, is useful in showing the occupancy of this separate space. A Z80 based example is given as Figure 10.13. A knowledge of the I/O map is essential in connecting further I/O devices to an existing system or in allocating I/O addresses during system design.

For the 68000 μP, I/O devices are treated as memory locations—I/O is "*memory mapped*"—and therfore forms part of a single memory map as shown. All 24 address lines are therefore used in I/O device addressing.

10.3.8 The Z80 hardware interface

It is possible to directly interface with the bus and control signals emanating from the microprocessor. The relevant data for the Z80 is set out as Figure 10.14.

It may be seen that a 16-bit (output) address bus is available and an 8- bit (bidirectional) data bus. The control signals of immediate interest are: MREQ', IORQ', RD', WR'. Further lines, INT', NMI', and M1' will be discussed with interrupts. Note that all control lines are active Low.

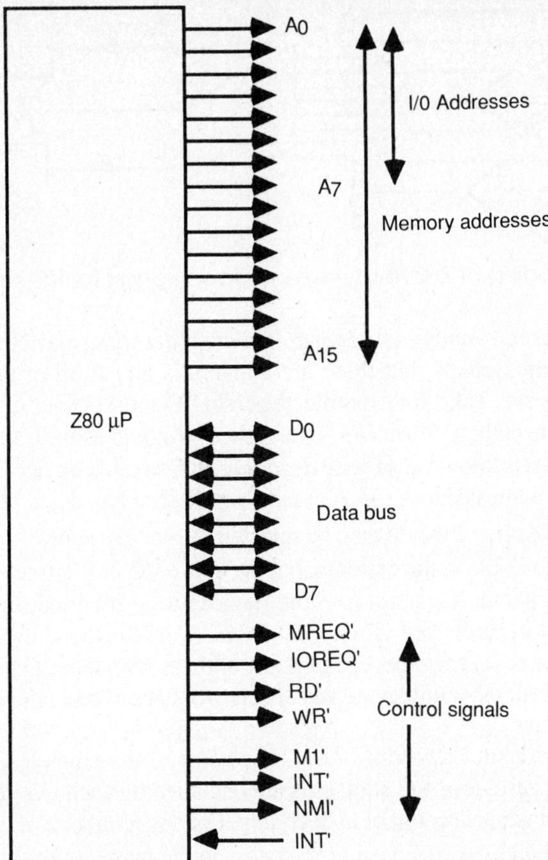

Figure 10.14 Z80 hardware interface signals

Perhaps the best way of illustrating the use of these facilities is to take an example of the way in which an input and an output device would be connected, meeting the three basic hardware interface requirements set out earlier.

A further example will illustrate the way in which memory would be interfaced.

In all cases the three basic requirements of address (device) selection, buffering of signals to the data bus and latching of data from the data bus must be allowed for.

Address selection—I/O device or memory selection

In the case of I/O, eight address lines, IORQ' and RD' and WR' are the signals to be decoded. Take for example the selection of an input device at, say, I/O address 23H. The necessary address decoding may be performed by simple gate logic as suggested in Figure 10.15.

In the case of a number of I/O devices, it is easier to use decoder chips, for example the 8205 or 74138 3-line to 8-line decoder as illustrated in Figure 10.16(a). This allows up to eight individual I/O devices to be connected which may be mixed input and output devices as in Figure10.16(b). Up to 16 may be connected if, as is often the case, an input and an output device may share one I/O address. In Figure 10.16(b), connections are shown for selecting eight mixed I/O devices in I/O addresses 30H to 37H.

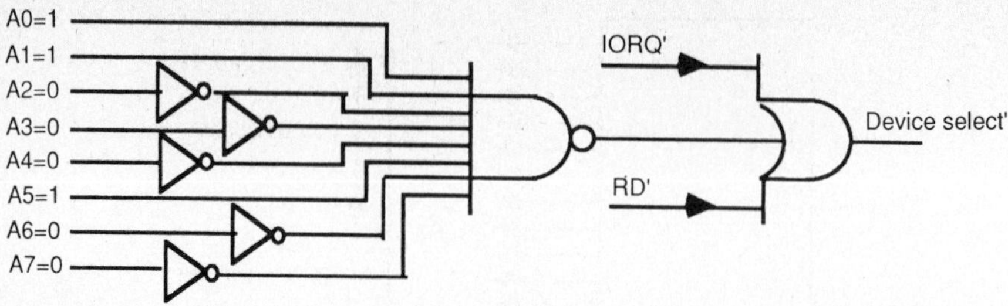

Figure 10.15 Decoding of Z80 hardware interface signals for I/O device 23H

Bitwise and other forms of ambiguous (partial) address decoding may be used for simplicity in some small cheap systems, but there are dangers in any form of incomplete address decoding as we may see. Take, for example, the eight I/O address lines of the Z80. We could argue that if there are eight or fewer I/O devices then we could use one address line to select each I/O device. This is known as *bitwise decoding* and, say, I/O device 0 could be selected by the Low state of address line A_0, I/O device 1 would then be selected by the Low state of address line A_1, I/O device 2 would then be selected by the Low state of address line A_2, etc. Decoding would be simple in the extreme, but the entire I/O map space would be occupied by just eight devices and, if a ninth or more devices were required, the whole decoding process would need to be revised. Every I/O device would respond to any one of 128 I/O addresses and worse is yet to come, because any address with more than one 0 would turn on more than one I/O device simultaneously. There would be a "real field day" if I/O address 00H was selected!

Figure 10.16(c) illustrates *partial address decoding* for the example taken earlier. It will be seen that, in this case, four I/O addresses are occupied by each I/O device because two address lines are not taken account of in decoding. For any number *n* of undecoded address lines, each decoded address will occupy 2^n spaces in the map.

For adding memory to a computer system, an equally straightforward decoding process is involved. Consider the memory map set out in Figure 10.11, and say we wished to connect 16K RAM in the unallocated area from 4000H to 7FFFH. Further, assume that the RAM is supplied in 4K byte packages, each package being as suggested in Figure 10.17(a).

We may conveniently use the decoder package set out in Figure 10.16(a) and use four of its outputs to select each RAM package in turn to cover the required 16K in the memory map.

To start with, we will find it convenient to consider the required state of each of the 16 address lines as follows.

A15 A0

0	1	S1	S0	a	a	a	a	a	a	a	a	a	a	a	a

|- FIXED- | DECODER |----------ADDRESS INPUTS OF RAM PACKAGES---------|

Note that the two most significant address lines are used to place the RAM area as required, the next two lines activate two select lines of the decoder and select the RAM in 4K blocks. The remaining lines address the 4K individual locations in each RAM when it is selected. The overall arrangement is shown in Figure 10.17(b).

Note that partial address decoding processes discussed above have similar and even more dramatic effects on the memory map.

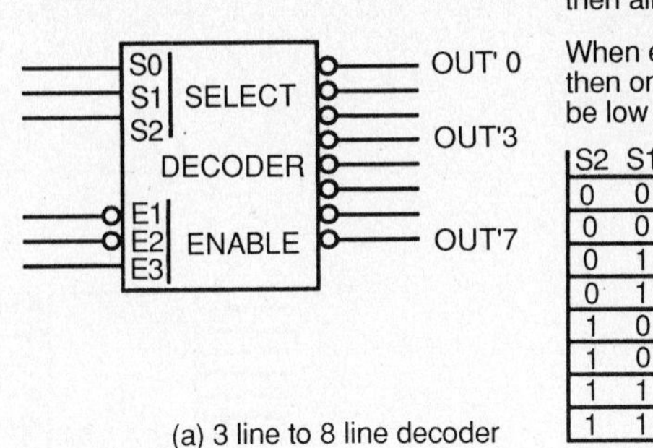

If enable inputs E1, E2, E3 ≠ 001 then all out lines will be high.

When enable inputs E1, E2, E3 = 001 then one and one only out line will be low as selected by select inputs.

S2	S1	S0	Out selected
0	0	0	OUT' 0
0	0	1	OUT' 1
0	1	0	OUT' 2
0	1	1	OUT' 3
1	0	0	OUT' 4
1	0	1	OUT' 5
1	1	0	OUT' 6
1	1	1	OUT' 7

(a) 3 line to 8 line decoder

Figure 10.16(a) 3-line to 8-line decoder

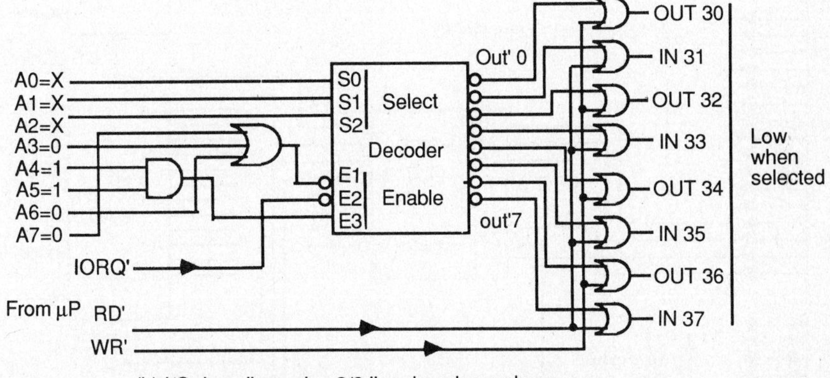

(b) I/O decoding using 3/8 line decoder package

Figure 10.16(b) I/O decoding using 3 to 8-line decoder package

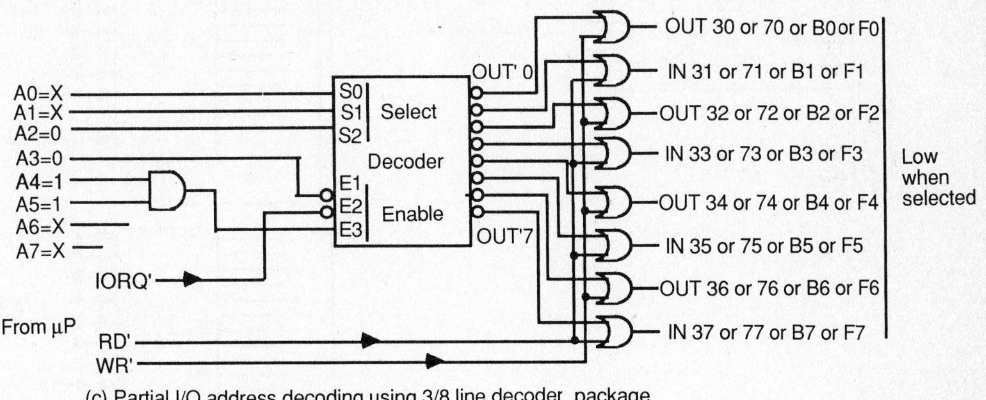

(c) Partial I/O address decoding using 3/8 line decoder package

Figure 10.16(c) Partial I/O address decoding using 3 to 8-line decoder package

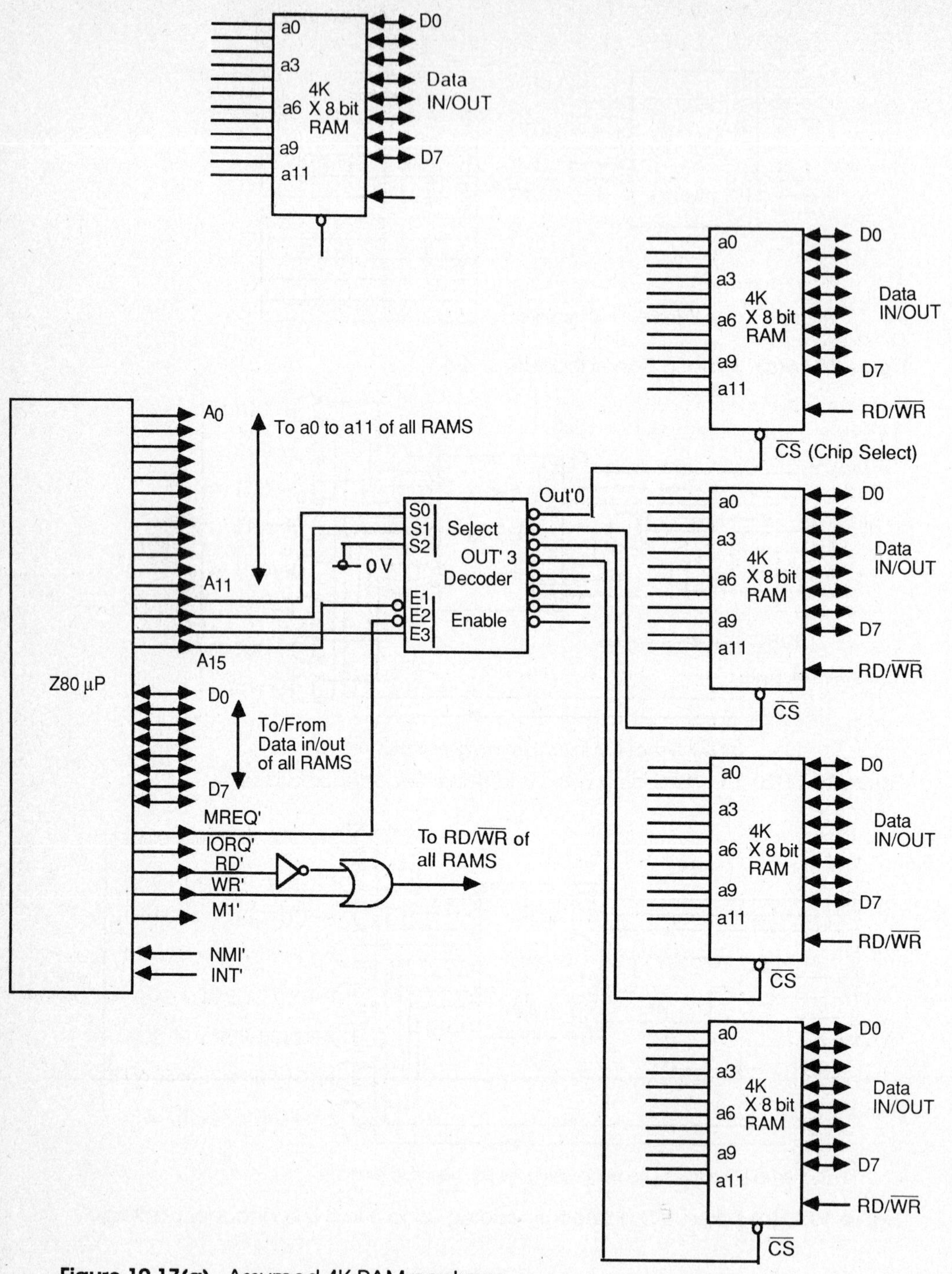

Figure 10.17(a) Assumed 4K RAM package
(b) Address decoding arrangement for 16K memory

To avoid problems, the message is:- *do not indulge in sloppy decoding practices unless you are absolutely forced to and even then, think twice.*

Buffering of inputs to the data bus

The second of the general requirements is to buffer all input device connections to the data bus so that,unless selected, a high impedance is presented to each line.

Tristate buffers allow this to be done quite simply. A tristate buffer has an enable facility which, if inhibited, will isolate input logic levels and present a high impedance at its output. The general mode of connection is to enable the buffers, one per input line, from the device select signal discussed in the previous section. Figure 10.18, hopefully,makes this clear.

For RAM and ROM and UART and PPI, etc., packages, buffers are usually built in to those lines providing inputs to the data bus.

Latching data from the data bus

This is the third general requirement and is associated with output devices. The problem of latching the momentary data sent by the μP is also readily tackled using D flip-flop(FF) circuits. Standard packages of a number of FFs known as, for example, quad (or octal) latches, may be used as suggested in Figure 10.19. Latches are usually built into RAMs, etc.

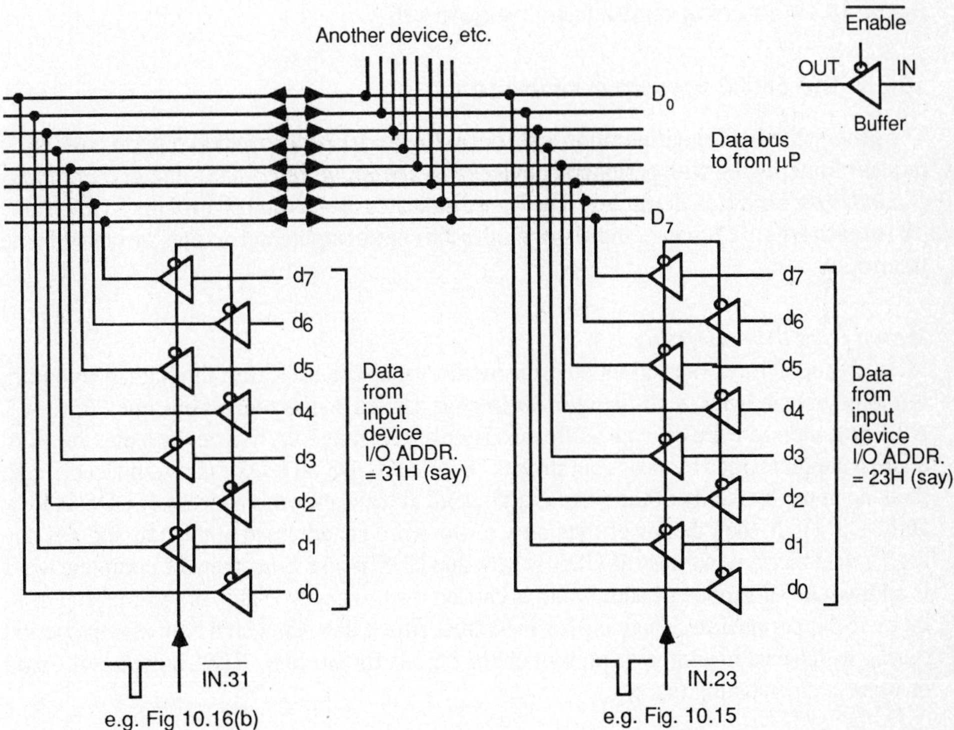

Figure 10.18 Buffering of inputs to the data bus

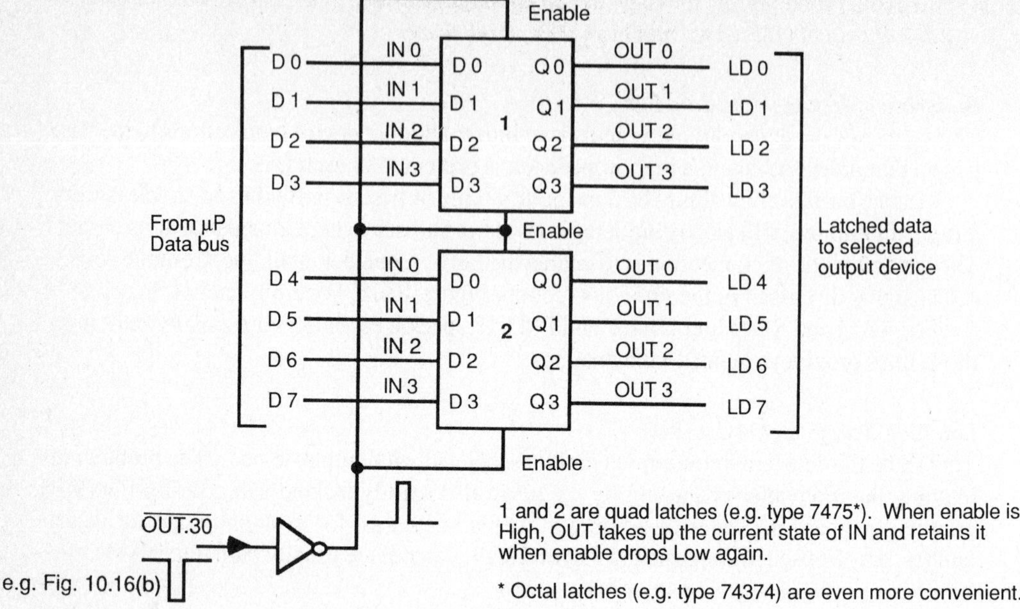

Figure 10.19 Latching output from the data bus

10.3.9 The 68000 hardware interface

The general hardware configuration is set out as Figure 10.20. The 68000 supports two basic modes of interfacing with peripheral devices—*synchronous interfacing* and *asynchronous ("handshake based") interfacing.* The figure indicates the bus and control lines which take part in each type of exchange and also includes the control inputs and outputs associated with Interrupts.

Memory and I/O Addressing

Memory and I/O are both addressed in the memory map and basically, although the memory is addressable in bytes, *a 16-bit word address* is carried on the address bus lines A_1 to A_{23}. All words therefore are on even addresses. Byte addressing is carried out by using the word address supplemented by two "data strobes" UDS' and LDS'. If UDS' is Low and LDS' High then the upper byte only of the word is addressed at the even word address. If LDS' is Low and UDS' High, then the lower byte only of the word is addressed at the odd address one higher than the word address. If UDS' is Low and LDS' is also Low, then the complete word is addressed. This mode of addressing is carried over to easily enable the addressing of 8-bit or 16-bit peripherals. There is also an address strobe line AS' which indicates the period during which a valid address is present on the bus. A further line, R/W', indicates if a read or write cycle is taking place.

Asynchronous interfacing

The signals utilized here are indicated in Figure10.20 and 10.21. Typical exchanges are as follows. (*Note* "assert" = enable; "negate" = disable or inhibit.):

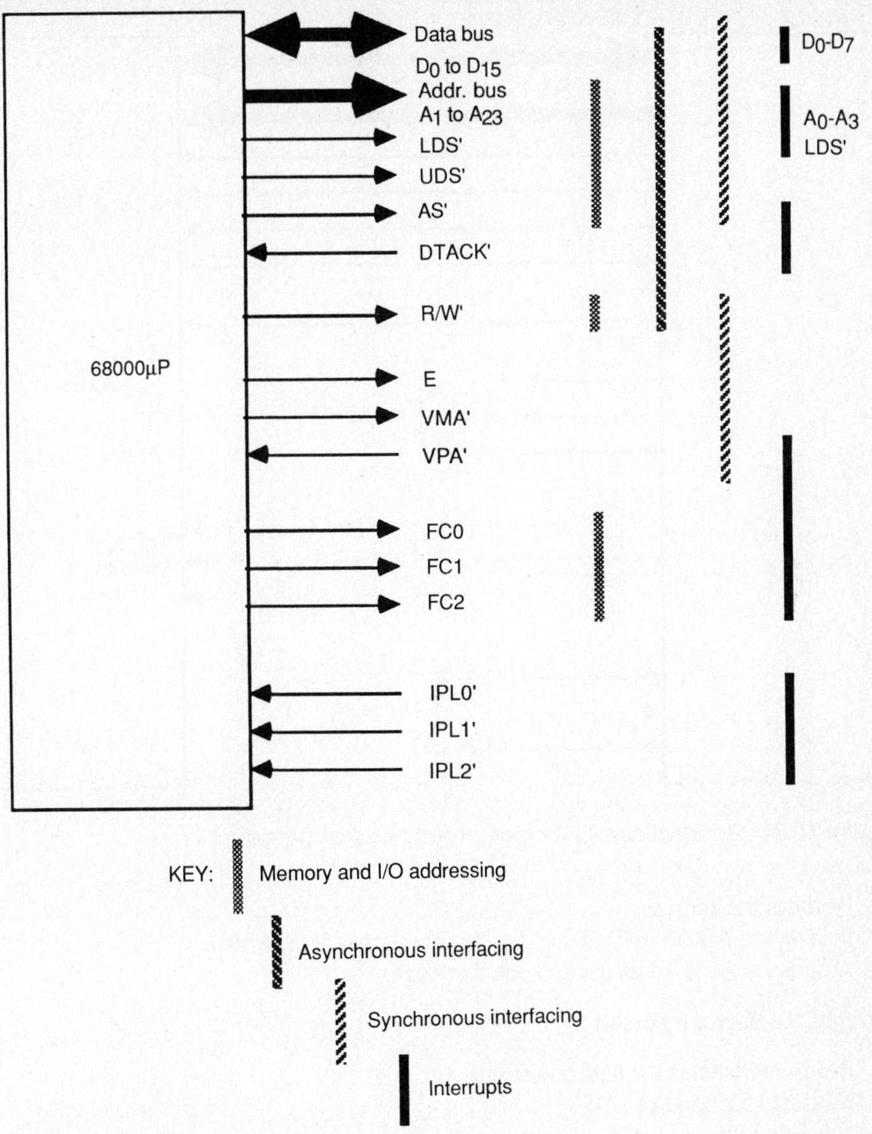

Figure 10.20 68000 hardware interface signals

For an input/read cycle, the microprocessor:

1. Sets R/W' to read.
2. Places address on bus (and also function code on FC0-FC2).
3. Asserts AS'.
4. Asserts UDS' or LDS' for byte or UDS' and LDS' for word data.

In reply, the peripheral device:

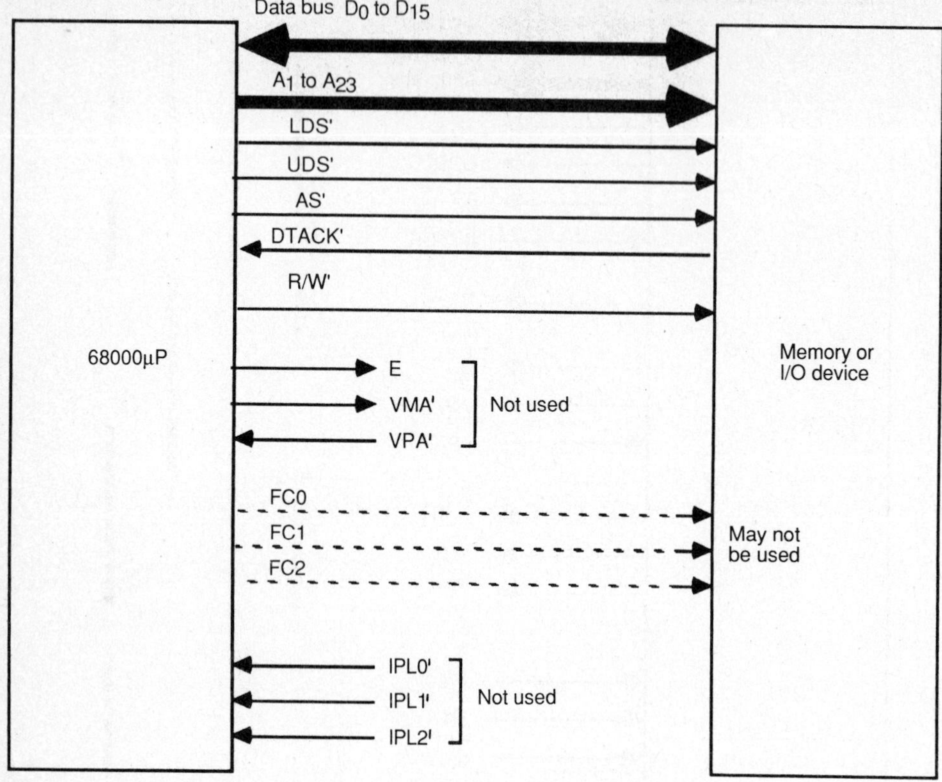

Figure 10.21 General asynchronous interface arrangement

1. Decodes the address.
2. Puts data on data bus D_0-D_7 or D_8-D_{15} (byte) or D_0-D_{15} (word).
3. Asserts DTACK' (data transfer acknowledge).

In reply, the microprocessor:

1. Reads and latches the data on the bus.
2. Negates UDS' and/or LDS'.
3. Negates AS'.

In reply, the peripheral device terminates the cycle by:

1. Removing data from the data bus; and
2. Negating DTACK'.

For an output/write cycle, the microprocessor:

1. Places address on bus (and also function code on FC0-FC2).
2. Asserts AS'.

3. Sets R/W' to write.
4. Puts data on data bus D_0-D_7 or D_8-D_{15}(byte) or D_0-D_{15}(word)
5. Asserts UDS' or LDS' for byte or UDS' and LDS' for word data

In reply, the peripheral device:

1. Decodes the address.
2. Latches data bus D_0-D_7 or D_8-D_{15} (byte) or D_0-D_{15} (word)
3. Asserts DTACK' (data transfer acknowledge).

In reply, the microprocessor:

1. Negates UDS' and/or LDS'.
2. Negates AS'.
3. Removes the data from the bus.
4. Sets R/W' to read.

In reply, the peripheral device terminates the cycle by negating DTACK'.

Note that in each case a "handshaking" exchange takes place and that *the peripheral device terminates the cycle.* This means that the processor operation will be hung up waiting if the DTACK' signal is not received. The general asynchronous interface connections are given in Figure 10.21 and an example showing the connection of eight 8-bit input peripheral devices in addresses 00A000 to 00A007 is given as Figure 10.22.

Synchronous interfacing

The signals utilized here are indicated in Figures 10.20 and 10.23 . Typical exchanges are as follows. (*Note* "assert" = enable; "negate" = disable or inhibit.):

For a read or write cycle, the microprocessor:

1. Sets R/W' to read or write.
2. Places address on bus (and also function code on FC0-FC2).
3. Asserts AS'.
4. Asserts UDS' or LDS' for byte or UDS' and LDS' for word data.

In reply, the peripheral device:

1. Decodes the address.
2. Asserts VPA' (valid peripheral address).

In reply, the microprocessor:

1. Monitors E until it is Low.
2. Asserts VMA' (valid memory address).

In reply, the peripheral device:

1. Waits until E is ACTIVE (High).
2. Transfers data over the data bus.

In reply, the microprocessor terminates the cycle by:

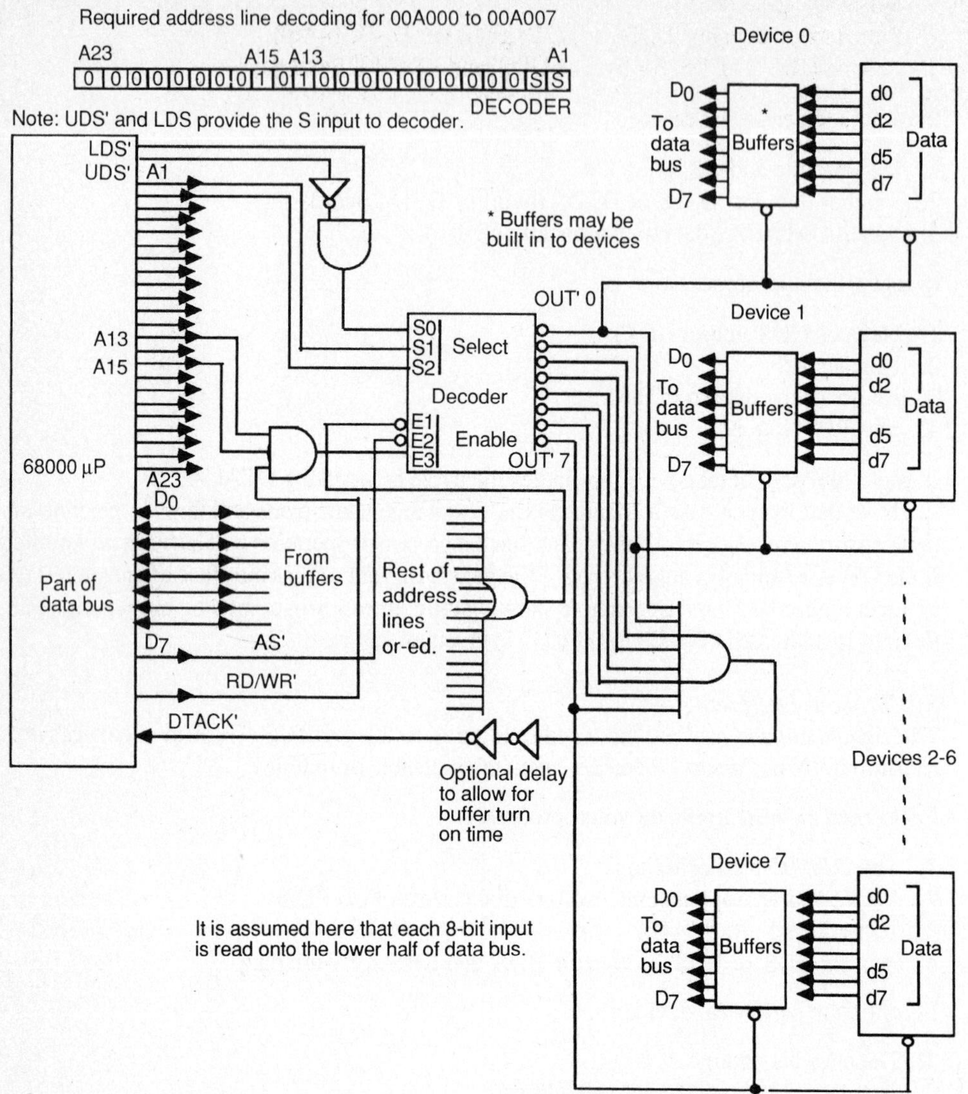

Figure 10.22 Asynchronous interfacing arrangements for input devices 00A000-00A007

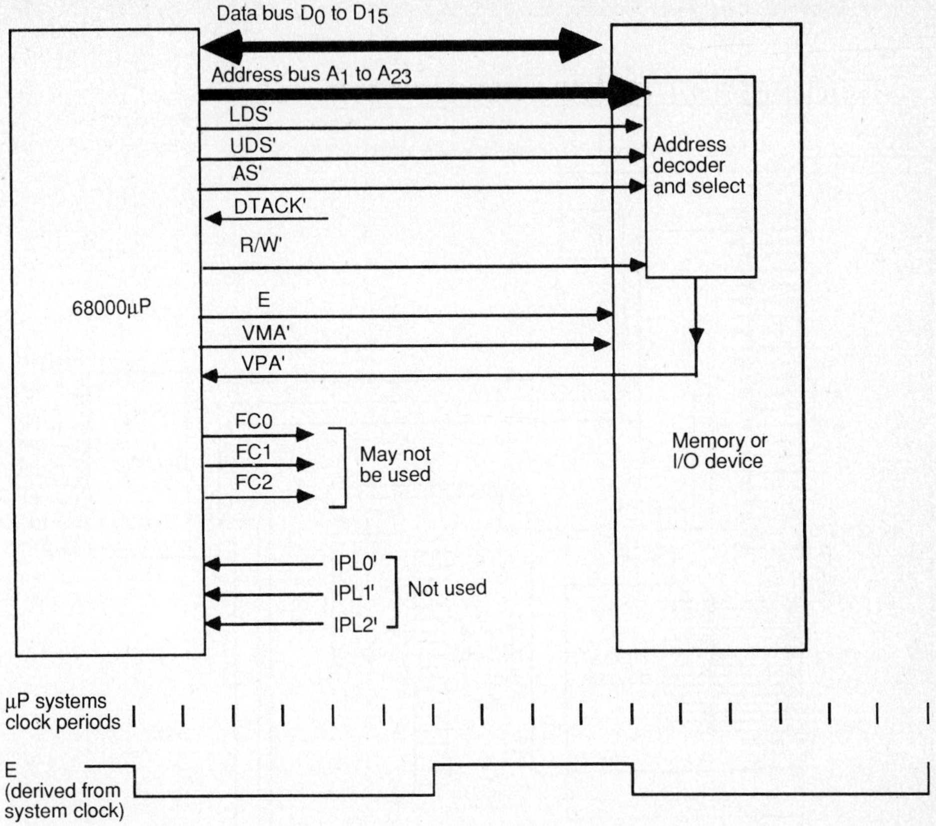

Figure 10.23 General synchronous interface arrangement

1. Wait for E to go Low—for read, data is latched on ∇E.
2. µP negates VMA', then negates AS', UDS', and LDS'.

An example showing connections to a block of synchronous peripherals at addresses 00A000-00A007 follows as Figure 10.24 (compare with Figure 10.22).

It should be noted that *DMA (direct memory access)* operations are also possible for which the reader is referred to the manufacturer's literature.

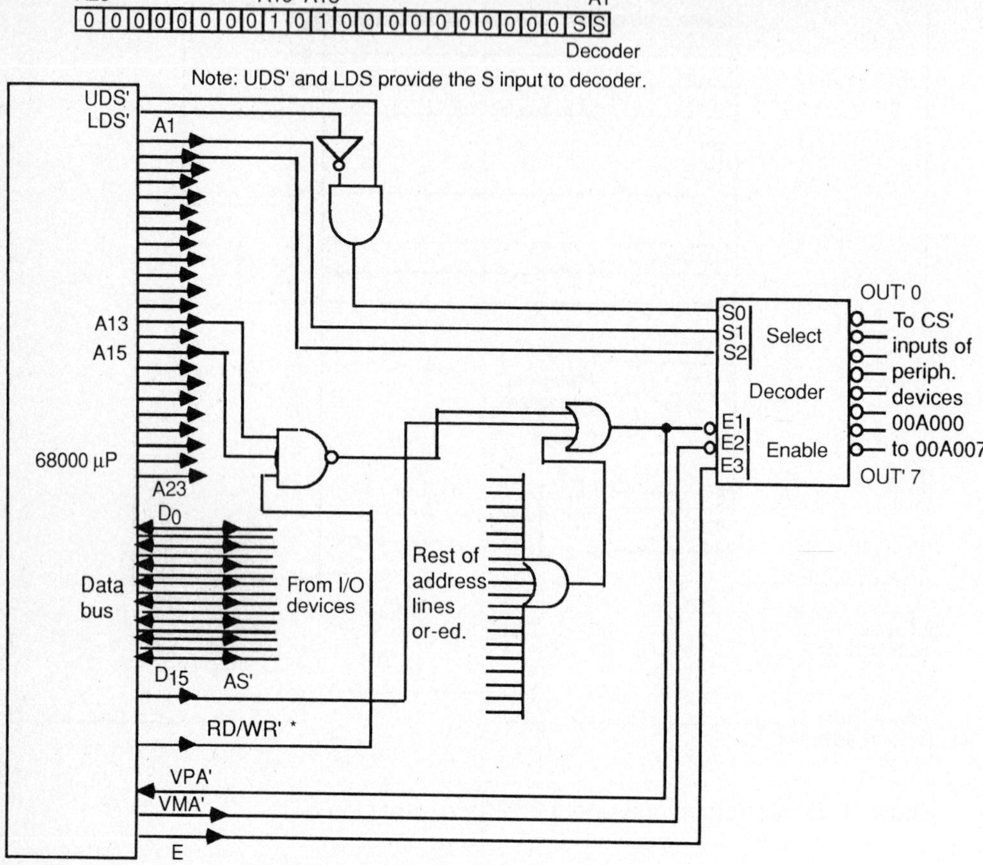

Figure 10.24 Synchronous interfacing arrangements for input devices 00A000-00A007

10.4 Interrupts

In many applications, the μP is interfaced with a number of peripheral devices, all of which must interact with it by sending or receiving data. Such exchanges may be of an urgent nature particularly in *real time* control or instrumentation applications. A way of attending to all peripheral needs is by *software driven polling* whereby the μP, as part of the normal program execution, interrogates each peripheral in turn for an indication of a need for service (e.g. the peripheral has data for or needs data from the μP). Service would normally occur through the execution of a service subroutine which would be entered only if the need was detected during interrogation.

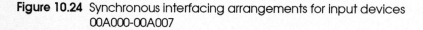

This process is quite effective but suffers from two big disadvantages:

1. Time is wasted by periodically interrogating all peripherals irrespective of whether or not a need exists.
2. The urgency of the need for attention by a fast peripheral device can only be taken into account by interrogating it first. Then there is a high probability that its need will arise immediately after it has been polled and it will have to wait until the next interrogation cycle.

A solution to these difficulties is to be found in a priority ordered interrupt system.

An *interrupt* is a signal from a peripheral which will cause the μP to break off the normal flow of program execution to execute an interrupt service routine for that peripheral.

Normal program execution should be resumed after completion of the service routine. This has distinct advantages:

1. Devices are only serviced when a need has arisen.
2. Response time is fast.

Facilities must exist in the μP architecture to handle interrupts. The simplest form of interrupt is *the single line interrupt* as in Figure 10.25(a).

This is a very simple facility and relies on subsequent software polling to identify the particular device which generated the interrupt.

At the other extreme is a *multiple input multilevel interrupt system* in which there is a separate interrupt input line to the μP for every peripheral device as indicated in Figure 10.25(b). The μP can thus directly identify the interrupting device. However, this is clearly impractical due to the potentially large number of separate inputs required.

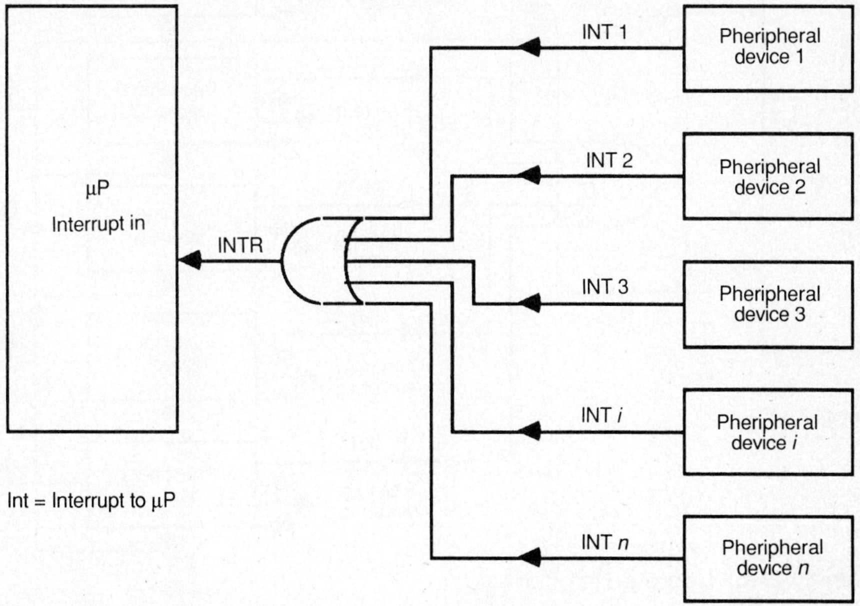

Figure 10.25(a) Single line interrupt

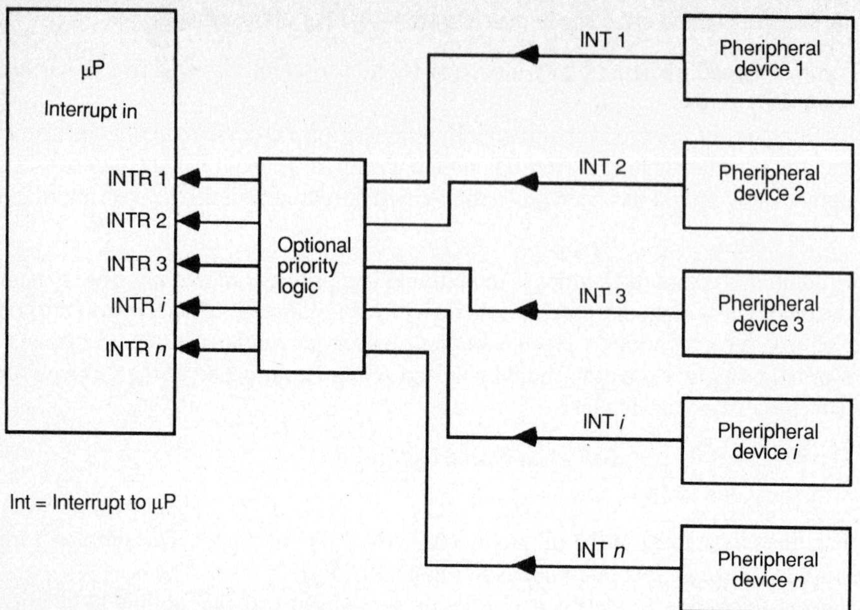

Figure 10.25(b) Multilevel interrupt

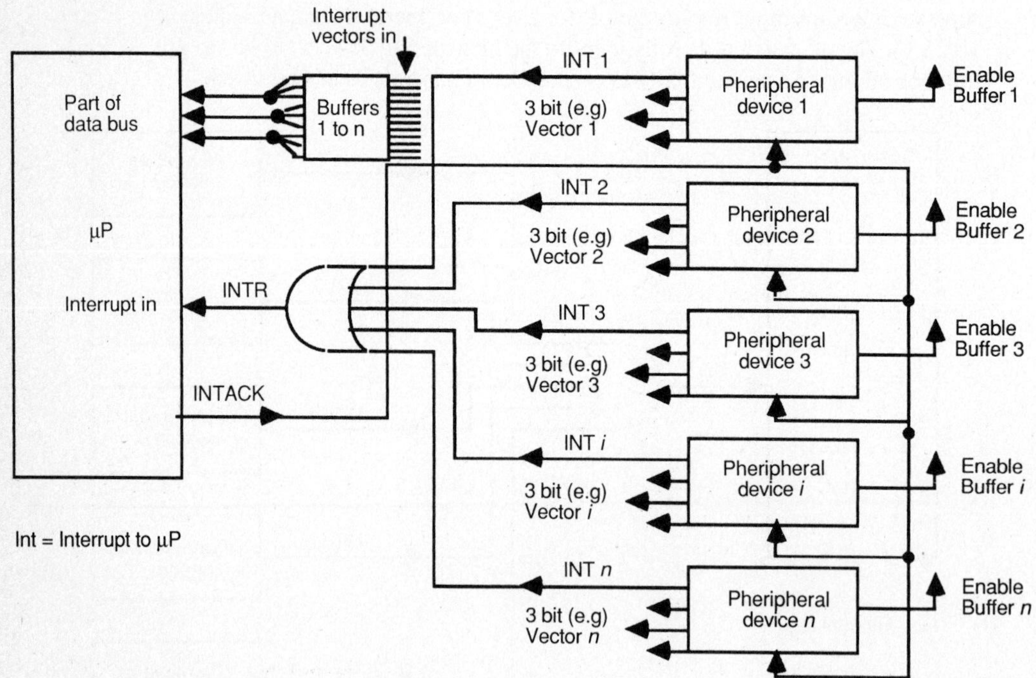

Figure 10.25(c) Vectored interrupt

As is often the case, a compromise is sought and a popular approach, and that available to both the Z80 and 68000, is to implement a *vectored interrupt system* as outlined in Figure 10.25(c).

On receiving a single line interrupt(e.g. INTR), the µP will issue an interrupt acknowledge signal (e.g. INTACK) which goes back to all peripheral devices having an interrupt input to the µP. The interrupting device then generates a coded signal of some sort which will identify it to the µP. For example, a vector could be placed on the data bus as suggested. This would be read by the µP in order to determine which particular service routine to execute.

Priority can be associated with such a system by interposing priority allocation logic, as discussed earlier in this text, and/or by "daisy chaining" the interrupt inputs. ["Daisy chaining" is a simple series logical interconnection of the interrupt lines of a group of interrupting devices in order of priority. The connection is such that the first device blocks the input from all other lines when active, the second device blocks the input from the third and beyond when active, etc.]

10.4.1 The Z80 interrupts

Referring back to Figure 10.15. we may see that four control lines are associated with the Z80 interrupt facilities:

Two input lines: INT' and NMI' (non maskable interrupt).
Two output lines: M1' and IORQ'

In the instruction set, there are five instructions which affect interrupts:

EI enable interrupts;
DI disable interrupts;
IM0, IM1 and IM2 (set interrupt modes 0,1,or 2 respectively.)

There are also two instructions associated with returns from interrupt service routines:

RETI (return from maskable interrupt), and
RETN (return from nonmaskable interrupt)

Z80 Non maskable interrupt input
The Z80 has a dedicated NMI' line which the processor cannot disable (with a DI instruction) so that it is always active and enjoys the highest priority. It may be used, for example, in entering "power fail" routines where the µP may have a few milliseconds to save data before the power drops below workable levels.

To generate an interrupt on this input:

1. Take NMI' to the active (Low) state.
2. µP will finish executing the current instruction.
3. Processor now executes an RST instruction, the first operation of which (a) pushes current PC contents onto the stack (to save a return address) and saves (internally) the current state of the interrupts and, (b) jumps to location 0066H in memory.

4. The NMI service routine must be written to start at address 66H and execution commences.
5. On completion, a return is made to the interrupted program via a RETN instruction which *must* be written at the end of the service routine.
6. The original state of the interrupt system is restored.

A simple example serves to illustrate the process:

MAIN PROGRAM

	Address in memory(e.g)	Instruction	Remarks
	——		
	2006	DEC BC	
	2007	LD A,B	**Assume NMI' occurs here.**
RETURN ADDRESS	2008	OR C	**OR C not executed and**
	2009	JR NZ, $-5	**2008 is put on stack.**
	——	etc.	**Internally, μP copies the state of IFF1(the maskable interrupt enable FF) to IFF2 (a temporary store). Clear IFF1. IFF1 clear means that the NMI cannot be interrupted Enter service routine via address 66H. e.g.**
	0066	**JMP 3000**	**Jump to usable RAM area**
	3000	**EX AF,AF'**	**"Save" current registers**
		——	
		service routine	
		——	
	30xx	**EX AF,AF'**	**Restore original registers.**
	30xx+1	**RETN**	**This returns to the main program via address (2008) from the stack and copies IFF2 to IFF1.**

Z80 Maskable interrupt input
Three modes are available, IM0, IM1 and IM2, each of which is selected via the instruction set and each of which can be enabled (EI) or disabled (DI). *These interrupt facilities are ineffective unless enabled.*

1. *Interrupt mode 0.* Assuming an EI instruction has been executed, i.e. IFF1 is set and that an IM0 instruction has followed:

   (a) Interrupting device puts a low level on INT'.
   (b) At the end of the current instruction, μP accepts the interrupt by responding with

IORQ' and M1' and disables interrupts.

(c) Interrupting device receives M1' and IORQ' and recognises this as an interrupt acknowledge.

(d) The interrupting device then places its 3-bit vector (000 [0] to 111 [7]) onto lines D_5 D_4 and D_3 of the data bus.

(e) The µP uses this vector as L for an RST nn instruction where nn = 8 X L and L will have a value from 0 to 7.

(f) The current PC value (return address) is put away on the stack and a jump is executed to memory location 8 X L. The service routine is written to start here.

(g) The service routine should save original register contents as appropriate and must end in a RETI (or RET) instruction. An EI instruction must precede the RETI (or RET) otherwise the interrupts will remain in the disabled state.

Interrupt mode 0 allows ready connection of eight interrupting devices but each must provide the external hardware to respond to "interrupt acknowledge" by placing a suitably buffered vector onto the data bus.

2. *Interrupt mode 1.* Assuming an EI instruction has been executed, that is, IFF1 is set and an IM1 instruction has followed, this facility allows the direct connection (no external hardware is needed) of one only interrupt.
   The action is as follows:

   (a) Interrupting device puts a low level on INT'.

   (b) At the end of the current instruction, µP accepts the interrupt (also issues IORQ' and M1') and disables interrupts.

   (c) The current PC value (return address) is put away on the stack and a jump is executed to memory location 38H. The service routine is written to start here.

   (d) The service routine should save original register contents as appropriate and must end in a RETI (or RET) instruction. An EI instruction must precede the RETI (or RET) otherwise the interrupts will remain in the disabled state.

3. *Interrupt mode 2.* Assuming an EI instruction has been executed, that is, IFF1 is set and that an IM2 instruction has followed, this facility allows the connection of up to 128 interrupting devices and uses the interrupt register "I" mentioned earlier.
   The heart of this system is the use of register I together with a 7-bit vector supplied by the device to the data bus. They are combined as a 16-bit address as follows:

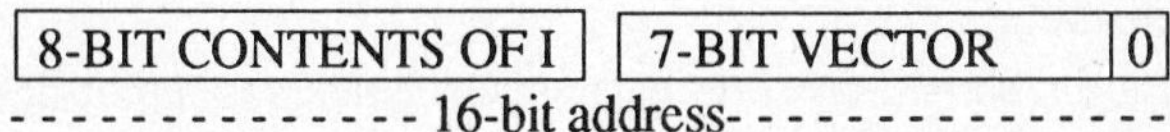

This address is used to access a table in memory comprising 16-bit addresses which point to the start of each service routine. Since 16-bit addresses must be held, each occupies two memory locations, hence only a 7-bit vector is needed on the data bus, the LSB always being taken as 0.

Thus, once again, the interrupting device must contain the necessary hardware to put the appropriate vector onto the data bus.

To illustrate this mode, assume that memory locations 2200H onwards are to be used for the interrupt service routine address table. Interrupt register I must, therefore, be loaded with 22H and this is done, for example by executing the following:

```
LD A,22H
LD I,A          (I now contains 22H)
```

Further, assume that service routine addresses are allocated in the table as follows:

Memory	*Address*
10	2200 Vector supplied by device 1 = 00
30	2201
50	2202 Vector supplied by device 2 = 02
30	2203
90	2204 Vector supplied by device 3 = 04
30	2205
D0	2206 Vector supplied by device 4 = 06
30	2207
etc.	

The service routine addresses, therefore, are as follows:

Device 1.....service routine starts at 3010H
Device 2.....service routine starts at 3050H
Device 3.....service routine starts at 3090H
Device 4.....service routine starts at 30D0H
etc.

Step by step the process is as follows:

(a) If IM2 is set and EI has been executed then the μP will accept the interrupt.
(b) After completing the current instruction, interrupts will be disabled and μP issues M1' and IORQ' (together forming INTACK).
(c) On receiving INTACK the device responds by placing its 7-bit vector with eighth bit (LSB) = 0 onto the data bus.
(d) This is merged with (I) as MS 8-bits to give the address of the appropriate entry in the table.
(e) The return address (PC) is put away on the stack.
(f) The address from (d) is accessed for the service routine start address which is then loaded into the PC.
(g) The service routine should save original register contents as appropriate and must end in a RETI (or RET) instruction so that execution of the interrupted program can resume. An EI instruction must precede the RETI (or RET) otherwise the interrupts will remain in the disabled state.

10.4.2 The 68000 interrupts

Referring back to Figure 10.20. we may see that eight control lines are associated with the interrupt facilities. They are:

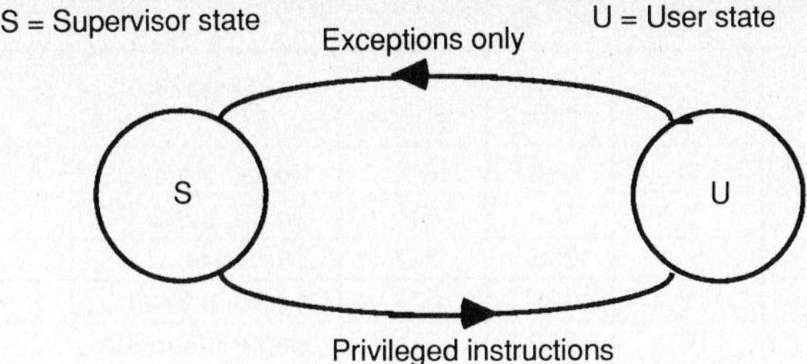

μP powers up into supervisor state.

Function code from μP FC2	FC1	FC0	*Significance*
0	0	0	Not used
0	0	1	User data
0	1	0	User program
0	1	1	Not used
1	0	0	Not used
1	0	1	Supvr. data
1	1	0	Supvr. program
1	1	1	Intr. acknowldg

Figure 10.26 Supervisor and user state and function codes

Five input lines: IPL0, IPL1, IPL2 and DTACK' and VPA'
Three output lines: FC0, FC1, FC2.(A0 to A3 are also affected).

To give it the widest range of applications the 68000 has two basic operating states. A user state in which certain instructions and, perhaps, through the use of the function code FC0-FC2, certain areas of memory, are not available. The second state is the supervisor state, which is the priviledged state, having access to all instructions. The transition between these two states is determined as shown in Figure 10.26 which also lists the significance of the code output on FC0-FC2.

The 68000 uses a vectored approach to interrupts with the table of vectors being in memory along with other *exception vectors.* The 68000 treats any departure from normal program flow as exceptions and all exception processing is carried out in the supervisor state of operation. The vector table is as set out in Table 10.4.

Interrupt service routine addresses can be generated in one of two ways:

(a) Auto vector generation, which is associated with a limited number (7) of, usually, synchronous peripheral devices; and
(b) Data bus vectored priority ordered generation.

Table 10.4 Exception vector assignment

Vector numbers	Dec	Address Hex	Space	Assignment
0	0	000	SP	Reset: Initial SSP[2]
	4	004	SP	Reset: Initial PC[2]
2	8	008	SD	Bus Error
3	12	00C	SD	Address Error
4	16	010	SD	Illegal Instruction
5	20	014	SD	Zero Divide
6	24	018	SD	CHK Instruction
7	28	01C	SD	TRAPV Instruction
8	32	020	SD	Privilege Violation
9	36	024	SD	Trace
10	40	028	SD	Line 1010 Emulator
11	44	02C	SD	Line 1111 Emulator
12[1]	48	030	SD	(Unassigned, Reserved)
13[1]	52	034	SD	(Unassigned, Reserved)
14[1]	56	038	SD	(Unassigned, Reserved)
15	60	03C	SD	Initialized Interrupt Vector
16-23[1]	64	040	SD	(Unassigned, Reserved)
	95	05F		-
24	96	060	SD	Spurious Interrupt[3]
25	100	064	SD	Level 1 Interrupt Autovector
26	104	068	SD	Level 2 Interrupt Autovector
27	108	06C	SD	Level 3 Interrupt Autovector
28	112	070	SD	Level 4 Interrupt Autovector
29	116	074	SD	Level 5 Interrupt Autovector
30	120	078	SD	Level 6 Interrupt Autovector
31	124	07C	SD	Level 7 Interrupt Autovector
32-47	128	080	SD	TRAP Instruction Vectors[4]
	191	08F		
48-63[1]	192	0C0	SD	(Unassigned, Reserved)
	255	0FF		-
64-255	256	100	SD	User Interrupt Vectors
	1023	3FF		-

Notes:

1. Vector numbers 12, 13, 14, 16 through 23, and 48 through 63 are reserved for future enhancements of Motorola. No user peripheral devices should be assigned these numbers.
2. Reset vector (0) requires four words, unlike other vectors which only require two words, and is located in the supervisor program space.
3. The spurious interrupt vector is taken when there is a bus error indication during interrupt processing.
4. TRAP #n uses vector number 32 + n.

Source: *Motorola*, published by Prentice Hall Australia

To make the differences plain and to explain the interrupt operation we may first consider the general interrupt servicing procedure as follows:

The device initiating an interrupt: Generates a potential interrupt by placing a 3-bit priority code on interrupt input lines IPL0', IPL1' and IPL2'. Highest priority is when all lines are active, descending in order to a "no interrupt" condition when all lines are negated.

IPL_2'	IPL_1'	IPL_0'	*Priority*	(highest number = highest priority)
A	A	A	7	Highest level
A	A	N	6	
A	N	A	5	
A	N	N	4	A = Assert or Active
N	A	A	3	N = Negated or Inactive.
N	A	N	2	
N	N	A	1	Lowest level
N	N	N	0	No interrupt

In reply, the microprocessor:

1. Compares incoming interrupt level with current level in the status register and:

   if incoming level $\leq$ current level, then servicing is postponed;
   if incoming level > current level, then servicing is commenced as follows.

2. μP finishes current instruction execution.
3. μP places interrupt level on address bus $A_3 A_2 A_1$.
4. Sets R/W' to High.
5. Generates interrupt acknowledge on FC_2 FC_1 FC_0.
6. Asserts AS'.
7. Asserts LDS'.

A vector number is then ascertained either:

(a) From the incoming level on IPL_2' IPL_1' IPL_0' (auto vector generation) if VPA' is asserted, or
(b) From an 8-bit vector placed on the data bus by the device which also asserts DTACK' (VPA' inactive).

Having ascertained the vector number, the microprocessor:

1. latches the vector number;
2. negates LDS';
3. negates AS';

In reply, the interrupting device: negates DTACK'.

Finally, the microprocessor commences interrupt processing.

The way in which the vector number is translated to an address in the exception vector table as follows:

1. *Auto vector generation* (i.e. VPA' asserted). The incoming priority level on IPL_2' IPL_1' IPL_0' is added to 18H (24_{10}) and then multiplied by 4 to obtain the address in the vector table.
2. *Data bus vectored* (i.e. DTACK' asserted, 8-bit vector on data bus). The 8-bit vector is multiplied by 4 by placing it in effect on address lines A_9 to A_2 with 0's on all other address lines.

Since priority level 7 is the highest level then an interrupt at this level may be used, for example, like the non-maskable interrupt of the Z80.

Thus it may be seen that interrupts are well catered for in both machines and that both allow easy connection of two (Z80) or seven (68000) individual interrupting devices with no external hardware requirement.

Where hardware is required, the interfacing techniques discussed earlier can be adapted to the needs.

10.5 Concluding remarks

It is appropriate to have ended this text with an overview of microprocessors.The advent of the μP was the result both of developments in microelectronics design and technology and the application of the fundamental principles of digital systems design. This text also intermixes these topics and it will be seen that this is a natural and logical mix. The applications of microelectronics technology has its major area of application in digital systems, and modern digital systems engineering relies on microelectronics technology for its components and implementation of its systems hardware. VLSI circuit and systems design is representative of the "state of the art" as far as applications are concerned and we have related logic to silicon using VLSI design techniques.

Overall, digital systems is an exciting area in which to work, particularly now that designers enjoy a great deal of scope and freedom. Packaged logic, semi-custom and full custom hardware is now readily and cheaply available and the full potential of microelectronics has not yet been realized so that considerable advances are yet to come. In order to fully take advantage of this situation, all that is required is a good knowledge of the relevant design techniques and available technologies together with creative ability. It is hoped that this text will prove useful in helping to establish a suitable background.

10.6 Tutorial 10

1. Micro processors are often used to accurately time events or signals, etc.
   Using the 68000 instruction set, write a series of instructions to form *a timing routine* which will take exactly 10msec to execute.

   You may assume (which is not the case*) that each 68000 instruction takes a constant time, say 5 μsec, to execute.

2. (a) Using the Z80 instruction set, write a program which will access from memory the contents of a table of 256 X 8-bit values (i.e.,1 word each) and output each value in turn to an output peripheral device at address 22H.

* Actual execution time varies from one type of instruction to another and with addressing modes specified and, of course with the system clock.

The table starts at location 3000H in memory and the entries occupy sequentially ascending memory addresses.

You may assume that your program is to start at location1000H.

(b) Repeat the requirements set out in part (a) but using the 68000 microprocessor instruction set.

3. (a) The Z80 μP presents the following lines to which a user may interface I/O devices:

Eight address lines (for I/O)	A_0 to A_7
Eight bidirectional data bus lines	D_0 to D_7
Three control lines (for I/O)	RD'
	WR'
	IORQ'

Design interface logic to accommodate an input from eight switches (each able to be set at 0 or1) and an output to eight LEDs (light emitting diodes) assumed to have drivers included so that each LED is turned on by a 1.

The allocated I/O addresses are to be:

63H for the eight switches
64H for the eight LEDs.

(b) Indicate how these arrangements would differ if we were interfacing with the 68000 μP.

4. Address decoding arrangements for the memory of a particular microcomputer outfit are given in Figure 10.27. Note that the memory comprises some RAM and 4K EPROM. Draw up a memory map for the entire memory space and indicate how much space is left for further memory expansion.

   Comment on your answer to the last part of the question.

5. Assume that a peripheral device is to use the auto vector generation facilities of the 68000 μP and that when interrupting, it has a priority level of 5. Draw a logic diagram of the necessary external (to the μP) hardware.

   Where will the start address of the service routine for this device be located in memory?

6. A peripheral device is to be able to interrupt the Z80 μP with which it is interfaced as an input device.

   Assume that *interrupt mode* 2 is to be used, and that a value of 34H is loaded in register "I". Explain how the device interrupt arrangements would be configured and connected so that the starting address 1234H of its interrupt service routine would be found in locations 3456H and 3457H. Assuming EI has been executed, outline the operations carried out in generating and starting to service an interrupt.

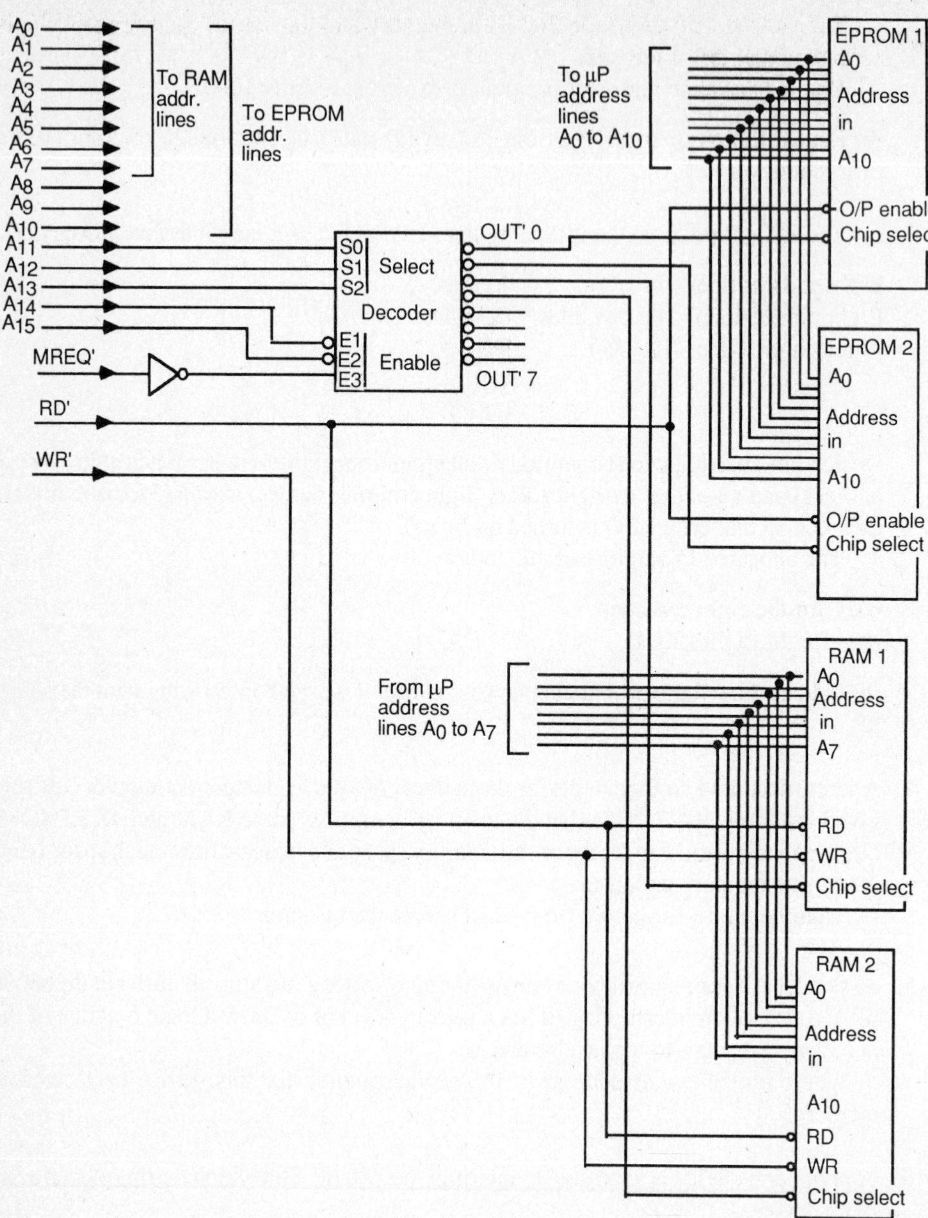

Figure 10.27 Memory addressing circuit

Appendix 1
The design of phase detectors using the logic signal flow graph

This appendix to the text describes the design of an asynchronous sequential circuit which is useful as a phase detector for recovering clock signals in digital communications or for recovering the carrier from an AM signal. The design procedure is illustrated with a new technique called a logic signal flow graph. This logic signal flow graph is a graphical technique which enables the operation of the asynchronous sequential circuit to be evaluated for all possible combinations of input signals.

A1.1 Introduction

Phase-locked loops are finding wide applications in communications. The block diagram of a phase-locked loop is shown in Figure A1.1. It consists of a voltage controlled oscillator (VCO), and a loop filter which primarily determines the transient behavior of the phase-locked loop, but which must also attenuate any high frequency components produced by the phase detector.

In digital communications, phase-locked loops are used in regenerative repeaters to recover timing information from a data stream. Most regenerative repeaters use the transitions in the data stream to extract timing information. Since these transitions occur at irregular intervals, the synchronization of the phase-locked loop is not straightforward and only certain types of phase detectors can be used.

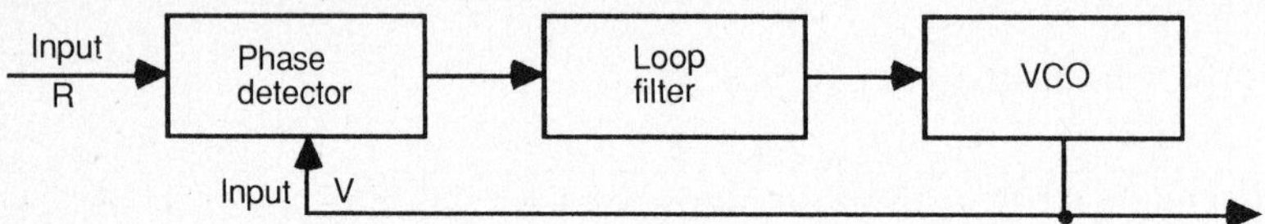

Figure A1.1 Phase-locked loop block diagram

A1.2 Phase detectors

A1.2.1 *Exclusive Or* gate

The simplest type of phase detector is the *Exclusive Or* gate which has the following disadvantages:

1. The mark space ratio of the VCO must be 1:1 otherwise phase errors will occur under normal operation and frequency drift will occur during the time when no transitions in the input data occur.

2. Under locked conditions, there is a large output signal at harmonics of the VCO. These frequency components must be filtered out in the loop filter otherwise FM noise will appear on the VCO output.

* The material set out in this appendix has been kindly provided by: Prof. C.J. Kikkert, Department of Electrical and Electronic Engineering, James Cook University of North Queensland.

3. The phase detector locks the input and VCO output 90° out of phase.

The asynchronous sequential logic phase detectors described below overcome these disadvantages.

A1.2.2 Basic sequential logic phase detector

An ideal phase detector provides no output signal if the phase difference between the inputs is zero, while generating positive and negative outputs if deviations in phase occur. In addition there should be as little high frequency energy as possible.

The waveform shown in Figure A1.2 satisfy these requirements.

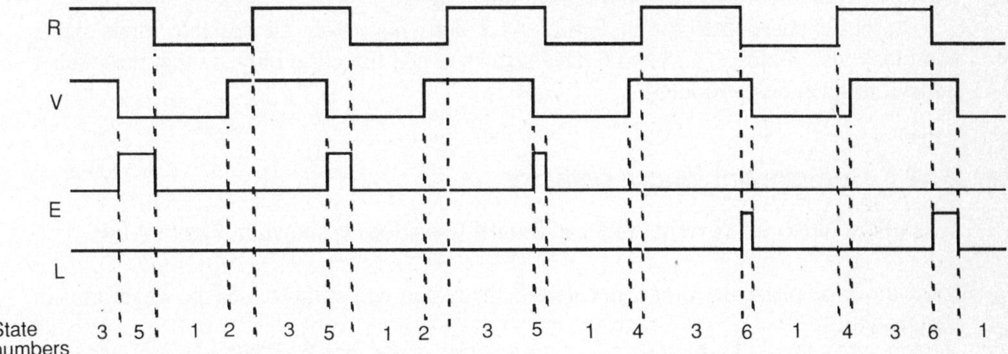

Figure A1.2 Phase detector waveforms

If V is earlier than R, pulses E appear which cause the phase-locked loop to slow down. If V is later than R, pulses L appear which cause the phase-locked loop to speed up.

A phase detector must now be designed using asynchronous sequential logic techniques to produce the outputs shown in Figure A1.2. The design waveforms can be represented by the logic signal flow graph of Figure A1.3.

The state numbers correspond to those in Figure A1.2. The logic signal flow graph has the following features.

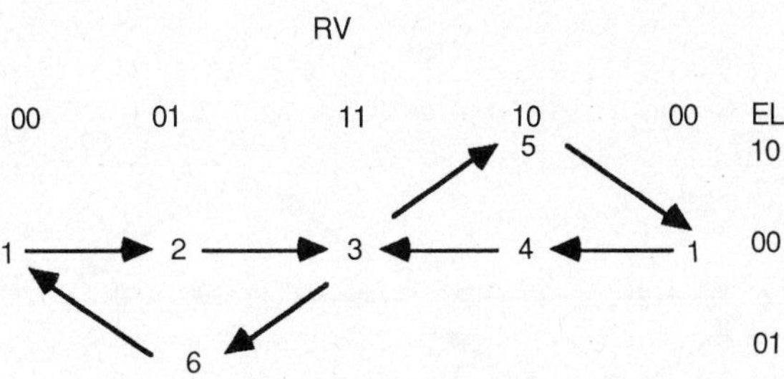

Figure A1.3 Logic signal flow graph for phase detector

1. It show the transitions that take place with the corresponding input and output signals.
2. It allows the output to be evaluated for any input sequence.
3. It enables the operation of the circuit to be visually checked regardless of the numbering of the states. For the above phase detector for instance, if V leads R the input sequence progresses from left to right and only E pulses occur. If V lags R the input sequence progresses from right to left and only L pulses occur. One can thus very simply check the operation of the circuit.

Two features of the diagram should be noted. First, the left-hand column corresponding to $RV = 00$ has been duplicated in the right-hand column. This keeps the diagram simple. Secondly, each state must have one only arrow to the left and one only arrow to the right, corresponding to both possible changes in input signal.

The basic phase detector in Figure A1.3 does not cover all possible input signal combinations. States 2, 4, 5 and 6 have arrows to one direction only, so that these other transitions must be considered.

A1.2.3 Combination phase detector

A simple solution for covering these additional transitions is shown in Figure A1.4.

Since the basic phase detector is not altered, the circuit will still produce the waveforms in Figure A1.2.

If $R = 1$ and V is a higher frequency, changes from state 3 to 5 and vice versa will occur, generating E pulses which cause the phase-locked loop to slow down the VCO output which is V. If $V = 1$ and R is a higher frequency, transitions between states 3 and 6 occur which give rise to L pulses which cause the phase-locked loop to speed V up and make it the same frequency as R. The phase detector will thus act as a phase and frequency detector if R or V are a 1.

If $R = 0$ and V is a higher frequency, transitions between states 1 and 2 occur and no output impulses are generated. Similarly if $V = 0$ and R is a higher frequency, transitions occur between states 1 and 4 and no output pulses are generated. The phase detector will thus work as a phase only detector when R or V is zero. This feature is useful in digital communications where transitions in the data stream do not always occur. In order to bias the phase detectors in the state $R = 0$, transitions in the binary data stream are used to fire a monostable

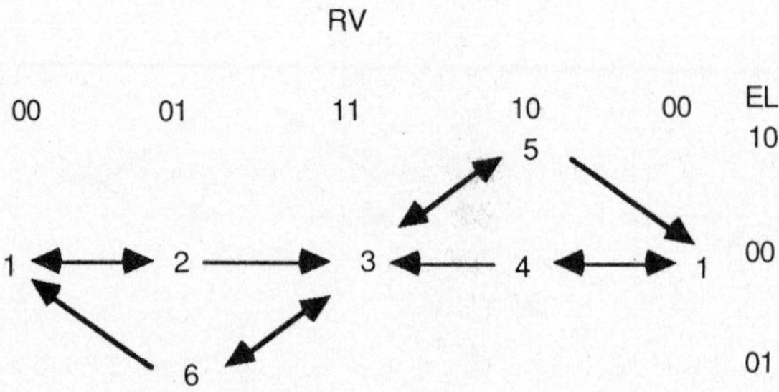

Figure A1.4 Modified logic signal flow graph for phase/frequency detector

multivibrator, generating a positive pulse which is used as the R input to the phase detector.

The phase detector is also suitable for the synchronous detection of AM, where the carrier amplitude becomes small. The AM input signal is biassed to $R = 0$ for small carrier signals which correspond to large modulation levels.

A1.2.4 Phase detector for negative pulses

If the input signal R and V contain negative going pulses only, the phase detector must be modified to allow for these possibilities. The resulting logic signal flow graph is shown in Figure A1.5.

It can be seen that the basic phase detector of Figure A1.3 is still incorporated in the circuit but that the states 7 and 8 have now been added to allow the circuit to respond to narrow negative going pulses. Similar to the combination phase detector above, this phase detector works as a phase only detector if R and V are zero and as a phase and frequency detector if R and V are a 1.

A1.2.5 Phase and frequency detector

If phase and frequency detection is required for all possible input signals, the logic signal flow graph must be modified to that of Figure A1.6.

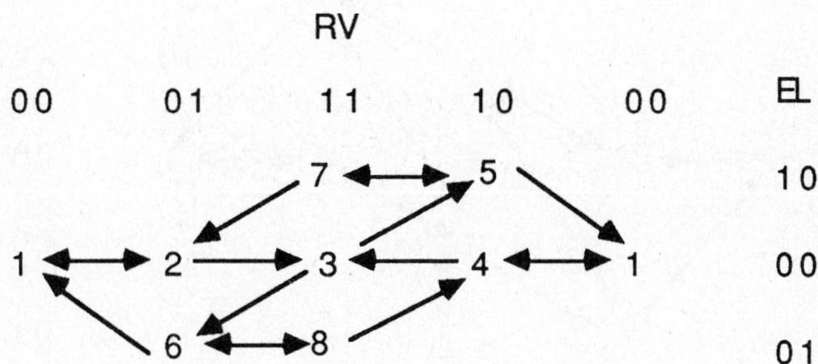

Figure A1.5 Logic signal flow graph modified for pulses

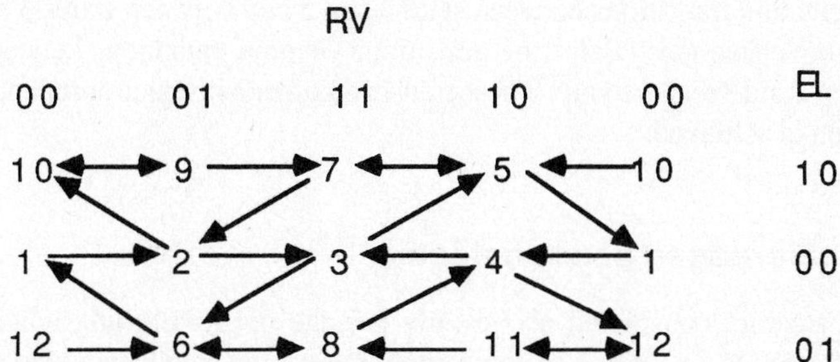

Figure A1.6 Logic signal flow graph for all possible input signals

It can be seen that the basic frequency detector of Figure A1.3 is still incorporated. If V is a higher frequency than R, the circuit gravitates to transitions between 5 and 7 and between 9 and 10 and an output E is produced. If R is a higher frequency than V, the circuit gravitates to transitions between 6 and 8 and between 11 and 12 and an output L is produced. Frequency as well as phase detection is thus obtained. The Motorola MC 4344, MC 4044 and MC 14046B phase/frequency detectors are realizations of this phase and frequency detector. Motorola, in their data sheet for the MC 14046B phase detector*, have drawn a diagram which shows the transitions between the logic states of their phase and frequency detector. Their diagram, however, does not give the intuitive understanding of the phase detector operation that the logic signal flow graph gives.

A1.2.6 Phase only detector

In some applications, frequency detection is undesirable and only phase detection is required. If the inputs to the phase detector are of different frequencies the phase only detector should produce no output. This can be achieved with the logic signal flow graph of Figure A1.7.

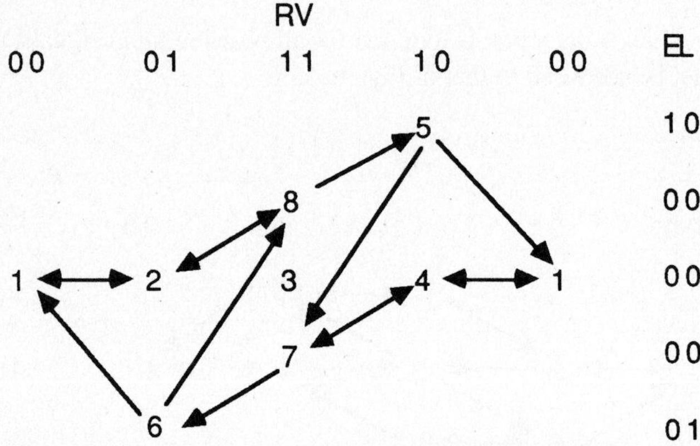

Figure A1.7 Logic signal flow graph for phase only detection

It can be seen that the state 3 has been split up into two states, 7 and 8, in order to avoid the frequency detecting transitions between state 3 and 5 and between states 3 and 6 of Figure A1.4. The above frequency detecting transitions become transitions between states 4 and 7 and between 2 and 8 respectively. These transitions do not produce output pulses and phase only detection is achieved.

A1.2.7 Double-edged phase detector

The phase detectors considered above only use the negative going edges of the input waveforms for phase detection. In some applications it is an advantage to generate E and

* Motorola Semiconductors Products Inc., Semiconductor Data Library, CMOS Integrated Circuits Data Book, Volume 5, Series B, 1976.

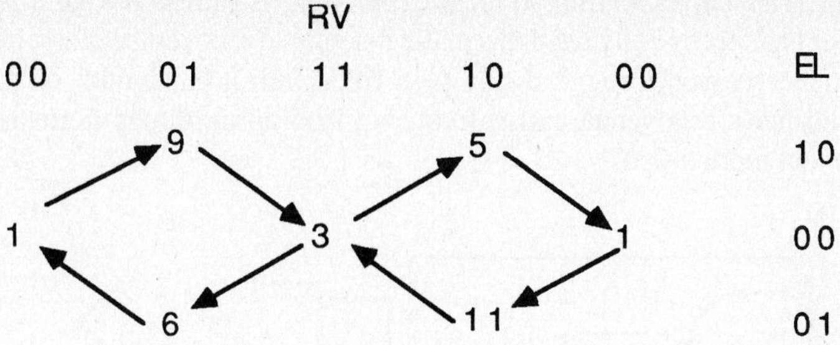

Figure A1.8 Logic signal flow graph for double edged phase detection

L pulses on the positive going edges as well as the negative going edges. The logic signal flow graph of Figure A1.8 obtains the desired result.

Figure A1.8 is a logic signal flow graph of a basic phase detector and like Figure A1.3 it does not include all input signals, phase and frequency detectors, phase detectors for pulsed inputs and phase only detectors can also be obtained.

A1.3 Phase detector design

In order to illustrate the design of a phase detector using the logic signal flow graph, the phase detector of Figure A1.4 will be realized using standard procedures*. The logic signal flow graph can be used directly to draw the primitive flow table of Figure A1.9.

RV 00	01	11	10	EL
①	2		4	00
1	②	3		00
	6	③	5	00
1		3	④	00
1		3	⑤	10
1	⑥	3		01

Figure A1.9 Primitive flow table

The primitive flow table can be merged to give a merged flow table which can then be used to obtain the following Boolean algebraic expressions for the circuit:

$$F = RV + fV + fR$$

$$E = F R V'$$

$$L = F R'V$$

* W.E. Wicks, *Logic Design with Integrated Circuits*, J. Wiley & Sons, 1968.

The circuit diagram corresponding to these expressions is shown in Figure A1.10.

The author has successfully used this phase detector to recover the clock from a pseudo-random sequence passed through a low pass filter with a bandwidth of half the clock frequency, and then to recover the carrier frequency from an amplitude modulated waveform with 100 percent modulation.

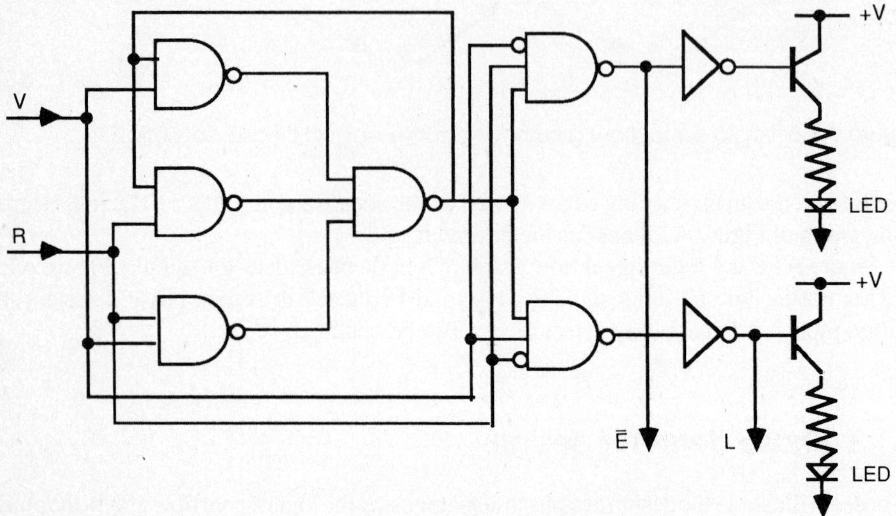

Figure A1.10 Phase detector for AM and digital communications

A1.4 Conclusion

The logic signal flow graph has been used to illustrate the operation of different types of phase detectors which all exhibit the following properties:

1. The phase detectors will operate without error with any mark space ratio.

2. Under perfectly locked conditions, there are no output signals from the phase detector, resulting in very little FM noise in the VCO output of a phase-locked loop.

3. The phase detector locks the input and VCO output in phase, with extremely small phase errors.

Appendix 2
Transition equations (TEs) and TE based design

A2.1 Introduction

In our consideration of clocked circuits, the nature of the clock activation has only been stated in words or indicated in the symbols used. Thus, in JK or D flip-flop based designs, for example, difficulties arise or design opportunities may be missed for the following reasons:

1. The clock signal does not appear in the characteristic tables or equations.
2. The *nature* of the clock activation does not appear, i.e. ΔT or ∇T. This information is critical for edge triggered circuits. For example, in a ripple through counter (dealt with in Chapter 8) as in Figure A2.1, the direction of count depends on the nature of the clock activation. ∇T clocking will give a forward (up) count while ΔT clocking will cause the circuit to count in reverse .
3. JK and D flip- flops usually have asynchronous Preset and/or Clear inputs which also do not appear in the characteristics.

A2.2 Transition characteristic equations

A solution to these shortcomings is to be found in a *transition equation* (TE)* approach. This requires the use of "*difference operators*"** the significance of which may be summarized as follows:

For any binary variable X, four possible events may occur at time t:

1. A transition of X from logic 0 to logic 1, denoted by ΔX.
2. A transition of X from logic 1 to logic 0, denoted by ∇X.
3. No change of X from logic 0, denoted by $\overline{\Delta} X$.
4. No change of X from logic 1, denoted by $\overline{\nabla} X$.

A further operator ∂ has the significance:

$$\partial X = \Delta X + \nabla X \text{ (any change in } X\text{) and ... } \bar{\partial} X = \overline{\Delta} X + \overline{\nabla} X \text{ (no change in } X\text{).}$$

With care, these operators can be manipulated by Boolean algebra provided that the significance of each operator is remembered. For example, the complementary variable will have the complementary transition:

$$\Delta X = \nabla X' \; ; \; \nabla X = \Delta X' \; ; \; \overline{\Delta} X = \overline{\nabla} X' \; ; \; \overline{\nabla} X = \overline{\Delta} X' \; .$$

Also note that:

$$(\Delta X)' \neq \nabla X \; ; \; (\nabla X)' \neq \Delta X, \text{ etc.}$$

* Pucknell, D.A. "Transition equations for the analysis and synthesis of sequential circuits", *IEE Electronic Letters*, 1970, 6, (23), pp. 731-3.
Smith, J.R. and Roth, C.H.,"Analysis and synthesis of asynchronous sequential networks using edge sensitive flip-flops," *IEEE Trans.*, 1971, C-20, pp. 847-55.
** Talantsev, A.D.,"On the analysis and synthesis of certain electrical circuits by means of special logical operators" *Avtom & Telemekh*, 1959, 20, pp. 895-907.

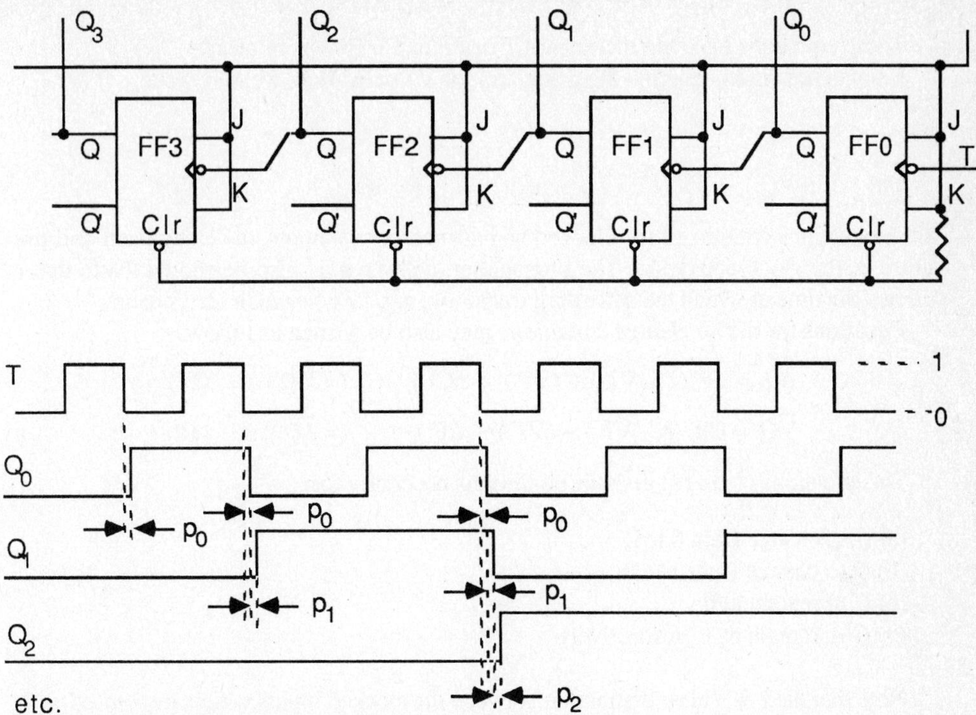

Figure A2.1 Ripple-through clocking (∇T clocking)

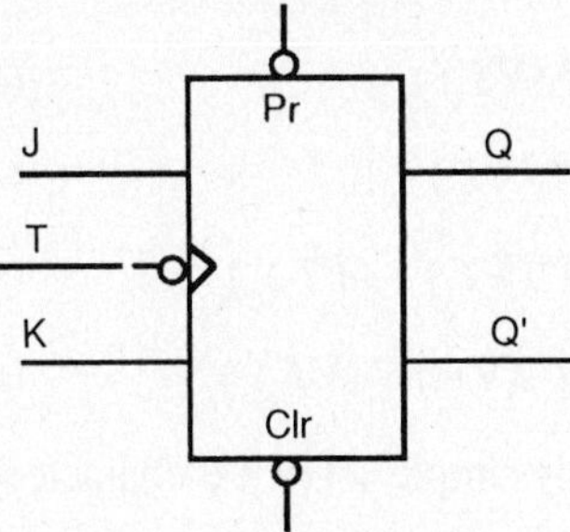

Figure A2.2 JK flip-flop with Preset (Pr′) and Clear (Clr′). Negative edge (∇T) activation

since there are four possible events. Thus, for example, we may interpret $(\Delta X)'$ as

$$(\Delta X)' = \nabla X + \overline{\nabla} X + \overline{\Delta} X$$

The operators may be used to advantage in setting out the transition equation (TE) characteristics of edge triggered circuits. Take, for example, the characteristic equations for a JK flip-flop as in Figure A2.2. Up to now we have written:

$$Q_{t+1} = (Q'J + Q.K')_t \text{ and } Q'_{t+1} = (Q'J' + Q.K)_t$$

These equations make no mention of T or Pr. or Clr (preset or clear).
Using a transition equation approach and for ∇T activation, we may write:

$$\Delta Q_{t+p} = \text{Clr}'.(J.(\nabla T) + \Delta\text{Pr})_t \qquad (= \nabla Q'_{t+p}) \qquad \textbf{(1a)}$$

and

$$\nabla Q_{t+p} = \text{Pr}'.(K.(\nabla T) + \Delta\text{Clr})_t \qquad (= \Delta Q'_{t+p}). \qquad \textbf{(1b)}$$

These equations contain all the clocked and asynchronous inputs, the clock itself and the nature of the clock activation. The propagation delay p may also be shown. Note that t denotes the time at which the activating transition, e.g. (∇T) or ΔClr etc., occurs.

Equations for the *no change conditions* may also be written as follows:

$$\bar{\Delta}Q = \text{Pr}'.(J'.(\nabla T) + (\nabla T)' + \Delta\text{Clr})_t \qquad (= \bar{\nabla}Q') \qquad \textbf{(1c)}$$

$$\bar{\nabla}Q = \text{Clr}'.(K'.(\nabla T) + (\nabla T)' + \Delta\text{Pr})_t \qquad (= \bar{\Delta}Q'). \qquad \textbf{(1d)}$$

The equations (1a)-(1d) give the conditions necessary for:

(a) Q to change from 0 to1;
(b) Q to change from 1 to 0;
(c) Q to remain at 0 ;
(d) Q to remain at 1, respectively.

Note that there is a clear distinction between the clocked inputs,which are *And*-ed with (∇T) in this case, and the asynchronous inputs which are effective alone. If the flip-flop has no asynchronous inputs or if they are not to be used then the TE characteristics become simply:

$$\Delta Q_{t+p} = (J.(\nabla T))_t \qquad (= \nabla Q'_{t+p}) \quad \textbf{(2a)}$$

$$\nabla Q_{t+p} = (K.(\nabla T))_t \qquad (= \Delta Q'_{t+p}). \quad \textbf{(2b)}$$

$$\bar{\Delta}Q = (J'.(\nabla T) + (\nabla T)')_t \qquad (= \bar{\nabla}Q') \quad \textbf{(2c)}$$

$$\bar{\nabla}Q = (K'.(\nabla T) + (\nabla T)')_t \qquad (= \bar{\Delta}Q'). \quad \textbf{(2d)}$$

For a D flip-flop a similarly simple set of TE characteristics follows:

$$\Delta Q_{t+p} = (D.(\nabla T))_t \qquad (= \nabla Q'_{t+p}) \quad \textbf{(3a)}$$

$$\nabla Q_{t+p} = (D'.(\nabla T))_t \qquad (= \Delta Q'_{t+p}). \quad \textbf{(3b)}$$

$$\bar{\Delta}Q = (D'.(\nabla T) + (\nabla T)')_t \qquad (= \bar{\nabla}Q') \quad \textbf{(3c)}$$

$$\bar{\nabla}Q = (D.(\nabla T) + (\nabla T)')_t \qquad (= \bar{\Delta}Q'). \quad \textbf{(3d)}$$

It may be seen that this type of TE characterisation is concise but full. We will now briefly apply it to the design of clocked sequential circuits.

A2.3 Example : The TE design of an 8421 BCD counter (up count only)

We have already examined the design of an up/down version of an 8421 BCD counter in Chapter 9, but the simpler up count only version will serve here to illustrate features of the TE design process. In this case we will assume that the input to be counted appears on line *W* and that the count is to advance on every falling edge (∇W).

We may start by setting out a state transition table as in Table A2.1 in which we have allocated vector *DCBA*, where *A* is taken as the LSB, to represent the state of the count.

Table A2.1 State table for BCD generator

BCD Count/ State	Counter D	C	Outputs B	A	Input W
0	0	0	0	0	— ∇W1
1	0	0	0	1	— ∇W2
2	0	0	1	0	— ∇W3
3	0	0	1	1	— ∇W4
4	0	1	0	0	— ∇W5
5	0	1	0	1	— ∇W6
6	0	1	1	0	— ∇W7
7	0	1	1	1	— ∇W8
8	1	0	0	0	— ∇W9
9	1	0	0	1	— ∇W10
0	0	0	0	0	

Let us further assume that we are to design the circuit using falling edge clocked JK flip-flops having an active Low Clr' input but no Preset input as shown in Figure A2.3. Equations (1a)-(1d) may be suitably restated as follows:

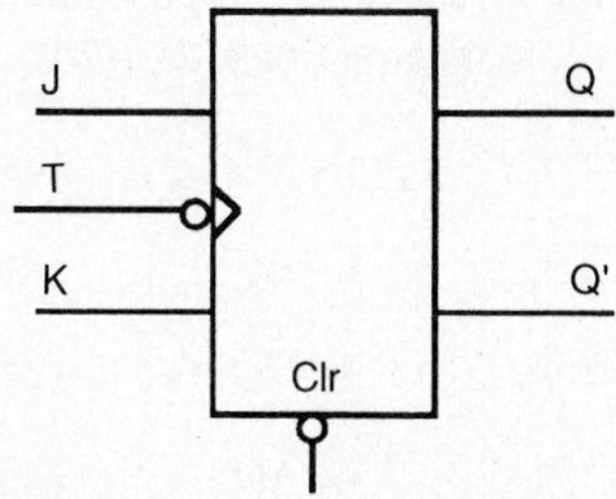

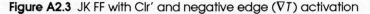

Figure A2.3 JK FF with Clr' and negative edge (∇T) activation

$$\Delta Q_{t+p} = \text{Clr}'.(J.(\nabla T))_t \qquad (= \nabla Q'_{t+p}) \qquad \textbf{(4a)}$$

$$\nabla Q_{t+p} = (K.(\nabla T) + \Delta \text{Clr})_t \qquad (= \Delta Q'_{t+p}) \qquad \textbf{(4b)}$$

$$\bar{\Delta} Q = (J'.(\nabla T) + (\nabla T)' + \nabla \text{Clr})_t \; (= \bar{\nabla} Q') \qquad \textbf{(4c)}$$

$$\bar{\nabla} Q = \text{Clr}'.(K'.(\nabla T) + (\nabla T)')_t \quad (= \bar{\Delta} Q'). \qquad \textbf{(4d).}$$

Inspection of these equations reveals that there are at least three separate ways in which the design may be pursued.

A2.3.1 Using the J and K inputs only

The asynchronous Clr input is confined to an overall reset to zero function. This is a "traditional" design which may have a state transition map set out as Table A2.2.

Table A2.2 State transition map

$DCBA_{t+p}$ / $(DC)_t$ \ $(BA)_t$	00	01	11	10
00	$\bar\Delta\ \bar\Delta\ \bar\Delta\ \Delta$	$\bar\Delta\ \bar\Delta\ \Delta\ \nabla$	$\bar\Delta\ \Delta\ \nabla\ \nabla$	$\bar\Delta\ \bar\Delta\ \bar\nabla\ \Delta$
01	$\bar\Delta\ \bar\nabla\ \bar\Delta\ \Delta$	$\bar\Delta\ \bar\nabla\ \Delta\ \nabla$	$\Delta\ \nabla\ \nabla\ \nabla$	$\bar\Delta\ \bar\nabla\ \bar\nabla\ \Delta$
11	Ø Ø Ø Ø	Ø Ø Ø Ø	Ø Ø Ø Ø	Ø Ø Ø Ø
10	$\bar\nabla\ \bar\Delta\ \bar\Delta\ \Delta$	$\nabla\ \bar\Delta\ \bar\Delta\ \nabla$	Ø Ø Ø Ø	Ø Ø Ø Ø

Noting that a Δ entry implies J=1, $\bar{\Delta}$ implies J=0, (K=Ø in both cases) and, a ∇ entry implies K=1, $\bar{\nabla}$ implies K=0, (J=Ø in both cases), we may readily extract expressions for J and K for each flip-flop as follows:

$$J_D{=}CBA\,;\; K_D{=}\,A\,;\; J_C{=}\,BA\,;\; K_C{=}\,BA\,;\; J_B{=}\,D'A\,;\; K_B{=}\,A\,;\; J_A{=}K_A{=}\,1$$

This leads to the realization shown in Figure A2.4.

A2.3.2 Using the Clock inputs "T" alone

Almost all the sequence may be generated with $J = K = 1$ for all flip-flops. Turning back to the TE based characteristics and noting that it is proposed to make $J = K = 1$ and not use the asynchronous Clr inputs we have, from equations (2a)-(2d):

$$\Delta Q_{t+p} = (\nabla T)_t \qquad (= \nabla Q'_{t+p}) \qquad \textbf{(2a')}$$

$$\nabla Q_{t+p} = (\nabla T)_t \qquad (= \Delta Q'_{t+p}). \qquad \textbf{(2b')}$$

$$\bar{\Delta} Q = (\nabla T)'_t \qquad (= \bar{\nabla} Q') \qquad \textbf{(2c')}$$

$$\bar{\nabla} Q = (\nabla T)'_t \qquad (= \bar{\Delta} Q'). \qquad \textbf{(2d').}$$

In words, the flip-flop will change state every time it is clock activated by (∇T) and will

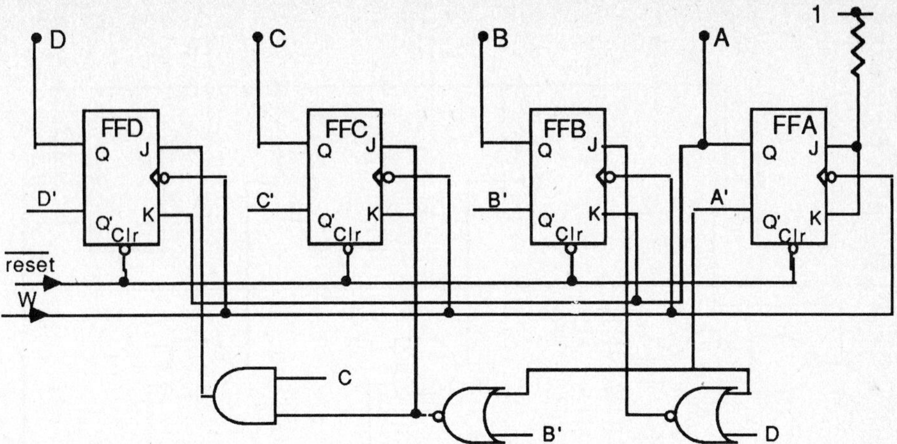

Figure A.2.4 Realization (i) for BCD generator (using J and K inputs) (Synchronous with ∇W)

not change state if no activation occurs-(∇*T*)'. Referring to Table A2.1, we may see that FFA must change state on every input ∇*W* so that a direct connection of *W* to the T_A input will effect this, that is:

$$T_A = W \qquad (\nabla T_A = \nabla W)$$

Now, for FFB, we see from the state table that four transitions of *B* are required coincident with ∇*W* and associated with:

$$D'C'B'A + D'C'BA + D'CB'A + D'CBA = D'A$$

Therefore, *And*-ing *D'A* with *W* will produce the required input to T_B, that is:

$$T_B = D'AW.$$

For FFC, two transitions only are needed associated with:

$$D'C'BA + D'CBA = D'BA.$$

but noting that *DC'BA* + *DCBA* do not occur and are therefore don't cares, we have:

$$T_C = BAW.$$

Finally, for FFD we have two transitions associated with:

$$D'CBA + DC'B'A$$

and to avoid the possibility of getting locked out in Illegal (unused) states we will not use the don't care states so that:

$$T_D = (D'CBA + DC'B'A).W.$$

The completed circuit is given as Figure A2.5.

Note: An overall "reset" line may be connected in realizations (i) and (ii) as shown, but this in no way interferes with the clocked inputs since reset is normally held at logical 1.

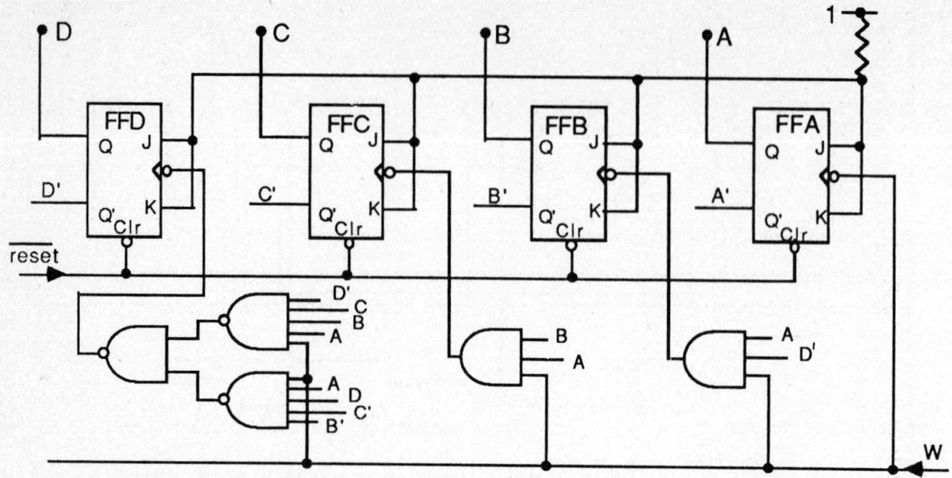

Figure A.2.5 Realization (ii) for BCD generator (using "clock" inputs only) (Synchronous with ∇W)

A2.3.3 A ripple through approach using *T* and also Clr inputs.

If we look again at the state transition table and at equations (2a')- (2d') we may see that with $J = K = 1$ for all flip-flops and leaving T_A = W, we can satisfy the remaining T input requirements by clocking each stage from the Q output of the preceding stage, so that:

for example: $\nabla T_B = \nabla A$

that is: $T_B = A$

similarly: $T_C = B$

$T_D = C$

and this will generate the required sequence of states up to and including state 9 but will not produce the required change from state 9 to state 0.

FFA will change to 0 as required since it is clocked by W. FFB would also be changed by the consequent ∇A transition of the Q output of FFA. This must be prevented and if we look to equations (1c):

$$\bar{\Delta}Q = \text{Pr}'.(J'.(\nabla T) + (\nabla T)' + \Delta\text{Clr})_t \qquad (= \bar{\nabla}Q') \quad \textbf{(1c)}$$

we see that this can be prevented by arranging for a ΔClr_B (i.e. a $\nabla\text{Clr}'_B$) input. If we thus prevent FFB from changing then FFC will not be clocked and will stay at 0 as required.

Finally for FFD we may see that the required ∇D transition may be achieved through the Clr_D input as in equation (1b):

$$\nabla Q_{t+p} = \text{Pr}'.(K.(\nabla T) + \Delta\text{Clr})_t \qquad (= \Delta Q'_{t+p}). \quad \textbf{(1b)}$$

This shows that a ΔClr_D (i.e. a $\nabla\text{Clr}'_D$) input is required.

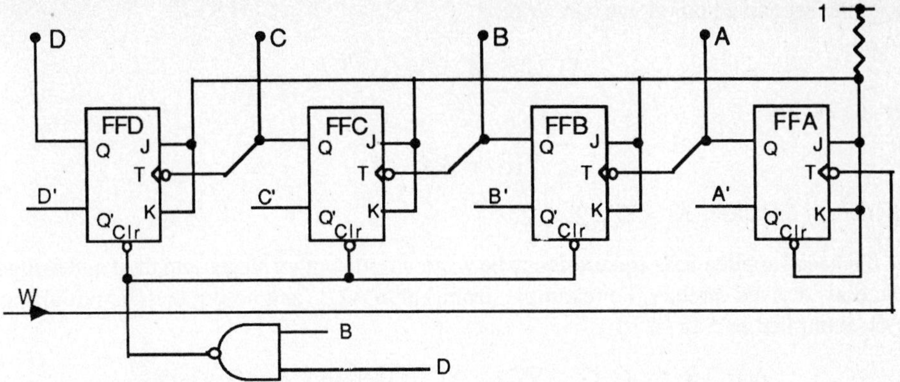

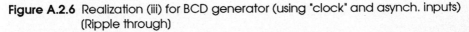

Figure A.2.6 Realization (iii) for BCD generator (using "clock" and asynch. inputs) (Ripple through)

Both $\nabla \mathrm{Clr'}_B$ and $\nabla \mathrm{Clr'}_D$ must be coincident with ∇W (i.e.$\Delta W'$) and with state 9 so that we may write:

$$\nabla \mathrm{Clr'}_B = \nabla \mathrm{Clr'}_D = \mathrm{DA}\Delta W' \quad \text{[or in conventional form: } \mathrm{Clr'_B} = \mathrm{Clr'_D} = DAW'\text{]}$$

which is a 3-input *Nand* gate.The overall circuit is given as FigureA2.6.

It is interesting to compare the three realizations (Figures A2.4-A2.6).

A2.4 Application equations revisited

We may present application equations in TE form, and as before, the input equations for JK flip-flops, for example, may be read directly from these equations.

Take, for example, a 4-bit Gray Code sequence as set out in Table A2.3. The conventional application equations for, say, stage Q_3 are as follows:

$$(Q_3)_{t+1} = \{Q'_3(Q_2.Q'_1.Q'_0) + Q_3(Q_2+Q_1+Q_0)\}_t \qquad \textbf{(5a)}$$

and,

$$(Q'_3)_{t+1} = \{Q_3(Q'_2.Q'_1.Q'_0) + Q'_3(Q'_2+Q_1+Q_0)\}_t \qquad \textbf{(5b)}$$

It may be seen that the first term on the right-hand side of each of equations (5a) and (5b) is the term associated with a transition of Q_3 (in this case). We can therefore write:

$$\Delta(Q_3) = Q_2.Q'_1.Q'_0(\nabla T)$$

Assuming we are to use JK FFs (FF3, FF2, FF1 and FFO) to realize the code generator outputs, we may compare this with the TE characteristic equation for the JK FF:

$$\Delta(Q) = J.(\nabla T).$$

We see that, for FF3 , we therefore have:

$$J_3 = Q_2.Q'_1.Q'_0.$$

From the second equation, we can write:

$$\nabla(Q_3) = Q'_2.Q'_1.Q'_0(\nabla T)$$

For the JK FF:

$$\nabla(Q) = K.(\nabla T).$$

whence, in this case, $K_3 = Q'_2.Q'_1.Q'_0$.

TE based applications equations can be written for the other stages and the J and K input equations derived directly. For example, from Table A2.3 (and noting that the initial state of Q_2 is implied in Δ or in ∇):

$$\Delta(Q_2) = Q'_3.Q_1.Q'_0.(\nabla T). \qquad \text{therefore } J_2 = Q'_3.Q_1.Q'_0$$

$$\nabla(Q_2) = Q_3.Q_1.Q'_0.(\nabla T). \qquad \text{therefore } K_2 = Q_3.Q_1.Q'_0 \text{ etc.}$$

Table A2.3

	Outputs				Clock
	Q_3	Q_2	Q_1	Q_0	Activation
	0	0	0	0	
t	0	0	0	1	$(\nabla T)_1$
$t+1$	0	0	1	1	$(\nabla T)_2$
	0	0	1	0	$(\nabla T)_3$
	0	1	1	0	
	0	1	1	1	
	0	1	0	1	
t	0	1	0	0	
$t+1$	1	1	0	0	$(\nabla T)_8$
	1	1	0	1	
	1	1	1	1	
	1	1	1	0	
	1	0	1	0	
	1	0	1	1	
	1	0	0	1	
	1	0	0	0	$(\nabla T)_{15}$
	0	0	0	0	$(\nabla T)_{16}$
	etc. Repeating				

As before, we see that application equations lead directly to input equations but an additional benefit arises from the use of a TE form. Since the specified transition, e.g. $\Delta Q3$, implies both the initial and final state of $Q3$, then only the state of the remaining variables is relevant. This means, for example, that a 5-variable problem can be mapped onto a 4-variable map.

It is also, very easy to directly write the TE form application equations from a state transition table or diagram.

Analysis is also readily carried out by reversing the processes of design. To further illustrate the design potential of a TE based approach, let us take a particular case study—a design we have worked through earlier in Chapter 8 of the text.

A2.5 A particular case study—the design of a successive approximation register (SAR) for an A/D converter

The overall arrangement of a successive approximation register was given as Figure 8.22. Once again, as in section 8.3.3 of the text, we will design a 3-bit SAR from which the requirements of larger systems may be readily deduced.

The overall specification of requirements may be stated as follows:

1. Number of A/D converter bits =3.
2. Conversion is to start on receipt of a Hi level on the "Start" input.
3. An EOC (end of conversion) output (Hi) is to be produced when the conversion process is completed. EOC to be Lo during conversion.
4. Negative edge clock activated JK flip-flops are to be used.
5. The comparator will generate an error signal *e* such that *e* is to change from Lo to Hi when $V_{out} \geq V_a + 0.5\partial V$ and change from Hi to Lo when $V_{out} \leq V_a - 0.5\partial V$, where ∂V is the quantizing increment or LSB.

A "conventional" design was carried out in Chapter 8 and the resulting circuit was presented as Figure 8.25. We now use a TE approach to achieve an alternative, simpler realization.

A2.5.1 A TE design approach

The state transition diagram (Figure 8.24) is repeated here for convenience as Figure A2.7. From the diagram we may set out a state transition table as in Table A2.4. This truly is a transition table as all transitions to the next states are mapped.

The design now proceeds using the TE form of the JK FF characteristics as follows:

$$\Delta Q_{t+p} = \text{Clr}'.(J.(\nabla T) + \Delta\text{Pr})_t \qquad (= \nabla Q'_{t+p}) \qquad \textbf{(1a)}$$

and

$$\nabla Q_{t+p} = \text{Pr}'.(K.(\nabla T) + \Delta\text{Clr})_t \qquad (= \Delta Q'_{t+p}). \qquad \textbf{(1b)}$$

Let us use the *J* and *K* and *T* inputs only for the register sequence with Clr inputs being confined to a reset on "start" operation (as in Figure A2.8).

Table A2.4 TE State transition table

	Present state				Transitions to next state								Activating clock pulse
	C	B	A	H(EOC)	E = 0				E = 1				
①	0	0	0	0	ΔC	-	-	-	ΔC	-	-	-	CA1
②	1	0	0	0	-	ΔB	-	-	∇C	ΔB	-	-	CA2
③	0	1	0	0	-	-	ΔA	-	-	∇B	ΔA	-	CA3
④	1	1	0	0	-	-	ΔA	-	-	∇B	ΔA	-	CA3
⑤	0	0	1	0	-	-	-	ΔH	-	-	∇A	ΔH	CA4
⑥	0	1	1	0	-	-	-	ΔH	-	-	∇A	ΔH	CA4
⑦	1	0	1	0	-	-	-	ΔH	-	-	∇A	ΔH	CA4
⑧	1	1	1	0	-	-	-	ΔH	-	-	∇A	ΔH	CA4
⑨ to ⑯	X	X	X	1	Transitions here to be independent of "E" but initiated by ΔS (or ΔH if "S" already HI). The transitions are to be $(\nabla C + \overline{\Delta}C)$. $(\nabla B + \overline{\Delta}B)$. $(\nabla A + \overline{\Delta}A)$. ∇H								ΔS or ΔH if S = 1

All changes here (states ① to ⑧) are independent of "Start" input "S".

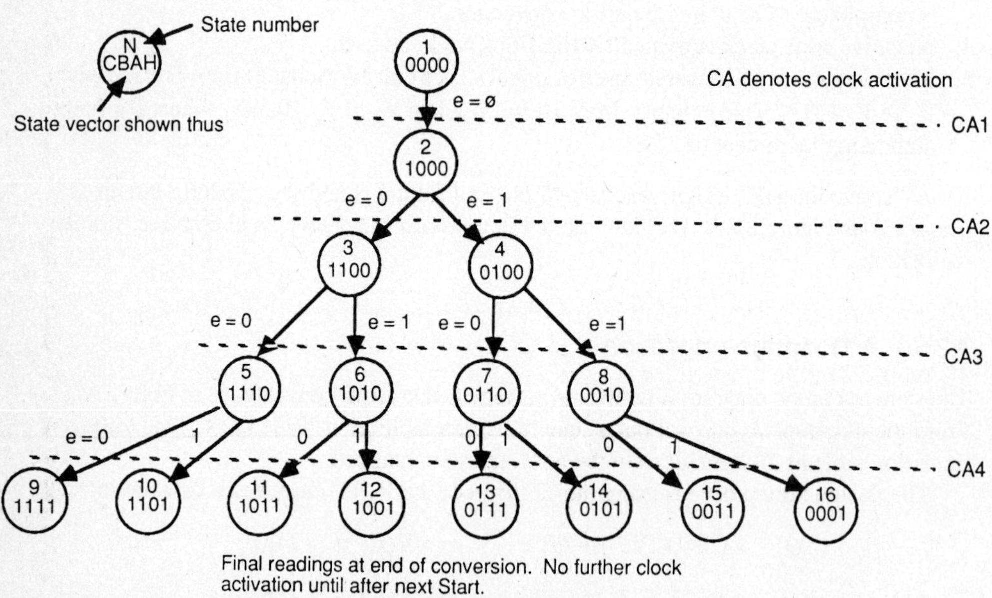

Figure A2.7 (Copy of Figure 8.24) State transition diagram for 3-bit SAR

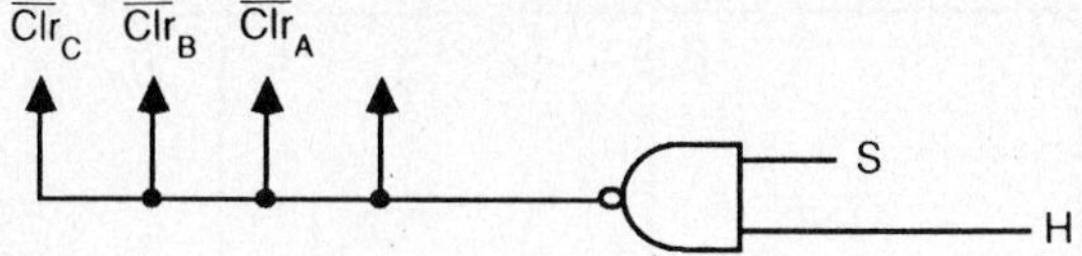

Figure A2.8 Start (reset) circuit

Considering possible use of the T inputs, we see that FFC is set on CA1 and reset or not on CA2 *and thereafter retains the condition following CA2*. If the clock signal is removed after CA2, then there can be no further disturbance to the setting of FFC. Similarly, the clock signal to FFB can be removed after CA3 and that to FFA and to FFH after CA4.

Formalizing this requirement, and with regard to Table A2.4, we may write:

$$T_H = T_A = T.H'$$

where T_A is the T input of FFA etc., and T is the clock input to the register.

Similarly: $T_B = T_A .A'$

and, $T_C = T_B .B'$.

The complete clock line gating circuit is thus very simple as shown by Figure A2.9 and clock line skew is not really important since, during conversion, stages operate sequentially and no output is available until EOC is generated.

Now in formulating J and K input requirements,we map don't care states wherever the clock signal is no longer present for the particular flip-flop.

It now remains to map 1's for J wherever Δ transitions of the output Q are required and 1's for K wherever ∇ transitions of the output Q are required.

Furthermore, recall that when mapping the input requirements of any flip-flop the variable represented by that FF need not be mapped since its state is implied in the operator Δ or ∇.

A final simplification in procedure may be made by recognizing that the error signal e can be omitted from the mapping since e=1 is always coincident with clocked ∇Q transitions of FFC, FFB and FFA, i.e. with K=1 input to these flip-flops. Thus the state of e may be omitted in mapping provided that e is later *and* -ed with all K input equations for FF C, B and A. To create the maps, entries are readily obtained from Table A2.4, 1 entries appearing for J where ΔQ is needed, 1 for K where ∇Q is required and "don't cares" appearing wherever

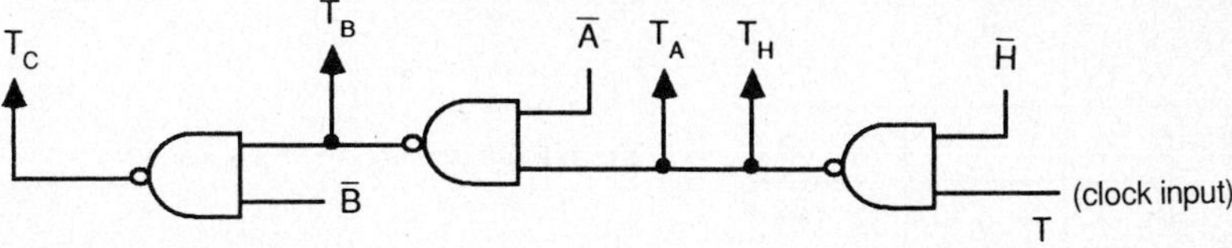

Figure A2.9 Clock line circuit

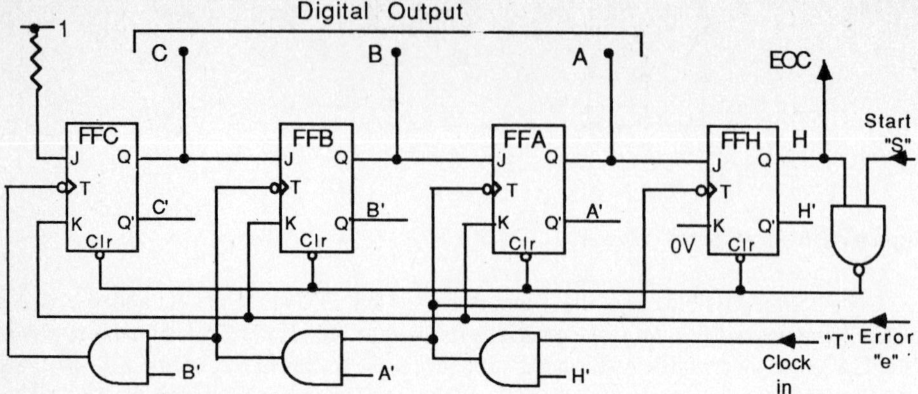

Figure A2.10 Circuit arrived at by the TE design approach—alternative design for a 3-bit successive approximation register

the clock input T_N is not present. Maps for all flip-flops appear as Tables A2.5(a)-(d) and the very simple J and K input equations are derived from those maps as shown.

The final circuit arrangement is shown as Figure A2.9 and will be seen to comprise four JK FFs, three 2I/P *And* gates and one 2I/P *Nand* gate as shown. The design is readily expanded to n bits by adding further stages requiring only one JK FF and one 2 I/P *And* gate per added bit.

A2.6 Conclusions

TE based design is flexible and easily carried out and may in some circumstances yield simpler, or at least alternative, implementations to the more conventional techniques. Specifying the transitions would appear to be quite a logical way to proceed with a sequential circuit design and analysis is also readily carried out through TE application equations.

It may also be shown that combinational logic can also be characterized using transition operators and we may infer from the brief studies here, that TEs provide a ready route to the design of edge triggered or "ripple through" circuits. It is further possible to approach the design of asynchronous sequential circuits this way but space does not permit a treatment here.

Table A2.5(a) Map of J and K inputs for flip-flop C

J_c K_c			H = 0		H = 1	
B.A	0	0	1	1	Ø	Ø
	0	1	Ø	Ø	Ø	Ø
	1	1	Ø	Ø	Ø	Ø
	1	0	Ø	Ø	Ø	Ø

From the map*: $J_c = 1$; $K_c = 1.e = e$

Table A2.5(b) Map of J and K inputs for flip-flop B

J_B K_B	H = 0		H = 1	
CA 0 0	0	1	Ø	Ø
0 1	Ø	Ø	Ø	Ø
1 1	Ø	Ø	Ø	Ø
1 0	1	1	Ø	Ø

From the map*: $J_B = C$; $K_B = e$

Table A2.5(c) Map of J and K inputs for flip-flop A

J_A K_A	H = 0		H = 1	
CB 0 0	0	1	Ø	Ø
0 1	1	1	Ø	Ø
1 1	1	1	Ø	Ø
1 0	0	1	Ø	Ø

From the map*: $J_A = B$; $K_A = e$

Table A2.5(d) Map of J and K inputs for flip-flop H (EDC)

J_H K_H	A = 0		A = 1	
CB 0 0	0	Ø	1	Ø
0 1	0	Ø	1	Ø
1 1	0	Ø	1	Ø
1 0	0	Ø	1	Ø

From the map: $J_H = A$; $K_H = Ø$

* Restoring *e* as noted in the text.

A2.7 Further reading

Pucknell, D.A., "Transition equations for the analysis and synthesis of sequential circuits", *IEE Electronic Letters*,1970, 6, (23), pp. 731-3.

Pucknell, D.A., "Sequential circuit characterisation and synthesis using a transition equation approach", *Proc. IEE*,Vol. 120, No. 5, May '73, pp. 551-6.

Pucknell, D.A., "Application equations, transition equations and the realisation of sequential networks of JK flip flops", *IREE Monitor*, Nov '86, pp. 339-42.

Pucknell, D.A., "T.E. design processes for successive approximation A/D converter registers", *Proc. IEE*, Vol. 128, Pt. E, No. 2, Mar '81, pp. 79-83.

Smith, J.R. and Roth, C.H.,"Analysis and synthesis of asynchronous sequential networks using edge sensitive flip-flops" *IEE Trans.*, 1971, C-20, pp. 847-55.

Talantsev, A.D.,"On the analysis and synthesis of certain electrical circuits by means of special logical operators" *Autom & Telemekh*, 1959, 20, pp. 895-907.

References for general reading

Bartee, T.C. *Digital computer fundamentals*, McGraw Hill, Singapore, 1985.

Booth, T.L. *Introduction to computer engineering—hardware and software design*, Wiley, USA, 1984.

Camenzind, H.R. *Circuit design for integrated electronics*, Addison-Wesley, USA, 1968.

Cavenor, M., Arnold, J. *Microcomputer interfacing—an experimental approach using the Z80*, Prentice Hall, Australia, 1989.

Dietmeyer, D.L. *Logic design of digital systems*, Allyn and Bacon, USA, 1988.

Eccles, W.J. *Microprocessor systems—a 16–bit approach*, Addison-Wesley, USA, 1985.

Fortino, A. *Fundamentals of computer-aided analysis and design (CAA/CAD) of integrated circuits—processes and devices*, Reston, USA, 1983.

Glasser, L.A., Dobberpuhl, D.W. *The design and analysis of VLSI circuits*, Addison-Wesley, USA, 1985.

Grove, A.S. *Physics and technology of semiconductor devices*, Wiley, USA, 1981.

Hill, F.J., Peterson, G.R. *Introduction to switching theory and logical design*, Wiley, USA, 1968.

Hill, F.J., Peterson, G.R. *Digital systems—hardware organisation and design*, Wiley, USA, 1973.

Hill, F.J., Peterson, G.R. *Digital logic and microprocessors*, Wiley, USA, 1984.

Kline, R.M. *Structured digital design including MSI/LSI components and microprocessors*, Prentice Hall, USA, 1983.

Krutz, R.L. *Microprocessors and logic design*, Wiley, USA, 1980.

Ledley, R.S. *Digital computer and control engineering*, McGraw Hill, USA, 1960.

Leventhal, L.A. *Introduction to microprocessors—software, hardware, programming*, Prentice Hall, UK, 1978

Lewin, D. *Design of logic systems*, Van Nostrand Reinhold, UK, 1985.

Lindmayer, J., Wrigley, C.Y. *Fundamentals of semiconductor devices*, Van Nostrand, USA, 1965.

Mano, M.M. *Digital design*, Prentice Hall, UK, 1984.

Mano, M.M. *Computer engineering—hardware design*, Prentice Hall, USA, 1988.

Marcus, M.P. *Switching circuits for engineers*, Prentice Hall, USA, 1967.

Mavor, J., Jack, M.A., Denyer, P.B. *Introduction to MOS LSI design*, Addison-Wesley, UK, 1983.

Mead, C.A., Conway, L.A. *Introduction to VLSI systems*, Addison-Wesley, USA, 1980.

Motorola, *M68000 16/32-bit microprocessor—programmer's reference manual*, Prentice Hall, USA, 1984.

Nashelsky, L. *Introduction to digital technology*, Wiley, USA, 1983.

Peatman, J.B. *Microcomputer-based design*, McGraw Hill, Japan, 1977.

Phister Jr., M. *Logical design of digital computers*, Wiley, USA, 1958.

Pucknell, D.A., Eshraghian, K. *Basic VLSI design—circuits and systems*, Prentice Hall, Australia, 1988.

Stone, H.S. *Microcomputer interfacing*, Addison-Wesley, USA, 1982

Triebel, W.A. Singh, A. *16-bit microprocessors—architecture, software, and interfacing techniques*, Prentice Hall, USA, 1985.

Wiatrowsky, C.A., House, C.H. *Logic circuits and microcomputer systems*, McGraw Hill, USA, 1980.

Wickes, W.E. *Logic design with integrated circuits*, Wiley, USA, 1968.

Wilcox, A.D. *68000 microcomputer systems—designing and troubleshooting*, Prentice Hall, USA, 1987.

Zilog, *Z80 assembly language programming manual*, Zilog Inc., USA, 1977.

Index

Abacus 2
adders 207
  bit-slice 211, 212
  carry look-ahead 214-216
  equations for 211, 212
  serial arrangement 216, 218, 219
  4-bit parallel arrangement 213, 217, 329
  CMOS adder element 330
Altera 207-209
analog/digital (A/D) converter 340
  successive approx process 341
  successive approx reg. 342
analysis of clocked sequential logic 371
Apple™ Computer 160
application equations 350
  design process using 352
  TE form 455
application oriented examples of design
  adders 207
  asynchronous counters 236
  clocked counters 317
    4-bit binary ripple through 317
    decimal count 319
    with zero detector 321
    4-bit U/D synchronous 323
    base-5 counter 357
      D FF based 358
      "one hot" (ring counter) design 361
      JK FF based 358
    BCD 8421 U/D synchronous 353
    BCD 8421 up only (3 versions) 451-455
    PLA version of for VLSI 366
    U/D (Incr/Decr) for VLSI 329
  phase detector 251, 440
  phase and frequency detector 443
  parity detector, clocked 271
  registers 331
    successive approx. reg. (SAR) 340, 457-460
    static 4-bit parallel 332
    mask layout for 333
    shift registers 334
    nMOS and CMOS cells 338
    mask layout for 339
    4-bit serial/parallel L/R shift 334
  sequence detector, clocked 278
  serial code detector 362
    PLA version 281
    CMOS, D FF version 284
  switch debounce circuit 259
  2-phase clock gen. circuit 330
ASICs 204
ASCII code 411, 412
  table 412
asynchronous sequential logic 228
  analysis of 230
    critical race 233, 247
    cycle 247
    excitation maps 230
    flow matrix 231
    stable states 231
  design procedure 245
    merger diagram 241
    output functions 243
    races 247
    row mergers 241
    specification of requirements 236
      importance of words 244
    state mergers/merging 240
    state transition diagram 242
    state transition table
      primitive 239
      merged 242
    state vector 231, 240
      allocation of states 237
  general model for 229
  hazards in 247
  logic signal flow graph 245, App 1
  memory property of 233
  synthesis of 235
  RS flip-flop 233
    characteristics of 234
Atanasoff 3

Babbage, Charles 2
  analytical engine 2
band-gap energy, E_g 7, 8
Baud rate 385
Berry 3
Boltzmann's constant, k 8
Boole, George 2
Boolean algebra 97, 98
  don't cares, Ø 123

Boolean algebra (cont.)
  duality 107
  expressions
    SOP form 107
    POS form 107
    conversion of form 110, 111, 112
    expansion of 110
    simplification of 115
      algebraic 115
      graphical 115
      Karnaugh map based 116
      prime implicants 120
  functions of 2 variables 103, 104
  functions 98
    Not 98, 99
    Or 99, 100
    Nor 100
    And 100, 101
    Nand 101, 102
    Exclusive Or 102
    Exclusive Nor 103
  Karnaugh mapping 112, 113, 114
    further uses of 122
    simplification 116
  literals 103
  maxterms 108
    notation 108
  minterms 108
    notation 109
  operators of 97
  product terms 107
  sum terms 107
  simplification 109, 115
  theorems, etc 104
    Demorgan's theorem 105
    Huntington's postulates 105
    Shannon's expansion 106
  truth tables 97, 98
  variables 97, 103

CISCs 405
clocked logic 190
  clocked complementary (C^2MOS) 197
  CMOS, 4 phase clocked 193
    rules of 194
  precharged single clock 196
clocked sequential logic 228, 262
  analysis of 371
  design using application equations 352
  design process 271
    summary 276
    FF programming 274
    secondary variable allocation 273
    state transition diagram 271
    state transition table 273
    state merging 272, 279

CMOS
  fabrication 20
  latch-up 21, 22
  parasitic components 22
capacitance
  area values for layers 34, 35
  peripheral C 34, 35
  parallel plate 15
  junction 16
Capilano Computing (LogicWorks™) 160
channel, MOS transistors 16
  gate oxide 16
  pinch-off 19
  resistance in resistive region 31
  resistance in saturation 32
  saturation 19
combinational logic 149
  random 150
  using MUXs 155-158, 159
    residual functions 158, 159
  using PLAs 158-161
  using ROM or RAMs 152-155
conduction band 5
conductivity
  extrinsic 8
  intrinsic 7
connections
  p-well and substrate 20
  contact cuts 26, 27
Conway, and Mead 23
computer hardware, generations of 3
counters
  asynchronous counters 236
  clocked counters 317
    4-bit binary ripple through 317
      decimal count 319
      with zero detector 321
    4-bit U/D synchronous 323
    base-5 counter 357
      D FF based 358
      "one hot" (ring counter) design 361
      JK FF based 358
    BCD 8421 U/D synchronous 353
    BCD 8421 up only (3 versions) 451-455
      PLA version of for VLSI 366
    U/D (Incr/Decr) for VLSI 329
crystal
  silicon 5
  lattice 6, 8
custom ICs

digtal/analog (D/A) converters 340
decoders 198
  CMOS 2-4 line 202
    as column select circuit 203

decoders (cont.)
mask layout 203, color plate 8
demultiplexers (DEMUXs) 198
arrangement of 199
based logic 198
CMOS 1-4 way 202
design rules 23
lambda based 23, 25 to 29, color plates 2 & 3
example layouts 29
D flip-flop 265, (and see under flip-flops)
difference equations 265
diffusion 16
doping
impurities 9
levels 9
level v resistivity 10
p-type 9
n-type 10
dynamic hazards 249

EPLD 207
Eniac 3
Eccles-Jordan trigger relay 3
Eckert 3
Edsac 3
electrical properties
of VLSI circuits 23
elements
group III, group IV, group V 9
atomic table 11
electron
charge 6
free 6
/hole concentration n_i 8
-hole pairs 7
mass m_e 8
orbit 5
per silicon atom 6
per shell 6
shells 6
velocity 12
energy gap 5
essential hazards 249
examples of design, application oriented
adders 207
asynchronous counter 236, 257
clocked counters 317
BCD 8421 U/D synchronous 353
BCD 8421 up only (3 versions) 451-455
base-5 counter 357
D FF based 358
"one hot" (ring counter) design 361
JK FF based 358
4-bit binary ripple through 317
decimal count 319
with zero detector 321
4-bit U/D synchronous 323
U/D (Incr/Decr)for VLSI 329
parity detector, clocked 271
phase detector 251, 440
phase and frequency detector 443
registers 331
successive approx. reg. (SAR) 340, 457-460
static 4-bit parallel 332
mask layout for 333
shift registers 334
nMOS and CMOS cells 338
mask layout for 339
4-bit serial/parallel L/R shift 334
sequence detector, clocked 278
PLA version 281
CMOS, D FF version 284
serial code detector 362
PLA version of for VLSI 366
switch debounce circuit 259
2-phase clock gen. circuit 330

fabrication
design rules for 23
nMOS summary of 19
masks 19
MOS layers 22
CMOS, p-well process 20
feature size 23
feature encoding 24
feedback in logic circuitry 228, 246
conditional 246
flip-flops
characteristic equations 265
common types 263
summary 268, 269, 350
D flip-flop 265, 303, 304
characteristics of 266, 267
VLSI +ve edge clocked 321
2-phase clocked,CMOS 284
mask layout for 286
JK flip-flop 263
characteristics of 264, 265
CMOS version of 269-271
design of -ve edge clocked 269
design of +ve edge clocked 304
RS flip-flop 233
characteristics of 234
layout for asynch. RS FF 235
T flip-flop 267
characteristics of 267, 268
design of edge sensitive FF 269-271
flow graph 252

free electrons 6
free electron-hole pairs 8
Fermi-level 7, 8
  energy E_F 10
  potential $Ø_F$ 10
gate arrays 204
  typical floorplans 205
  routing channels 205, 206
Gray code 279, 455

hole-electron pairs 7
Hollerith 2
  punched cards 2

I.C.s, integrated circuits
  LSI 4
  MSI 4
  SSI 4
  TTL logic 4
  VLSI 4
I_{ds} v V_{ds} relationship MOS transistors
  beta factor ß 30
  resistive region 27, 30, 31
  saturation region 31
  technology factor K 30
illegal (not used) states 354
impurities
  acceptor 9
  donor 10
  doping 8
  n-type 10
  p-type 9
Intel 8080µP™ 162
interrupts 426
  multilevel 427
  priority allocation 429
  software driven polling 426
  single line 427
  vectored 429
inverter
  complementary CMOS 41, 42, 181
    logic based on 181, 182
  delays 42, 43
  nMOS and pMOS 38, 39, 40
    current and dissipation 40, 41, 44, 45
    general model for 184
    logic based on 182
    ratio rules 39, 40, 182, 183
  noise margins 44, 45, 46
    effect of Zpu/Zpd ratio 45, 46
    effect of ßp/ßn ratio 45, 46
  pair delay 44, 45
  pseudo-nMOS 40, 41
    current and dissipation 41, 44, 45
    general model for 184
    inverter based logic
      Nor gate 185
      Nand gate 186
        mask layout 187
    ratio rules 40, 41
  switching times 43, 44
I/O maps 412
  for Z80 414

junction capacitance,
  effective dielectric width 15
  expression for 16
junction, p-n, n-p 13
JK flip-flop 263, (and see under flip-flops)

Karnaugh maps 111, 112, 113, 114
  for 5 and 6 variables 124-130
Kikkert, C.J. 245, 252, App 1

lambda based design rules 23
latches/latching circuits 246, 265
layers
  electrical properties of 23
  encoding schemes 24
  MOS fabrication 22
  diffusion 22
  polysilicon 22
  metal 22
  stick diagrams for 22
LogicWorks™ software 160
logic
  active low 162
  Boolean algebra 97
    don't cares, Ø 123
    duality 107
    expressions
      SOP form 107
      POS form 107
      conversion of form 110-112
      expansion of 110
    simplification of 115
      algebraic 115
      graphical 115
      Karnaugh map based 116
      Quine-McCluskey method 130-137
      prime implicants 120
    functions of 2 variables 103, 104
    functions 98
      multiple output 150, 151
      Not 98, 99
      Or 99, 100
      Nor 100
      And 100, 101
      Nand 101, 102

functions (cont.)
Exclusive Or 102
Exclusive Nor 103
truth tables 97
variables 97, 103
literals 103
maxterms 108
notation 108
minterms 108
notation 109
operators of 97
product terms 107
sum terms 107
simplification 109
switch based 171
theorems, etc 104
Demorgan's theorem 105
Huntington's postulates 105, 106
Shannon's expansion 106
combinational logic 150
in silicon 171
Karnaugh mapping 112, 113, 114
further uses of 122
simplification 116
level representation 97
random logic 150-152
universal logic module (ULM) 199, 202
logic circuits, static current flow 175
logic in silicon 171
custom design of 171
complementary switch based 173
alternative structures 178
bridging switches, use of 187, 189
CMOS *Nand* gate 182, 183, color plate 7
CMOS *Nor* gate 177, 181-183
design procedure 176
Exclusive Or gate 176 -178
switching time estimation 173, 180
general model 176, 177
multiplexer (MUX) based 198
nMOS and pMOS ratio logic 38, 39, 40
current and dissipation 40, 41, 44
design procedure 184
general model for 184
general rules for 182, 183
logic based on 182
Nand gate 182, color plate 6
Nor gate 182, 185, color plate 6
ratio rules 39, 40, 182, 183
PLA based 189, 190
general arrangement 191
PLA cell, mask layout 192
precharged and clocked logic 190
clocked complementary (C^2MOS) 197
CMOS, 4-phase clocked 193-196
rules of 194
precharged single clock 196
pseudo-nMOS
Nor gate 185
Nand gate 186
Nand gate mask layout 187
switch based 171
3I/P *And* gate 173
3I/P *Or* gate 174, 175
switching times 171, 172
logic level transmission 172
logic signal flow graph 245, App 1
Macintosh plus™ 160
masks
design rules for 23, color plates 2 & 3
for fabrication 19
layout of 23
layout for asynchronous RS FF 235
layout for CMOS compl. cell 302
layouts for CMOS 2I/P *Nor*, 3I/P *Nand* color plate 6
layout for D FF 286
layout example 30
layouts for inverters color plate 4
layout for nMOS static latch 299
layout for nMOS and CMOS shift reg cells 339
layout for PLA cell 192
layout for 1 transistor cell 297
layouts for 2I/P *Nor* color plate 7
layout for 2-4 line decoder 203, color plate 8
layout for 3 transistor cell 295
layout for 3 I/P *Nand* 187
layout for 4-way MUX 200, 201
Mauchly 3
Mead and Conway 23
Mealy and Moore models 287
memory
general considerations 290
flip-flop circuits 291
charge storage 291
inverter based cells 292
3-transistor cell 294
mask layout for 295
1-transistor cell 296
mask layout for 297
static storage circuits 298
nMOS pseudo static latch 298
mask layout for 299
CMOS complementary cell 299
mask layout for 302

memory maps 412
for 68000 413
for Z80 413
merger diagram 243
merging 240
microcomputers 382
architecture and organisation 383
general arrangement of 382
microprocessors 382
accumulator 385
address selection (decoding) 415
bitwise 416
memory 418
partial 416
ALU 385
as a system component 383
buffering to the data bus 419
latching data from data bus 419
buses 382, 384, 386
control unit 385
flags/status register 387
hardware interface 407
interface model 388
interfacing, handshake based 420
interrupt facilities 407
I/O 385, 408
memory mapped 414
serial and parallel 409, 411
PPI 409
UARTs 411
control word 409
more detailed architecture 384
Motorola 68000™ 383
addressing modes 399
examples of 401
asynchronous interfacing 420
DMA 425
data structure 399
exceptions 433
vector table 434
hardware interface signals 421
instruction set 400
interrupts 432
priority 435
memory and I/O addressing 420
memory map 412
programmer's model 397
synchronous interfacing 423
program counter (PC) 386
stack pointer (SP) 384, 387
status reg/flags 387
word length 384
working registers 385
Zilog Z80™ 383
addressing modes 392
hardware interface 414
interface signals 415
interrupts 429
non-maskable (NMI) 429
maskable (IM0-IM2) 430
I register 431
memory map 413
programmer's model 389
instruction set 390
I/O map 414
MOS technology 16
basic transistor structures 16
layers 22
mobility μ 12
bulk 12,13
surface 13
Moore and Mealy models 287
Motorola 68000™ micro proc. 383
addressing modes 399
examples of 401
asynchronous interfacing 420
DMA 425
data structure 399
exceptions 433
vector table 434
hardware interface signals 421
instruction set 400
interrupts 432
priority 435
memory and I/O addressing 420
memory map 412
programmer's model 397
synchronous interfacing 423
multiplexers (MUXs) 155-158, 199
arrangement of 199
mask layout for 200
stick diagram for 200
transmission gate based 201
multiplexer-based logic 198
Murphy (of the law) 251

numbers
arithmetic, operations on 62
overflow in 70, 74
binary system 63, 69
division of fractions 78, 79
division of integers 79, 80
signed multiplication 76, 77, 78
unsigned multiplication 75, 76
floating point 83
arithmetic operations 85, 86
format & representation 83, 84, 85
hexadecimal system 65, 66
multiplier architecture 76
octal system 64, 65

numbers (cont.)
quadernary 81
residue system 81
serial transmission of 278
ternary 80, 81
fractions, general form 62
integers, general form 61
radix (base) conversion 66
fractions 67
integers 66
representation of 61
signed repesentations 67
diminished radix form 72
nines complement form 72
ones complement form 73, 74
radix complement form 68
tens complement form 68, 69
twos complement form 68-71
rules 71
sign/magnitude form 67, 68

oscillators 246, 247

PAL 207
phase detectors 440
.phase locked loops 440
phase and frequency detector 443
PLA based logic 189, 190
analysis of 373
form of transition table 282
general arrangement 191
PLA cell,mask layout 192
dimension reduction of 368
PLD 207
pass transistor switches
limitations in use of 36, 37
properties of 35-37
switch-based logic 171
switching times 171, 172
logic level transmission 172
Pascal, Blaire 2
permittivity
of free space 15
relative for silicon 12, 14
Planck's constant, h 8
phase detector 251
polysilicon (POLY) 16
programmable logic arrays (PLAs) 158-161
in silicon 189, 190
dimensions 159
reduction of 161
general arrangement 191
cell, mask layout 102

Quine-McCluskey simplification 130-137

radix (base) conversion of numbers 66
registers 331
successive approx. reg.(SAR) 340
design of 341
static 4-bit parallel 332
mask layout for 333
shift registers 334
nMOS and CMOS cells 338
mask layout for 339
4-bit serial/parallel L/R shift 334
regular design 254
resistance
of contacts, typical values 34
of transistor channel 31, 32, 34
sheet values, R_s 33, 34
resistivity
expressions for 13
intrinsic 6
RISCs 405
RS flip-flop 233

saturation
transistor channel 19
resistance in 32
secondary variables 242, 248
sequence dependant logic 228
sequential logic/circuits 228
clocked 228, 262
analysis of 371
design using application
equations 352
design process 271
summary 276
FF programming 274
secondary variable allocation 273
state transition diagram 271
state transition table 273
PLA form 282
state merging 272, 279
asynchronous 228
analysis of 230
critical race 233
cycle 247
design procedure 245
excitation maps 230
flow matrix 231
general model for 229
hazards 247
logic signal flow graph 245, App 1
memory property of 233
merger diagram 241
output functions 243
races 247
RS flip-flop 233
characteristics of 234

sequential logic/circuits
asynchronous (cont.)
row mergers 241
specification of requirements 236
importance of words 244
stable states 231
state mergers/merging 240
state transition diagram 242
state transition table
primitive 239
merged 242
state vector 231, 240
allocation of states 237
synthesis of 235
seven segment display 163
Shannon, C.E. 3
sheet resistance R_s 33, 34
signed numbers 67
silicon
atom 5
crystal 5
monocrystalline 5
relative permittivity 12, 15
some properties of 11, 12
simplification of logic functions 115
algebraic 115
graphical 115
Karnaugh map based 116
Quine-McCluskey method 130-137
prime implicants 120
Sperry Rand 3
static hazards 249
stick diagrams
color/monochrome encoding of 23, 24, color plate 1
successive approx A/D conv 341
successive approx reg. 342
substrate
body effect 32
bias, effect of 32, 33
switching algebra (Boolean) 97

T flip-flop 267
theorems, etc 104
Demorgan's theorem 105
Huntington's postulates 105, 106
Shannon's expansion 106
threshold voltage 16
typical values of 32, 33
transmission gates
properties of 35, 37, 38, 39
limitations in use of 37, 38
switching times 171, 172
logic level transmission 172
transistor
channel, MOS devices 16
depletion mode 17, 19
enhancement mode 17, 18, 19
fabrication
CMOS, n-well process 20
CMOS, p-well process 20
CMOS, twin-tub process 21
nMOS summary of 19
masks 19
invention of 4
MOS, structure of 16
parameters for, MOS 25
regions of operation 19
saturation 19
transition equations (TEs) 448
TE based characteristic equations 448
TE based design 451
two-phase clock generator circuit 330

UCIC 207
Univac 3

vacuum tubes 4
valence
band 5
bonding 5
electrons 5
VLSI 3
alternative structures in design 178
regular design 254
Von Leibnitz 2

Xilinx 207, 210

Zilog Z80™ microprocessor 383
addressing modes 392
hardware interface 414
interface signals 415
interrupts 429
non-maskable (NMI) 429
maskable (IM0-IM2) 430
I register 431
memory map 413
programmer's model 389
instruction set 390
I/O map 414